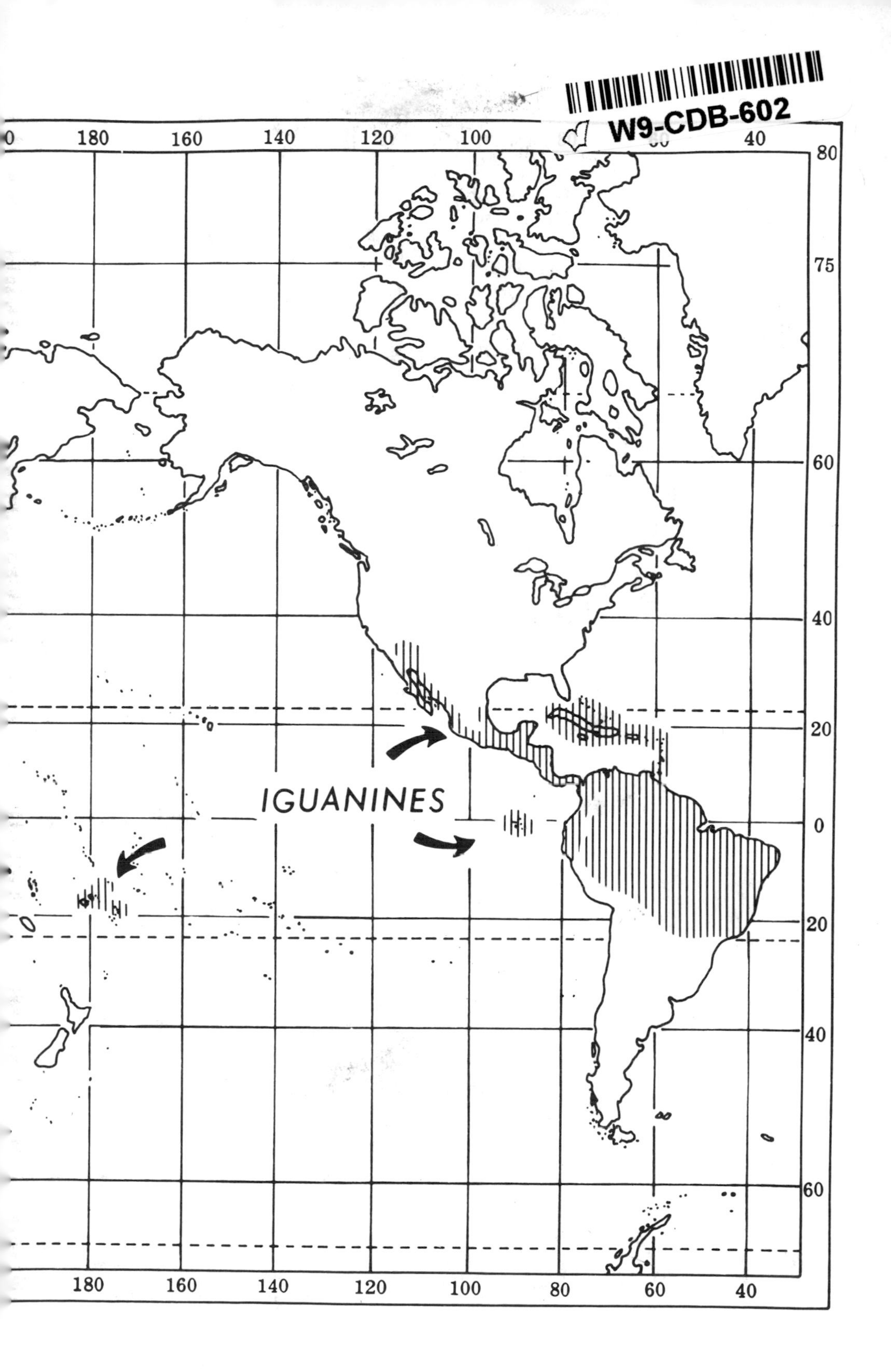

180
160
140
120
100
80
60
40
80
75
60
40
20
0
IGUANINES

1. *Amblyrhynchus cristatus*

2. *Brachylophus* (new species)

3. *Conolophus subcristatus*

4. *Ctenosaura quinquecarinata*

5. *Ctenosaura similis*

6. *Cyclura cornuta stejnegeri*

7. *Cyclura rileyi cristata*

8. *Dipsosaurus dorsalis*

9. *Iguana iguana*

10. *Sauromalus obesus obesus*

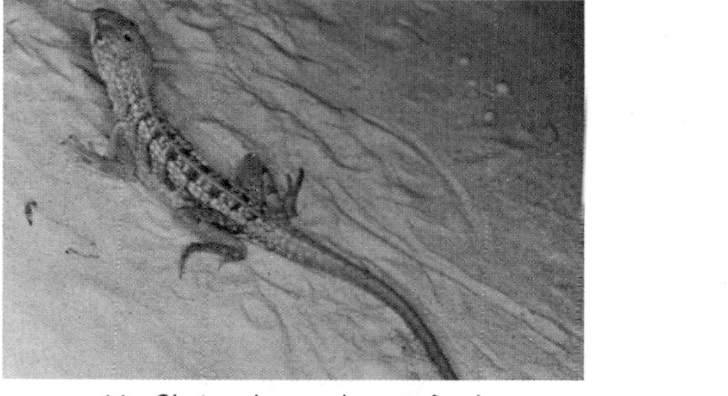

11. *Chalarodon madagascariensis*

12. *Oplurus fierinensis*

13. Female Iguana at Nesting Site

IGUANAS OF THE WORLD

NOYES SERIES
IN
ANIMAL BEHAVIOR, ECOLOGY,
CONSERVATION AND MANAGEMENT

A series of professional and reference books in ethology devoted to the better understanding of animal behavior, ecology, conservation, and management.

WOLVES OF THE WORLD: Perspectives of Behavior, Ecology, and Conservation.
Edited by *Fred H. Harrington and Paul C. Paquet*

IGUANAS OF THE WORLD: Their Behavior, Ecology, and Conservation.
Edited by *Gordon M. Burghardt* and *A. Stanley Rand*

HORSE BEHAVIOR: The Behavioral Biology and Adaptations of Domestic and Wild Horses, Including Ponies.
By *George H. Waring*

GAZELLES AND THEIR RELATIVES: A Study in Territorial Behavior
By *Fritz R. Walther, Elizabeth Cary Mungall,* and *Gerald A. Grau*

THE MANAGEMENT AND BIOLOGY OF AN EXTINCT SPECIES: *PÈRE DAVID'S DEER*
Edited by *Benjamin B. Beck* and *Christen Wemmer*

IGUANAS OF THE WORLD

Their Behavior, Ecology, and Conservation

Edited by

GORDON M. BURGHARDT

University of Tennessee
Knoxville, Tennessee

and

A. STANLEY RAND

Smithsonian Tropical Research Institute
Balboa, Republic of Panama

NOYES PUBLICATIONS
Park Ridge, New Jersey, USA

Library of Congress Catalog Card Number: 82-7932
ISBN: 0-8155-0917-0
Printed in the United States

Published in the United States of America by
Noyes Publications
Mill Road, Park Ridge, New Jersey 07656

10 9 8 7 6 5 4 3 2 1

Library of Congress Cataloging in Publication Data
Main entry under title:

Iguanas of the world.

(Animal behavior, ecology, conservation, and
management)
Bibliography: p.
Includes index.
1. Iguanidae. I. Burghardt, Gordon M., 1941-
II. Rand, A. Stanley (Austin Stanley) III. Series.
QL666.L25I37 1982 597.95 82-7932
ISBN 0-8155-0917-0 AACR2

Preface

The giant iguanas not only play important cultural and nutritional roles, but are increasingly endangered, true even of the most ubiquitous species, the green iguana. Scientifically, too, these animals are important; as the main line of herbivorous lizards their ecology and social organization have some intriguing parallels to herbivorous mammals, including primates.

For a decade and a half we have been studying green iguanas in Panama. Several years ago we set out to compare what we were finding with what was known about iguanas elsewhere. We discovered that little had been published, but that we have colleagues scattered from the West Indies through Central America to the Galápagos Islands and Fiji who had been working on similar problems in other species.

To tap this wealth of as yet unpublished information and bring together these scientists we organized a symposium on the world's iguanas that was held at the joint meeting of the Society for the Study of Amphibians and Reptiles and The Herpetologist's League in August, 1979, at the University of Tennessee, Knoxville. As most species of iguanas are restricted to islands we included studies on the unique but distantly related group of iguanid lizards found only on Madagascar (Malagasy Republic).

We were gratified at the large acceptance rate and the interest expressed in preparing a comprehensive "state of the art" book on iguanas including manuscripts by those unable to attend the symposium and additional authors added later, for iguana research is an active field. Much of the work discussed herein is provisional or still in progress. Exciting new findings will still be made. At the Knoxville meeting, for example, discovery of a new giant *Brachylophus* species was announced and its behavior and ecology described.

Many persons aided in the organization of the symposium, the extensive correspondence, and the production of this book. The contributors themselves deserve first mention; they were generally prompt, good natured, and full of excellent suggestions. Many also aided in the review of manuscripts and provided treasured color photos for reproduction herein. Others that reviewed papers included George Bartholomew, Jon Mamau, Henry Molina, Neil Smith, Kathy Troyer, and E.E. Williams. The following also helped in many ways: Barbara Allen, Bruce Batts, Karen Easter, A.C. Echternacht, Carl Gans, Lisa Holbert, Sharon Lane, Maxine McSwords, Debbie Myers, Argelis Ruiz, Roy Scudder, Paul Weldon, and Dan York. Brian Bock deserves special mention for organizing the preparation of the Species Indexes with only short notice.

Knoxville, Tennessee
August, 1982

Gordon M. Burghardt
A. Stanley Rand

Contributors

Walter Auffenberg
Florida State Museum
University of Florida
Gainesville, Florida

Charles P. Blanc
Laboratoire de Zoogéographie
Université Paul Valéry
Montpellier Cedex, France

P. Dee Boersma
Institute for Environmental Studies
University of Washington
Seattle, Washington

Gordon M. Burghardt
Department of Psychology
University of Tennessee
Knoxville, Tennessee

Charles C. Carpenter
Department of Zoology
University of Oklahoma
Norman, Oklahoma

Ted J. Case
Department of Biology
University of California at San Diego
La Jolla, California

Keith A. Christian
Department of Zoology and Entomology
Colorado State University
Ft. Collins, Colorado

Hansjürgen Distel
Max-Planck-Institut für Psychiatrie
Munich, Federal Republic of Germany

Hugh Drummond
Institute of Biology
Department of Zoology
Universidad Nacional Autónoma de México
México D.F., México

Beverly Dugan
Department of Psychology
University of Tennessee
Knoxville, Tennessee

Richard E. Etheridge
Department of Zoology
San Diego State University
San Diego, California

Henry S. Fitch
Biological Sciences
University of Kansas
Lawrence, Kansas

John R.H. Gibbons
School of Natural Resources
University of the South Pacific
Suva, Fiji

Neil Greenberg
Department of Zoology
University of Tennessee
Knoxville, Tennessee

Harry W. Greene
Museum of Vertebrate Zoology
University of California
Berkeley, California

Dennis M. Harris
Museum of Zoology
University of Michigan
Ann Arbor, Michigan

Robert W. Henderson
Vertebrate Division
Milwaukee Public Museum
Milwaukee, Wisconsin

David M. Hillis
Department of Herpetology
Museum of Natural History
University of Kansas
Lawrence, Kansas

John B. Iverson
Department of Biology
Earlham College
Richmond, Indiana

Thomas A. Jenssen
Department of Biology
Virginia Polytechnic Institute
and State University
Blacksburg, Virginia

Richard H. McBee
Department of Microbiology
Montana State University
Bozeman, Montana

Virginia H. McBee
Department of Microbiology
Montana State University
Bozeman, Montana

Kenneth A. Nagy
Laboratory of Biomedical and
Environmental Sciences
University of California
Los Angeles, California

A. Stanley Rand
Smithsonian Tropical Research
Institute
Balboa, Republic of Panama

Michael J. Ryan
Neurobiology and Behavior
Langmuir Laboratory
Cornell University
Ithaca, New York

C. Richard Tracy
Department of Zoology and
Entomology
Colorado State University
Fort Collins, Colorado

Robert Wayne Van Devender
Department of Biology
Appalachian State University
Boone, North Carolina

J. Veazey
Max-Planck-Institut für Psychiatrie
Munich, Federal Republic of
Germany

Ivy F. Watkins
Orchid Island Fijian Cultural
Centre
Suva, Fiji

Dagmar I. Werner
Charles Darwin Research Station
Isla Santa Cruz
Galápagos, Ecuador

Thomas A. Wiewandt
Department of Ecology and
Evolutionary Biology
University of Arizona
Tucson, Arizona

Contents

List of Color Plates

Adult males (Plates 1-12) were chosen for consistency and their more dramatic appearance. Plate 13 is a female iguana at a nest site.

1. *Amblyrhynchus cristatus*
   Hood Island, Galápagos
   Charles C. Carpenter
2. *Brachylophus* (new species)
   Yaduataba, Fiji
   John R.H. Gibbons
3. *Conolophus subcristatus*
   Fernandina Island, Galápagos
   Dagmar Werner
4. *Ctenosaura quinquecarinata*
   Honduras
   John B. Iverson
5. *Ctenosaura similis*
   Great Corn Island, Zelaya, Nicaragua
   Robert W. Henderson
6. *Cyclura cornuta stejnegeri*
   Mona Island, Puerto Rico
   Thomas A. Wiewandt
7. *Cyclura rileyi cristata*
   White Cay, Exuma Islands, Bahamas
   John B. Iverson
8. *Dipsosaurus dorsalis*
   Dallas Zoo
   Wayne Van Devender
9. *Iguana iguana*
   Finca La Pacifica, Guanacaste Costa Rica
   Wayne Van Devender
10. *Sauromalus obesus obesus*
    San Bernadino Co., California
    Wayne Van Devender
11. *Chalarodon madagascariensis*
    Malagasy Republic
    Charles Blanc
12. *Oplurus fierinensis*
    Malagasy Republic
    Charles Blanc
13. Female iguana at nesting site
    A. Stanley Rand

Introduction

Iguanas are large, primeval appearing animals that have always attracted considerable attention, if not scientific study. The 30 species of iguanas are today collectively referred to as iguanines, or the subfamily Iguaninae, of the lizard family Iguanidae. The word iguana is the Spanish version of the carib *iwana*. Some confusion exists because it has been used to refer to the entire family (e.g., Woodward, 1911). Complicating matters even more is the association of iguanid lizards with the extinct herbivorous dinosaur Iguanodon, which had similar peculiar teeth (round at the base and blade-like at the tip). Burghardt (1977) has argued, however, that data from iguanas and other living reptiles can be used in reconstructing the behavioral repertoires of dinosaurs.

The family Iguanidae is large, containing many species and genera. It includes 4 genera: *Anolis, Sceloporus, Liolaemus* and *Leiocephalus* that have radiated extensively and complexly. The iguanines have not.

Most iguanids are insectivorous, some eat other lizards, and our special subset the iguanines, are herbivorous. The family has a wide ecological range in both temperate and tropical areas from extreme deserts to tropical rainforest interiors, from sea level to above tree line. The fossil record is scant but shows that the family has been in existence in North and South America since the Cretaceous.

With a few critical exceptions, iguanid lizards today occur only in the western hemisphere. Here they are widespread in North, Central, and South America as well as in the Antilles and the Galapagos. Iguanids occur in the old world only in one genus on Fiji and Tonga, and the two genera on Madagascar. The former are iguanines, the latter are not; but all the old world forms of iguanids are considered in this book.

Generally, in the old world, iguanid niches are ecologically occupied by agamids (family Agamidae). This complementation is not on a simple "replacement" basis. For example there are large herbivorous agamids such as *Uromastyx* and its relatives in northern Africa and arid Asia, and *Hydrosaurus* in the

Philippines, but there are many areas, such as tropical Africa, that lack herbivorous lizards completely. We should really not be surprised at this. The ways in which communities are organized are flexible enough that the environment is divided differently in different places. The absence of a large herbivorous lizard, in tropical Africa say, does not necessarily imply that there is any empty lizard-shaped niche but rather that the relevant resources are divided differently.

The classic view (Darlington, 1957) is that the iguanids originated in the old world tropics and spread to the new world. The agamids arose from them, also in the old world tropics, and spread, displacing the more primitive iguanids except in the island refuges of Madagascar and possibly Fiji and Tonga. A more recent suggestion is that iguanids are of South American origin (Estes and Price, 1973). Estes (1970) sees similarities between iguanids and teids. Blanc in Chapter 2, on the iguanids of Madagascar, suggests dispersal through Antarctica, when the southern continents were closer together, as an alternative explanation of present distributions (see also Laurent, 1979).

The iguanines are a subfamily of the iguanids containing the largest iguanids and those that are strictly herbivorous. The group appears to be monophyletic and there are only 8 genera (see Table on page 6). While Etheridge (Chapter 1) points out that the Iguaninae is not a formally recognized subfamily, it is clear that many of our authors feel that this is indeed a mere formality. Iguanine distribution centers in Mexico where 4 out of the 8 genera occur. Throughout the range there are usually only one or two, occasionally three, iguanine species sympatric at any locality. Where two of similar size occur together they usually differ in habitat preferences, frequently one species occurring in more moist and the other in more dry environments. Species which occur over extensive ranges, particularly if they occur on a set of islands, such as the Galápagos, the Bahamas, or the Fiji archipelago, vary geographically.

Most iguanas occur in dry environments that are either rainless or physiologically dry. There are exceptions. *Brachylophus* occurs in wet forests in the Fiji archipelago and *Iguana* that invades wet forests but is most common along edges.

Most iguanines refuge in holes, in trees (some *Ctenosaura*), among rocks (some *Ctenosaura* and *Amblyrhynchus*), and in burrows in the ground (*Conolophus* and *Cyclura,* even *Iguana* does this in some places). Most iguanas forage on ground and in bushes and trees. *Amblyrhynchus* is unique in feeding on marine algae.

For all iguanines for which we have information, eggs are laid in a burrow dug in the ground, a burrow deep enough for the animal to get completely inside. Many iguanines such as *Dipsosaurus, Cyclura* and *Conolophus* regularly refuge in burrows and dig their own when necessary. That even the arboreal species dig full burrows rather than just a deep pit for their eggs argues that the ancestral iguanine may have been terrestrial, refuging in burrows, though probably also climbing to feed and forage.

All iguanines are diurnal and, of course, ectothermic ("cold-blooded"), but also as far as is known, showing behavioral thermoregulation by moving into and out of the sun. A large lizard must have easy access to the sun for substantial periods to function effectively as a heliotherm. This is easiest if it lives in lowland dry open habitats.

Iverson (Chapter 4) shows that iguanines are primarily herbivorous, some opportunistically taking animal food when they can get it, but surviving and growing on a strictly vegetable diet. He argues that although some *Ctenosaura* may eat considerable insect material when young, this is unusual in the subfamily. *Brachylophus* is partly insectivorous but it is not clear to what extent in the wild. Many iguanas, including *Iguana*, will sometimes eat insects, chopped meat, and canned dog food in captivity.

It is important in discussions of herbivory to distinguish between herbivores that eat primarily fruit, flowers and buds, and those who eat leaves and get a substantial proportion of their needs from structural carbohydrates. Leaves generally have adequate protein levels, but are frequently high in fiber, low in easily digested energy sources, and frequently protected chemically against predation by herbivores. Accounts of the diet of *Dipsosaurus, Sauromalus, Cyclura, Brachylophus, Conolophus,* even *Iguana* all stress the selectivity of the lizards in the plants eaten; fruit, flower buds and young leaves are repeatedly mentioned. Mature leaves are mentioned less often.

The large size of iguanines has been correlated with their herbivorous diet. It can be argued that large animals need less energy per gram of tissue and so are metabolically more efficient and thus can survive on lower quality food. It is also true that large animals can eat larger fruit and can travel longer distances between food resources than can small animals. Perhaps also important, larger animals can crop more of an ephemeral food source that is available during brief periods, can store more food during periods of abundant food, and probably endure longer periods of food scarcity than could a smaller animal.

As ectotherms, lizards have much lower metabolic rates and so are probably better than homeotherms such as birds and mammals at exploiting an environment where food is intermittently and unpredictably abundant only for short periods. Birds can exploit intermittently available resources by their high mobility, but they are less effective at coping with unpredictable resources. This discussion suggests that iguanines' basic adaptations are to exploit intermittently and perhaps unpredictably abundant, high quality plant resources such as fruit, flowers, and buds, *Iguana* feeding on mature leaves may be a more recent and further specialization. This may explain, paradoxically, why large lizards (1.5 meters long and up) are also found among the monitor lizards (Varanidae). These are often considered to be virtually entirely carnivorous. Some may, however, eat fruit opportunistically (e.g., Gray's monitor, *Varanus grayi,* Affenberg, 1979), although this may not differ from carnivorous mammals, such as foxes, that also eat fruit. None the less, monitors and iguanas are the two groups of lizards that reach large size and this could relate to a breaking away from the typical lizard tendency to eat arthropods almost exclusively. It is thus important to eventually compare the activities, social and reproductive behavior, and other aspects of monitors and iguanas. The differences between the groups, which seem so great currently, may turn out not to be so fundamental, or at least not as obvious, when detailed field work is completed.

As might be expected of large animals living on intermittently available resources, iguanines are generally slow growing, long lived animals. That at least some of their food is unpredictable would add further importance to their repetitive (iterative) breeding over many years. Wiewandt (Chapter 7) points out that iguanines show two strategies. *Ctenosaura* and *Iguana* seem to grow more quickly and to reach maturity more rapidly than most iguanines, and also to lay larger clutches of smaller eggs. Wiewandt argues that this is a strategy to be expected in the predator filled continental environments with seasonally abundant food where these lizards occur; the alternate strategy of small clutches of very large eggs is to be expected in the groups that live on small islands or in deserts where predation risks are (or were until recently) lower but where lesser food resources are also found. Note that the very widespread *Iguana iguana* has a large clutch but *Iguana delicatissima,* restricted to the Lesser Antilles, seems to lay bigger and fewer eggs (Rand, pers. obs.).

Most iguanines seem to have a restricted breeding season with the possible exception of *Brachylophus. Sauromalus* and perhaps *Dipsosaurus* are also exceptional in that they may produce several clutches in succession in a favorable year. Most of the world, including most tropical areas, is more or less seasonal. Uniformity or truly unpredictable changes seem restricted to very few locations in the tropics. The dry, but not desertic, part of the tropics with its predictable and well marked dry seasons is the kind of region in which one might expect the evolution of a breeding system such as we see in iguanines.

In summary, iguanines seem basically adapted in size, thermoregulation, refuging, and nesting habits to feeding on intermittently available high quality plant parts in a dry seasonal environment in tropical North America. From this they have spread to islands, to the wet tropics and to temperate deserts. In this volume information is gathered that summarizes what we know, what is beginning to be known, and what we need to know. These animals, popular in zoos, folklore, and illustrations from the time they were first discovered, deserve study, protection, and perhaps even reverence.

Section I

Systematics and Biogeography

It is appropriate to begin with a section that characterizes and classifies the animals we are concerned with and considers their biogeographic origins. In order to write about any organism, or any phenomenon for that matter, one must first have a name for it; for readers to understand what has been written they must share the author's referents for the name. The checklist by Etheridge, (Chapter 1) which covers the iguanine lizards as well as the Malagasy iguanids, provides such a list of lizard names and, in the descriptions referenced, a way to find out precisely to what the names apply. Already it has proved useful, for in editing the chapters in this volume we repeatedly found authors differing in their usage of names.

We have inserted a list of common names and used them in a standard way throughout the book. In contrast to scientific names, there is no accepted way of determining the correct common name for an animal. Indeed, we feel that the view that there *is* a "correct" common name is misguided; in contrast, we rather treasure local common names and do not want to surpress them with the book names. Still, people talking together must agree on their terminology and after consulting with our contributors, we have assigned English common names for all of the recognized genera of iguanines and used them here.

A checklist is much more than just a list of names. By the species recognized and the genera to which they are assigned it represents a set of conclusions about phylogenetic relationships. Classification is a continuing process; Etheridge's conclusions will be changed as new information becomes available and is analyzed. For example, Gibbon's discovery of a new species of *Brachylophus* in Fiji makes it highly likely that additional discoveries will be made in the South Pacific. The relationships among the spiny-tailed iguanines in Mexico and Central America are under study and it may be that the genus *Ctenosaura,* to which they are all assigned here, will be subdivided, perhaps bringing back on a more solid footing *Enyaliosaurus,* a genus not recognized as separate from *Ctenosaura* in this volume. Etheridge's checklist will be revised in years to come but it represents our present knowledge accurately and therefore it is useful now and will be useful for a long time. The table below includes a breakdown of the 30 iguanine species and their distribution. The numerous subspecies listed for *Cyclura* indicate that this island group's taxonomy is unstable, and it may never be resolved as extinction closes in on these remarkable animals.

The Iguanine Genera

Genus	Common Name	Number of Species[1]	Recognized Subspecies[1]	Geographic Range[1]
Amblyrhynchus	Marine Iguanas	1	3	Galápagos Islands
Brachylophus	Banded Iguanas	2[2]	0	Fiji and Tonga Island groups
Conolophus	Land Iguanas	2	0	Galápagos Islands
Ctenosaura	Spiny-tailed Iguanas	9	7	Mexico to Panama, Colombian Islands
Cyclura	Ground Iguanas	7	14	Greater Antilles and Bahamas
Dipsosaurus	Desert Iguana	1	3	Southwestern U.S., Mexico, islands in the Gulf of California
Iguana	Green Iguanas	2	0	Mexico to southern Brazil and Paraguay, Lesser Antilles
Sauromalus	Chuckawallas	6	7	Southwestern U.S., Mexico, islands in the Gulf of California
		30		

[1] Based on Etheridge, Chapter 1

[2] See also Gibbons and Watkins, Chapter 23

Blanc's contribution on the biogeography of the Malagasy lizards (Chapter 2) bears directly on the origin of the family Iguanidae and its past distribution, and thus on the early history of the iguanines. As Blanc points out, the presence of iguanids in Madagascar and their absence in Africa and Asia is one of the paradoxes in classic biogeography. Calling on the currently accepted concepts of plate tectonics and shifting continents, Blanc suggests that iguanids evolved in South America, reached Madagascar via Antarctica when they were much closer together, and may never have occurred in Asia, Africa and Australia at all.

Representatives of all the iguanine and Malagasy iguanid genera are depicted in the color plates of the frontispiece. The end paper map may prove useful in quickly locating geographically the various kinds of iguanas.

1

Checklist of the Iguanine and Malagasy Iguanid Lizards

Richard E. Etheridge
Department of Zoology
San Diego State University
San Diego, California

INTRODUCTION

A checklist is included in this volume to provide those interested in or working on iguanine and Malagasy iguanid lizards with a ready guide to the taxonomic literature of the group, and reasonably detailed distributions for each taxon. The group of genera referred to here as iguanines is not now recognized as a formal taxonomic unit, i.e., subfamily. Nevertheless iguanines are easily distinguished from other iguanid lizards, and very probably form a natural, monophyletic group. They share a large number of derived anatomical and behavioral characteristics, the most notable being specializations of the skull, dentition, and gut adaptive for an herbivorous diet, as well as a distinctive condition of the caudal vertebrae, presumably not functionally related to herbivory, which is unique within the family. The relationships of the iguanine genera, however, remain unclear, as does the position of the group within the family as a whole. There is much work to be done on the evolutionary and biogeographic history of the group.

Cope (1886) was apparently the first to use the term Iguaninae in a more or less formal sense. He included the genera *Cyclura, Ctenosaura, Brachylophus, Cachryx (= Enyaliosaurus), Iguana, Conolophus* and *Amblyrhynchus*. The North American genera *Dipsosaurus* and *Sauromalus* were not included. Previously Boulenger (1885) in his *Catalogue of the Lizards in the British Museum* had listed these same genera in sequence (including *Metapoceros* for *Cyclura cornuta*), implying perhaps his belief in their affinity. In 1890 Boulenger provided a brief description of the skulls of these genera, again considering *Metapoceros* as distinct but failing to recognize the validity of *Cachryx*. In 1900 Cope presented a revised formal classification of iguanid subfamilies, but in this work his Iguaninae was quite different from the group he recognized under that name in 1886. Cope proposed the subfamilies Anolinae, Basiliscinae and

Iguaninae, the latter being defined merely as lacking the diagnostic features of the first two subfamilies, and thus including all of the iguanid genera not assigned to the Anolinae and Basiliscinae.

Subsequently various authors have offered comments on the possible relationships of iguanine genera, but without proposing their formal recognition as a taxonomic unit. On the basis of musculature Camp (1923:416) stated that "*Holbrookia* and other North American iguanids are regarded as primitive, *Iguana, Cyclura, Dipsosaurus* and *Amblyrhynchus* as central, *Basiliscus* and *Anolis* as offshoots of the latter," and that "the Fijian *Brachylophus* is closely related to *Ctenosaura* and *Cyclura*." In his review of the genus *Urosaurus* Mittleman (1942: 112-113; Fig. 1) briefly considered the phylogeny and relationships of all North American iguanid genera except *Anolis,* implying that they form a monophyletic group, with *Ctenosaura* basal to the remaining genera. He stated: "*Dipsosaurus* is probably the most primitive of the North American Iguanidae (excepting *Ctenosaura,* which is properly a Central and South American form), and possesses several points in common with *Ctenosaura,* most easily observed of which is the dorsal crest; the genera further show their relationship in the similarity of cephalic scutellation. *Sauromalus* is considered a specialized offshoot of *Crotaphytus,* or more properly pre-*Crotaphytus* stock, by reason of its solid sternum, as well as the five-lobed teeth; the simple type of cephalic scalation indicates its affinity with the more primitive *Dipsosaurus-Ctenosaura* stock." Smith (1946: Fig. 92) apparently accepted Mittleman's view that all North American iguanids except *Anolis* form a monophyletic group, in which he also included the West Indian *Leiocephalus*. Following Mittleman, he placed *Ctenosaura, Dipsosaurus* and *Sauromalus* near the base of the group. However, Smith (1946: 101) apparently recognized the broader affinities of the North American iguanines, for under the heading "The Herbivore Section," he included *Iguana, Amblyrhynchus, Conolophus, Dipsosaurus, Sauromalus, Ctenosaura* and *Cyclura.*

Savage (1958: 48-49) questioned the assumption that North American iguanids form a natural, inter-related group, stating: "Insofar as can be determined at this time, the so-called Nearctic iguanids form two diverse groups that can be only distantly related. These two sections are distinguished by marked differences in vertebral and nasal structures, and include several genera not usually recognized as being allied to Nearctic forms." Savage then defined as one of the primary divisions of the Iguanidae a group of ten genera, informally referred to as the "Iguanine group": *Amblyrhynchus, Brachylophus, Conolophus, Crotaphytus, Ctenosaura, Cyclura, Dipsosaurus, Enyaliosaurus, Iguana* and *Sauromalus.* He characterized the group by the presence of zygosphenes and zygantra on each dorsal vertebra, and a nasal organ of the relatively simple S-shaped type with a concha present (= *Dipsosaurus*-type of Stebbins 1948: 209). Other skeletal and integumentary characters present in the majority of these genera were also listed. In a study of sceloporine lizards Etheridge (1964: 628-629) pointed out that zygosphenes and zygantra are poorly developed in *Crotaphytus,* but well developed in several non-iguanine genera, apparently their presence being correlated with large body size, and also that the *Dipsosaurus*-type nasal structure may be found in non-iguanine genera as well. It was also shown, however, that if *Crotaphytus* is removed from the list of iguanine genera, then the latter all share a number of skeletal and integumentary characters, some of which are clearly derived, implying a natural, monophyletic group. Thus the iguanine group as redefined by Etheridge includes the genera *Amblyrhynchus, Brachylophus, Conolophus, Ctenosaura, Cyclura, Dipsosaurus, Enyaliosaurus, Iguana* and *Sauromalus.*

Subsequent to Etheridge's redefinition of the iguanine group there has been but one attempt to determine the evolutionary relationships within the group, that of Avery and Tanner (1971). Their work included the Malagasy genera *Oplurus* and *Chalarodon* along with the iguanines, and contains descriptions and measurements of parts of the skeleton, the musculature of the head and neck, the tongue and hemipenes. Length-width measurements of bones and bone shapes were utilized to analyze the osteological relationships between the genera, those genera sharing the most characters in common being considered the most closely related. No consideration was given to ontogenetic changes in proportions due to allometric growth, and there was no discussion of the polarities of character states. A "phylogenetic chart" was provided, but without an indication as to how the chart was constructed. It is not certain why *Chalarodon* and *Oplurus,* rather than some other iguanid genera, were chosen for comparison with iguanines. The Malagasy genera themselves form an anatomically distinctive group of iguanids, but one not especially close to the iguanines.

The individual iguanine genera have received unequal treatment in the taxonomic literature. Little attention has been paid to *Conolophus,* and Van Denburgh and Slevin's (1913) brief descriptions of the two species seem to be the most recent. Eibl-Eibesfeldt (1962) described the geographic variation in *Amblyrhynchus,* and provided figures and descriptions of the subspecies, but no key. Gibbons and Watkins (this volume) discuss the distribution of *Brachylophus fasciatus,* and describe, informally, a new species (subsequently described as *B. vitiensis* Gibbons, 1981). They provide evidence for the restriction of the type locality of *B. fasciatus* to Tongatapu in the Tonga Island Group, and for the synonymy of *B. brevicephalus* with *B. fasciatus.*

Lazell (1973) studied the two species of *Iguana* as they occur in the Lesser Antilles, where *I. delicatissima* is confined to that region while *I. iguana* occurs over a vast area on the mainland as well. Lazell included in his studies of geographic variation in *Iguana iguana* samples from mainland populations as well, and found no geographically consistant pattern, either on the islands or mainland, of the enlargement and alignment of the median scales on the snout, a character used by Dunn (1934) and other to distinguish the subspecies *I. i. iguana* from *I. i. rhinolopha.* Accordingly I have placed the latter in the synonymy of the former in this list. In the same work, Lazell (1973) designated a neotype for *Lacerta iguana* Linnaeus 1758, and restricted the type locality to the island of Terre de Haut. However, Hoogmoed (1973) pointed out that the specimens upon which Linnaeus based his description are still extant in Stockholm and Uppsala, and furthermore there is a good reason to believe these specimens came from Paramaribo, at the confluence of the Cottica River and Perica Creek in Surinam.

The most recent taxonomic revision and key for the genus *Ctenosaura* is that of Bailey (1928), but several important papers on individual species or groups of species have appeared subsequently. Bailey recognized 13 species, including those forms with a relatively small body size and a short, strongly spinose tail referred by some authors to *Enyaliosaurus*. Following Gray's (1845) description of *Enyaliosaurus* the name was seldom used until its revival by Smith and Taylor (1950: 75). In this work the species *clarki, defensor, erythromelas, palearis* and *quinquecarinata* were allocated to *Enyaliosaurus,* but no justification was provided for the revival of the genus. Duellman (1965: 599), followed Smith and Taylor in recognizing the validity of *Enyaliosaurus,* placed *erythromelas* in the synonymy of *defensor,* provided a key to the species, and suggested that: "*Enyaliosaurus* doubtless is a derivative of *Ctenosaura,* all species of which are larger and have relatively longer tails and less well-developed spines than *Enyaliosaurus*." Meyer and Wilson (1973) referred *Ctenosaura bakeri* to *Enyaliosaurus,* but Wilson and Hahn (1973: 114-5) returned *bakeri* to *Ctenosaura,* commenting that: "John R. Meyer is currently studying the problems of the relationship of the species now grouped in *Enyaliosaurus* to those now grouped in *Ctenosaura*. He (pers. comm.) advised us that he considers the two genera inseparable, and that *bakeri* appears to be closely related to both *palearis* (now in *Enyaliosaurus*) and *similis* (now in *Ctenosaura*)." In addition, Ernest Williams of Harvard University has informed me (pers. comm.) that based on an unpublished study of the group by him and Clayton Ray, he does not believe the recognition of *Enyaliosaurus* is warranted. At the present time, the problem of the relationships of *Ctenosaura* and *Enyaliosaurus* are under study by Diderot Gicca of the Florida State Museum. With one exception, the remaining long-tailed species of *Ctenosaura* have received little attention in recent years; Smith (1972) reviewed geographic variation in *Ctenosaura hemilopha,* discussed the zone of overlap between this species and *C. pectinata,* thought by Smith to be ancestral to *hemilopha,* and provided a key to *C. pectinata* and the subspecies of *C. hemilopha.*

Figure 1.1: The spiny tail of *Ctenosaura* is clearly evident in this *Ctenosaura defensor* captured in Piste, Yucatan, Mexico by Charles H. Lowe (photo by T.A. Wiewandt).

Schwartz and Carey (1977) have published a thorough revision of the genus *Cyclura,* together with a key to the species and subspecies and a discussion of their possible relationships and biogeographic history. *Cyclura* is said to be most similar to *Ctenosaura,* from which it differs in having short series, or "combs," of enlarged, fused, ventrolateral subdigital scales on the hind feet, and it was suggested that *Cyclura* probably originated on Hispaniola following an invasion of pre-*Ctenosaura* stock from the mainland. *Cyclura carinata* and *C. ricordii* are considered basal members of the genus, most closely related to *Ctenosaura* by virtue of their possession of enlarged scales setting off the caudal verticils, absent in other species of *Cyclura*. Following a study of the fossil lizards of Puerto Rico, Gregory Pregill (unpubl.) now considers the extinct *Cyclura mattea* from St. Thomas and *C. portoricensis* to be synonyms of *C. pinguis,* now living on Anegada Island.

As yet no comprehensive study of the genus *Dipsosaurus* has been published. Van Denburgh (1922: 71) provides a good description of the genus, followed by detailed descriptions of the forms recognized at that time. In their study of the lizards on the islands in the Gulf of California, Soulé and Sloan (1966) listed *Dipsosaurus catalinensis* as a subspecies of *D. dorsalis,* and implied the synonymy of *D. carmenensis* with *D. dorsalis lucasensis* by listing the island of Carmen within the range of the latter. I have followed these arrangements although no formal justification for them was provided by the authors.

In his revision of *Sauromalus*, Shaw (1945) recognized two species groups: a coarse-scaled group consisting of *hispidis, slevini, klauberi, ater* and *australis,* and another group with relatively less spinose scalation including *varius* and *obesus* (including *townsendi* and *tumidus*). Shaw felt (1945: 277) that the separation of the Gulf of California was the original factor in separating the genus, and that *hispidus* and *varius* were the oldest forms of the genus, "their differentiation occurring before that of other species, through early breaking off, by faulting, of the islands upon which they now occur, *hispidus* being derived from a peninsular population, while *varius* was derived from a Sonoran population." Shaw also included a diagram of his ideas of the phylogeny of the genus. Subsequently Cliff (1958) described *S. shawi,* Tanner and Avery (1964) described a new subspecies of *S. obesus,* and Soulé and Sloan (1966) listed *S. klauberi* and *S. shawi* as subspecies of *S. ater.* (See also Robinson, 1972.)

The citations provided here include the original descriptions of all taxa, all references that involve nomenclatorial changes, restricted type localities, and the designations of neotypes and lectotypes. In addition the most recent taxonomic revisions, as well as published illustrations, are included. The orthography of scientific names is reproduced here as used in the reference cited, and all type localities have been copied as originally stated. Some of the type specimens listed here as "not located" may eventually be identified. Specimens formerly in the collection of Thomas Bell were deposited in the British Museum, and in the zoological collections of Oxford and Cambridge Universities. It is possible that the specimens used by Gray (1831) in his descriptions of *Ctenosaura armata, C. belli, C. similis, Amblyrhynchus ater,* and *Conolophus subcristatus* may eventually be discovered among these specimens. The material upon which Laurenti (1768) based *Iguana delicatissima, I. minima,* and *I. tuberculata* may have at one time been in the Museum of Zoology of the University of Torino, but a search of that collection by me in 1969 failed to discover them. The types of *Brachylophus fasciatus* (Brongniart 1800) and *Cyclura cornuta* (Bonnaterre 1789) may eventually be identified in the Natural History Museum in Paris. Harlan's (1824) type of *Cyclura carinata* may have at one time been in the Academy of Natural Sciences of Philadelphia, but if so is almost certainly now lost. Merrem (1820) referred to earlier published accounts, but listed no specimens in his description of *Oplurus cyclurus;* quite possibly no type has ever existed.

In most cases the arrangement of species and subspecies and their synonymies follows their most recent taxonomic revisions, but there are a few exceptions.

Museum Abbreviations. *Acad. Nat. Sci. Philad.:* Academy of Natural Sciences of Philadelphia, Nineteenth and the Parkway, Philadelphia, Pennsylvania 19103, U.S.A.; *Amer. Mus. Nat. Hist.:* American Museum of Natural History, Central Park West at 79th Street, New York, New York 10002, U.S.A.; *Brig. Young Univ.:* Brigham Young University, Provo, Utah 84103, U.S.A.; *Brit. Mus. Nat. Hist.:* British Museum (Natural History), Cromwell Road, London S.W. 7, England; *Calif. Acad. Sci.:* California Academy of Sciences, Golden Gate Park, San Francisco, California 94118, U.S.A.; *Eibl-Eibesfeldt private coll.:* private collection of Irenäus Eibl-Eibesfeldt, location not known; *Field Mus. Nat. Hist.:* Field Museum of Natural History, Roosevelt Road at Lake Shore Drive, Chicago, Illinois 60605, U.S.A.; *Mus. Comp. Zool.:* Museum of Comparative Zoology, Harvard University, Cambridge, Massachusetts 02138, U.S.A.; *Mus. Hist. Nat. Paris:* Muséum National d'Histoire Naturelle, 57, rue Cuvier, 75005 Paris, France; *Nat. Ricksmus. Stockholm:* Naturhistoriska Ricksmuseet, Vertebratavdeiningen, Stockholm 50, Sweden; *Oxf. Univ. Mus.:* The Zoological Collections, University Museum, Oxford University, Park Road, Oxford, England; *San Diego Soc. Nat. Hist.:* Natural History Museum, The San Diego Society of Natural History, P.O. Box 1390, San Diego, California 92112, U.S.A.; *Senck. Mus. Frankfurt:* Natur-Museum und Forschungs-Institut Senckenberg der Senckenbergischen Naturforschenden Gesellschaft, 6 Frankfurt, 1, Senckenberg Anlage 25, Germany; *Stanford Univ. Mus.:* Natural History Museum, Stanford University, Palo Alto, California 94305, U.S.A.; *U.S. Natn. Mus.:* United States National Museum of Natural History (Smithsonian Institution), Washington, D.C. 20560, U.S.A.; *Univ. Colo. Mus.:* University of Colorado Museum, Boulder, Colorado 80302, U.S.A.; *Zool. Inst. Univ. Uppsala:* Zoological Institute, University of Uppsala, Uppsala, Sweden; *Zool. StSamm. München:* Zoologisches Sammlung des Bayerischen Staates, Schloss Nymphenburg, Nordflügel, München 19, Germany; *Zool. Mus. Berlin:* Institut für Spezielle Zoologie und Zoologisches Museum der Humboldt-Universität zu Berlin, 104 Berlin, den Invalidenstr. 43, Democratic Republic of Germany; *Zool. Mus. Hamburg:* Zoologisches Staatsinstitut und Zoologisches Museum, 2000 Hamburg, den Von-Melle-Park, Germany; *Zool. Mus. Torino:* Istituto e Museo di Zoologia della Universita de Torino, Via Accademia Albertina, 17, Torino, Italy.

IGUANINES

Amblyrhynchus Bell

1825 *Amblyrhynchus* Bell, Zool. Jr., London, *2:* 206. – **Type species** (by monotypy): *Amblyrhynchus cristatus* Bell 1825.

1843 *Hypsilophus (Amblyrhynchus)* – Fitzinger, Syst. Rept., Wien, *1:* 55.
1845 *Oreocephalus* Gray – Cat. Spec. Liz. Coll. Brit. Mus., London, 189. – **Type species** (by monotypy): *Amblyrhynchus cristatus* Bell 1825.
1885 *Amblyrhynchus* – Boulenger, Cat. Liz. Brit. Mus., London, *2:* 185.
Range: Rocky coasts of various islands of the Galápagos Archipelago, Ecuador.

Amblyrhynchus cristatus Bell

1825 *Amblyrhynchus cristatus* Bell, Zool. Jr., London, *2*: 206. – **Type locality:** Mexico (**Holotype:** Oxf. Univ. Mus. Ref. No. 6176). – **Restricted type locality** (Eibl-Eibesfeldt 1956): Narborough (Fernandina).
1831 *Iguana (A. [mblyrhynchus]) Cristatus* – Gray *in* Cuvier *edit.* Griffith, Anim. Kingd., London, *9:* 37.
1831 *Iguana (A. [mblyrhynchus]) Ater* Gray (*syn. fide* Gray 1845), *in* Cuvier *edit.* Griffith, Anim. Kingd., London, *9:* 37 – *Type locality:* Galápagos (**Holotype:** not located).
1843 *Hypsilophus (Amblyrhynchus) cristatus* – Fitzinger, Syst. Rept., Wien, *1:* 55.
1843 *Hypsilophus (Amblyrhynchus) ater* – Fitzinger, Syst. Rept., Wien, *1:* 55.
1845 *Oreocephalus cristatus* – Gray, Cat. Spec. Liz. Coll. Brit. Mus., London, 189.
1876 *Amblyrhynchus cristatus* – Steindachner, Festshr. zool.-bot. Ges., Wien, 316; Pl. 3, 7.
1885 *Amblyrhynchus cristatus* – Boulenger, Cat. Liz. Brit. Mus., London, *2:* 185.
Range: Galápagos Archipelago.

Amblyrhynchus cristatus cristatus Bell

1956 *Amblyrhynchus cristatus cristatus* – Eibl-Eibesfeldt, Senckenberg., Biol., Frankfurt a. M., *37:* 88; Pl. 9, Fig. 1, 2a-b; Fig. 1a, 2.
Range: Narborough (= Fernandina) Island, Galápagos Archipelago.

Amblyrhynchus cristatus albemarlensis Eibl-Eibesfeldt

1962 *Amblyrhynchus cristatus albemarlensis* Eibl-Eibesfeldt, Senckenberg., Biol., Frankfurt a. M., *43* 3: 184; Pl. 14, Fig. 2; Fig. 2f. – **Type locality:** Insel Albemarle (Isabella) (**Holotype:** Eibl-Eibesfeldt private coll.).
Range: Albermarle (= Isabella) Island, Galápagos Archipelago.

Amblyrhynchus cristatus hassi Eibl-Eibesfeldt

1962 *Amblyrhynchus cristatus hassi* Eibl-Eibesfeldt, Senckenberg., Biol., Frankfurt a. M., *43* 3: 181; Pl. 15, Fig. 4; Fig. 2e, 3b. – **Type locality:** Indefatigable Südküste, westliche Akademiebucht . . . .Indefatigable (Santa Cruz), Galápagos-Inseln (**Holotype:** Senck. Mus. Frankfurt No. 57407).
Range: Indefatigable (= Santa Cruz) Island, Galápagos Archipelago.

Amblyrhynchus cristatus mertensi Eibl-Eibesfeldt

1962 *Amblyrhynchus cristatus mertensi* Eibl-Eibesfeldt, Senckenberg., Biol., Frankfurt a. M., *43* 3: 185; Fig. 3c-d, 3d-e, – **Type locality:** etwa 3 km südwestlich der Wrack-Buct der insel Chatham (S. Cristobal), Galápagos Insel. (**Holotype:** Senck. Mus. Frankfurt No. 57430).
Range: Chatham (= San Cristobal) and James (= Santiago) Islands, Galápagos Archipelago.

Amblyrhynchus cristatus nanus Garman

1892 *Amblyrhynchus nanus* Garman, Bull. Essex Inst., Salem, *24*: 8. – **Type locality:** Tower Island (**Holotype:** Brit. Mus. Nat. Hist. No. 99.5.4 [RR 1946.8.30.20]).

1962 *Amblyrhynchus cristatus nanus* – Eibl-Eibesfeldt, Senckenberg., Biol., Frankfurt a. M., *43* 3: 189; Pl. 15, Fig. 6; Fig. 2b, 3g.

Range: Tower (= Genovesa) Island, Galápagos Archipelago.

Amblyrhynchus cristatus sielmanni Eibl-Eibesfeldt

1962 *Amblyrhynchus cristatus sielmanni* Eibl-Eibesfeldt, Senckenberg., Biol., Frankfurt a. M., *43* 3: 188; Fig. 2h, 3f. – **Type locality:** Westküst der Insel Abington (**Holotype:** Senck. Mus. Frankfurt No. 57417).

Range: Abington (= Pinta) Island, Galápagos Archipelago.

Amblyrhynchus cristatus venustissimus Eibl-Eibesfeldt

1956 *Amblyrhynchus cristatus venustissimus* Eibl-Eibesfeldt, Senckenberg., Biol., Frankfurt a. M., *37:* 89. – **Type locality:** Nordküste der Insel Hood (Española) (**Holotype:** Senck. Mus. Frankfurt No. 49851).

Range: Hood (= Española) and Gardner Islands, Galápagos Archipelago.

Brachylophus Cuvier

1829 *Brachylophus* Cuvier *in* Guérin-Ménville, Icon. Règ. Anim., Paris, 1; Pl. 9, Fig. 1. – **Type species** (by monotypy): *Iguana fasciata* Brongniart 1800.

1831 *Iguana (Brachylophus)* – Gray *in* Cuvier *edit.* Griffith, Anim. Kingd. London, *9:* 37.

1862 *Chloroscrates* Günther, Proc. Zool. Soc. Lond., 189. – **Type species** (by monotypy): *Chloroscartes fasciatus* Günther 1862 (*non* Brongniart 1800).

1885 *Brachylophus* – Boulenger, Cat. Liz. Brit. Mus., London, *2*: 192.

Range: Numerous islands of the Fiji Island Group, Tongatapu and perhaps other islands of the Tonga (= Friendly) Island Group, and Îles Wallis northeast of Fiji, all in the southwestern Pacific Ocean.

Brachylophus fasciatus (Brongniart)

1800 *Iguana fasciata* Brongniart, Bull. Soc. Philom., Paris, *2:* 90; Pl. 6, Fig. 1. – **Type locality:** none given; probably Tongatapu in the Tonga Islands *fide* Gibbons 1981 (**Holotype:** none designated).

1802 *Agama fasciata* – Daudin, Hist. Nat. Rept., Paris, *3:* 352.

1829 *Brachylophus fasciatus* – Cuvier *in* Guérin-Ménville, Icon. Règ. Anim., Paris, *1:* 9, Pl. 9, Fig. 1, 1a-c.

1831 *Iguana (Brachylophus) fasciatus* – Gray *in* Cuvier *edit.* Griffith, Anim. Kingd., London, *9:* 37.

1843 *Hypsilophus (Brachylophus) fasciatus* – Fitzinger, Syst. Rept., Wien, *1:* 55.

1862 *Chloroscartes fasciatus* Günther, Proc. Zool. Soc. Lond., 189; Pl. 25. – **Type locality:** Feegee Islands (**Syntypes:** Brit. Mus. Nat. Hist. No. 55.8.13.1-2 [RR 1946.8.3.83-84]).

1885 *Brachylophus fasciatus* – Boulenger, Cat. Liz. Brit. Mus., London, *2:* 192.

1970 *Brachylophus brevicephalus* Avery & Tanner (*syn. fide* Gibbons 1981).

Gt. Basin Nat., Provo, *30* 3: 167. – **Type locality:** Nukalofa, Tongatabu Island, Friendly Islands (**Holotype:** Brig. Young Univ. No. 32662).
1981 *Brachylophus fasciatus* – Gibbons, J. Herpet., *15* 3: 255.
Range: In the Fiji Island Group recorded from the islands of Viti Levu, Wakaya, Ovalau, Moturiki, Gau, Beqa, Vatuele, Kandavu Ono, Dravuni, Taveuni, Nggamea, Vanua, Balavu, Avëa, Vatu Vara, Lakeba, Aiwa, Oneata, Vanua Levu, Vanua Vatu, Totoya, Kabara and Fulaga; records from Cikobia, Koro, Naviti and Yasawa, as well as other records from Viti Levu are likely to be of this species; in the Tonga Island Group known definitely only from the island of Tongatapu, but likely to occur elsewhere; also recorded from Îles Wallis northeast of Fiji, and from Efate Island in the New Hebrides, where it may have recently been introduced.

Brachylophus vitiensis (Gibbons)

1981 *Brachylophus vitiensis* Gibbons, J. Herpet., *15* 3: 257; Pl. I, IIa, c-d; Fig. 2, 4a, 5a. – **Type locality:** Yaduataba island (16°50′S; 178°20′E), Fiji (**Holotype:** Mus. Comp. Zool. No. 157192).
Range: Known only from the type locality in the Fiji Island Group.

Conolophus Fitzinger

1834 *Hypsilophus (Conolophus)* Fitzinger, Syst. Rept., Wien, *1:* 55. – **Type species** (by original designation): *Amblyrhynchus demarlii* Duméril & Bibron 1837 = *Amblyrhynchus subcristatus* Gray 1831.
1845 *Trachycephalus* Gray, Cat. Spec. Liz. Coll. Brit. Mus., London, 188. – **Type species** (by monotypy): *Amblyrhynchus subcristatus* Gray 1831.
1885 *Conolophus* – Boulenger, Cat. Liz. Brit. Mus., London, *2*: 186.
Range: The Galápagos Archipelago.

Conolophus pallidus Heller

1903 *Conolophus pallidus* Heller, Proc. Wash. Acad. Sci., Washington, D.C. *5:* 87. – **Type locality:** Barrington Island, Galápagos (**Holotype:** Stanford Univ. Mus. No. 4749).
1913 *Conolophus pallidus* – Van Denburgh & Slevin, Proc. Calif. Acad. Sci., San Francisco, *2:* 190.
Range: Barrington (= Santa Fe) Island, Galápagos Archipelago.

Conolophus subcristatus (Gray)

1831 *Amb. [lyrhynchus] subcristatus* Gray, Zool. Misc., London, *1831:* 6. – **Type locality:** Galápagos? (**Holotype:** not located).
1837 *Amblyrhynchus Demarlii* Duméril & Bibron (*syn. fide* Gray 1845), Erpét. Gén., Paris, *4:* 197. – **Type locality:** inconnue (**Holotype:** not located).
1843 *Hypsilophus (Conolophus) demarlii* – Fitzinger, Syst. Rept., Wien, *1*: 55.
1845 *Trachycephalus subcristatus* – Gray, Cat. Spec. Liz. Coll. Brit. Mus., London, 188.
1876 *Conolophus subcristatus* – Steindachner, Festr. zool.-bot. Ges. Wien, 22; Pl. 4; P. 7, Fig. 5.

1899 *Conolophus subcristatus pictus* Rothschild & Hartert (*syn. fide* Van Denburgh & Slevin 1913) Novit. Zool., London, *6:* 102. – **Type locality:** Narborough (**Syntypes:** Brit. Mus. Nat. Hist. No. 99.5.6.41-44).

1913 *Conolophus subcristatus* – Van Denburgh & Slevin, Proc. Calif. Acad. Sci., San Francisco, *2:* 188.

Range: The islands of James (= Santiago), Indefatigable (= Santa Cruz), Albermarle (= Isabella), Narborough (= Fernandina), and South Seymour, Galápagos Archipelago.

Ctenosaura Wiegmann

1828 *Ctenosaura* Wiegmann, Isis (von Oken), Leipzig, *21*: 371. – **Type species** (subsequent designation by Fitzinger 1843): *Ctenosaura cycluroides* Wiegmann 1828 = *Lacerta acanthura* Shaw 1802.

1845 *Enyaliosaurus* Gray, Cat. Spec. Liz. Coll. Brit. Mus., London, 192. – **Type species** (by monotypy): *Cyclura quinquecarinata* Gray 1842.

1866 *Cachryx* Cope, Proc. Acad. Nat. Sci. Philad., *18:* 124. – **Type species** (by monotypy): *Cachryx defensor* Cope 1866.

1928 *Ctenosaura* – Bailey, Proc. U.S. Natn. Mus., Washington, *73* 12: 7.

Range: Arid and subhumid lowlands of México and Central America, from southeastern Baja California and the vicinity of Hermosillo, Sonora, in western México, and from near the Tropic of Cancer in Tamaulipas, eastern México, southward along both the Atlantic and Pacific versants through most of Central America to Panamá, in the vicinity of Colón on the north, and Panamá on the south, as well as on various off-shore islands and the Colombian islands of the western Caribbean.

Ctenosaura acanthura (Shaw)

1802 *Lacerta Acanthura* Shaw, Gen. Zool., London, *3* 1: 216. – **Type locality:** not given (**Holotype:** Brit. Mus. Nat. Hist. No. xxii 20a [RR 1946.8.30.19]). – **Restricted type locality** (Bailey 1928): Tampico, Tamaulipas, Mexico.

1820 *Uromastyx acanthurus* – Merrem, Tent. Syst. Amphib., Marburg, 56.

1825 *Cyclura teres* Harlan (*syn. fide* Boulenger 1885), Proc. Acad. Nat. Sci. Philad., *4*: 250; Pl. 16. – **Type locality:** Tampico (**Holotype:** Acad. Nat. Sci. Philad., now lost).

1828 *Ct.[enosaura] cycluroides* Wiegmann (*syn. fide* Boulenger 1885), Isis (von Oken), Leipzig, *21:* 371. – **Type locality:** Tampico (**Syntypes:** Zool. Mus. Berlin No. 576, 578; Mus. Comp. Zool. No. 2253). – **Restricted type locality** (Smith & Taylor 1950): Veracruz, Veracruz.

1831 *Iguana (Ctenosaura) Cycluroides* – Gray *in* Cuvier *edit.* Griffith, Anim. Kingd., London, *9:* 37.

1831 *Iguana (Ctenosaura) Acanthura* – Gray *in* Cuvier *edit.* Griffith, Anim. Kingd., London, *9:* 38.

1831 *Cyclura Shawii* Gray (substitute name for *Lacerta acanthura* Shaw 1802), *in* Cuvier *edit.* Griffith, Anim. Kingd., London, *9:* 38.

1831 *Iguana (Ctenosaura) Armata* Gray (*syn. fide* Boulenger 1885) *in* Cuvier *edit.* Griffith, Anim. Kingd., London, *9:* 38. – **Type locality:** not given (**Holotype:** not located). – **Restricted type locality** (Smith & Taylor 1950): Tampico, Tamaulipas.

1831 *Iguana (Ctenosaura) Lanceolata* Gray (*syn. fide* Boulenger 1885) in *Cuvier edit.* Griffith, Anim. Kingd., London, *9:* 38. – **Type locality:** not given (**Holotype:** not located). – **Restricted type locality** (Smith & Taylor 1950): Tampico, Tamaulipas.
1831 *Iguana (Ctenosaura) Bellii* Gray (*syn. fide* Boulenger 1885) *in* Cuvier *edit.* Griffith, Anim. Kingd., London, *9:* 38. – **Type locality:** not given (**Holotype:** not located). – **Restricted type locality** (Smith & Taylor 1950): Tampico, Tamaulipas.
1831 *Iguana (Cyclura) Teres* – Gray *in* Cuvier *edit.* Griffith, Anim. Kingd., London, *9:* 39.
1834 *C.[yclura] articulata* Wiegmann (substitute name for *Iguana (Ctenosaura) armata* Gray 1831), Herp. Mex., Saur. Spec., Berlin, *1:* 43.
1834 *C.[yclura] denticulata* Wiegmann (substitute name for *Ctenosaura cycluroides* Wiegmann 1828), Herp. Mex., Saur. Spec., Berlin, *1:* 43; Pl. 3.
1843 *Cyclura (Ctenosaura) denticulata* – Fitzinger, Syst. Rept., Wien, *1:* 56.
1843 *Cyclura semicristata* Fitzinger (substitute name for *Cyclura denticulata* Wiegmann 1834), Syst. Rept., Wien, *1:* 56.
1843 *Cyclura (Ctenosaura) articulata* – Fitzinger, Syst. Rept., Wien, *1:* 56.
1843 *Cyclura (Ctenosaura) Shawii* – Fitzinger, Syst. Rept., Wien, *1:* 56.
1843 *Cyclura (Ctenosaura) Bellii* – Fitzinger, Syst. Rept., Wien, *1:* 56.
1845 *Ctenosaura acanthura* – Gray, Cat. Spec. Liz. Coll. Brit. Mus., London, 191.
1855 *Cyclura denticulata* – Hallowell, J. Acad. Nat. Sci. Philad., (2) *3:* 36.
1869 *Cyclura (Ctenosaura) acanthura* – Cope, Proc. Am. Philos. Soc., Philadelphia, *6:* 161.
1874 *Ctenosaura teres* – Bocourt *in* Duméril & Bocourt, Miss. Sci. Mex., Paris, *3:* 142.
1885 *Ctenosaura acanthura* – Boulenger, Cat. Liz. Brit. Mus., London, *2:* 195.
1886 *Ctenosaura multispinis* Cope (*syn. fide* Bailey 1928), Proc. Am. Philos. Soc., Philadelphia, *23:* 267. – **Type locality:** Dondomingovillo, in the state of Oaxaca (**Holotype:** U.S. Natn. Mus. No. 72737).
1928 *Ctenosaura acanthura* – Bailey, Proc. U.S. Natn. Mus., Washington, D.C. *73* 12: 9; Pl. 1-4.
1950 *Ctenosaura acanthura* – Smith & Taylor, Bull. U.S. Natn. Mus., Washington, D.C. *199:* 74.

Range: Eastern México, from Liera and Tepehuaje de Arriba near the Tropic of Cancer in Tamaulipas southward to the Isthmus of Tehuantepec in southeastern Veracruz and eastern Oaxaca, at elevations below 360 meters.

Ctenosaura bakeri Stejneger

1901 *Ctenosaura bakeri* Stejneger, Proc. U.S. Natn. Mus., Washington, D.C. *23:* 467. – **Type locality:** Utilla Island, Honduras (**Holotype:** U.S. Natn. Mus. No. 26317).
1928 *Ctenosaura bakeri* – Bailey, Proc. U.S. Natn. Mus., Washington, D.C. *73* 12: 38; Pl. 21-22.
1973 *Enyaliosaurus bakeri* – Meyer & Wilson, L. A. Co. Nat. Hist. Mus. Contrib. Sci., Los Angeles, *244:* 24.
1973 *Ctenosaura bakeri* – Wilson & Hahn, Bull. Fla. St. Mus., Biol. Sci., Gainesville, *17* 2: 114.

Range: Isla de Utilla, Isla de Roatán and Isla de Santa Elena of the Islas de la Bahía off the northern coast of Honduras.

Ctenosaura clarki Bailey

1928 *Ctenosaura clarki* Bailey, Proc. U.S. Natn. Mus., Washington, *73* 12: 44; Pl. 27. – **Type locality:** Ovopeo, Michoacan, Mexico (**Holotype:** Mus. Comp. Zool. No. 22454). – **Corrected type locality** (Duellman & Duellman 1959): Oropeo . . . ., at an elevation of about 1000 feet in the Tepalcatepec Valley about 8 miles south of La Huacana.

1959 *Enyaliosaurus clarki* – Duellman & Duellman, Occ. Pap. Mus. Zool. Univ. Mich., Ann Arbor, 598: 1 Fig. 1; Pl. 1.

Range: Western México in the valley of the Río Tepalcatepec, Michoacán, between 200 and 510 meters.

Ctenosaura defensor (Cope)

1866 *Cachryx defensor* Cope, Proc. Acad. Nat. Sci. Philad., *18:* 124. – **Type locality:** Yucatan (**Syntypes:** U.S. Natn. Mus. No. 12282 [3]). – **Restricted type locality** (Bailey 1928): Chichén Itzá, Yucatan, Mexico.

1886 *Ctenosaura erythromelas* Boulenger (*syn. fide* Duellman 1965), Proc. Zool. Soc. London, *1886:* 241; Pl. 23. – **Type locality:** not given (**Holotype:** Brit. Mus. Nat. Hist. No. 86.8.9.1 [RR 1946.8.30.18]). – **Restricted type locality** (Smith & Taylor 1950): Balchacaj, Campeche.

1887 *Cachryx erythromelas* – Cope, Bull. U.S. Natn. Mus., Washington, D.C. *32:* 43.

1890 *Ctenosaura defensor* – Günther, Biol. Cent. Amer., Rept. & Batr., 58.

1911 *Ctenosaura (Cachryx) annectens* Werner (*syn. fide* Bailey 1928), Jb. Hamb. Wiss. Anst., Hamburg, *27* 2: 25. – **Type locality:** not given (**Holotype:** Zool. Mus. Hamburg, destroyed).

1928 *Ctenosaura erythromelas* – Bailey, Proc. U.S. Natn. Mus., Washington, D.C. *73* 12: 46; Pl. 28-29.

1928 *Ctenosaura defensor* – Bailey, Proc. U.S. Natn. Mus., Washington, D.C. *73* 12: 48; Pl. 30.

1950 *Enyaliosaurus erythromelas* – Smith & Taylor, Bull. U.S. Natn. Mus., Washington, D.C. *199:* 77.

1965 *Enyaliosaurus defensor* – Duellman, Univ. Kans. Publ. Mus. Nat. Hist., Lawrence, *15* 12: 598.

Range: Southern México in the states of Campeche and Yucatán.

Ctenosaura hemilopha Cope

1863 *Cyclura (Ctenosaura) hemilopha* Cope, Proc. Acad. Nat. Sci. Philad., *15*: 105. – **Type locality:** Cape St. Lucas (**Syntypes:** U.S. Natn. Mus. No. 529 [4]).

1866 *Ctenosaura hemilopha* – Cope, Proc. Acad. Nat. Sci. Philad., *18*: 312.

1928 *Ctenosaura hemilopha* – Bailey, Proc. U.S. Natn. Mus., Washington, D.C. *73* 12: 17; Pl. 5.

1969 *Ctenosaura hemilopha* – Hardy & McDiarmid, Univ. Kans. Publ. Mus. Nat. Hist., Lawrence, *18* 3: 119.

Range: Northwestern México, including southeastern Baja California, various islands in the Gulf of California, and on the mainland from the vicinity of Hermosillo, Sonora, southward to the northern third of Sinaloa, and inland at low elevations to extreme western Chihuahua.

Ctenosaura hemilopha hemilopha Cope

1882 *Ctenosaura interrupta* Bocourt (*syn. fide* Boulenger 1885), Le Naturaliste, Paris, *2:* 47. – **Type locality:** California (**Syntypes:** Mus. Hist. Nat. Paris No. 2243, 2245, 2843; Brit. Mus. Nat. Hist. No. 85.11.2.1 [RR 1946.8.3.85]). – **Restricted type locality** (Smith & Taylor 1950): Cape San Lucas.

1972 *Ctenosaura hemilopha hemilopha* – Smith, Gt. Basin Nat., Provo, *32* 2: 104.

Range: The southern part of the Peninsula of Baja California south of La Paz, extending northward on the lower eastern slopes of the Sierra de Giganta at least as far as Loreto, and perhaps as far north as Santa Rosalía.

Ctenosaura hemilopha conspicuosa Dickerson

1919 *Ctenosaura conspicuosa* Dickerson, Bull. Am. Mus. Nat. Hist., New York, *41* 10: 461. – **Type locality:** San Esteban Island, Gulf of California, Mexico (**Holotype:** Amer. Mus. Nat. Hist. No. 5027).

1955 *Ctenosaura hemilopha conspicuosa* – Lowe & Norris, Herpetologica, *11*: 89.

Range: Isla San Esteban, and possibly Isla Lobos just south of Isla Tiburon, in the Gulf of California, México.

Ctenosaura hemilopha insulana Dickerson

1919 *Ctenosaura insulana* Dickerson, Bull. Am. Mus. Nat. Hist., New York, *41* 10: 462. – **Type locality:** Cerralvo Island, Gulf of California, Mexico (**Holotype:** Amer. Mus. Nat. Hist. No. 2694).

1955 *Ctenosaura hemilopha insulana* – Lowe & Norris, Herpetologica, *11*: 90.

Range: Isla Cerralvo, just east of La Paz, in the southern Gulf of California, México.

Ctenosaura hemilopha macrolopha Smith

1972 *Ctenosaura hemilopha macrolopha* Smith, Gt. Basin Nat., Provo, *32* 2: 104. – **Type locality:** La Posa, San Carlos Bay, 10 mi NW Guaymas, Sonora (**Holotype:** Field Mus. Nat. Hist. No. 108705).

Range: Northwestern México from the vicinity of Hermosillo, Sonora, southward through the northern third of Sinaloa, and inland at low elevations to extreme western Chihuahua.

Ctenosaura hemilopha nolascensis Smith

1972 *Ctenosaura hemilopha nolascensis* Smith, Gt. Basin Nat., Provo, *32* 2: 107. – **Type locality:** Isla San Pedro Nolasco, Sonora (**Holotype:** Univ. Colo. Mus. No. 26391).

Range: Isla San Pedro Nolasco in the Gulf of California, México.

Ctenosaura palearis Stejneger

1899 *Ctenosaura palearis* Stejneger, Proc. U.S. Natn. Mus., Washington, D.C. *21*: 381. – **Type locality:** Gualan, Guatemala (**Holotype:** U.S. Natn. Mus. No. 22703).

1928 *Ctenosaura palearis* – Bailey, Proc. U.S. Natn. Mus., Washington, D.C. *73* 12: 40; Pl. 22-23.

1963 *Enyaliosaurus palearis* – Stuart, Misc. Publ. Mus. Zool. Univ. Mich., Ann Arbor, *122:* 68.

Range: Valley of the Río Motagua in southeastern Guatemala, and the valley of the Río Aguan in northern Honduras, at elevations from 150 to 250 meters.

Ctenosaura pectinata (Wiegmann)

1834 *Cyclura pectinata* Wiegmann, Herp. Mex., Saur. Spec., Berlin, 42; Pl. 2. – **Type locality:** not given (**Holotype:** Zool. Mus. Berlin No. 574). – **Restricted type locality** (Bailey 1928): Colima, Colima, Mexico.

1845 *Ctenosaura pectinata* – Gray, Cat. Spec. Liz. Coll. Brit. Mus., London, 191.

1886 *Ctenosaura brevirostris* Cope (*syn. fide* Smith 1949), Proc. Am. Philos. Soc., Philadelphia, *23*: 268. – **Type locality:** Colima, in Western Mexico (**Holotype:** U.S. Natn. Mus. No. 24709).

1886 *Ctenosaura teres brachylopha* Cope (*syn. fide* Smith 1949), Proc. Am. Philos. Soc., Philadelphia, **23**: 269. – **Type locality:** Mazatlan, Sinaloa, Mexico (**Syntypes:** U.S. Natn. Mus. No. 7180-7183).

1928 *Ctenosaura brachylopha* – Bailey, Proc. U.S. Natn. Mus., Washington, D.C. *73* 12: 22; Pl. 6.

1928 *Ctenosaura pectinata* – Bailey, Proc. U.S. Natn. Mus., Washington, D.C. *73* 12: 24; Pl. 7-11.

1928 *Ctenosaura brevirostris* – Bailey, Proc. U.S. Natn. Mus., Washington, D.C. *72* 12: 27; Pl. 12, 13, 15.

1928 *Ctenosaura parkeri* Bailey (*syn. fide* Smith 1949), Proc. U.S. Natn. Mus., Washington, D.C. *73* 12: 29; Pl. 14, 15. – **Type locality:** Barranca Ibarra, Jalisco, Mexico (**Holotype:** U.S. Natn. Mus. No. 18967).

1949 *Ctenosaura pectinata* – Smith, J. Wash. Acad. Sci., Washington, D.C. *39* 1: 36.

Range: Western México, from just north of Culiacan, Sinaloa, southward at elevations below 1000 meters to the Isthmus of Tehuantepec in southeastern Oaxaca; also the Islas de las Tres Marías in the Pacific Ocean west of Nayarit.

Ctenosaura quinquecarinata (Gray)

1842 *Cyclura quinquecarinata* Gray, Zool. Misc., London, *1842:* 59. – **Type locality:** Demarara? (**Holotype:** Brit. Mus. Nat. Hist. No. 41.3.5.61 [RR 1946.8.30.48]). – **Restricted type locality** (Bailey 1928): Tehuantepec, Oaxaca, Mexico.

1845 *Enyaliosaurus quinquecarinatus* – Gray, Cat. Spec. Liz. Coll. Brit. Mus., London, 192.

1869 *Cyclura (Ctenosaura) quinquecarinata* – Cope, Proc. Am. Philos. Soc., Philadelphia, *11:* 161.

1874 *Ctenosaura (Enyaliosaurus) quinquecarinata* – Bocourt *in* Duméril & Bocourt, Miss. Sci. Mex., Paris, *3:* 138.

1928 *Ctenosaura quinquecarinata* – Bailey, Proc. U.S. Natn. Mus., Washington, D.C. *73* 12: 42; Pl. 24-26.

Range: Isolated populations in arid lowland forests in southern México and northern Central America: in México known from the vicinity of Tehuantepec westward to Puerto Escondido and eastward to Juchitán, Oaxaca; in Honduras from near La Paz and Yaro; in Nicaragua from the states of Chontales, Matagalpa, Jinotega, Boaco, Managua, and Granada; in San Salvador from no specific locality; and in Costa Rica from the state of Guanacaste.

Ctenosaura similis Gray

1831 *Iguana (Ctenosaura) Similis* Gray *in* Cuvier *edit.* Griffith, Anim. Kingd., London, *9:* 38. – **Type locality:** not given (**Holotype:** not located). – **Restricted type locality** (Bailey 1928): Tela, Honduras, Central America.

1928 *Ctenosaura similis* – Bailey, Proc. U.S. Natn. Mus., Washington, D.C. *73* 12: 32; Pl. 16-20.

Range: Low to moderate elevations on both Pacific and Atlantic versants from the Isthmus of Tehuantepec, México, southward to Panamá, and the Colombian Caribbean Islands.

Ctenosaura similis similis Gray

1874 *Ctenosaura completa* Bocourt (*syn. fide* Bailey 1928) *in* Duméril & Bocourt Miss. Sci. Mex., Paris, *3*: 145. – **Type locality:** Guatemala. . . .[and] Union (**Syntypes:** Mus. Hist. Nat. Paris No. 2252, 2256, 6499, 6500; Mus. Comp. Zool. No. 22662). – **Restricted type locality** (Smith & Taylor 1950): La Unión.

1934 *Ctenosaura similis similis* – Barbour & Shreve, Occ. Pap. Boston Soc. Nat. Hist., *8:* 197.

1950 *Ctenosaura similis similis* – Smith & Taylor, Bull. U.S. Natn. Mus., Washington, 199: 73.

Range: Southern México from the Isthmus of Tehuantepec southward along both Atlantic and Pacific versants below 800 meters through Central America to the sandy beaches of Panamá at least as far as Colón in the north and Panamá in the south; also Isla Mujeres, Isla del Carmen, and Isla Aguada off the Yucatan Peninsula, Isla de Utilla and Isla de Guanaja off the north coast of Honduras, Isla San Miguel in the eastern Golfo de Panamá, and the Colombian Isla San Andres in the southwestern Caribbean.

Ctenosaura similis multipunctata Barbour & Shreve

1934 *Ctenosaura similis multipunctata* Barbour & Shreve, Occ. Pap. Boston Soc. Nat. Hist., *8:* 197. – **Type locality:** Old Providence Island (**Holotype:** Mus. Comp. Zool. No. 36830).

Range: Colombian Isla Providencia in the Southwestern Caribbean.

Cyclura Harlan

1824 *Cyclura* Harlan, J. Acad. Nat. Sci. Philad., *4*: 250. – **Type species** (subsequent designation by Fitzinger 1843): *Cyclura carinata* Harlan 1824.

1830 *Metapoceros* Wagler, Natür Syst. Amphib., München, 147. – **Type species** (by monotypy): *Iguana cornuta* Bonnaterre 1789.

1837 *Aloponotus* Duméril & Bibron, Erpét. Gén., Paris, *4:* 189. - **Type species** (by monotypy): *Aloponotus ricordii* Duméril & Bibron 1837.
1843 *Hypsilophus (Aloponotus)* - Fitzinger, Syst. Rept., Wien, *1:* 54.
1843 *Hypsilophus (Metapoceros)* – Fitzinger, Syst. Rept., Wien, *1*: 54.
1843 *Cyclura (Cyclura)* – Fitzinger, Syst. Rept., Wien, *1*: 56.
1977 *Cyclura* – Schwartz & Carey, Stud. Faun. Curaçao & Carib. Is., Utrecht, *53* 173: 21.
Range: The Bahama Islands, Cuba and nearby islets and archipelagos, the Cayman Islands, Navassa Island, Mona Island, Hispaniola and its satellite islands, Jamaica and its satellite islands, and Anegada Island.

Cyclura carinata Harlan

1824 *Cyclura carinata* Harlan, J. Acad. Nat. Sci. Philad., *4*: 250. – **Type locality:** Turk's Island (**Holotype:** not located).
1831 *Iguana (Cyclura) Carinata* – Gray *in* Cuvier *edit.* Griffith, Anim. Kingd., *9:* 39.
1843 *Cyclura (Cyclura) carinata* – Fitzinger *(partim),* Syst. Rept., Wien, *1:* 48.
1916 *Cyclura carinata* – Barbour & Noble, Bull. Mus. Comp. Zool. Harv. Cambridge, *60* 4: 157; Pl. 8, Fig. 3,4; Pl. 13, Fig. 3,4.
1977 *Cyclura carinata* – Schwartz & Carey, Stud. Faun. Curaçao & Carib. Is., Utrecht, *53* 173: 68.
Range: The Caicos Islands, Turk's Islands, and Booby Cay off Mayaguana Island in the eastern Bahama Islands.

Cyclura carinata carinata Harlan

1935 *Cyclura carinata carinata* – Barbour, Zoologica, New York, *19* 3: 118.
1977 *Cyclura carinata carinata* – Schwartz & Carey, Stud. Faun. Curaçao & Carib. Is., Utrecht, *53* 173: 69; Fig. 17.
Range: The Caicos Islands (Pine Cay, Ft. George Cay, North Caicos, Big Iguana Cay off East Caicos, Big Ambergris Cay, Little Water Cay, and Fish Cay), and the Turk's Islands (Big Sand Cay, Long Cay) in the eastern Bahama Islands.

Cyclura carinata bartschi Cochran

1931 *Cyclura carinata bartschi* Cochran, J. Wash. Acad. Sci., Washington, D.C. *21* 3: 39. – **Type locality:** Booby Cay, east of Mariguana Island, Bahamas (**Holotype:** U.S. Natn. Mus. No. 81212).
1977 *Cyclura carinata bartschi* – Schwartz & Carey, Stud. Faun. Curaçao & Carib. Is., Utrecht, *53* 173: 72.
Range: Booby Cay off Mayaguana Island in the eastern Bahama Islands.

Cyclura collei Gray

1845 *Cyclura Collei* Gray, Cat. Spec. Liz. Coll. Brit. Mus., London, 190. – **Type locality:** Jamaica (**Holotype:** Brit. Mus. Nat. Hist. 1936.12.3.108).
1848 *Cyclura lophoma* Gosse (*syn. fide* Grant 1940), Proc. Zool. Soc. London, *1848:* 99. – **Type locality:** Jamaica (**Holotype:** Brit. Mus. Nat. Hist. No. 47.12.27.101).
1916 *Cyclura collei* – Barbour & Noble, Bull. Mus. Comp. Zool. Harv., Cambridge, *60* 4: 158; Pl. 9, Pl. 15, Fig. 5, 6.

1940 *Cyclura collei* – Grant, Bull. Inst. Jamaica, Sci. Ser., Kingston, 1 (2): 97.
1977 *Cyclura collei* – Schwartz & Carey, Stud. Faun. Curaçao & Carib. Is., Utrecht, *53* 173: 56; Fig. 14.
Range: Jamaica (close to extinction; may still occur in the Hellshire Hills.), Goat Island and Little Goat Island.

Cyclura cornuta (Bonnaterre)

1789 *Lacerta cornuta* Bonnaterre, Tab. Encycl. Méth. Règ. Nat., Erpét., Paris, 40; Pl. 4, Fig. 4. – **Type locality:** Sainte-Domingue. . . .dans les mornes de l'Hôpital, entre l'Artibonite & les Gonaves (**Holotype:** not located).
1789 *Iguana cornuta* – Lacépède, Hist. Nat. Quad. Ovip. et Serp., Paris, *2*: 493.
1830 *Metapoceros cornutus* – Wagler, Natür. Syst. Amphib., München, 147.
1843 *Hypsilophus (Metapoceros) cornutus* – Fitzinger, Syst. Rept., Wien, *1:* 54.
1886 *Cyclura cornuta* – Cope, Proc. Am. Phil. Soc., Philadelphia, *23* 122: 263.
1977 *Cyclura cornuta* – Schwartz & Carey, Stud. Faun. Curaçao & Carib. Is., Utrecht, *53* 173: 47.
Range: Hispaniola, including the islands of Ile Petite Cayemite, Isla Saona, and Isla Cabritos in Lago Enriquillo, and the off-shore islands of Isla Beata, Ile de la Petite Gonave, Ile de la Tortue, and Ile Grand Cayemite, as well as the islands of Mona and Navassa.

Cyclura cornuta cornuta (Bonnaterre)

1937 *Cyclura cornuta cornuta* – Barbour, Bull. Mus. Comp. Zool. Harv., Cambridge, *82* 2: 132.
1941 *Cyclura cornuta cornuta* – Cochran, Bull. U.S. Natn. Mus., Washington, D.C. 177: 195; Fig. 92.
1977 *Cyclura cornuta cornuta* – Schwartz & Carey, Stud. Faun. Curaçao & Carib. Is., Utrecht, *53* 173: 48; Fig. 12.
Range: Xeric areas of Hispaniola in Haiti and southwestern República Dominicana, including the islands of Ile Petite Cayemite, Isla Saona and Isla Cabritos in Lago Enriquillo, and the Hispaniolan satellite islands of Isla Beata, Ile de Petite Gonave, Ile de Tortue and Ile Grande Cayemite.

Cyclura cornuta onchiopsis Cope

1885 *C.[yclura] onchiopsis* Cope, Am. Nat., Lancaster, *19* 10: 1006. – **Type locality:** unknown (**Syntypes:** U.S. Natn. Mus. No. 9977, 12239; Mus. Comp. Zool. No. 4717). – **Restricted type locality** (Cope 1886): Navassa Island.
1885 *C.[yclura] nigerrima* Cope (*syn. fide* Schwartz & Thomas 1975), Am. Nat., Lancaster, *19* 10: 1006. – **Type locality:** Navassa (**Holotype:** U.S. Natn. Mus. No. 9974).
1886 *Cyclura onchiopsis* – Cope, Proc. Am. Philos. Soc., Philadelphia, *23:* 264.
1975 *Cyclura cornuta onchiopsis* – Schwartz & Thomas, Carnegie Mus. Nat. Hist. Spec. Publ., Pittsburgh, 1: 112.
1977 *Cyclura cornuta onchiopsis* – Schwartz & Carey, Stud. Faun. Curaçao & Carib. Is., Utrecht, *53* 173: 54.
Range: Navassa Island, probably extinct.

Cyclura cornuta stejnegeri Barbour & Noble

1916 *Cyclura stejnegeri* Barbour & Noble, Bull. Mus. Comp. Zool. Harv., Cambridge, *60* 4: 163; Pl. 12. – **Type locality:** Mona Island (**Holotype:** U.S. Natn. Mus. No. 29367).

1975 *Cyclura cornuta stejnegeri* – Schwartz & Thomas, Carnegie Mus. Nat. Hist. Spec. Publ., Pittsburgh, 1: 112.

1977 *Cyclura cornuta stejnegeri* – Schwartz & Carey, Stud. Faun. Curaçao & Carib. Is., Utrecht, *53* 173: 51.

Range: Isla Mona between Hispaniola and Puerto Rico.

Cyclura cychlura (Cuvier)

1829 *Iguana cychlura* Cuvier, Règ. Anim., Ed. 2, Paris, *2:* 45. – **Type locality:** Carolina (**Holotype:** Mus. Hist. Nat. Paris No. 2367). – **Restricted type locality** (Schwartz & Thomas 1975): Andros Island, Bahama Islands.

1975 *Cyclura cychlura* – Schwartz & Thomas, Carnegie Mus. Nat. Hist. Spec. Publ., Pittsburgh, 1: 112.

1977 *Cyclura cychlura* – Schwartz & Carey, Stud. Faun. Curaçao & Carib. Is., Utrecht, *53* 173: 37.

Range: Andros Island and the Exuma Cays (except White Cay) in the western Bahama Islands.

Cyclura cychlura cychlura (Cuvier)

1862 *Cyclura baelopha* Cope (*syn. fide* Schwartz & Thomas 1975), Proc. Acad. Nat. Sci. Philad., (1861) *13:* 123. – **Type locality:** Andros Island, one of the Bahamas (**Holotype:** Acad. Nat. Sci Philad. No. 8120).

1975 *Cyclura cychlura cychlura* – Schwartz & Thomas, Carnegie Mus. Nat. Hist. Spec. Publ., Pittsburgh, 1: 112.

1977 *Cyclura cychlura cychlura* – Schwartz & Carey, Stud. Faun. Curaçao & Carib. Is., Utrecht, *53* 173: 39; Fig. 10.

Range: Andros Island in the western Bahama Islands.

Cyclura cychlura figginsi Barbour

1923 *Cyclura figginsi* Barbour, Proc. New Engl. Zool. Club, Cambridge, *8:* 108. – **Type locality:** Bitter Guana Cay, near Great Guana Cay, Exuma Group, Bahama Islands (**Holotype:** Mus. Comp. Zool. No. 17745).

1975 *Cyclura cychlura figginsi* – Schwartz & Thomas, Carnegie Mus. Nat. Hist. Spec. Publ., Pittsburgh, 1: 112.

1977 *Cyclura cychlura figginsi* – Schwartz & Carey, Stud. Faun. Curaçao & Carib. Is., Utrecht, *53* 173: 44.

Range: The Exuma Cays, including Guana Cay, Prickly Pear Cay, Allan Cay, Guana Cay off the north end of Norman's Pond Cay, Bitter Guana Cay, Gaulin Cay, and possibly Ozie Cay, all in the western Bahama Islands.

Cyclura cychlura inornata Barbour & Noble

1916 *Cyclura inornata* Barbour & Noble, Bull. Mus. Comp. Zool. Harv., Cambridge, *60* 4: 151; Pl. 14 – **Type locality:** U Cay in Allan's Harbor, near Highborn Cay, Bahamas (**Holotype:** Mus. Comp. Zool. No. 11602).

1975 *Cyclura cychlura inornata* – Schwartz & Thomas, Carnegie Mus. Nat. Hist. Spec. Publ., Pittsburgh, 1: 112.
1977 *Cyclura cychlura inornata* – Schwartz & Carey, Stud. Faun. Curaçao & Carib. Is., Utrecht, *53* 173: 42.
Range: U Cay (= Southwest Allan's Cay) and Leaf Cay in Allan's Cay at the northern extreme of the Exuma Cays in the western Bahama Islands.

Cyclura nubila Gray

1831 *Iguana (Cyclura) Nubila* Gray *in* Cuvier *edit.* Griffith, Anim. Kingd., London, *9:* 39. – **Type locality:** South America? (**Holotype:** Brit. Mus. Nat. Hist. No. xxii.18.a [RR 1946.8.29.88]). – **Restricted type locality** (Schwartz & Thomas 1975): Cuba.
1977 *Cyclura nubila* – Schwartz & Carey, Stud. Faun. Curaçao & Carib. Is., Utrecht, *53* 173: 23.
Range: Cuba and the Isla de Pinos, and nearby islets and archipelagos, and the Cayman Islands.

Cyclura nubila nubila Gray

1837 *Cyclura Harlani* Duméril & Bibron (*partim; syn. fide* Schwartz & Thomas 1975), Erpét. Gén., Paris, *4:* 218. - **Type locality:** Caroline (**Syntypes:** Mus. Hist. Nat. Paris No. A661, 2367; **Lectotype** (Schwartz & Carey 1977): Mus. Hist. Nat. Paris No. A661).
1845 *Cyclura MacLeayii* Gray (*syn. fide* Schwartz & Thomas 1975), Cat. Spec. Liz. Coll. Brit. Mus., London, 190. – **Type locality:** Cuba (**Holotype:** Brit. Mus. Nat. Hist. No. xx.17.a [RR 1946.8.4.28]).
1916 *Cyclura macleayi* – Barbour & Noble, Bull. Mus. Comp. Zool. Harv., Cambridge, *60* 4: 145; Pl. 1, 2; Pl. 13, Fig. 5, 6.
1975 *Cyclura nubila nubila* – Schwartz & Thomas, Carnegie Mus. Nat. Hist. Spec. Publ., Pittsburgh, 1: 113.
1977 *Cyclura nubila nubila* – Schwartz & Carey, Stud. Faun. Curaçao & Carib. Is., Utrecht, *53* 173: 24; Fig. 7.
Range: Cuba and the Isla de Pinos, and numerous islets of the Archipiélago de los Canarreos, Cayos de San Felipe, Jardin de la Reina, and Archipiélago de Sabana-Camagüey, and presumably other nearby islets and cays; introduced on Isla Magueyes off the southwestern coast of Puerto Rico.

Cyclura nubila caymanensis Barbour & Noble

1916 *Cyclura caymanensis* Barbour & Noble, Bull. Mus. Comp. Zool. Harv., Cambridge, *60* 4: 148; Pl. 3. – **Type locality:** Cayman Islands, probably Cayman Brac (**Holotype:** Mus. Comp. Zool. No. 10534).
1940 *Cyclura macleayi caymanensis* – Grant, Bull. Inst. Jamaica, Sci. Ser., Kingston, *2*: 29; Pl. 1, Fig. 1, 2.
1975 *Cyclura nubila caymanensis* – Schwartz & Thomas, Carnegie Mus. Nat. Hist. Spec. Publ., Pittsburgh, 1: 113.
1977 *Cyclura nubila caymanensis* – Schwartz & Carey, Stud. Faun. Curaçao & Carib. Is., Utrecht, *53* 173: 30.
Range: Cayman Brac and Little Cayman Island; introduced on Grand Cayman Island.

Cyclura nubila lewisi Grant

1940 *Cyclura macleayi lewisi* Grant, Bull. Inst. Jamaica, Sci. Ser., Kingston, *2:* 35, Pl. 2, Fig. 3, 4. – **Type locality:** Battle Hill, east end of Grand Cayman (**Holotype:** Brit. Mus. Nat. Hist. No. 1939.2.3.68 [RR 1946.8.9.32]).
1975 *Cyclura nubila lewisi* – Schwartz & Thomas, Carnegie Mus. Nat. Hist. Spec. Publ., Pittsburgh, 1: 113.
1977 *Cyclura nubila lewisi* – Schwartz & Carey, Stud. Faun. Curaçao & Carib. Is., Utrecht, *53* 173: 33.
Range: Grand Cayman Island.

Cyclura pinguis Barbour

1917 *Cyclura pinguis* Barbour, Proc. Biol. Soc. Wash., Washington, D.C. *30:* 100. – **Type locality:** Anegada, British Virgin Islands (**Holotype:** Mus. Comp. Zool. No. 12082).
1977 *Cyclura pinguis* – Schwartz & Carey, Stud. Faun. Curaçao & Carib. Is., Utrecht, *53* 173: 60; Fig. 15.
Range: Anegada Island on the Puerto Rican Bank.

Cyclura ricordii Duméril & Bibron

1837 *Aloponotus Ricordii* Duméril & Bibron, Erpét. Gén., Paris, *4:* 190; Pl. 38. – **Type locality:** Sainte-Domingue (**Holotype:** Mus. Hist. Nat. Paris. No. 8304).
1843 *Hypsilophus (Aloponotus) Ricordii* – Fitzinger, Syst. Rept., Wien, *1*: 54.
1924 *Cyclura ricordii* – Cochran, Proc. U.S. Natn. Mus., Washington, D.C. *66* 6: 5.
1977 *Cyclura ricordi* – Schwartz & Carey, Stud. Faun. Curaçao & Carib. Is., Utrecht, *53* 173: 64; Fig. 16.
Range: Hispaniola, from the Valle de Neiba and the Península de Barahona south of the Sierra de Baoruco, and Isla Cabritos in Lago Enriquillo, southwestern República Dominicana; presumably also in the Cul de Sac Plain of Haiti.

Cyclura rileyi Stejneger

1903 *Cyclura rileyi* Stejneger, Proc. Biol. Soc. Wash., Washington, D.C. *16:* 130. – **Type locality:** Watlings Island, Bahamas (**Holotype:** U.S. Natn. Mus. No. 31969).
1977 *Cyclura rileyi* – Schwartz & Carey, Stud. Faun. Curaçao & Carib. Is., Utrecht, *53* 173: 74.
Range: Central Bahama Islands, including San Salvador (= Watlings), the Exuma Cays, and islands of the Crooked-Acklins group.

Cyclura rileyi rileyi Stejneger

1975 *Cyclura rileyi rileyi* – Schwartz & Thomas, Carnegie Mus. Nat. Hist. Spec. Publ., Pittsburgh, 1: 114.
1977 *Cyclura rileyi rileyi* – Schwartz & Carey, Stud. Faun. Curaçao & Carib. Is., Utrecht, *53* 173: 75; Fig. 18.
Range: San Salvador (= Watlings), Man Head Cay, and Green Cay in the central Bahama Islands.

Cyclura rileyi cristata Schmidt

1920 *Cyclura cristata* Schmidt, Proc. Linn. Soc. New York, *33*: 6. – **Type locality:** White Cay, Bahama Islands (**Holotype:** Amer. Mus. Nat. Hist. No. 7238). – **Corrected type locality** (Schmidt 1936): White Cay, Exuma Cays, Bahamas.

1936 *Cyclura cristata* – Schmidt, Field Mus. Nat. Hist., Zool. Ser., Chicago, *20* 16: 128.

1975 *Cyclura rileyi cristata* – Schwartz & Thomas, Carnegie Mus. Nat. Hist. Spec. Publ., Pittsburgh, 1: 114.

1977 *Cyclura rileyi cristata* – Schwartz & Carey, Stud. Faun. Curaçao & Carib. Is., Utrecht, *53*, 173: 80.

Range: White Cay, at the southern end of the Exuma Cays, central Bahama Islands.

Cyclura rileyi nuchalis Barbour & Noble

1916 *Cyclura nuchalis* Barbour & Noble, Bull. Mus. Comp. Zool. Harv., Cambridge, *60* 4: 156, Pl. 8, Fig. 1, 2. – **Type locality:** Fortune Island, Bahamas (**Holotype:** Acad. Nat. Sci. Philad. No. 11985).

1975 *Cyclura rileyi nuchalis* – Schwartz & Thomas, Carnegie Mus. Nat. Hist. Spec. Publ., Pittsburgh, 1: 114.

1977 *Cyclura rileyi nuchalis* – Schwartz & Carey, Stud. Fauna Curaçao & Carib. Is., Utrecht, *53* 173: 78.

Range: Fortune Island, Fish Cay and North Cay in the Crooked-Acklin's group, central Bahama Islands.

Dipsosaurus Hallowell

1854 *Dipso-saurus* Hallowell, Proc. Acad. Nat. Sci. Philad., *7*: 92. – **Type species** (by monotypy): *Crotaphytus dorsalis* Baird & Girard 1852.

1855 *Dipsosaurus* – Boulenger, Cat. Liz. Brit. Mus., London, *2*: 201.

Range: Desert regions of southwestern United States in southeastern California, extreme southern Nevada and southwestern Utah and western Arizona, southward in northwestern México through western Sonora and extreme northwestern Sinaloa, and to the southern tip of Baja California, as well as on various islands in the Gulf of California.

Dipsosaurus dorsalis (Baird & Girard)

1852 *Crotaphytus dorsalis* Baird & Girard, Proc. Acad. Nat. Sci. Philad., *6*: 126. – **Type locality:** Desert of Colorado, Cal. (**Holotype:** U.S. Natn. Mus. No. 2699). – **Restricted type locality** (Smith & Taylor 1950): Winterhaven (= Fort Yuma), Imperial County.

1854 *Dipso-saurus dorsalis* – Hallowell, Proc. Acad. Nat. Sci. Philad., *7*: 92.

1950 *Dipsosaurus dorsalis* – Smith & Taylor, Bull. U.S. Natn. Mus., Washington, D.C. 199: 78.

Range: Desert regions of southwestern United States in southeastern California, extreme southern Nevada and southwestern Utah and western Arizona, southward in northwestern México through western Sonora and extreme northwestern Sinaloa, and to the southern tip of Baja California, as well as various islands in the Gulf of California.

Dipsosaurus dorsalis dorsalis (Baird & Girard)

1920 *Dipsosaurus dorsalis dorsalis* – Van Denburgh, Proc. Calif. Acad. Sci., San Francisco, (4) *10* 4: 33.

1922 *Dipsosaurus dorsalis dorsalis* – Van Denburgh, Occ. Pap. Caif. Acad. Sci., *10* 1: 73; Pl. 2.

Range: The lower levels of the Colorado and Mojave Deserts in southern California, extending northward into Owen's, Panamint, Death, Mesquite, and Amargosa valleys, southern Nevada and extreme southwestern Utah, and western Arizona, extending southward in México into extreme northwestern Sonora, and down the peninsula of Baja California, except for the region west of the Sierra de Juarez and the Sierra de San Pedro Martir and north of Punta Santa Rosalia, to at least as far south as Bahía Magdalena; in the Gulf of California on the islands of Encantada Grande, Angel de la Guarda, and San Marcos, and on Magdalena off the Pacific coast.

Dipsosaurus dorsalis catalinensis Van Denburgh

1922 *Dipsosaurus catalinensis* Van Denburgh, Occ. Pap. Calif. Acad. Sci., San Francisco, *10* 1: 83. – **Type locality:** Santa Catalina Island, Gulf of California, Mexico (**Holotype:** Calif. Acad. Sci. No. 50505).

1966 *Dipsosaurus dorsalis catalinensis* - Soulé & Sloan, Trans. San Diego Soc. Nat. Hist., *14* 11: 140.

Range: Isla Santa Catalina in the Gulf of California, México.

Dipsosaurus dorsalis lucasensis Van Denburgh

1920 *Dipsosaurus dorsalis lucasensis* Van Denburgh, Proc. Calif. Acad. Sci., San Francisco, (4) *10* 4: 33. – **Type locality:** San Jose del Cabo, Lower California, Mexico (**Holotype:** Calif. Acad. Sci. No. 46090).

1922 *Dipsosaurus carmenensis* Van Denburgh (*syn. fide* Soulé & Sloan 1966), Occ. Pap. Calif. Acad. Sci., San Francisco, *10* 1: 81. – **Type locality:** Near Puerto Bellandro, Carmen Island, Gulf of California (**Holotype:** Calif. Acad. Sci. No. 50504).

1966 *Dipsosaurus dorsalis lucasensis* – Soulé & Sloan, Trans. San Diego Soc. Nat. Hist., *14* 11: 140.

Range: The southern part of the peninsula of Baja California, and in the Gulf of California on the islands of Carmen, Coronados, Monserrate, San José, and Cerralvo, western México.

Iguana Laurenti

1768 *Iguana* Laurenti, Spec. Med., Synop. Rept., Wien, 47. – **Type species** (by tautonymy): *Lacerta iguana* Linnaeus 1758.

1828 *Prionodus* Wagler, Isis (von Oken), Leipzig, *21* 8/9: 860. – **Type species** (by monotypy): *Lacerta iguana* Linnaeus 1758.

1830 *Hypsilophus* Wagler, Natür Syst. Amphib., München, 147. - **Type species** (by monotypy): *Lacerta iguana* Linnaeus 1758.

1843 *Hypsilophus (Hypsilophus)* – Fitzinger, Syst. Rept., Wien, *1*: 16.

1885 *Iguana* – Boulenger, Cat. Liz. Brit. Mus., London, *2:* 189.

1973 *Iguana* – Lazell, Bull. Mus. Comp. Zool. Harv., Cambridge, *145* 1: 1.

Range: On the American mainland from Sinaloa and Veracruz, México, southward at low elevations through Central America and South America to southern Brazil and Paraguay; in the Caribbean northward through the lesser Antilles to the Virgin Islands.

Iguana delicatissima Laurenti

1768 *Iguana delicatissima* Laurenti, Spec. Med., Synop. Rept., Wien, 48. – **Type locality:** Indiis (**Holotype:** Zool. Mus. Torino, not located). – **Restricted type locality** (Lazell 1973): island of Terre de Bas, Les Iles de Saintes, Departement de la Guadeloupe, French West Indies.

1820 *Iguana nudicollis* Merrem (substitute name for *Iguana delicatissima* Laurenti 1768), Tent. Syst. Amphib., Marburg, 48.

1830 *Amblyrhynchus delicatissima* - Wagler, Natür Syst. Amphib., München, 148.

1885 *Iguana delicatissima* – Boulenger, Cat. Liz. Brit. Mus., London, *2*: 191.

1973 *Iguana delicatissima* – Lazell, Bull. Mus. Comp. Zool. Harv., Cambridge, *145* 1: 18; Fig. 2.

Range: The lesser Antilles on the islands of Anguilla, St. Martin, Ile Fourchue, Les Iles Frégates, Ile Chevreau, St. Barthélemy, St. Eustatius, Nevis (presence now uncertain), Antigua, the Grande-Terre portion of Guadeloupe, La Desirade, Les Iles de Saintes (Terre-de-Bas and Terre-de-Haut), Dominica, and Martinique.

Iguana iguana (Linnaeus)

1758 *Lacerta iguana* Linnaeus, Syst. Nat., Ed. 10, *1:* 206. – **Type locality:** Indiis (**Syntypes:** Nat. Ricksmus. Stockholm, number unknown; Zool. Inst. Univ. Uppsala, number unknown). – **Restricted type locality** (Lazell 1973): island of Terre de Haut, Les Iles de Saintes, Departement de La Guadeloupe, French West Indies; (Hoogmoed 1973): confluence of the Cottica River and Perica Creek, Surinam.

1768 ? *Iguana minima* Laurenti (*syn. fide* Fitzinger 1843), Spec. Med., Synop. Rept., Wien, 48. – **Type locality:** not given (**Holotype:** Zool. Mus. Torino, not located).

1768 *Iguana tuberculata* Laurenti (*syn. fide* Dunn 1934), Spec. Med., Synop. Rept., Wien, 48. – **Type locality:** not given (**Holotype:** Zool. Mus. Torino, not located).

1802 *Iguana coerulea* Daudin (*syn. fide* Fitzinger 1843), Hist. Nat. Rept., Paris, *3:* 286. – **Type locality:** l'ile Formose (**Holotype:** based upon Seba, 1734, Locupl. v. natur. thesaur., *1:* 44; Fig. 4-5).

1806 *I.[guana] vulgaris* Link (substitute name for *Lacerta iguana* Linnaeus 1758), Beschr. Natural.-Samml. Univ. Rostock., *2*: 58.

1820 *Iguana sapidissima* Merrem (substitute name for *Lacerta iguana* Linnaeus 1758), Tent. Syst. Amphib., Marburg, 47.

1825 *Iguana squamosa* Spix (*syn. fide* Gray 1831), Spec. Nov. Lacert. Brazil, Monachii, *1*: 5; Pl. 5. – **Type locality:** Bahiae, Parae (**Syntypes:** Zool. StSamm. München No. 520/0, 537/0).

1825 *Iguana viridis* Spix (*syn. fide* Gray 1831), Spec. Nov. Lacert. Brazil., Monachii, *1*: 6; Pl. 6. – **Type locality:** supra ripam Rio St. Francisci et Itapicuru (**Holotype:** Zool. StSamm. München No. 540/0).

1825 *Iguana coerulea* Spix (*non* Daudin 1802; *syn. fide* Fitzinger 1843), Spec. Nov. Lacert. Brazil., Monachii, *1*: 7; Pl. 7. – **Type locality:** in locis ripariis vel humidis Rio St. Francisci (**Syntypes:** Zool. StSamm. München No. 71/0 (2), destroyed).

1825 *Iguana emarginata* Spix (*syn. fide* Gray 1831), Spec. Nov. Lacert. Brazil., Monachii, *1*: 7; Pl. 8. – **Type locality:** ad flumen St. Francisci (**Syntypes:** Zool. StSamm. München No. 535/0 (2)).

1825 *Iguana lophyroides* Spix (*syn. fide* Fitzinger 1843), Nov. Spec. Lacert. Brazil., Monachii, *1*: 8; Pl. 9. – **Type locality:** in sylvis Rio de Janeiro, Bahiae (**Syntypes:** Zool. StSamm. München No. 546/0 (2)).

1826 *Iguana tuberculata* – Fitzinger, Neu Class. Rept., Wien, *1:* 48.

1828 *Prionodus iguana* – Wagler, Isis (von Oken), Leipzig, *21*: 860.

1830 *Hypsilophus tuberculatus* - Wagler, Natür. Syst. Amphib., München, 147.

1831 *Iguana (Iguana) tuberculata* – Gray *in* Cuvier *edit.* Griffith, Anim. Kingd., London, *9:* 36.

1834 *Iguana (Hypsilophus) rhinolophus* Wiegmann (*syn. fide* Lazell 1973), Herp. Mex., Saur. Spec., Berlin, 44. – **Type locality:** not given (**Syntypes:** Zool. Mus. Berlin No. 571 (2)). – **Restricted type locality** (Smith & Taylor 1950): Córdoba, Veracruz.

1843 *Hypsilophus (Hypsilophus) Rhinolophus* – Fitzinger, Syst. Rept., Wien, *1:* 55.

1843 *Hypsilophus (Hypsilophus) tuberculatus* – Fitzinger, Syst. Rept., Wien, *1:* 55.

1845 *Iguana tuberculata* – Gray, Cat. Spec. Liz. Coll. Brit. Mus., London, 186.

1845 *Iguana rhinolophus* – Gray, Cat. Spec. Liz. Coll. Brit. Mus., London, 186.

1857 ? *Iguana Hernandessi* Jan (*nomen nudum fide* Smith & Taylor 1950), Indice Sistem. Rett. e. Anfib. Medesimo, Milano, 58.

1885 *Iguana tuberculata* – Boulenger, Cat. Liz. Brit. Mus., London, *2*: 189.

1885 *Iguana tuberculata* var. *rhinolopha* – Boulenger, Cat. Liz. Brit. Mus., London, *2*: 190.

1898 *Iguana iguana rhinolopha* – Van Denburgh, Proc. Acad. Nat. Sci. Philad., (1897) *49:* 461.

1934 *Iguana iguana iguana* – Dunn, Copeia, 1: 1.

1934 *Iguana iguana rhinolopha* – Dunn, Copeia, 1: 1.

1950 *Iguana iguana rhinolopha* – Smith & Taylor, Bull. U.S. Natn. Mus., Washington, D.C. 199: 72.

1973 *Iguana iguana* – Lazell, Bull. Mus. Comp. Zool. Harv., Cambridge, 145 1: 7; Fig. 2, 12.

1973 *Iguana iguana iguana* – Hoogmoed, Biogeographica, The Hague, *4:* 148; Fig. 23; Pl. 16.

Range: On the American mainland from northern México (the town of Costa Rica in Sinaloa on the west, and Laguan de Tamiahua, Veracruz on the east) southward, excluding most of the Yucatan Peninsula, through Central America and South America at least to the Tropic of Capricorn in Paraguay and southeastern Brazil; altitude records include to 800 meters in Michoacán, México, to 500 meters in Surinam, and to 1000 meters in Colombia. Pacific island records include the Archipiélago de las Perlas in the Golfo de Panamá, and Isla Gorgona off the coast of Colombia. Island records in the western and southern Caribbean include Isla Cozumel off Quintana Roo, México, Las Islas de la Bahía (Isla de Utilla, Isla de Roatán, and Isla de Guanaja), Honduras, the Corn Islands, Providencia and San Andres, and the coastal South American islands of Margarita, Los Testigos, Los Frailes, Los Hermanos, La Blanquilla, La Tortuga, Isla Orchilla, Los Roques, Isla Aves, Bonaire, Klein Bonaire, Curaçao, Aruba, Trinidad, and Tobago. In the Lesser Antilles records include the Virgin Islands (St. Thomas and its satellites Water Island, Patricia Cay, and Hassel Island, St. John, St. Croix, and Tortola and its satellites Peter Island and Guana Island), Saba, Montserrat, the Guadeloupe Bank (the Basse-Terre portion of Guadeloupe and the Iles de Pigeon ou Goyave), Les Iles des Saintes (La Coche, Grand Ilet, Terre-de-Haut, and Ilet à Cabrit), the St. Lucia Bank (southern tip of the larger Maria Island and the northern coast of St. Lucia), the St. Vincent Bank (St. Vincent and all coastal cays that support trees), and the Grenada Bank (Grenada and on most adjacent cays, Bequia Island, Ile Quatre, Battowia Island, Petit Mustique Island, Savan Island, Cannouan Island, the Tobago Cays, Union Island, Frigate Cay, Petite St. Vincent, Mabuya Cay, Carriacou Island, Kick-'em-Jenny, and Ile-a-Caille).

Sauromalus Duméril

1856 *Sauromalus* Duméril, Arch. Mus. Hist. Nat. Paris, *8:* 535 – **Type species** (by monotypy): *Sauromalus ater* Duméril 1856.
1859 *Euphryne* Baird, Proc. Acad. Nat. Sci. Philad., (1858) *10:* 253. – **Type species** (by monotypy): *Euphryne obesus* Baird 1859.
1885 *Sauromalus* – Boulenger, Cat. Liz. Brit. Mus., London, *2*: 202.
1945 *Sauromalus* – Shaw, Trans. San Diego Soc. Nat. Hist., *10* 15: 269.
Range: Desert regions of southwestern United States in southern California, extreme southern Utah and Nevada, and western and central Arizona, and western México in western Sonora, various islands in the Gulf of California, and the eastern part of southern Baja California.

Sauromalus ater Duméril

1856 *Sauromalus ater* Duméril, Arch. Mus. Hist. Nat. Paris, *8:* 536, Pl. 23, Fig. 3, 3a – **Type locality:** not given (**Holotype:** Mus. Hist. Nat. Paris No. 813). - **Restricted type locality** (Smith & Taylor 1950): Espíritu Santo Island.
1919 *Sauromalus interbrachialis* Dickerson (*syn. fide* Schmidt 1922), Bull. Am. Mus. Nat. Hist., New York, *41* 10: 463. – **Type locality:** La Paz, Lower California (*en error fide* Schmidt 1922) (**Holotype:** U.S. Natn. Mus. No. 64443).
1922 *Sauromalus ater* – Schmidt, Bull. Am. Mus. Nat. Hist., New York, 46 11: 640 (part).
1945 *Sauromalus ater* – Shaw, Trans. San Diego Soc. Nat. Hist., *10* 15: 284.
1950 *Sauromalus ater* – Smith & Taylor, Bull. U.S. Natn. Mus., Washington, 199: 80.

Range: The islands of Espíritu Santo, Partida, San José, San Francisco, San Diego, Santa Cruz, San Marcos, and Santa Catalina in the Gulf of California, México.

Sauromalus ater ater Duméril

1966 *Sauromalus ater ater* – Soulé & Sloan, Trans. San Diego Soc. Nat. Hist., *14* 11: 141.

Range: The islands of Espíritu Santo, Partida, San José, San Francisco, San Diego, and Santa Cruz in the Gulf of California, México.

Sauromalus ater klauberi Shaw

1941 *Sauromalus klauberi* Shaw, Trans. San Diego Soc. Nat. Hist., *9* 28: 285. – **Type locality:** Santa Catalina Island, Gulf of California, Mexico (**Holotype:** San Diego Soc. Nat. Hist. No. 6859).

1966 *Sauromalus ater klauberi* – Soulé & Sloan, Trans. San Diego Soc. Nat. Hist., *14* 11: 141.

Range: Santa Catalina Island in the Gulf of California, México.

Sauromalus ater shawi Cliff

1958 *Sauromalus shawi* Cliff, Copeia, *1958* 4: 259. – **Type locality:** San Marcos Island (**Holotype:** Stanford Univ. Mus. No. 16120).

1966 *Sauromalus ater shawi* – Soulé & Sloan, Trans. San Diego Soc. Nat. Hist., *14* 11: 141.

Range: San Marcos Island in the Gulf of California, México.

Sauromalus australis Shaw

1945 *Sauromalus australis* Shaw, Trans. San Diego Soc. Nat. Hist., *10* 15: 286. – **Type locality:** San Franciscito Bay, Baja California, Mexico (**Holotype:** San Diego Soc. Nat. Hist. No. 30170).

Range: The eastern part of southern Baja California, from Punta San Gabriel southward to La Paz, México.

Sauromalus hispidus Stejneger

1891 *Sauromalus hispidus* Stejneger, Proc. U.S. Natn. Mus., Washington, D.C. *14* 864: 409. – **Type locality:** Angel de la Guarda Island, Gulf of California (**Holotype:** U.S. Natn. Mus. No. 8563).

1922 *Sauromalus hispidus* – Van Denburgh, Occ. Pap. Calif. Acad. Sci., San Francisco, *10* 1: 99; Pl. 5, 6.

1945 *Sauromalus hispidus* – Shaw, Trans. San Diego Soc. Nat. Hist., *10* 15: 279.

Range: The islands of Angel de la Guarda, Smith, Pond, Granite, Mejía, San Lorenzo Norte, and San Lorenzo Sur in the Gulf of California. México.

Sauromalus obesus (Baird)

1859 *Euphryne obesus* Baird, Proc. Acad. Nat. Sci. Philad. (1858) *10:* 253. – **Type locality:** Fort Yuma (**Holotype:** U.S. Natn. Mus. No. 4172).

1922 *Sauromalus obesus* – Schmidt, Bull. Am. Mus. Nat. Hist., New York, *46* 11: 641 (part).

Range: Southwestern United States in southern California, extreme southern Nevada and Utah, western and central Arizona, and extreme northwestern México in western Sonora.

Sauromalus obesus obesus (Baird)

1945 *Sauromalus obesus obesus* – Shaw, Trans. San Diego Soc. Nat. Hist., *10* 15: 295.

Range: Desert regions of southwestern United States in southern California east of the mountains, extreme southern Nevada and southwestern Utah, and western and central Arizona.

Sauromalus obesus multiforminatus Tanner & Avery

1964 *Sauromalus obesus multiforminatus* Tanner & Avery, Herpetologica, *20* 1: 38. – **Type locality:** North Wash, 11 miles northwest of Hite, Garfield County, Utah (**Holotype:** Brig. Young Univ. No. 11376).

Range: The Colorado River area from Glenn Canyon Dam in northern Arizona, northward and eastward to just north of Hite in southern Utah.

Sauromalus obesus townsendi Dickerson

1919 *Sauromalus townsendi* Dickerson, Bull. Am. Mus. Nat. Hist., New York, *41* 10: 464. – **Type locality:** Tiburon Island, Gulf of California, Mexico (**Holotype:** U.S. Natn. Mus. No. 64442).

1922 *Sauromalus townsendi* – Schmidt, Bull. Am. Mus. Nat. Hist., New York, *46* 11: 643; Fig. 3c, 3d.

1945 *Sauromalus obesus townsendi* – Shaw, Trans. San Diego Soc. Nat. Hist., *10* 15: 290.

Range: Tiburon Island in the Gulf of California, and the adjacent coast of western Sonora southward at least as far as Guaymas, and inland to the vicinity of Hermosillo, northwestern México.

Sauromalus obesus tumidus Shaw

1945 *Sauromalus obesus tumidus* Shaw, Trans. San Diego Soc. Nat. Hist., *10* 15: 292. – **Type locality:** Telegraph Pass, Gila Mountains, Yuma County, Arizona (**Holotype:** San Diego Soc. Nat. Hist. No. 27323).

Range: Southwestern Arizona and adjacent extreme northwestern Sonora, northwestern México.

Sauromalus slevini Van Denburgh

1922 *Sauromalus slevini* Van Denburgh, Occ. Pap. Calif. Acad. Sci., San Francisco, *10* 1: 97. – **Type locality:** South end of Monserrate Island, Gulf of California, Mexico (**Holotype:** Calif. Acad. Sci. No. 50503).

1945 *Sauromalus slevini* – Shaw, Trans. San Diego Soc. Nat. Hist., *10* 15: 280.

Range: The islands of Monserrate, Carmen and Coronados in the Gulf of California, western México.

Sauromalus varius Dickerson

1919 *Sauromalus varius* Dickerson, Bull. Am. Mus. Nat. Hist., New York, *41* 10: 464. – **Type locality:** San Esteban Island, Gulf of California, Mexico (**Holotype:** U.S. Natn. Mus. No. 64441).

1922 *Sauromalus varius* – Schmidt, Bull. Am. Mus. Nat. Hist., New York, *46* 11: 641; Pl. 48.

1945 *Sauromalus varius* – Shaw, Trans. San Diego Soc. Nat. Hist., *10* 15: 288.

Range: The Islands of San Estebán, Lobos, and Pelicano in the Gulf of California, western México.

THE MALAGASY IGUANIDS

Chalarodon Peters

1837 *Tropidogaster* Duméril & Bibron (Official Index [Invalid], Op. 955, 1971). Erpét. gén., Paris, *4:* 329. – **Type species** (by monotypy): *Tropidogaster blainvillii* Duméril & Bibron 1837.
1843 *Ptychosaurus (Tritropis)* Fitzinger (Official Index [Invalid], Op. 955, 1971), Syst. Rept., Wien, *1*: 59. – **Type species** (by original designation): *Tropidogaster blainvillii* Duméril & Bibron 1837.
1854 *Chalarodon* Peters, Mber. K. Akad. Wiss., Berlin, 616. – **Type species** (by monotypy): *Chalarodon madagascariensis* Peters 1854.
1885 *Chalarodon* – Boulenger, Cat. Liz. Brit. Mus., London, *2*: 128.
Range: Arid and semiarid regions of southwestern Madagascar, throughout most of Tulear Province, extending into southwestern Fianarantsoa and extreme southwestern Majunga Provinces.

Chalarodon madagascariensis Peters

1837 *Tropidogaster Blainvillii* Duméril & Bibron (Official Index [Invalid], Op. 955, 1971), Erpét. gén., Paris, *4*: 330. – **Type locality:** inconnue (**Holotype:** Mus. Hist. Nat. Paris No. 6869).
1843 *Ptychosaurus (Tritropis) Blainvillii* – Fitzinger (Official Index [Invalid], Op. 955, 1971), Syst. Rept., Wien, *1*: 59.
1854 *Chalarodon madagascariensis* Peters, Mber. K. Akad. Wiss., Berlin, 616. – **Type locality:** Madagascar (St. Augustins Bay) (**Syntypes:** Zool. Mus. Berlin No. 4360 (2), 5617, 9214) (2)).
1885 *Chalarodon madagascariensis* – Boulenger, Cat. Liz. Brit. Mus., London, *2:* 128.
1885 *Tropidurus ? blainvillii* – Boulenger, Cat. Liz. Brit. Mus., London, *2*: 178.
1933 *Tropidurus blainvillii* – Burt & Burt, Trans. Acad. Sci. St. Louis, *27* 1: 45.
1942 *Chalarodon madagascariensis* – Angel, Mem. l'Akad. Malag., Tananarive, *36:* 89; Pl. 13, Fig. 3.
1977 *Chalarodon madagascariensis* – Blanc, Faune de Madagascar, Paris, *20* 59; Pl. 7.
Range: Arid and semiarid regions of southwestern Madagascar, throughout most of Tuléar Province, extending into southwestern Fianarantsoa and extreme southwestern Majunga Provinces.

Oplurus Cuvier

1829 *Oplurus* Cuvier, Règ. Anim., Paris, Ed. 2, *2:* 47. – **Type species** (by monotypy): *Oplurus torquatus* Cuvier 1829; subsequent invalid designation by Fitzinger 1843: *Oplurus sebae* Duméril & Bibron 1837.
1843 *Hoplurus* – Fitzinger (invalid emendation of *Oplurus* Cuvier 1829), Syst. Rept., Wien, *1:* 76.
1843 *Doryphorus* Fitzinger (*non* Cuvier 1829), Syst. Rept., Wien, *1:* 77. – **Type species** (by original designation): *Oplurus maximiliani* Duméril & Bibron 1837.
1885 *Hoplurus* – Boulenger, Cat. Liz. Brit. Mus., London, *2*: 129.

1942 *Hoplurus* – Angle, Mem. l'Akad. Malag., Tananarive, *36*: 82.
1952 *Oplurus* – Savage, Copeia, *1952* 3: 182.
Range: Western and central Madagascar and Grand Comore Island.

Oplurus cuvieri (Gray)

1829 *Oplurus torquatus* Cuvier (secondary homonym for *Tropidurus torquatus* Wied 1821), Règ. Anim., Ed. 2, Paris, *2*: 48. – **Type locality:** Brésil (**Holotype:** Mus. Hist. Nat. Paris No. 1433).
1831 *Trop.[idurus] Cuvieri* Gray *in* Cuvier *edit.* Griffith (substitute name for *Oplurus torquatus* Cuvier 1829), Anim. Kingd., London, *9:* 41.
1837 *Oplurus Sebae* Duméril & Bibron (*syn. fide* Savage 1952), Erpét. gén., Paris, *4*: 361. – **Type locality:** Brésil (**Holotype:** Mus. Hist. Nat. Paris No. 1433).
1843 *Hoplurus Sebae* – Fitzinger, Syst. Rept., Wien, *1:* 76.
1845 *Hoplurus Barnardi* Peters (part; *syn. fide* Boulenger 1885), Mber. K. Akad. Wiss., Berlin, 616. – **Type locality:** Madagascar (Bombatuka, St. Augustins-Bay) (**Syntypes:** Zool. Mus. Berlin No. 674 (2), 3951, 5393).
1885 *Hoplurus sebae* – Boulenger, Cat. Liz. Brit. Mus., London, *2:* 129.
1952 *Oplurus cuvieri* – Savage, Copeia, *1952* 3: 182.
1977 *Oplurus cuvieri* – Blanc, Faune de Madagascar, Paris, *45*: 28.
Range: Northwestern Madagascar and Grand Comore Island.

Oplurus cuvieri cuvieri (Gray)

1942 *Hoplurus sebae [sebae]* – Angel, Mem. l'Akad. Malag., Tananarive, *36:* 83; Pl. 3, Fig. 4; Pl. 12, Fig. 1.
1952 *Oplurus cuvieri cuvieri* – Savage, Copeia, *1952* 3: 182.
1977 *Oplurus cuvieri cuvieri* – Blanc, Faune de Madagascar, Paris, *45:* 28; Pl. 1.
Range: Subhumid regions of northwestern Madagascar from western Diego-Suarez Province southward into northern Tuléar Province and inland to the western slopes of the Tananarive Province, with an isolated population in northwestern Fianarantsoa Province.

Oplurus cuvieri comorensis Angel

1942 *[Hoplurus sebae]* var. *comorensis* Angel, Mem. l'Akad. Malag., Tananarive, *36:* 84. – **Type locality:** Grand Comore (**Syntypes:** Mus. Hist. Nat. Paris No. 22-298, 22-299).
1952 *O.[plurus] cuvieri comorensis* – Savage, Copeia, *1952* 3: 182.
Range: Grand Comore Island (westward of northern Madagascar in the northern Mozambique Channel).

Oplurus cyclurus (Merrem)

1820 *Uromastyx cyclurus* Merrem, Tent. Syst. Amphib., Marburg, 56. – **Type locality:** Brasilia (**Holotype:** not located).
1837 *Oplurus Maximiliani* Duméril & Bibron (*syn. fide* Boulenger 1885), Erpét. gén., Paris, *4:* 365. – **Type locality:** Brésil (**Holotype:** Mus. Hist. Nat. Paris No. 1431).
1843 *Dorphorus Maximiliani* – Fitzinger, Syst. Rept., Wien, *1:* 77.
1885 *Hoplurus cyclurus* – Boulenger, Cat. Liz. Brit. Mus., London, *2:* 130.

1942 *Hoplurus cyclurus* – Angel, Mem. l'Akad. Malag., Tananarive, *36:* 84; Pl. 3, Fig. 30; Pl. 11, Fig. 4.
1952 *Oplurus cyclurus* – Savage, Copeia, *1952* 3: 182.
1977 *Oplurus cyclurus* – Blanc, Fanue de Madagascar, Paris, *45:* 20, Pl. 1.
Range: Southwestern Madagascar throughout most of Tuléar Province and extending northward into extreme southwestern Majunga Province.

Oplurus fierinensis Grandidier

1869 *Oplurus Fierinensis* Grandidier, Rev. et. Mag. Zool., Paris, (2) *21*: 341. – **Type locality:** Mafale (**Syntypes:** Mus. Hist. Nat. Paris No. 7638 (4)).
1942 *Hoplurus fierinensis* – Angel, Mem. l'Akad. Malag., Tananarive, *36:* 87; Pl. 13, Fig. 1.
1952 *Oplurus fierinensis* – Savage, Copeia, *1952* 3: 182.
1977 *Oplurus fierinensis* – Blanc, Faune de Madagascar, Paris, *45*: 42.
Range: Arid regions of southwestern Madagascar in the lower reaches of the Fiherenana and Onilahy river valleys, and on the Mahafly Plateau, western Tuléar Province.

Oplurus grandidieri Mocquard

1900 *Hoplurus Grandidieri* Mocquard, Bull. Soc. Philomath., Paris (9) *2* 1: 105; Pl. 2. – **Type locality:** Vananitalo (Forêt d'lkongo) (**Holotype:** Mus. Hist. Nat. Paris. No. 99-359).
1942 *Hoplurus grandidieri* – Angel, Mem. l'Akad. Malag., Tananarive, *36*: 85; Pl. 12, Fig. 2.
1952 *Oplurus grandidieri* – Savage, Copeia, *1952* 3: 182.
1977 *Oplurus grandidieri* – Blanc, Faune de Madagascar, Paris, *45*: 47; Pl. 6.
Range: Semiarid regions of southern central Madagascar in eastern central Tuléar Province and western central Fianarantsoa Province.

Oplurus quadrimaculatus Duméril & Bibron

1851 *Oplurus quadrimaculatus* Duméril & Bibron *in* Duméril & Duméril, Cat. Meth., Paris, 83. – **Type locality:** Madagascar (**Holotype:** Mus. Hist. Nat. Paris No. 1404).
1856 *Centrura quadrimaculatus* – Duméril, Arch. Mus. Hist. Nat., Paris, *8:* 558; Pl. 22, Fig. 4.
1869 *Oplurus montanus* Grandidier (*syn. fide* Boulenger 1885), Rev. et Mag. Zool., Paris, (2) *21*: 340. – **Type locality:** Fiérin (**Syntypes:** Mus. Hist. Nat. Paris No. 95-173, 95-175).
1885 *Hoplurus quadrimaculatus* – Boulenger, Cat. Liz. Brit. Mus., London, *2:* 131.
1942 *Hoplurus quadrimaculatus* – Angel, Mem. l'Akad. Malag., Tananarive, *36:* 88; Pl. 13, Fig. 2.
1952 *Oplurus quadrimaculatus* – Savage, *1952* 3: 182.
1977 *Oplurus quadrimaculatus* – Blanc, Faune de Madagascar, Paris, *45*: 34; Pl. 3.
Range: Arid to humid regions of central and southern Madagascar from the coast of southern Tuléar Province through the higher parts of western Fianarantsoa Province and northern central Tananarive Province.

Oplurus saxicola Grandidier

1869 *Oplurus saxicola* Grandidier, Rev. et Mag. Zool., Paris (2) *21:* 340 – **Type locality:** Fiérin [**Syntypes:** Mus. Hist. Nat. Paris No. 7637 (2)].

1942 *Hoplurus saxicola* – Angel, Mem. l'Akad. Malag., Tananarive, *36*: 86; Pl. 12, Fig. 3.

1952 *Oplurus saxicola* – Savage, Copeia, *1952* 3: 182.

1977 *Oplurus saxicola* – Blanc, Faune de Madagascar, Paris, *45:* 53; Pl. 5.

Range: Semiarid regions of southern Madagascar in southern central Tuléar Province.

2

Biogeographical Aspects of the Distribution of Malagasy Iguanids and Their Implications

Charles P. Blanc
Laboratoire de Zoogéographie
Université Paul Valéry
Montpellier Cedex
France

INTRODUCTION

The unique nature of the Malagasy fauna is elegantly illustrated by the presence of the Iguanidae, a family that is absent today, as well as in the fossil records, from Africa, Asia and Australia. Such a problematic distribution has always excited biogeographers who have referred to it as a "biogeographic enigma" and an "irritating problem."

This question has only been analysed from a purely geographical and numerical angle. The vast distribution, the enormous generic and specific diversity of the new world iguanids, have been opposed to the two genera and seven species of Madagascar and the genus *Brachylophus* of the Fiji Islands. Indeed, J. Cracraft (1973 : 382) did not hesitate to write: "Most evolution of the Iguanidae was in South America but they must have been present in Africa at one time to have left descendents on Madagascar." The lack of any fossil remains in Africa does not support this hypothesis and forces us to look for evidence of the origin of the Malagasy Iguanidae using strictly neontological data. Actually, moreover, the comparative study of recent species permits certain approaches (for instance, scalation, pattern of coloration, etc.) which are impossible when only fossils are examined.

The present study will first deal with the characteristics of the distribution of all of the lizards of Madagascar, and then will focus on the Iguanidae beginning with the available biological data, and incorporating an analysis of the distributional possibilities and the taxonomic implications, as well as data about climate, soil and vegetation.

GENERAL CHARACTERISTICS OF THE DISTRIBUTION OF MALAGASY LIZARDS

Structure

The endemism of Malagasy lizards is 95% on the specific and 62% on the generic level (Blanc, 1971). Its main distributional character is its insularity. In spite of the enormous species richness (more than 173 species of lizards are known), the faunal composition is unbalanced when compared with that of adjacent continents and even islands, such as the Seychelles. Western Indian Ocean islands have no faunistic unity and the convenient term "Malagasy Region" should not be used except in a purely geographical sense.

This disequilibrium in the Malagasy lizard fauna may be explained by the survival of a small group of archaic lineages, constituting phylogenetic, geographic, and often numerical relicts, as well as the occurrence of certain groups resulting from very rare immigration and subjected to strong adaptive radiation. Madagascar behaves as an island refuge for the first category which includes endemic genera, generally monospecific and corresponding to the last survivors of remote lineages, present in Madagascar since the time of the first stages of separation of Gondwana. Many of the scincids, as well as turtles and boids, apparently belong here. In contrast, the second group includes some non-endemic genera such as *Scelotes, Lygodactylus,* and *Chamaeleo* which have markedly diversified so that one may recognize respectively around 25, 16 and 30 endemic species. The chance arrival of certain founder forms explains another characteristic of the fauna, namely its heterochronism (presence at the same time of very ancient and new lineages). Madagascar acts as a small continent in terms of its size and the diversity of its biotopes. These grade from the oriental tropical humid forest to the dry bush of the southwest and have favored speciation characterized by the survival of extremes in size, both the largest and the smallest species known of the same genus being present on Madagascar. This suggests the occurrence of limited competition or empty ecological niches. Also having arrived on Madagascar since the fragmentation of Gondwana are endemic genera of Gerrhosauridae: *Zonosaurus* (10 sp.) and Gekkonidae: *Phelsuma* (12 sp.), *Uroplatus* (6 sp.) and *Millotisaurus* (1 sp.), indicative of supra-specific evolution that most likely occurred in situ.

Affinities

I have shown (Blanc, 1971) that, with the exception of rare cosmopolitan forms which generally have a coastal distribution (3 species of gekkonids and the skink *Ablepharus boutonii*), the Malagasy herpetofauna is fundamentally African. This is true whether this continent is believed to represent the center of differentiation of the genus, i.e., *Lygodactylus* (Pasteur, 1964) or *Chamaeleo* (Hillenius, 1959), or of the family, at the generic level, i.e., Gerrhosauridae, Scincidae, etc.

PECULIARITIES OF THE MALAGASY IGUANIDS

The peculiar nature of the distribution of Malagasy iguanids stems on the one hand from the original position of the family, and on the other hand, from intergeneric relations which suggest the existence of only one phylum in Madagascar.

Position of the Iguanids

The position of the Iguanidae among lizards has two aspects, biogeographic and evolutionary.

The absence of any known ancestor or related form on the adjacent continents, particularly in Africa, sets the Iguanidae apart from the other lizards that lack a pantropical distribution. Also, neither of the two fundamental models put forward for the lizard distribution of Madagascar may account for the different speciation patterns of two iguanid genera, *Chalarodon* (monotypic) and *Oplurus* (six species).

Intergeneric Relations

Multiple arguments of varied origin lead to the conclusion that the Malagasy iguanids form a natural group. Among these are: the arrangement of their ventral ribs (Etheridge, 1965), the specificity of their serum albumin (Gorman et al. 1971 : 185), and in the innervation of their zeugopodes (Renous, 1978). All these characters separate the Malagasy forms from the rest of the family, including the Fijian forms of the genus *Brachylophus*. Furthermore, a multidimensional analysis of 105 variables shows that the phenetical distance between *Chalarodon* and *Oplurus* is of the same order of magnitude as the maximal interspecific distance within the genus *Oplurus* (Blanc, Blanc and Rouault, in prep.). Indeed, in many characters such as nuchal and caudal scales, color pattern, pterygoid teeth, displays, macrochromosome characteristics, antigenic similarities and electrophoretic identity of two esterases, *Chalarodon* is indeed closer to one or another species of *Oplurus* than these species are to each other (Blanc, 1977 : 178).

In discussing the distribution of Malagasy iguanids, the grouping within one taxon, the considerable age and Gondwanian origin of the infraorder Iguania (Agamidae, Chamaeleonidae, Iguanidae) must be kept in mind.

HYPOTHESES REGARDING DISTRIBUTION

Two hypotheses are proposed to explain the current distribution of these Iguania. The Malagasy representatives might be descendents of a group of Pre-iguania that were an integral part of the primitive fauna of Madagascar during the earliest stages of the breakup of Gondwanaland.

This viewpoint is supported by the contribution of Renous (1978) who distinguished four patterns of limb innervation : L and V in forelegs according to the path of the ulnar nerve, and A and B in legs, yielding four possible combinations (LA, LB, VA and VB). The breakup of Gondwanaland leads to the evolution of Pre-iguania into true Iguania in South America, which are characterized by an innervation of type LA or LB whereas Indian Agamidae (the Asian ones are much more diversified than the African ones) show type LB, the chameleons of Africa type VB and the Malagasy so-called iguanids type VB. As the latter have not yet acquired an acrodont dentition, they deserve a family status.

Another version of this Gondwanian hypothesis is the origin of the Malagasy phylum from iguanias living in a large area on the continent of Gondwana. In spite of the fact that no iguanian fossils have yet been found in Africa, it may be postulated that Malagasy iguanias have a history analogous to that of the Malagasy turtle *Erymnochelis madagascariensis*. The ancestors of this species are known from the Eocene of Egypt and from the Senonian of In Beceten, on the Niger (Buffetaut and Taquet, 1975), and the most closely related forms are South American (genus *Podocnemis*). Another example is provided by the boids of the extinct genus *Madtsoia* that were discovered in the Paleocene-Eocene of South America and in the Senonian of In Beceten and Madagascar. Darlington (1957) has suggested a replacement of the Iguanids by the more advanced Agamids on the Old World Continents, notably in Africa, but not on the isolated island of Madagascar.

The second hypothesis is based on the possibility of an accidental invasion of Madagascar during one of the stages of the breakup of Gondwanaland. Their absence in Africa would then be due to their chance inability to reach, or establish successfully, on this continent. Similarly, the genus *Phelsuma* is present in the Andaman islands which are located further away from the Malagasy region than the Laquadives where it is lacking.

Although the chronology of the breakup of Gondwanaland (Figure 2.1) is still uncertain, several stages may be characterized:

1. Separation of Africa and Madagascar : the opening of the Mozambique channel may be considered to date from the end of the Upper Cretaceous (Hoffstetter, 1976). However, this separation is now considered by some authors to have occurred much earlier, perhaps in the Upper Jurassic (McHeleny, 1973 : 267) or even in the Triassic to judge from stratigraphic arguments based on the marine sediments of the Upper Triassic and Lower Jurassic (Bathonian) on the coast of Kenya and the north of Madagascar, as well as from paeleomagnetic data (Andriamirado, 1976). The latter author considers that Madagascar became completely separated from Africa during the Triassic, in other words about 180 million years ago. The zone of separation that he, in accord with many authors, places on the latitude of Kenya and Tanzania, suggests a Malagasy shift southways from Africa during the Triassic and, subsequently, from the beginning of the Jurassic to the Upper Cretaceous, Madagascar maintained a latitude approximating that of Salisbury in a position very close to the current one.

2. Separation of Africa and South America : this separation occurred considerably later and ended only at the beginning of the Upper Cretaceous *(Turonien)*, that is, between 105 to 90 million years ago according to absolute dating (Tarling and Runcorn, 1973).

3. Separation of South America from the southern block : this separation did not occur before the beginning of the Tertiary, approximately 75 million years ago. The southern block split into Antartica and Australia plus New Guinea during the Eocene, in other words, between 45 and 43 million years ago.

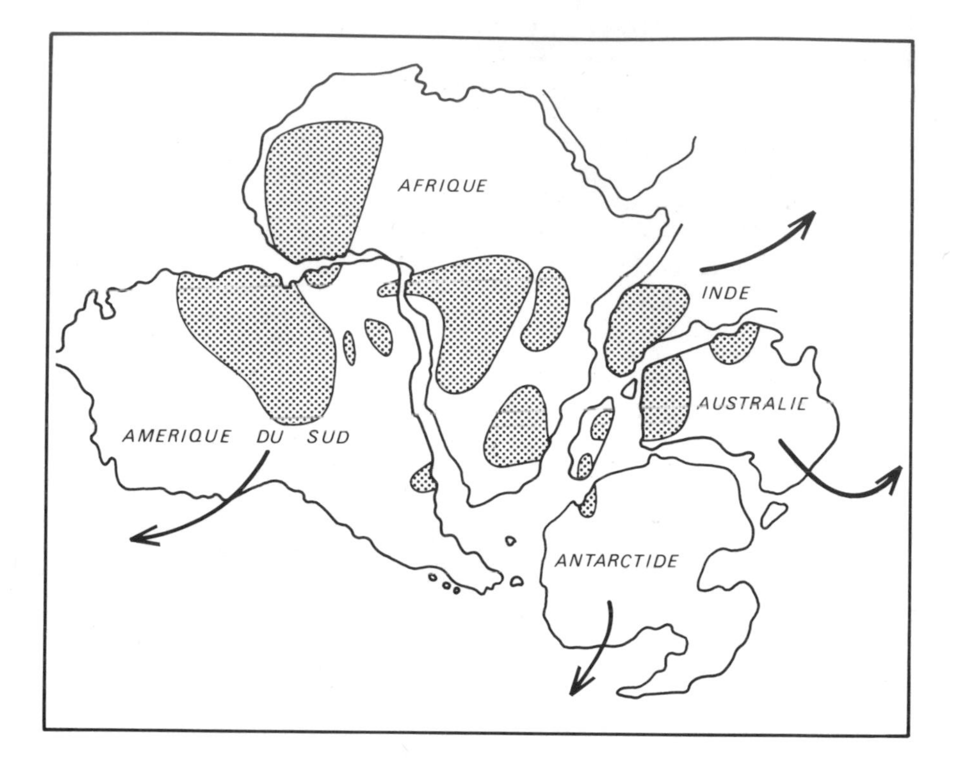

Figure 2.1: The breakup of the Gondwana continent. The geological regions older than 1.6 million years correspond to the hatched zones. Adapted from S. Moorbath.

According to this hypothesis, the Malagasy iguanids are true iguanids derived from the stock of the family differentiating in South America and that they then migrated via Antartica at a period when it was probably, but not necessarily, separated from Australia. Only one founder population is required to explain the actual Malagasy iguanids. It may have reached Madagascar from Antartica on a floating raft (as did the ancestor of the Comoran population of *Oplurus cuvieri* from Madagascar, in a more recent past).

Multiple arguments speak in favor of this hypothesis : although paleoclimatic data are still imprecise (Frakes and Kemps, 1973), these continents are known to have favoured life since the path proposed was that taken successfully by marsupials, cheliid turtles, leptodactylid amphibians, and the tree *Nothophagus* between South America and Australia (Hoffstetter, 1976).

Absence of iguanids from Australia (and the beech tree *Nothofagus,* etc., from Madagascar) does not argue against the hypothesis that iguanids might have been living in Antarctica and, from there, have reached Madagascar. We must keep in mind that the Malagasy iguanid phylum is now living in the driest part of Madagascar i.e., in the past, in a probably different area of Antarctica than that peopled by *Nothafagus,* cheliids and leptodactylids (the driest part of a mainland is always located on the western side). So, their areas of distribution may have been separated when the Southern block split into Antarctica and Australia. Nowadays, many African animals are only living on the East part of the Rift Valley and may be isolated in the future from the other part of Africa.

The Malagasy iguanids are true iguanids. I have shown (Blanc, 1965) that the skeleton of *Chalarodon* has affinities with that of American forms such as *Dipsosaurus* and, according to Presh (1969), it furnishes the fundamental type of axial skeleton for small and moderate sized iguanids. The dentition is pleurodont. The antigenic specificity of its albumin suggests that *O. cuvieri* is a true iguanid but of a very ancient divergence (Gorman et al., 1971).

We have difficulty explaining how the poorly diversified African agamids could have succeeded in totally supplanting the eventual iguanids. Their currently allopatric distributions appears to be more the geographical result of general historical consequences than of competition.

The organization of the Malagasy iguanids into a monophyletic group agrees better with a chance immigration than with the hypothesis of the survival of a very ancient group.

The slow separation of fragments of Gondwanaland must have produced during long periods ocean channels narrow enough to permit immigration of a founder population, in the southwest part of the island where the group became diversified.

According to these hypotheses all of the Malagasy iguanids should represent a single monophyletic unit perhaps deserving a family or subfamily rank.

IMPLICATIONS FOR ENVIRONMENTAL FACTORS

Climatic Implications

The paleoclimatologic circumstances underlying the current distribution have been determined for *Chalarodon madagascariensis* (Blanc, 1969). Those of the other species have been deduced from their distribution by comparison with the preceeding form and by observation of their behavior in the field.

The Malagasy iguanids are essentially localized in an arid or semi-arid bioclimate. Their distribution does not seem to be limited by increasing aridity; on the other hand, only a single species *(Oplurus quadrimaculatus)* invades the "humid" bioclimatic zone and none of them enters the "very humid" one.

The high summer temperatures are not limiting, but the minimal winter temperatures apparently keep the majority of the species from reaching the High Plateau of the region.

All of the Malagasy iguanids are highly heliophilic. They feed in sunny places whereas other sympatric lizard species forage in the vicinity of thickets (Gerrhosauridae).

The limitation to a dry environment is correlated with a strong insolation and with both reduced rainfall and unclouded skies. Periods of about 18 months without significant rainfall may occur.

The homogeneity of environmental requirements for these two genera seems to indicate that these conditions were prevalent at the time the group became established. The ubiquitous species *O. quadrimaculatus* may have extended its distribution to the north and east during a more arid period. Residual populations that may have benefited from locally less unfavorable climates and stretched their climatic limits, now mark the remmants of an ancient peneplane.

It should be noted that these species share the same reproductive phenology: oogenesis takes place during the rainy season (abundance of insects) and young are born at the beginning of the long dry season.

Implications of the Nature of Soil

Chalarodon is associated with loose sands (coastal dunes, main beds of rivers and portions of erosion slopes), while the speciation of the rupicolous *Oplurus* implies an enduring habitat of diverse rocky outcrops: limestone and gneiss with diaclases and with sub-horizontal fissures of various diameters.

Implications of Vegetation

Two notions may be proposed: The heliophilism of Malagasy iguanids in which movements occur only on the ground, trunks and large branches of trees, implies a bushy type of vegetation (Mahafaly and Androy districts) or dry forests on the western slopes (West district) affording sufficient sunshine either through discontinuity of the cover (isolated rocky outcrops, dunes) or reduction of the above-ground vegetation.

Oplurus selects tree shelters with a specific shape and architectural system that favours rotting of the trunk when the dead branches have fallen. Also, tree *Oplurus* do not invade smooth trunks, such as those occupied by the equally-sized sympatric gecko *Phelsuma standingi,* but prefer trees with rough bark such as *Poupartia* (Sakoa) and *Tamarindus* (Kily) from which they hang by their toes, head downward.

The climatic and botanic characteristics required by Malagasy iguanids agree with the nature of the southern flora of this country, rich in endemic forms such as the family of Didieraceae, many species of baobabs, euphorbias and aloes. These plants tend to be thorny, reviviscent or with a water-filled parenchyma, suggesting long-term adaptations to an enduring arid climate.

SUMMARY AND CONCLUSIONS

I have discussed two possible hypotheses for the occurrence of iguanids on Madagascar, keeping their close affinities in mind. The comparative study of the (South) American forms, particularly of the most primitive forms from which the Malagasy group might have been derived will possibly provide a clue as to which of these two hypotheses should be retained.

The occurrence of closely related American forms would support the hypothesis that the distribution results from chance dispersal, probably via the Antarctic. In contrast, the diversity of entirely original characteristics among Malagasy iguanids suggests a need for a comparative study of the primitive forms of chamaeleons and agamids and favours the second hypothesis of an original evolution or immigration via Africa from the primitive Pre-iguanian stock. It is also possible that the discovery of fossils may still provide some information regarding the history of this group. Such a collaborative study to define the present phylogeny of all iguanids as well as the presumptive stages of their differentiation would be of great interest.

Acknowledgments

I would like to thank Carl Gans and Neil Greenberg for their help in translating my manuscript from the original French into English. I also thank A. Stanley Rand, Gordon M. Burghardt, and outside reviewers for their critical reading of the manuscript.

Section II

Food and Energetics

All animals must eat in order to survive; certainly this is a truism if ever there was one. Yet ethology and behavioral ecology are just awakening to the impact that metabolic requirements and feeding strategies have on all aspects of an organism's existence. Social behavior has been the focus of most ethology and virtually all of sociobiology. Given the name of the latter this is appropriate *until* it leads to the ignoring of factors that also have consequences for social behavior and behavioral evolution. Thus matters pertaining to the obtaining, ingesting, and metabolizing of foodstuffs are finally receiving attention (e.g., Curio, 1976; Kamil and Sargent, 1981).

The interest in iguanine feeding, however, goes beyond the above because iguanines are among the few lizards that are largely, if not exclusively, herbivores. Thus the four chapters in this section tell us some things about how these largely plant eating animals forage, select, and process food. In their food habits iguanines, like some turtles and tortoises, are far more similar to various terrestrial and arboreal mammals than they are to other reptiles.

Nagy (Chapter 3) treats the metabolism of iguanid lizards, and compares it to mammals. Maintaining a fairly constant body temperature by endogenous physiological means is generally considered the major biological difference between reptiles, on the one hand, and birds and mammals on the other. Of course this is simplistic and the "inferior" status of reptiles is in large part due to anthropomorphic biases (see Regal, 1977; Pough, 1980 for alternative perspectives). Be that as it may, reptiles do not need to divert large amounts of food to heat production (or cooling) and hence need to eat less in order to survive and reproduce. And regardless of the type of animal, herbivores need to eat much more bulk than carnivores. Although not many iguanines have been looked at, Nagy concludes that iguanines are not metabolically different from other iguanid lizards.

Iverson (Chapter 4) considers the structural modifications in the digestive system that are found in herbivorous lizards regardless of family. The subdivision of the colon into sections by septa and valves are apparently necessary adaptations for a lizard to effectively enter upon a life "feasting" on leaves, fruits, and flowers. While iguanines are the most prevalent herbivorous lizards there are also a few scattered in other families (Agamidae, Scincidae). Iverson also postulates that the large nematode worm populations found in these animals' hindguts are involved in the digestion of plant matter. Altogether these factors warrant detailed comparative treatment.

McBee and McBee (Chapter 5) deal with the actual physiology of digestion of plant matter in green iguanas. How are lizards able to make use of difficult to digest materials? McBee and McBee establish here that an important process is the hindgut fermentation by bacteria. This may explain the quiet basking by lizards at close to maximum tolerable temperatures. Also of interest is how the bacteria first get inside newly hatched lizards. We now know that most iguanines are herbivorous from hatching. Transfer of bacteria from the mother via some embryonic process seems unlikely. Speculation more generally invokes the ingestion of the bacteria fortuitously via the frequent tongue extensions, flicking and licking, either underground in the nest cavity or in the frequent substrate and sibling tongue touches seen in the newly emerged neonates. In the green iguana Troyer (1982) has documented neonatal acquisition of the requisite microbes from feces of conspecifics.

With Auffenberg's chapter (6) this section closes with a detailed consideration of the actual foraging and plant selection behavior of the Caicos ground iguana, now one of our better known iguanines. When we have to actually confront an animal in the field dealing with seasonal food items, with a variety of costs and rewards associated with each, our amazement at the sophisticated systems – behavioral, physiological, morphological, even psychological – involved should be great indeed. Although after this section the focus of the remaining chapters is on reproductive and social matters, it is quite clear that these cannot be considered in isolation from the mechanisms used to procure, process, and utilize foodstuffs.

3

Energy Requirements of Free-Living Iguanid Lizards

Kenneth A. Nagy
Laboratory of Biomedical and Environmental Sciences
University of California
Los Angeles, California

INTRODUCTION

The concept of energy demand and supply has been quite valuable in unifying many aspects of the environmental physiology and ecology of animals. Studies of animal energetics have increased understanding of how single species manage their energy budgets in various seasons and habitats (see for example Brett, 1970; Chaplin, 1974; King, 1974; Turner et al., 1976), as well as how populations interact while partaking of the energy flow through their ecosystems (see chapters in Golley et al., 1975; Ricklefs, 1976; Wiegert, 1976).

A question of central importance in this regard is: How much energy do animals require to exist in their natural habitats? Information of this kind provides the means for estimating rates of food consumption by individuals and populations, and subsequent assessment of the impact and role animals have in the ecosystems they occupy. Unfortunately, energy requirements (metabolic rates) of free-living vertebrates have previously been difficult to measure. Methods such as heart rate telemetry and excretion rates of various radioactive tracers such as iodine and several metals have been used, with marginal success (Gessaman, 1973). The most reliable method is the doubly labeled water technique (Lifson and McClintock, 1966), but its high cost and analytical difficulty have limited its use in the past. However, the recent availability of less expensive isotopic oxygen and new developments in isotope analysis (Wood et al., 1975) have made feasible the wider application of this method.

It is not reasonable to try to measure the field metabolic rate (FMR) of every animal that might be of interest in terms of, say, ecosystem analysis. Thus, it should be of value to have a model, based on measurements of FMR in animals of varying sizes, which can be used to predict FMR in animals that haven't been studied. Correlations between basal or standard metabolic rate and body mass (allometric regressions) have been remarkably useful for predicting standard metabolism in many species of birds, mammals, reptiles and amphibians for which data are unavailable (Bartholomew, 1977). Field metabolic rates are now known for enough species of iguanid lizards to permit calculation of an allometric regression for this family of reptiles. This chapter includes a summary of the doubly labeled water measurements of FMR in reptiles that are currently available, an allometric regression analysis of iguanid lizard results along with comparisons of this relationship with similar regressions for birds and mammals, and discussion of how feeding rates in free-living iguanids might be estimated.

METHODOLOGY

Rates of CO_2 production can be measured in free-living vertebrates by using doubly labeled water (Lifson and McClintock, 1966). The hydrogen isotope can be deuterium (^2H or D) or tritium (^3H or T) (tritium is easier to measure and is used in most studies), while heavy oxygen (^{18}O) serves as the oxygen isotope. In an animal whose body water is labeled with tritium, the specific activity of T will decline exponentially through time because of the loss of isotope as water via evaporation and excreta, and the simultaneous dilution of the isotope through gain of unlabeled water via food, drink and metabolic water production. The rate of decline of the specific activity of T is a measure of water flux through the animal. The oxygen-18 in the body water of a labeled animal is in isotopic equilibrium with the oxygen of CO_2 in the blood due to the action of carbonic anhydrase (Lifson et al., 1949). Thus, the specific activity of O-18 in body water declines faster than that of T because O-18 is lost in the form of CO_2 as well as in the form of H_2O. It follows that the difference between rates of decline of O-18 and T is a measure of CO_2 production rate (metabolic rate).

The doubly labeled water method has been checked in the laboratory by comparing it with simultaneous measurements or estimates of CO_2 output using other techniques in a variety of vertebrates, and it has proved accurate to within 8% in all species tested thus far (Nagy, 1980). In field studies, there are two potential sources of error. If a labeled animal breathes air containing a relatively high CO_2 concentration, then ambient CO_2 will diffuse into the animal, equilibrate with O-18 and then diffuse out, yielding a falsely high FMR measurement. The size of the error should depend primarily on ambient CO_2 concentration. This situation may occur if a labeled animal enters a communal burrow or nest. If a labeled animal is alone in a confined space and rebreathes its own CO_2, then no error will result because the ambient CO_2 will be labeled at the same specific activity as in the animal's body water. The second error can occur in animals that have high water flux rates relative to their rates of CO_2 production (e.g., amphibians). In this case, the washout rate for O-18 is only slightly greater than that for tritium, and small errors in isotope measurements, or the slight differential evaporation rates of HTO vs. H_2O will cause a large error in estimated CO_2 production, which is calculated as the difference between the washout rates of the two isotopes (Nagy, 1980).

In the field, reptiles are captured, marked for identification, weighed, sexed and injected intraperitoneally with 3 ml per kg body mass of water containing about 95% oxygen-18 and 0.3 mCi T. After waiting 1-4 hours (depending on body mass) for complete mixing of the labeled water in the animals, an initial blood sample is taken for analysis of isotope levels. The animals are then released where captured. Following an appropriate time interval (2 to 20 or so days depending on predicted isotope loss rates for each species), the animals are recaptured, weighed, and a second sample of blood or urine is taken. Total body water volume is calculated from the dilution volume of injected isotopic water, since this value is needed in calculating metabolic rates. Body fluid samples are distilled, and the pure water thus obtained is measured for T by liquid scintillation counting, and for O-18 by cyclotron-generated proton activation (Wood et al., 1975). These data are then used to calculate FMR (in milliliters of carbon dioxide produced per gram of body mass per hour, abbreviated ml CO_2/g h) with the equations given by Nagy (1975).

The amount of heat (in joules, J) liberated per unit CO_2 produced differs for the oxidation of carbohydrates, fats and proteins (Schmidt-Nielson, 1975). Since the isotopic water method yields results in terms of CO_2, and since reptilian diets range from herbivory to carnivory, it is necessary to convert FMRs to units of energy (J) before comparing between species. Feeding experiments, in which assimilation of carbohydrate, fat and protein were estimated, have been done on a herbivorous lizard (Nagy and Shoemaker, 1975) and an insectivorous lizard (Nagy, unpublished). These results were used to calculate the following conversion factors: for herbivores, respiratory quotient (R.Q.) = 0.93, energy equivalent = 21.7 J/ml CO_2, and for insectivores, R.Q. = 0.75, energy equivalent = 25.7 J/ml CO_2 (to convert from joules to calories, divide J by 4.184). These energy equivalents were used to convert FMRs for the various species in Table 3.1 to units of energy for purposes of allometric analysis.

RESULTS

Field metabolic rates (FMRs) of the reptiles studied to date are summarized in Table 3.1. Of the twelve species listed, nine are iguanid lizards, and three of these are iguanines. FMR values have been reported for adults, or for males and females separately, and for juveniles of a few species. Also, several studies included metabolic rate measurements during various seasons. All these parameters can influence FMR. This is illustrated in Figure 3.1, which shows monthly measurements of FMR in male, female and juvenile *Uta stansburiana* living in the Mojave Desert in Nevada (Nagy, unpubl.). At different times of the year, either males or females tended to have the higher FMR. However, the differences between sexes were not significant in most months. Bennett and Nagy (1977) measured FMR in male and female *Sceloporus occidentalis* during spring and fall, and found no significant differences between sexes or seasons. However, Merker and Nagy (unpubl.) found that FMR in male *Sceloporus virgatus* in Arizona was significantly higher than in females during spring, but both

Table 3.1: Summary of Metabolic Rates in Free-Living Reptiles

Taxon	Body Mass (g)	Metabolic Rate [ml CO_2 $(g\text{-}h)^{-1}$]	Season[a]	Cohort[b]	Figure Number[c]	Reference[d]
Sauria						
Iguanidae						
Amblyrhynchus	1,481	0.124	D	A	1	1
cristatus	69	0.135	D	J	2	1
Callisaurus	4.6	0.238	Su, F	F	3	2
draconoides	6.3	0.318	Su, F	M	4	2
Dipsosaurus	50.3	0.152	Sp	A	5	3
dorsalis	3.5	0.350	Sp	J	6	3
Sauromalus obesus	167	0.18	Sp	A	7	4
	148	0.04	Su	A	–	4
Sceloporus jarrovi	12.1	0.224	Su	A	8	5
	19.5	0.224	F	A	9	5
	23.9	0.194	F	M	10	5
	17.4	0.133	F	F	11	5
	8.5	0.278	F	J	12	5
	11.0	0.079	F, W	A	–	6
	11.9	0.031	W, Sp	A	–	6
Sceloporus	11.9	0.220	Sp	A	13	7
occidentalis	11.5	0.198	F	A	14	7
	11.4	0.252	Sp	A	15	8
Sceloporus virgatus	5.3	0.139	Su	F	–	5
	5.5	0.303	Sp	M	16	9
	7.6	0.254	Sp	F	17	9
	5.7	0.249	Su	M	18	9
	6.7	0.290	Su	F	19	9
Tropidurus	43.7	0.155	D	M	20	10
albemarlensis	12.7	0.186	D	F	21	10
Uta stansburiana	3.3	0.34	Sp	A	22	10
	2.8	0.30	Su	A	23	10
	3.0	0.14	F	A	–	10
	2.5	0.08	W	A	–	10
	0.5	0.45	Su	J	24	10
	1.6	0.26	F	J	25	10
Lacertidae						
Aporosaura anchietae	3.1	0.238	Su, W	F	26	11
	5.4	0.272	Su, W	M	27	11
Teiidae						
Cnemidophorus	3.9	0.356	Su	A	28	2
hyperythrus	4.2	0.540	Su	A	29	2
Testudines						
Testudinidae						
Gopherus agassizii	900	0.15	Sp	A	30	12
	900	0.08	Su	A	–	12

[a]Season: Sp = spring, Su = summer, F = fall, W = winter, D = dry.
[b]Cohort: A = adults, M = males, F = females, J = juveniles.
[c]Number corresponding to those in Figures 2 or 3.
[d]References: (1) Nagy and Shoemaker (unpubl.), (2) W. Karasov and R. Anderson (pers. comm.), (3) Mautz and Nagy (unpubl.), (4) Nagy and Shoemaker (1975), (5) J. Congdon (pers. comm.), (6) Congdon et al. (1979), (7) Bennett and Nagy (1977), (8) Bickler and Nagy (unpubl.), (9) Merker and Nagy (unpubl.), (10) Nagy (unpubl.), (11) P. Cooper and M. Robinson (pers. comm.), (12) Nagy and Medica (unpubl.).

sexes had the same metabolic rate during summer. If FMR in *Sceloporus* spp. varies with season as it does in *Uta stansburiana* (Figure 3.1), it is not difficult to understand how one can find either no differences or significant differences between sexes or seasons. It depends heavily on the precise segment of time over which FMR measurements are made.

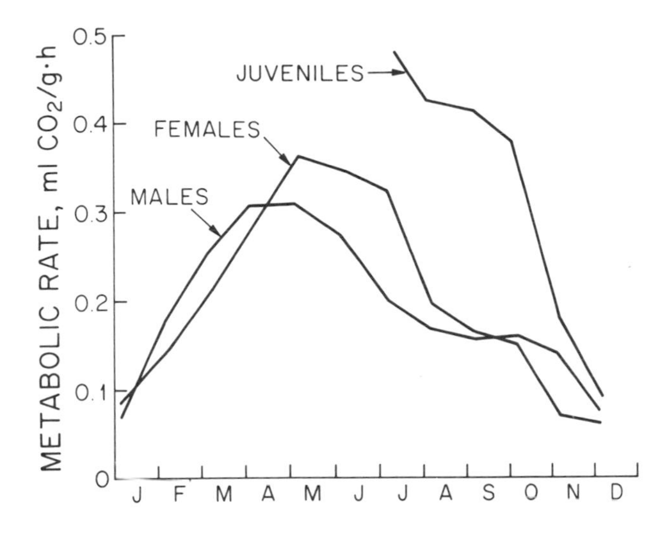

Figure 3.1: Seasonal changes in the metabolic rates of male, female and juvenile lizards *(Uta stansburiana)* living in the Mojave Desert in Nevada (Nagy, unpubl.).

Reptiles living in strongly seasonal habitats tend to be relatively inactive when the climate is harsh, such as during hot, dry periods (e.g., *Sauromalus obesus*) and during cold periods (e.g., *Uta stansburiana* and *Sceloporus jarrovi*), and their FMRs at these times are lower than during periods of normal activity (Figure 3.1, Table 3.1). This additional source of variation in FMR complicates an examination of the relationship between body mass and FMR. In an attempt to minimize seasonal influences on the allometric analysis, I selected from Table 3.1 only those measurements that were made during the animal's normal activity season (spring and summer in most cases). The relationship between field metabolic rate (in kilojoules per day) during the activity season and body mass (in grams) in iguanid lizards is shown in Figure 3.2. The regression line is described by:

$$\text{kJ/day} = 0.224\ \text{g}^{0.80} \qquad \text{(eq. 1)}$$

with the correlation coefficient (r) = 0.991, standard error of the slope = 0.023, N = 25, df = 23 and $P<0.001$ via an F test for significance of the regression. When converted to units of calories, this equation becomes:

$$\text{cal/day} = 53.5\ \text{g}^{0.80} \qquad \text{(eq. 2)}$$

$$\text{or} \quad \text{cal/h} = 2.23\ \text{g}^{0.80} \qquad \text{(eq. 3)}$$

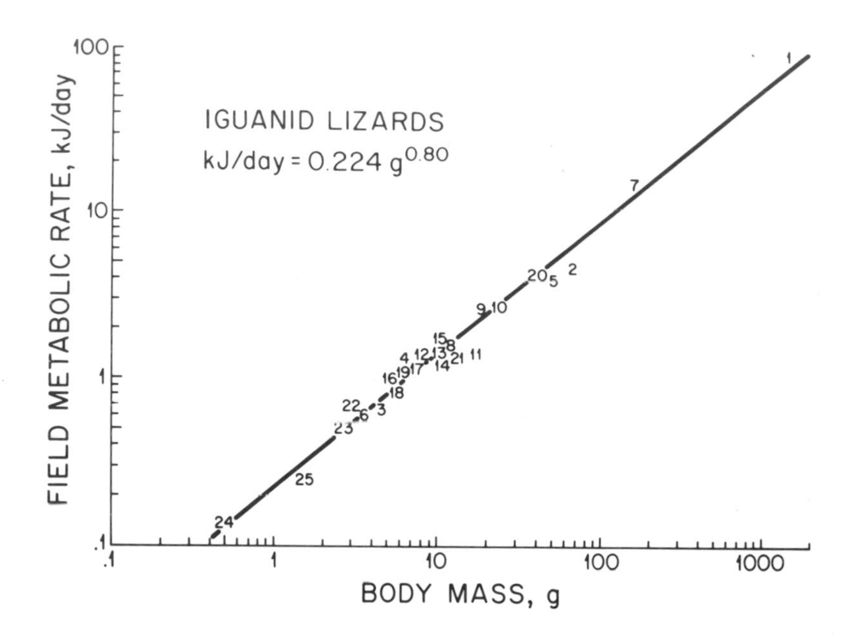

Figure 3.2: Relationship between field metabolic rate and body mass in iguanid lizards during their activity season. The coordinates are logarithmic and the line is the least squares regression for the logarithmically transformed data. Numbers refer to Table 3.1.

DISCUSSION

Several questions arise in connection with this analysis: (1) are the FMRs of juvenile iguanids higher than expected from size alone, because of possible additional costs of rapid growth?; (2) do herbivorous lizards have different FMRs than carnivorous lizards, possibly because of differing costs of foraging along with differing costs of processing food in the digestive tract?; and (3) is the cost of living for iguanine lizards different from that of other iguanids? Inspection of Figure 3.2 suggests that the field metabolic rates of juveniles, herbivores and iguanines do not tend to differ much from those expected on the basis of body mass alone. The results for juveniles (points 2, 6, 12, 24 and 25) and for herbivores and iguanines (points 1, 2, 5, 6 and 7–the herbivores studied were all iguanines and the iguanines studied were all herbivorous) do not appear to diverge from the regression line any farther than the other points. It should be emphasized here that logarithmic transformations (as in Figure 3.2) visually and statistically reduce differences that may be quite large in the untransformed results, so the statements above can be no more than suggestions at this time.

The slope of the FMR regression for iguanid lizards (0.80) is virtually identical to the slopes calculated by Bennett and Dawson (1976) for standard metabolic rate in lizards at different temperatures (0.80 at 20°C, 0.83 at 30°C and 0.82 at 37°C). This indicates that daily energy expenditure of free-living iguanid lizards during their activity season, and their standard metabolic rates are related to body mass in the same way. In other words, the additional energy cost of living unrestrained in their natural habitats, above that for simply remaining alive in a resting, starved state at a constant temperature in a laboratory, requires the same percentage increase above resting levels for a small as well as a large lizard. This conclusion has some fascinating ecological implications, for the following reasons. In *Sceloporus occidentalis,* 85% of total 24 hour energy expenditure occurs during the 8 hours when this lizard is abroad and active during the daytime (Bennett and Nagy, 1977), so total daily cost (FMR) of this lizard is primarily influenced by (1) body temperature while active, (2) the duration of the activity period and (3) the intensity of activity while abroad (as a multiple of resting metabolic rate). All of the species represented by FMR regression in Figure 3.2 are active during the daylight hours only, and most maintain body temperatures between 35° to 38°C (*Dipsosaurus,* at about 40°C, is an exception), so body temperatures are roughly the same, regardless of body mass. This means that the energetic cost of being active (the product of hours spent active and the intensity of activity while abroad) accounts for about the same increment above resting requirements, regardless of body size. This is surprising. Small lizards are generally more short-lived than large lizards, so they should be more constrained by time (especially in a seasonal habitat), and would be expected to work relatively harder, or longer, or both each day in order to obtain the needed extra food to provide for the next generation. Large, long-lived lizards can afford to skip a year or two, and still insure reproduction of their species by breeding in subsequent years. This argument suggests that small lizards should work relatively harder each day than large lizards to insure reproductive success; this would yield a slope of log FMR on log body mass that is lower than the slope for resting metabolic rates. Such is not the case (Figure 3.2). It would be interesting to see if reproductive effort in iguanid lizards also scales with body mass to the 0.80 power.

The available results for FMR in non-iguanid reptiles are compared with the iguanid lizard regression in Figure 3.3. The desert tortoise (*Gopherus agassizi,* point 30) and the desert lacertid *(Aporosaura anchietae* point 26, females and point 27, males) have activity season FMRs that are similar to those in iguanid lizards. In the teiid lizard *Cnemidophorus hyperythrus,* the average metabolic rate of a population living in a thin thorn forest (point 28) was lower than that in a population occurring in a thick thorn forest a few kilometers away (point 29). The reason for this difference is not clear. Teiid lizards are generally more active while abroad than are iguanids, so they might be expected to have higher FMRs.

Allometric regressions for FMR in birds and rodents have been calculated from compliations of doubly labeled water measurements, time-energy budgets and laboratory-based estimates of field metabolism (King, 1974). These are shown in Figure 3.3 for comparison with iguanid lizards. For a 100 g animal (a fairly representative size for these three taxa), these regressions yield the following predicted FMRs: for a bird, 268 kJ/day; for a rodent, 162 kJ/day; for an iguanid lizard, 8.9 kJ/day. Thus, the cost of free-existence for a 100 g iguanid lizard during its activity season is about 6% that of a rodent and only 3% that of a bird having the same body mass.

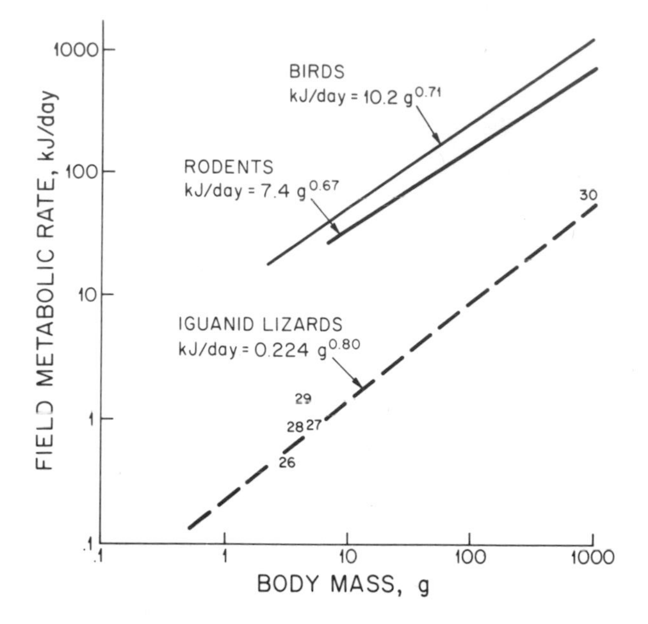

Figure 3.3: Comparison of FMR allometric regressions for free-living birds, rodents and iguanid lizards. The coordinates are logarithmic. Numbers refer to non-iguanid reptiles, as listed in Table 3.1.

As might be expected, the differences between lizards and endotherms are even greater when examined in terms of yearly energy expenditure. Annual field metabolic rates have been estimated from labeled water measurements for two iguanid lizards: *Uta stansburiana* (2.5 g), which were active about 10 or 11 months of the year (Nagy, unpubl.), and *Sauromalus obesus* (167 g), as estimated for animals that were active about four months of the year (Nagy and Shoemaker, 1975). For comparison, results are available for jackrabbits (*Lepus californicus,* 1.8 kg), which are closely related to rodents, and were active throughout the year (Shoemaker et al., 1976). These three species were all studied in the Mojave Desert. To correct for differences in body mass, annual FMRs (in kJ/yr) were divided by body masses (in g) raised to the 0.80 power (for *Uta* and *Sauromalus*) or 0.67 power (for *Lepus*), and the resulting expression was solved for an animal weighing 100 g. The size-corrected, annual FMRs predicted for a 100 g animal are: 2.4 MJ/yr, based on *Uta* results; 1.2 MJ/yr for *Sauromalus;* and 60 MJ/yr for *Lepus.* The annual energy requirements of the lizards are about 2 to 4% those of the mammal.

The low field energy requirements of lizards compared to birds and mammals are due to at least three phenomena. First, the resting metabolic rates of lizards are much lower than in mammals. A 100 g lizard resting at 37°C has a predicted metabolic rate that is only 15% that of a 100 g placental mammal at rest in its thermal neutral zone (Bennett and Dawson, 1976; Bartholomew, 1977). Second, lizards spend half or more of each 24 hour period resting at comparatively low body temperatures, which dramatically reduces their metabolic rates (Bennett and Nagy, 1977). Third, annual FMR in temperate zone lizards is reduced because they are usually inactive during the cold times of the year. It would be interesting to examine the annual energy requirements of a tropical lizard that is active year-round, and a rodent or a bird that uses daily or seasonal torpor.

The differences in FMR between lizards and endotherms suggest that important differences may exist in the ecology of these taxa. If food supply is important in determining the density and biomass of vertebrates in a given habitat, and if lizards and endotherms are equally good food gatherers, then one would predict that, in a given trophic level, the biomass of endotherms should be but a few percent of the biomass of reptiles. A variety of studies have indicated that this trend is evident in many habitats. Another prediction can be made from this comparison. Lizards and mammals are made of similar kinds of tissues, and their anatomies are generally comparable, so a lizard and a mammal of the same size should be able to consume about the same amount of food at one time. Because of its intrinsically higher metabolic rate, the mammal should be able to process its food faster than the lizard, but even after allowing for differences in resting metabolic rate and daily body temperature regimes, it seems that the lizard should be capable of obtaining far more surplus energy (relative to that required by FMR) than the mammal. This suggests that a greater fraction of assimilated energy can be invested in growth and reproduction by lizards. In fact, ectotherms in general channel a much larger fraction of their assimilated energy into production of new biomass (Turner, 1970; Ricklefs, 1976). The low metabolic rates of ectotherms constitute an important aspect of this ability.

There are many potential uses for a model that could be used to predict food consumption rates in free-living animals. The feeding rates necessary to satisfy metabolic rate requirements for iguanid lizards can be predicted from their FMRs provided that the amount of food required to yield a unit of metabolizable energy is known. Estimates of the average amounts of energy in leaves and insects have been reported (Golley, 1961), and metabolic efficiency:

$$\frac{\text{energy ingested} - \text{energy voided as feces and urine}}{\text{energy ingested}}$$

is known for a few herbivorous and insectivorous lizards (Nagy and Shoemaker, 1975; Nagy, unpublished data for *Uta;* also see summaries by Pough, 1973 and Bennett and Dawson, 1976). Based on the few results available, the diet of plant-eating lizards contains about 17 kJ/g dry matter, of which about 55% is metabolizable, and a diet of insects contains about 23 kJ/g dry matter, and about 80% of this energy is metabolizable. Thus, the approximate metabolizable energy yield to lizards eating plants is 17 kJ/g dry matter x 0.55 kJ metab./kJ ingested = 9.4 kJ metab./g dry food, and for insectivores, 23 kJ/g dry food x 0.80 kJ metab./kJ ingested = 18.4 kJ metab./g dry food. Dividing these values into equation 1 yields the following relationships. For herbivorous iguanid lizards during their activity season,

$$\text{mg dry food ingested/day} = 24\ g^{0.80} \qquad \text{(eq. 4)}$$

and for active insectivorous iguanids,

$$\text{mg dry food ingested/day} = 12\ g^{0.80} \qquad \text{(eq. 5)}$$

It should be emphasized that equations 4 and 5 are based on just a few measurements of metabolic efficiency in lizards, and may yield predicted feeding rates that are in error by more than 30 or 40% if the above assumptions do not apply to the species in question. Metabolic efficiency in herbivorous lizards may be highly variable, depending on the composition of the diet, so equation 4 is probably of less predictive value than equation 5.

Free-living lizards grow and reproduce so their feeding rates are actually higher than would be predicted from metabolic rates alone (equations 4 and 5), because extra energy is required to build new tissue. Since the FMR measurements upon which equations 4 and 5 are based include lizards that were growing and reproducing, any additional metabolic costs of creating new biomass should already be accounted for in these equations, but the chemical potential energy from the food that ends up as new tissue is not. Reproductive effort, measured as the ratio of egg mass or energy content to mass or energy content of the female's body, can be as high as 0.5 in some lizards (Vitt and Congdon, 1978). However, the fraction of metabolizable energy that is invested in offspring biomass has been measured in only one lizard, *Uta stansburiana* (Nagy, unpubl.). Female *Uta* allocated about 23% of their metabolizable energy to eggs during the breeding season, and on an annual basis, this value was about 12%. The energy devoted to gametes by male *Uta* was too low to be measured. Since *Uta* is a small, essentially annual species and produces several clutches per year, its reproductive investment may be among the higher values for iguanid lizards in general. Thus, feeding rates predicted directly from equations 4 and 5, and feeding rates corrected upward by 25% to account for production, may bracket actual feeding rates in the females of many iguanid lizard species during their activity season. Food consumption of breeding males should be about as predicted directly from the equations.

SUMMARY AND CONCLUSIONS

Measurements of field metabolic rate (FMR) in reptiles, made by means of doubly labeled water, are summarized and examined using allometric regression analysis. For nine species of iguanid lizards during their activity season, FMR is significantly correlated with body mass (in grams), and is described by: kJ metabolized/day = $0.224\ g^{0.80}$. Field metabolic rates of juveniles, herbivores or iguanine lizards do not appear to differ consistently from those of adults, carnivores or other iguanid lizards, respectively. A desert tortoise and a desert lacertid lizard had metabolic rates that were similar to iguanids, but one population of teiid lizards had much higher daily energy expenditures. The slope of the iguanid FMR regression is essentially identical to those for resting metabolism in lizards at 20°, 30° and 37°C. The daily cost of free existence for an active, 100 g lizard is only 6% that of a free living 100 g rodent, and 3% that of a 100 g bird in the field. On an annual basis, these percentages are even lower. This is probably part of the basis for the higher reproductive efforts in lizards as compared to endotherms. Equations are derived for estimating feeding rates of active herbivorous and insectivorous iguanid lizards from their predicted FMR, and adjustments to account for the greater food intake needed to produce new biomass are discussed.

Acknowledgements

Preparation of this review and many of the studies cited herein were supported by Contract DE-AM03-76-SF00012 between the U.S. Department of Energy and the University of California. I am grateful to R. Anderson, P. Bickler, J. Congdon, P. Cooper, W. Karasov, G. Merker and M. Robinson for generously allowing me to use unpublished results, and to G.M. Burghardt and A.S. Rand for valuable comments on the manuscript.

4

Adaptations to Herbivory in Iguanine Lizards

John B. Iverson
Department of Biology
Earlham College
Richmond, Indiana

INTRODUCTION

Totally herbivorous lizards are found in three disjunct regions: (1) the New World tropics and subtropics northward into the Mohave Desert of the SW United States (all the Iguaninae but two species), (2) the Near and Middle East from North Africa to Southwest Asia (spiny tailed agamids *Uromastyx);* and (3) the tropical Far East in the Fiji Islands (the banded iguanas *Brachylophus),* the Philippines and Indonesia (water lizards *Hydrosaurus*), and the Solomon Islands (giant skinks *Corucia*). They are found on islands as well as continents (commonly on the former), in predator-filled to nearly predator-free environments (predator species and herbivorous lizard species diversities negatively correlated), in xeric (most), mesic, and hydric habitats, and they include both oviparous and viviparous (only *Corucia*) forms. Although individual species are usually allopatric, two species are occasionally microsympatric. However, to my knowledge, no record yet exists for microsympatry of more than two species, even though more than two species may have broadly overlapping ranges in Mexico and Central America.

These lizards are generally considered to be unspecialized (Pough, 1973) in comparison to more obviously specialized lizard groups such as the Teiidae, Chamaeleonidae, Varanidae, Anguidae, and Helodermatidae. However, I believe that these herbivorous lizards, in particular the iguanines, are much more specialized than is generally realized.

Perhaps the most unique characteristic of the lizards of the sub-family Iguaninae is their success at herbivorousness. Unfortunately, complete, reliable data on diet and feeding habits for most of the iguanines are not available; most of the published information is only anecdotal, and, in some cases (see below), clearly misleading. Intrigued by my observations that the ground iguana,

Cyclura carinata, is almost totally herbivorous from hatching through adulthood (a weight range of 15 to 1900 g) (Iverson, 1979), I began examining lizard herbivory in terms of diet, feeding habits, feeding strategies, and adaptational correlates (both morphological and physiological). By looking simultaneously at all of these aspects of herbivory, I hoped to better understand the mechanisms involved in its evolution in lizards.

DIET

One of the problems inherent to discussions of herbivory lies with the definition of "herbivorous." Many workers have called lizards herbivorous based only on a few records of plant food in their diets. In reality each species probably lies somewhere along a carnivory-omnivory-herbivory continuum, and may even fluctuate along that continuum depending on such factors as season, size, or resource availability. For purposes of this discussion I consider as 'truly herbivorous' (i.e., on the far right of the continuum) only those species whose diets include essentially only vegetation (whether fruits, flowers, seeds, *or* foliage) throughout the year.

Many species generally called 'herbivorous' are probably facultative herbivores at best, and more likely, simple omnivores. Based on the literature and my own dissections of several hundred lizard species, such forms as *Anolis equestris, Basiliscus* spp., *Agama* spp., *Physignathus leseuri, Angolosaurus skoogi, Gerrhosaurus* spp., *Egernia* spp., *Phymaturus* spp., *Tiliqua* spp., *Macroscincus coctei,* and *Trachydosaurus rugosus* (review in Pough, 1973; Fleet and Fitch, 1974; Cogger, 1975; Van Devender, 1975; Greer, 1976; Mares and Hulse, 1978), although often termed herbivorous in the literature, are clearly not true herbivores. In fact, by my definition, the only totally herbivorous extant lizards are the iguanines (±30 species; but see below) among the Iguanidae (Hirth, 1963; Henderson, 1965; Wiewandt, 1977; and Iverson, 1979); the genera *Uromastyx* (Henke, 1975; Grenot, 1976) and *Hydrosaurus* (Taylor, 1922) among the Agamidae (ca 17 species); and *Corucia zebrata* (Greer, 1976) in the Scincidae.

I also found no basis for earlier speculations that the iguanines *Amblyrhynchus cristatus* (Wilhoft, 1958), *Cyclura nubila* (Carey, 1966), *Iguana iguana* (Mertens, 1960; Pough, 1973), and *Dipsosaurus dorsalis* (Pough, 1973), and the agamid *Uromastyx hardwicki* (Mertens, 1960; Minton, 1966) exhibit an ontogenetic shift from carnivory to herbivory. Most of these suggestions were based: (1) on diet information from captive lizards, or (2) on anecdotal field observations. In fact, of all the true herbivores I have dissected, only the iguanine *Ctenosaura similis* showed any indication of an omnivorous juvenile diet (Kathleen Smith, pers. comm.; Montanucci, 1968; Iverson, unpublished data; Van Devender, this volume; among others). Further field study will be necessary to quantitatively establish the extent of this omnivory by size and season. I thus conclude that an ontogenetic shift from carnivory to herbivory is not usual in lizards truly herbivorous as adults, and further, that the documentation of such a transition (as appears to be the case for *C. similis*) will at best be the very rare exception rather than the rule.

Surprisingly few data are available on the specific diets of most herbivorous lizards (for example, see review of natural foods for species of *Cyclura* in Iverson, 1979). Most of the attention has been directed to species in the American southwest, the desert iguana *(Dipsosaurus dorsalis)* and the chuckawalla *(Sauromalus obesus)*, with some consideration given to the Neotropical *Iguana iguana.* Although some authorities provide significant lists of plant foods (Minnich and Shoemaker, 1970, 21 plant species for *Dipsosaurus dorsalis;* Nagy, 1973, 22 species for *Sauromalus obesus*), most references refer to but a few of the plant species eaten, and rarely categorize them by plant part. Thus the true diversity of diet in these lizards is generally unappreciated. For example, it has recently been shown that *Cyclura carinata* includes at least 58 plant species in its diet (Iverson, 1979; Auffenberg, this volume); *Cyclura cornuta stejnegeri,* at least 71 species (Wiewandt, 1977); *Sauromalus varius,* about 60 genera (Richard Hansen, pers. comm.); and *Uromastyx acanthinurus,* at least 45 species (Dubuis, et al., 1971). These high diversities imply that foraging strategies in these lizards are probably much more complex than generally realized. Auffenberg's study (this volume) is the first attempt to qualitatively and quantitatively analyze these strategies in an herbivorous lizard *(Cyclura carinata).* His results suggest some very fruitful directions for further study of herbivorous species.

FEEDING ECOLOGY

Other than Auffenberg's (this volume), studies of the ecological aspects of feeding in herbivorous lizards have received little attention. However, it is known that herbivorous lizards spend the majority of their activity cycle resting, not feeding or foraging as do most carnivorous lizards (e.g., Andrews, 1971) or mammalian herbivores. It apparently takes very little time to fill the digestive tract (Auffenberg, 1979; Auth, 1980), especially in relation to normal total food passage time (96+ hours; Harlow et al., 1976; Grenot, 1976; Auth, 1980; personal observation). Moberly (1968) estimated that *Iguana iguana* spends 90% of its time resting, and Beverly Dugan (personal communication) estimated that the same species spent 96% of the day inactive and only 1% feeding. Auffenberg (this volume) has calculated that *Cyclura carinata* spends only 18% of its daily activity period on a typical summer day actually involved in feeding and/or foraging behavior; and Minnich and Shoemaker (1970), Nagy (1973), and Grenot (1976) have commented on the reduced time spent feeding by *Dipsosaurus dorsalis, Sauromalus obesus,* and *Uromastyx acanthinurus,* respectively. Several herbivorous lizard species also inhabit regions temperate enough to necessitate the suspension of activity for the winter months: *Dipsosaurus dorsalis* (Norris, 1953), *Sauromalus obesus* (Berry, 1974; Case, 1976a), and *Uromastyx acanthinurus* (Grenot, 1976). Yet although no feces may be produced and the rest of the gastrointestinal tract may be empty, the proximal colon apparently always contains digesta (Grenot, 1976; Case, 1976a; pers. observ.).

Another neglected aspect of the feeding ecology of herbivorous lizards concerns food limitation. Despite the apparent abundance of plant food, availability of food resources may be the primary limiting factor for populations of many iguanine lizard species. For example, in Colombia, Mueller (1968, 1972) has shown that green iguanas, *Iguana iguana,* inhabiting strongly seasonal habitats are smaller than those in less seasonal habitats. Similarly, Case (1976b; see also 1978) has shown that chuckawallas, *Sauromalus obesus,* grow faster and larger in habitats with more diverse and more abundant food resources. Nagy (1973) and Case (this volume) have even documented cessation of reproduction in adult *Sauromalus* in harsh years. *Cyclura carinata* unquestionably grew faster and increased its fecundity in captivity with unlimited food (Iverson, 1979). *Dipsosaurus dorsalis* likewise grew faster with unlimited food in captivity than it did in the field (Mayhew, 1971). Insufficient food resources (in quantity and/or quality or availability and/or useability) appear to impose restraints on growth and fecundity in these lizards.

Rand (1978) has suggested that food may be limiting to *Iguana iguana* in highly seasonal tropical habitats only during part of the year. During the winter, food resources for *Cyclura carinata* are not only restricted to items more difficult to digest (primarily leaves) and of lower caloric content than at other times of the year (Auffenberg, personal communication), but environmental temperatures may also physiologically limit the effectiveness (efficiency?) of the lizard's use of those resources which *are* available (see Harlow et al., 1976). Further, even during times of maximum primary productivity, i.e., times when high quality foods (e.g., fruits) are most abundant, lizards can only eat and assimilate as much as their digestive machinery can process. Because of their low relative metabolic rates and daily fluctuations in body temperatures, this machinery may well limit energy intake even at maximum efficiency. Thus food limitation in individual iguanine lizards (but not populations) may operate via resource useability (or processability) in addition to simply availability. Again we lack the critical ecological and physiological studies to test this hypothesis.

DIGESTIVE PHYSIOLOGY

Other physiological aspects of lizard herbivory have received some attention. For example, digestive efficiencies of carnivorous lizards are known to typically vary from 70 to 90% and to significantly exceed those of herbivorous lizards, which normally range from 30 to 70% (literature review in Iverson, 1979). But Throckmorton (1973) and Hansen and Sylber (1979) have reported efficiencies high in the carnivore range for the herbivores *Ctenosaura pectinata* fed sweet potato tubers and *Sauromalus varius* fed dandelion flowers, shredded carrot roots and "chick starter." However, those two species, like most herbivorous lizards, probably don't have such easily digestible foods available in nature for most of the year. For example, during at least parts of the year, the herbivorous iguanines *Dipsosaurus dorsalis* (Minnich, 1970), *Cyclura carinata* (Iverson, 1979), and *Iguana iguana* (Rand, 1978) each primarily rely on leaves which are fibrous and presumably difficult to digest. Ingested leaves often pass through the entire gastrointestinal tract of *Cyclura carinata* (Figure 4.1) and *Cyclura cornuta* (Wiewandt, 1977) nearly intact. We badly need data on the relationship of digestive efficiencies to variables such as diet (and age and colon anatomy).

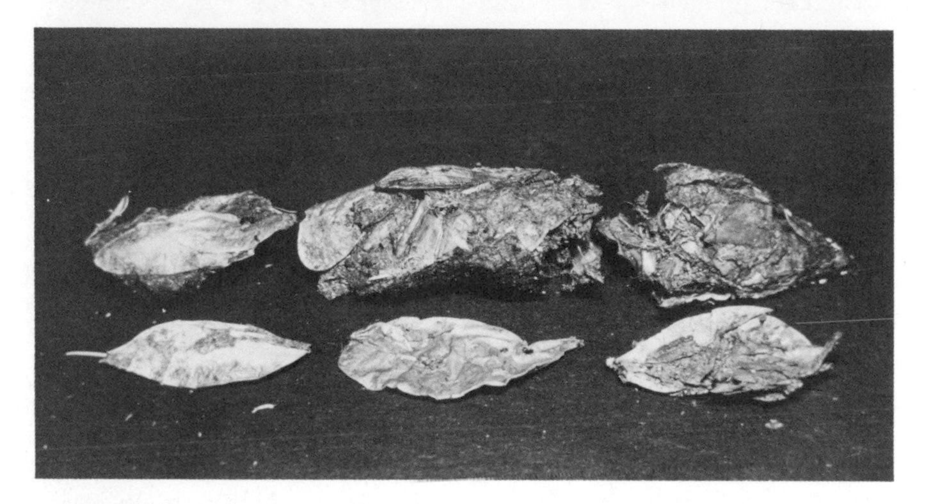

Figure 4.1: Typical fecal pellet of *Cyclura carinata* as deposited (top) and after partial dissection (bottom).

Freeland and Janzen (1974) have suggested that the detoxification of plant secondary compounds by herbivores may be metabolically very expensive. Thus herbivorous lizards must either avoid toxic vegetation (see Swain, 1976) or expend energy detoxifying plant compounds. It is therefore not surprising that *Dipsosaurus dorsalis* apparently eats only the flowers of the creosote bush, *Larrea divaricata,* leaving the fruits and leaves (Norris, 1953; Pianka, 1971; Mares and Hulse, 1978). It is noteworthy that *Cyclura cornuta* (Wiewandt, 1977) and *C. carinata* (Iverson, 1979) both feed commonly on the fruits and leaves of the manchineel, *Hippomane mancinella,* and the fruits of the poisonwood tree, *Metopium toxiferum,* species which contain strong alkaloids. Specific feeding strategies of herbivorous lizards have not been studied well enough to shed light on the relative importance of plants such as these to lizard nutrition and energetics, but evidence reviewed by Swain (1976) indicates that secondary plant compounds may be among the most important factors determining food preferences in herbivores.

Many aspects of the diet, feeding behavior, and digestive physiology of herbivorous lizards obviously remain to be investigated. The information is especially critical to our understanding of the functional and evolutionary significance of their morphological specializations. Because they number so few among extant lizard species and because their numbers are so rapidly declining (Fitch et al., this volume) it is important that attention be focused on them without delay.

TROPHIC ADAPTATIONS

All true herbivorous lizards, regardless of family, are specialized in that they all share one significant morphological adaptation (and a suite of associated physiological and ecological ones) found in no other living lizards; all have a distinctly enlarged, partitioned colon (Iverson, in press).

All iguanine lizards but *Amblyrhynchus cristatus* possess from one (*Dipsosaurus dorsalis;* Figure 4.2) to eleven (some *Cyclura cornuta;* Figure 4.3) transverse valves in the proximal colon (Table 4.1). Valves are of two kinds, circular (sometimes with sphincter) or semilunar (Figures 4.2 and 4.3), and circular valves (if present) always occur proximally to semilunar valves. Intraspecific variation in the number and type of valves is small, but greater in species with higher modal numbers of valves. There is no significant ontogenetic change in the number or the kind of valves.

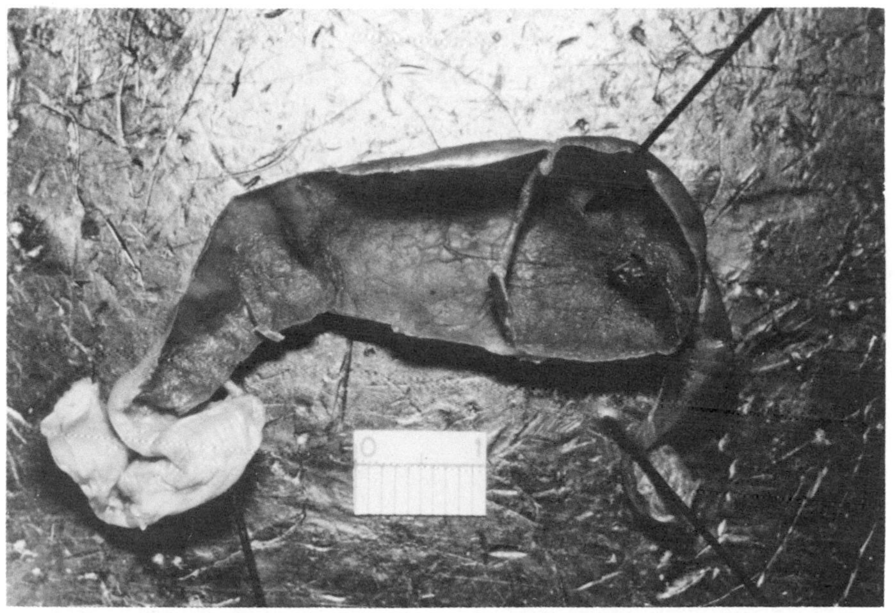

Figure 4.2: Ventral view of frontal section through the colon of *Dipsosaurus dorsalis* (UF 40726). Note ileocecal valve and single circular valve. Anterior to right. Scale = 1 cm.

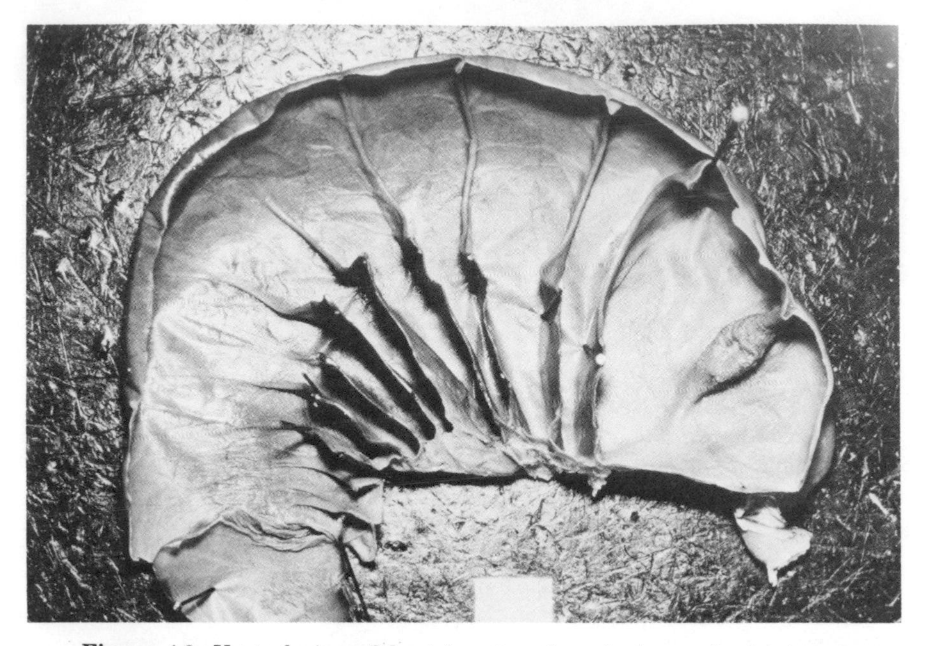

Figure 4.3: Ventral view of frontal section through the proximal colon of *Cyclura cornuta* (UF 33554). Note four circular valves (CV's) and seven semilunar valves (SLV's). Anterior to right. Scale = 1 cm.

Table 4.1: Modal Number of Colic Valves in Iguanine Lizards with Sample Sizes Greater than Five

Ranges appear in parentheses below modes. See Iverson (in press) for complete data.

Species	% of Total Sample with Modal Condition	Modal Number CV's	Modal Number SLV's	Species Range
Dipsosaurus dorsalis	100	1	0	SW U.S./NW Mexico
Ctenosaura clarki	88	1 (0-2)	1 (1-2)	Southern Mexico
Ctenosaura quinquecarinatus	100	1	1	SE Mexico to Nicaragua
Ctenosaura palearis	86	1	2 (1-2)	Guatemala
Ctenosaura similis	95	1	2 or 3 (2-4)	Mexico to Panama
Ctenosaura acanthura and *C. pectinata*	100	1	3	Mexico
Brachylophus fasciatus	100	0	4	Fiji Islands
Cyclura carinata	93	1	4 (4-5)	Turks & Caicos Islands
Iguana iguana	48	1	6 (3-6)	Mexico to South America; West Indies
Cyclura cychlura figgensi	43	4 (2-6)	3 (2-6)	Bahamas
Cyclura cornuta	38	3 (3-4)	6 (5-7)	Hispaniola

The colon of *Amblyrhynchus cristatus* differs from that of other iguanines only in the height of the valves (Figure 4.4); the infolded tissue layers involved in them are the same (Iverson, in press). The valves in all iguanines are formed by the infolding of the mucous membrane, the submucosa, and at least the circular (internal) muscular component of the muscularis externa. The serosa is not involved in the valvular structure. The valves may have evolved as simple infoldings (or creases) along the medial colic wall due to the increased bulk of digesta commensurate with an increasingly herbivorous diet and limited abdominal space (Iverson, 1980). I visualize this process as functionally similar to crease formation when rigid tubing is bent.

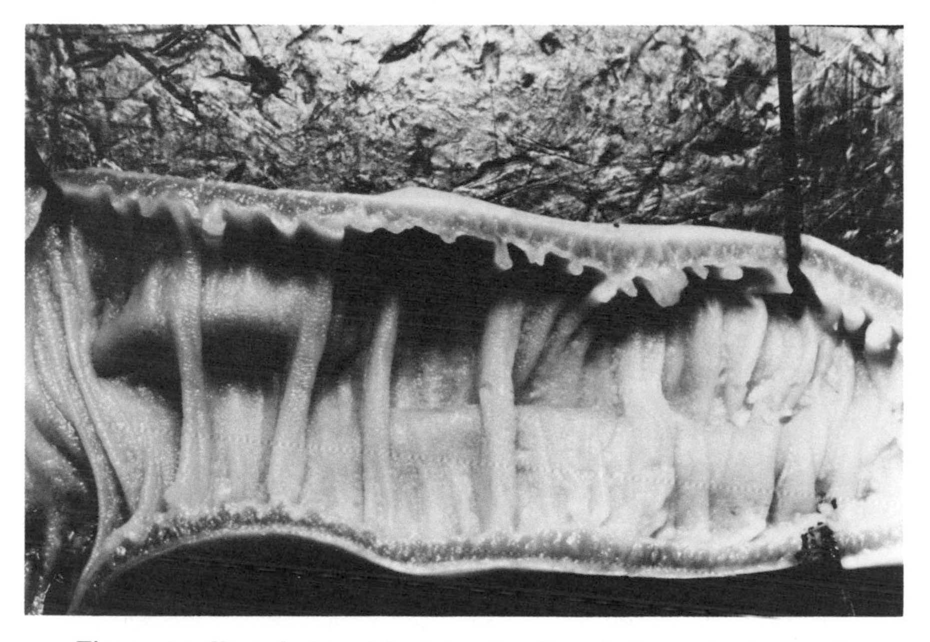

Figure 4.4: Ventral view of frontal section through the proximal colon of *Amblyrhynchus cristatus* (UF 41565). Anterior to right.

The variation in colon valve morphology in the iguanines seems to be of little value in phylogenetic comparisons. For example, modal number of valves varies within the genus *Cyclura* from 5 to 9. In addition, as mentioned above, the colon of *Amblyrhynchus* (with only folds, not complete valves) differs internally from all other iguanines, including *Conolophus* (with 1 circular and 4 semilunar valves) to which it is supposedly most closely related. However, because the number, type, and size of valves is so constant within a given iguanine species, colon structure is an important taxonomic adjunct. Colons of unknown species can nearly always be allocated at least to genus, based solely on morphology of that organ (Iverson, in press). In fact, this level of constancy suggests the existence of an undescribed iguanine taxon in the West Indies. The colon of *Iguana iguana* from the northern Lesser Antilles differs radically from that of the remainder of the species' range. Four total valves were present in each of three individuals from Montserrat and St. Croix, whereas all other specimens examined by me bore five, six (modal), or seven valves. No other iguanine species exhibits such extensive geographic variation. In light of Lazell's (1973) comments on the slightly different pattern and scalation characteristics in the same northern Antilles population, a systematic reappraisal seems warranted.

Perhaps the most intriguing thing about iguanine colic variation is the significant linear relationship between number of valves and mean body size for interspecific comparisons (Figure 4.5). The larger the species, the more complex is the colon (i.e., the more colic compartments present).

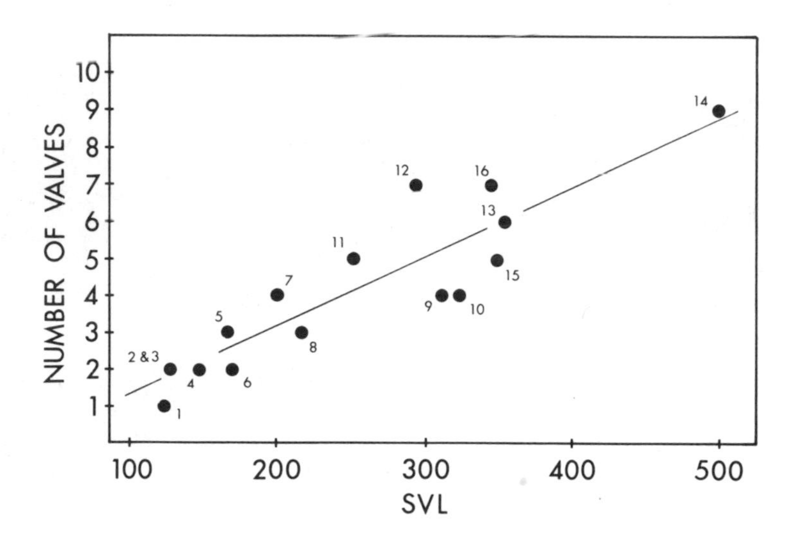

Figure 4.5: Relationship between modal number of colic valves and body size (mm snout-vent length) for the following iguanine species: *Dipsosaurus dorsalis* (1), *Ctenosaura defensor* (2), *C. quinquecarinatus* (3), *C. clarki* (4), *C. palearis* (5), *Sauromalus obesus* (6), *Brachylophus fasciatus* (7), *Ctenosaura hemilopha* (8), *C. similis* (9), *C. acanthura (=pectinata)* (10), *Cyclura carinata* (11), *C. cychlura* (12), *C. ricordi* (13), *C. cornuta* (14), *Conolophus subcristatus* (15), and *Iguana iguana* (16). Data from Iverson (1979, 1980, and unpublished). Least squares regression is: $y = 0.01895X - 0.6196$; $r = 0.908$; $p << 0.01$.

Although not as complexly modified, the colon of *Uromastyx* and *Hydrosaurus* is also partitioned; *Uromastyx* proximally (Figure 4.6) and *Hydrosaurus* distally (Figures 4.7 and 4.8). In being partitioned, the colon of *Corucia* differs radically from all other skinks I have examined (Figure 4.9), including species of the closely related genera *Egernia* and *Tiliqua*. Thus a partitioned colon has apparently evolved independently at least three times (once in each of the three families) in the Lacertilia. At least in regard to their gastrointestinal tract anatomy, these lizards can hardly be considered unspecialized.

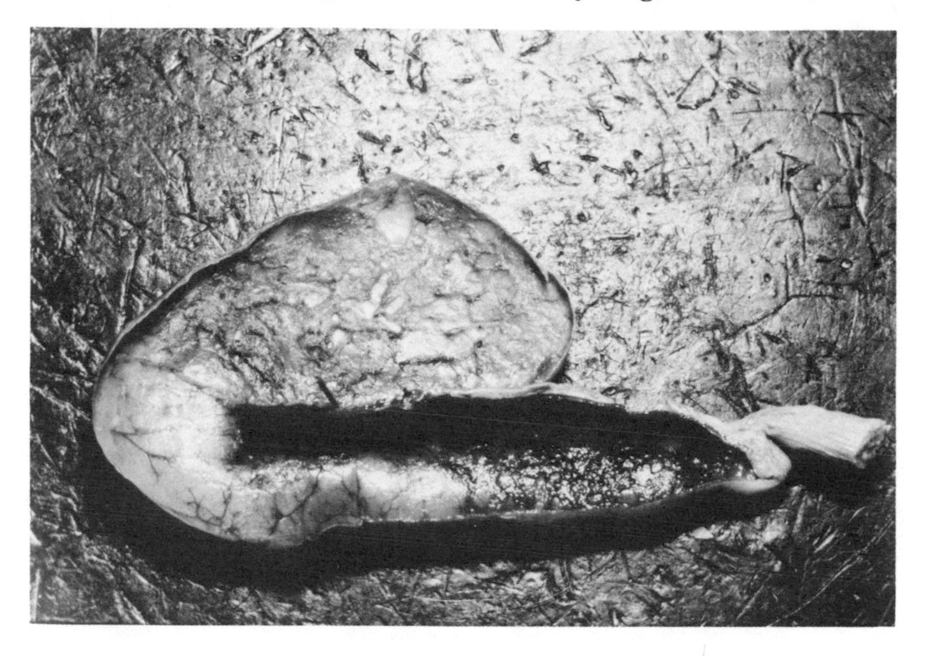

Figure 4.6: Ventral view of frontal section through the proximal colon of *Uromastyx hardwicki* (UF 20179). Anterior to left.

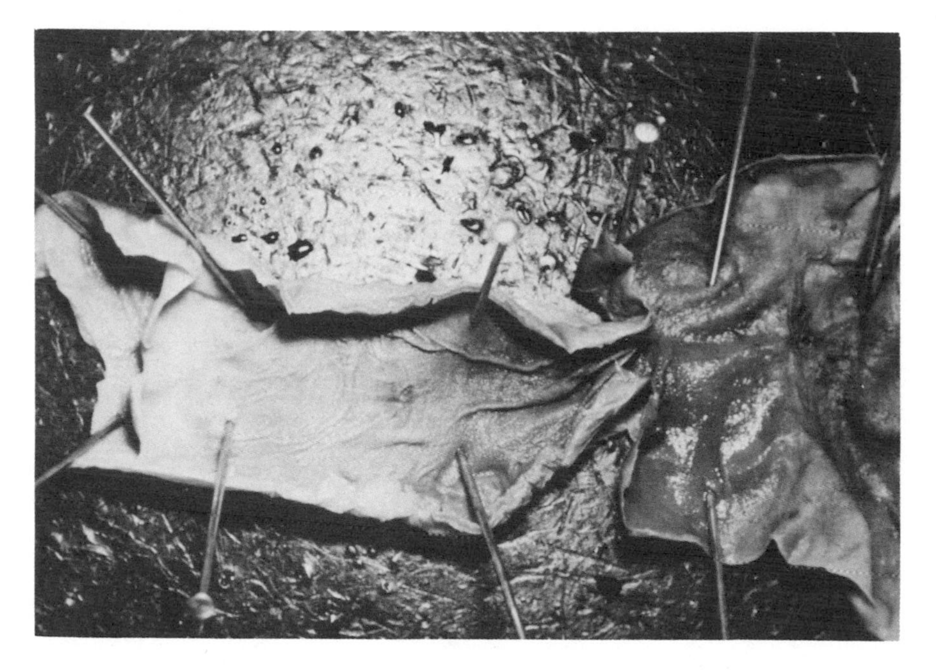

Figure 4.7: Ventral view of frontal section through the distal colon of *Hydrosaurus amboinensis* (FMNH 142331). Anterior to right.

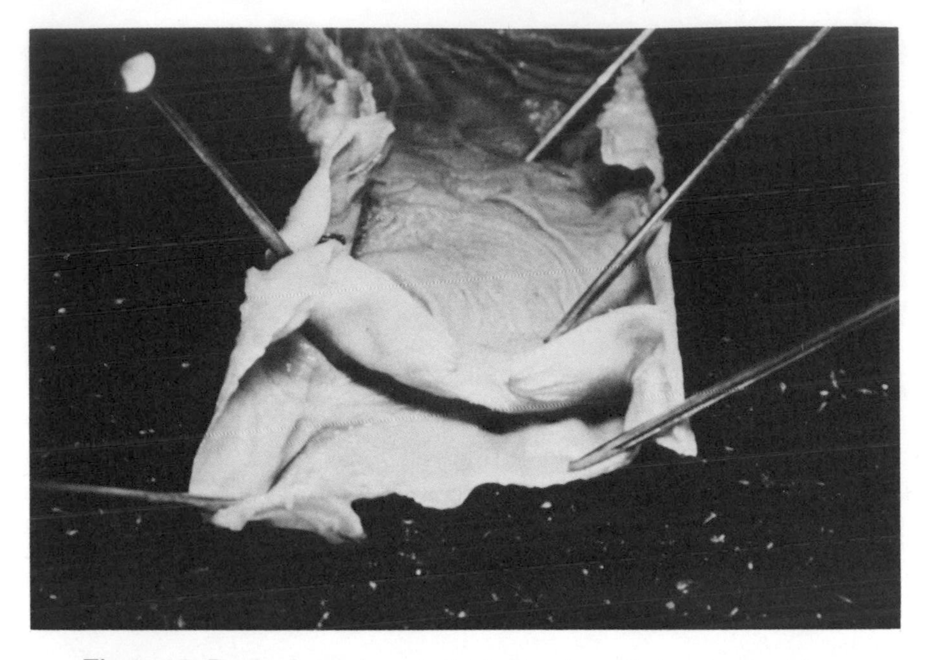

Figure 4.8: Proximal view through distal colic valve of *Hydrosaurus amboinensis* (FMNH 142331).

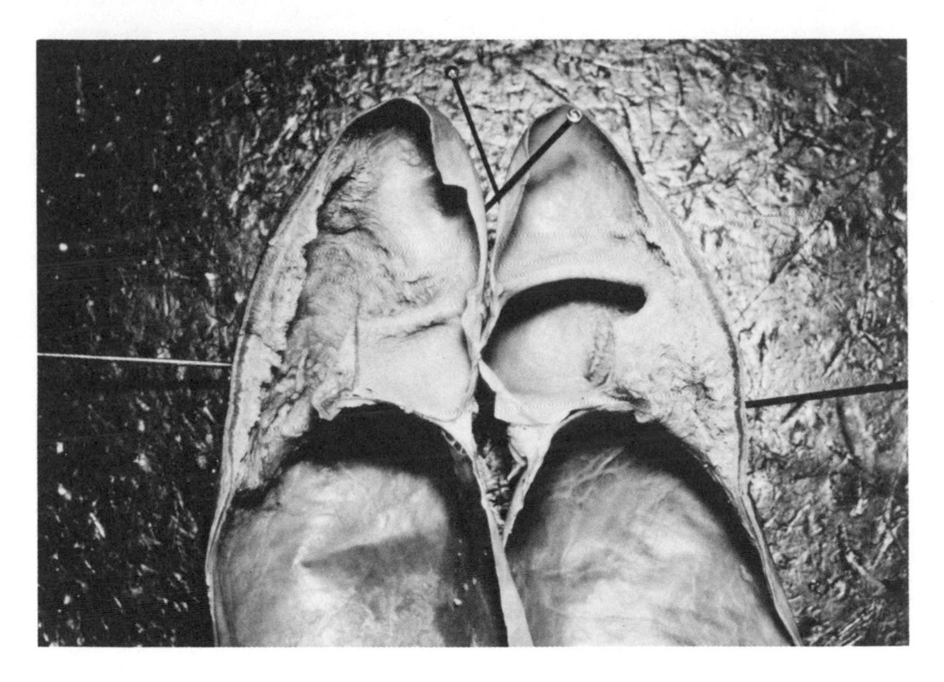

Figure 4.9: Ventral (right) and dorsal halves of the proximal colon of *Corucia zebrata* (UF 41309). Anterior to top.

Though I have established that all truly herbivorous lizards possess a partitioned colon, I am still uncertain about its functional significance. The partitioned colon surely slows the passage of digesta through the gut, and relative absorptive surface area (for water and nutrients) is certainly increased. But the presence of tremendously dense nematode faunas (and presumably bacterial and protozoan populations) in the normal cecum of all of these herbivorous lizards suggests that a much more significant advantage of these partitions is that they provide important microhabitats for colic (cellulytic?) symbionts. For example, McBee and McBee (this volume) have documented an excess of 10^{10} bacterial cells or clumps per gram of colic material, and I estimated the population of nematodes in the colon of single healthy adult *Cyclura carinata* to be in excess of 15,000 (Iverson, 1979). Juveniles of *C. carinata* begin accumulating colic nematodes soon after hatching, and worm populations are usually about 100 by age 3 months, continuing to increase with age. These nematodes (families Atractidae and Oxyuridae) have direct life cycles, and eggs are likely ingested during substrate licking, geophagy, or coprophagy–behaviors frequently observed in these lizards (Sokol, 1967, 1971; and review in Iverson, 1979).

Figure 4.10 illustrates most of the nematodes removed from the proximal colon (i.e., valve region) of a single adult Cuban ground iguana *(Cyclura nubila)*. Significantly, these heavy worm burdens are typical of herbivorous lizards, whereas such burdens are not found in omnivorous or carnivorous lizards (Bowie, 1974; among others; pers. obs.).

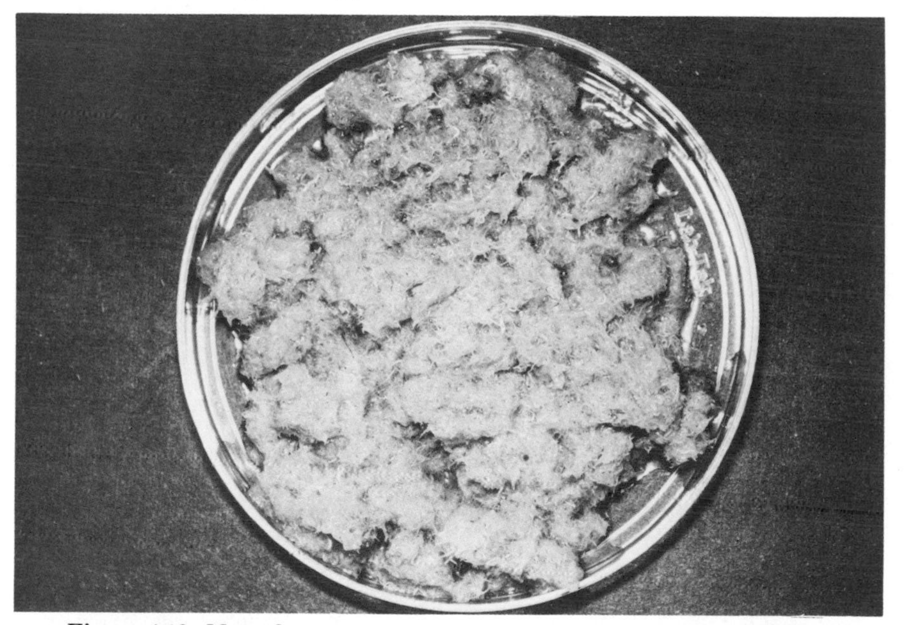

Figure 4.10: Most of nematodes removed from the proximal colon of an adult *Cyclura nubila* (USNM 37908) in 10 cm Petri dish.

The question remains as to just what the relationship is between these organisms and their hosts. Hypothesizing that colic compartmentalization permits the proliferation of nematodes, bacteria, and perhaps protozoa (if by no other means than by reducing the likelihood of egestion due to peristalic flow of digesta), I compared the number of colic nematode species described from each lizard species (tabulated in Iverson, 1979) with number of colic valves for the 11 best-studied iguanine lizard species. The linear regression relating those two variables is highly significant (Figure 4.11). Clearly increased colic partitioning allows an increase in diversity (and surely abundance) of at least nematodes, and probably bacteria and protozoa as well.

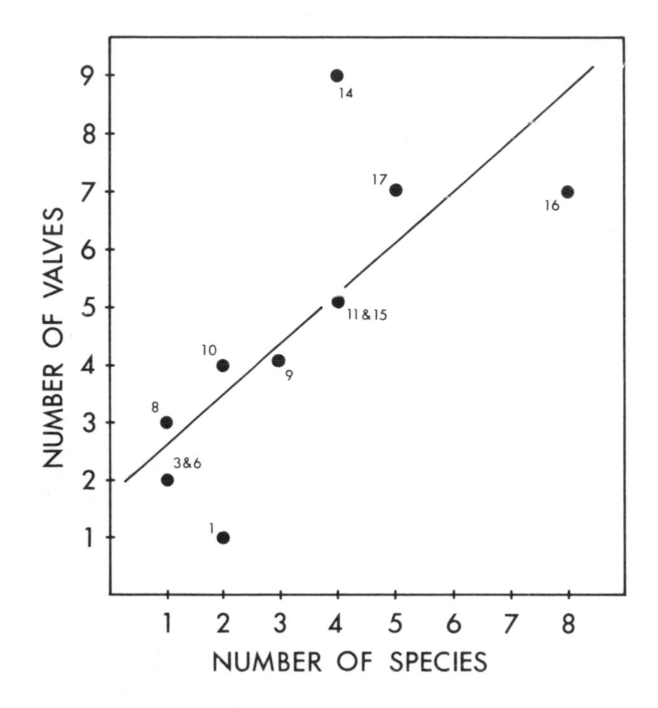

Figure 4.11: Intestinal nematode species richness (referenced in Table 26 of Iverson, 1979) versus colic modification (modal number of colic valves from Iverson, 1980) for the 11 best-studied iguanine species. Species are coded as in Figure 4.5, except *Cyclura nubila* (17). Least squares regression is plotted: $y = 0.8785X + 1.659$; $r = 0.762$; $p < 0.01$.

A positive linear correlation is also suggested by a comparison of nematode species richness and mean body length for the ten iguanine species for which data are available (Figure 4.12). The simple linear regression is not quite significant, whereas a log-log comparison is ($r = 0.70$; $p < 0.05$). No such relationship for cecal nematodes is identifiable in carnivorous lizards (data from Pearce and Tanner, 1973 and Bowie, 1974). In fact, larger lizard species often harbor fewer nematode species than smaller lizard species. I would be surprised if colic bacteria and perhaps protozoa did not exhibit these same relationships.

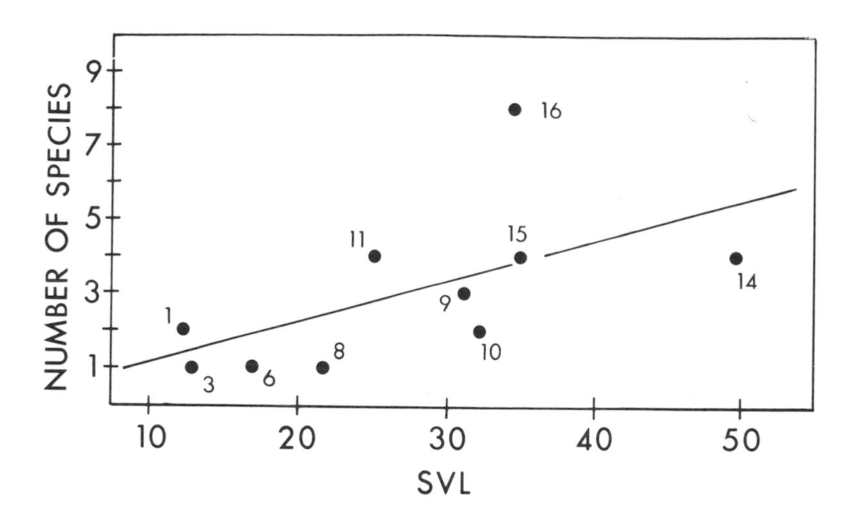

Figure 4.12: Relationship between body size (cm snout-vent length) and nematode species richness for ten iguanine lizards. Species are coded as in Figure 4.11. Least squares regression is plotted: $y = 0.1099\ X + 0.01718$; $r = 0.592$; $r_{0.05} = 0.632$. Data sources as in Figures 4.5 and 4.11.

The tremendous nematode densities in healthy lizards suggest they are not parasitic, but rather commensalistic, or perhaps even mutualistic. Potential roles for these nematodes include (1) the simple mixing and mechanical breakdown of vegetation, effectively increasing the surface area of digesta particles; (2) the production of useable waste products (vitamins, cellulase, volatile fatty acids?); and/or (3) the regulation of the composition and/or abundance of colonic microbes (on which some nematodes are known to feed; Schad, 1963) However, in a study of digestion in the yellow chuckawalla *(Sauromalus varius),* Hansen and Sylber (1979) treated their lizards for intestinal worms, assumed none were present, and yet found very high digestibility coefficients for low to mid-fiber diets. This seems to indicate a reduced importance of colonic nematodes, but only if it can be certainly established that the worms were, in fact, eliminated.

The importance of bacteria and protozoa to digestion in other herbivorous organisms (primarily higher vertebrates and insects) is well-established (McBee, 1971, 1978; Gasaway, 1976; Baldwin, 1970; among others). Their significance in herbivorous reptile digestion should be intuitive, especially since it has recently been established that there is considerable cellulytic activity in the colon of herbivorous lizards (Nagy, 1977; Hansen and Sylber, 1979), and also that colic fermentation products (volatile fatty acids) are assimilated by the green iguana, *Iguana iguana* (Mcbee and McBee, this volume) and yellow chuckawalla, *Sauromalus varius* (Hansen and Sylber, in press) (and even provide at least 15.2% of the daily energy budget in the herbivorous green turtle, *Chelonia mydas* [Bjorndal, 1979]). This last point contradicts Nagy's (1977) conclusion that fermentation products are not assimilated by *Sauromalus obesus.* However, because Nagy's experiments were run at only 23°C, his data are suspect. Thus, contrary to earlier speculations, hindgut fermentation appears to be very important to the digestive physiology of herbivorous reptiles.

Gastrointestinal tract modification for symbiont culture is also the norm in herbivorous organisms that have previously been studied (Vaughan, 1978; Romer and Parsons, 1977; Barnes, 1969). Colonic partitioning in herbivorous lizards is thus by no means novel. What is surprising is the lack of attention it has received, especially since it appears to be the one adaptation essential for a lizard's completely herbivorous existence. However, numerous studies now in progress by several investigators are addressing the physiology and biochemistry of herbivory in reptiles.

BODY SIZE

True herbivorous lizards share another obvious character. Excluding the varanids (which are very specialized carnivores), they are the largest extant lizards. This relationship has been frequently mentioned in the literature (review in Pough, 1973), and several hypotheses offered to explain the adaptive significance of the correlation between large body size and herbivory. These include the advantage of greater mechanical strength for reducing vegetation for consumption (Sokol, 1967) (although there is no mastication; Throckmorton, 1976); the advantage of reduced predation and competition (Ellis and Ross, 1978; among others); the advantage of eating easily obtainable foods of medium to poor quality (i.e., plants) rather than higher quality items (e.g., insects) more energetically costly to obtain (since metabolic rates are relatively lower in large lizards: Pough, 1973); and the advantage of greater thermal inertia, allowing the maintenance of elevated body temperatures and facilitating the digestion of vegetation (Ellis and Ross, 1978). Each of these theories relates selective advantages which no doubt accrue to large, herbivorous lizards, but none appears to fully explain the body size-herbivory relationship in these lizards. I here extend another hypothesis which I think more adequately explains the body size-herbivory correlation and permits speculation on the evolution of these characters in lizards. I believe that the anatomical, physiological, and ecological characteristics of the gastrointestinal tract are the most important determinants, not only of herbivorous capabilities, but also of body size in these lizards.

I have already mentioned here the positive relationship of colic complexity to body size in these lizards, and the presumed benefit of that modified colon to digestion. Previous studies on both lizards (Case, 1976b) and turtles (Parmenter, 1978) have shown that body size may be significantly related to resource availability. Analogously, I believe that the evolutionary increase in colon complexity has increased resource useability (not necessarily *availability*!), and thereby energetically permitted increased body size. Thus by modifying their colons (and diversifying and increasing their intestinal flora and fauna), these herbivores have been able to grow to larger sizes, and thus more fully gain other selective advantages, such as reduced predation, metabolic (Pough, 1973) and thermoregulatory (Ellis and Ross, 1978) benefits.

DISCUSSION

I believe that the iguanines, *Uromastyx, Hydrosaurus,* and *Corucia* are much more specialized than generally realized. Each has a relatively large body size and a modified colon with large nematode and microbe populations–a combination of characters unique among the lizards. In addition, although I have not mentioned them, these lizards all possess dentitions modified for herbivory (Edmund, 1969; Avery and Tanner, 1971; Greer, 1976; and Throckmorton, 1976) and (apparently) nasal salt glands for extrarenal salt secretion (Peaker and Linzell, 1975; Dunson, 1976). These last two characters are shared by lizards which are not entirely herbivorous, however they nevertheless facilitate an herbivorous existence; teeth for more efficient cropping and salt glands for dealing with the increased potassium load.

In summary, the evolution of lizard herbivory probably had its roots in xeric habitats (evidenced by the significant inverse correlation between percentage of lizards exhibiting at least some herbivory and annual rainfall in North American communities; Keeling, in Rand, 1978). In order to inhabit those areas, lizards would have been faced with: (1) problems of ionic balance (regardless of their diets), and (2) seasonal difficulties in obtaining adequate animal food (probably insects). The evolution of extrarenal nasal salt glands in these lizards would presumably have reduced the osmotic problems, and simultaneously may also have allowed the lizards to utilize small amounts of plant resources during seasonal animal food shortages without being susceptible to additional ionic problems from ingesting potassium rich plant parts. The acquisition by these proto-herbivores of small populations of symbiotic (cellulytic) nematodes and/or bacteria would have even further facilitated the efficient use of ingested vegetation. As the diet of these lizards became more omnivorous, the bulk of the digesta in the gastrointestinal tract would presumably have increased. Among other things, this presumably would have resulted in increased gastrointestinal tract capacity, cramped abdominal space and even medial crimping of intestinal loops where they folded back on themselves. If, as I believe, these crimps (1) slowed the movement of digesta (permitting more time for digestion or absorption), or (2) reduced the likelihood of egestion of gut symbionts by providing microhabitats for them, selection would have favored their maintenance. Thus evolution of these "protovalves" in conjunction with salt glands and the acquisition of gut symbionts may finally have made it energetically feasible to subsist primarily on vegetation. Both before and (more intensely) after the appearance of these valves, selection would also have been operating on the dentition to permit efficient handling of first an omnivorous, and later an herbivorous diet.

Finally, those species in which the evolution of valves was more rapid, would, because they had more valves and more compartments, also have had more symbionts and hence more efficient food processing. The energy benefit from this apparatus would result in increased body size, the size increase in direct proportion to the efficiency of the gastrointestinal machinery. When measured in terms of number of partitions, this is exactly the relationship that is obtained for the iguanines (Figure 4.5). Body size increases would, of course, then have imparted additional advantages to these lizards, specifically the ecological, metabolic, and thermoregulatory benefits already mentioned.

Thus, it appears to me that the evolution of colic compartmentalization is the key character complex both necessary for and permitting total herbivory in lizards. In addition, because there is no ontogenetic change in colon structure I expect both juveniles and adults of these lizards to be primarily herbivorous. Deviations from this scheme (as apparently is the case for small juvenile *Ctenosaura similis*) are presumed to be secondary adaptations to omnivory or carnivory (for example, as a response to atypically abundant alternate animal food sources).

It is hoped that this volume will stimulate more interest in studies of the ecology and physiology of herbivory in this group of lizards. It is time that the uniqueness of these lizards be generally accepted and the appropriate scientific attention be directed to them.

SUMMARY AND CONCLUSIONS

The iguanine lizards, with the agamid genera *Uromastyx* and *Hydrosaurus,* and the scincid genus *Corucia* comprise the only true lizard herbivores. They share the following characteristics, considered adaptive to an herbivorous existence: modified dentition, generally large body size, nasal salt glands, a partitioned colon (unique among lizards), and huge populations of colic nematodes and, presumably, colic bacteria and protozoa. Although each of these character complexes is considered adaptive in permitting the increased use of plant material for food by lizards ancestrally carnivorous, the evolution of colic modification (and the associated symbiotic microinhabitants) appears to be the key character complex permitting total herbivory in lizards. The larger body size of these lizards is explained in terms of this model, and the general applicability of the theory on an ontogenetic shift from carnivory to herbivory to these lizards is questioned.

Acknowledgments

The support of the New York Zoological Society, the American Philosophical Society, and the Earlham College Faculty Development Fund made my field work on iguanines possible.

The comments, criticisms, and suggestions of Walter Auffenberg, David Auth, Karen Bjorndal, Bill Buskirk, Ted Case, Bev Dugan, Dick Franz, Diderot Gicca, Richard Hansen, Hank Harlow, Bob Henderson, Dan Janzen, Bill Mautz, Richard McBee, Peter Meylan, Ken Nagy, Harvey Pough, Stan Rand, Rick Shine, Kathy Smith, Otto Sokol, and Tom Wiewandt are all acknowledged. The Florida State Museum, the University of Florida, and Earlham College provided support and space for the duration of these studies.

5

The Hindgut Fermentation in the Green Iguana, *Iguana iguana*

Richard H. McBee
Virginia H. McBee
Department of Microbiology
Montana State University
Bozeman, Montana

INTRODUCTION

The green iguana, *Iguana iguana,* is one of several species of lizards living chiefly on a diet of vegetation. It does not appear to be highly selective in the type of vegetation that it consumes except that it is usually a tree dweller and thus consumes the leaves of trees and vines. It also eats a variety of small fruits when they are available.

Because of its dentition the iguana does not chew its food. Leaves are perforated and sections torn out. Fruits observed from iguana stomachs during the Panama dry season had not been chewed. Most of them remained intact in the stomach and small intestine.

An examination of the iguana digestive system revealed a tubular acidic stomach several times as long as its diameter. It was not empty in any of the iguanas captured between 9 A.M. and midnight. The small intestine appeared to be fairly simple and discharged rather large boluses of poorly digested leafy material or intact fruits into the mid-section of an enlarged hindgut, one end of which was sac-like and the other end considerably expanded before it connected with the large intestine. The hindgut had a series of septa that extended partially across it, dividing it into approximately seven chambers as also observed by Iverson (this volume). On the basis of visual observation, greater changes took place in the leafy ingesta while it was in the hindgut than had occurred in the stomach and the small intestine combined.

This hindgut was examined as a possible site of a microbial fermentation that would aid in the digestion of its contents and produce fermentation products absorbable and metabolizeable by the iguana. This would be comparable to the fermentations found in the ceca (McBee, 1971) and other hindgut modifications of some insects, birds, rodents and many other animals (McBee, 1977).

EXPERIMENTAL DESIGN

Iguanas of various sizes were caught at different times of the day for a study of hindgut fermentation and the bacteria involved in it. A reliable source of iguanas was established by employing Panamanian boys to hunt on Perico Island at the Pacific entrance to the Panama Canal. The animals studied were largely restricted to males, since during the period of January to April in the Canal Zone, females were gravid and had little room for food in their digestive systems. Males ranging in total length from 86 to 103 cm and weighing between 420 and 1356 g were used for most examinations since they were large enough to have sufficient hindgut contents for study (44 to 170 g) and did not contain as many fruits as the larger iguanas that were caught. Fruits reached the hindgut essentially intact and only began to break up in that site. Their presence prevented effective sampling.

Animals were anaesthetized with chloroform, opened with a midline incision and the content of the hindgut removed to a plastic bag flushed with oxygen-free carbon dioxide to maintain what was believed to be an anaerobic condition. The content was mixed by kneading the bag and then incubated at approximately the rectal temperature of the animal, 27° to 36°C. Samples of gut content were removed from the bag immediately following kneading and at 15 minute intervals for 45 minutes. These samples were weighed and preserved with measured amounts of 1 N sulfuric acid for subsequent fermentation product analyses. In some instances blood samples of both arterial and venous blood from the main vessels supplying the hindgut were collected (Bailey and McBee, 1964) and preserved with acid for the determination of whether fermentation products were absorbed from the hindgut.

Most hindgut contents were also cultured anaerobically using the roll-tube technique (Hungate, 1950) and a glucose, cellobiose and starch agar culture medium (Caldwell and Bryant, 1966) to ascertain the size of the culturable bacterial populations and the dominant bacterial types involved in the fermentation. Formalin preserved gut content was used to make a direct microscopic clump-count for the numbers of bacteria present and also to search for other microorganisms such as protozoa.

Changes in the pH of the gut content were monitored during the 45 minute incubation period with indicator papers.

Fermentation products in all gut samples were measured quantitatively by gas chromatography and the fermentation rates were determined by the O-time method (Carroll and Hungate, 1954).

Bacterial culture counts were made from triplicate roll-tubes containing fewer than 100 colonies following a five day incubation at 35°C. Dominant types of bacteria were determined by picking 20 colonies at random from a high dilution (10^{-8} or 10^{-9}) roll-tube into anaerobic agar slants which were later checked for purity and species by the use of appropriate identification procedures.

A total of 21 animals was examined, 11 were cultured. Fourteen were large males, one a large female, and six sexually immature iguanas not sexed.

RESULTS

The fermentation products found in the hindgut were chiefly acetic, propionic and butyric acids with lesser amounts of lactic acid and some of the higher volatile fatty acids (VFAs) sometimes present. Ethanol was not found in measureable quantities.

Although the pH dropped rapidly from an inital value of about 6.8 to 7.0 in many incubated samples of gut content, the samples were inadequately mixed to give meaningful fermentation rates in most instances. Due to the relatively dry state of the gut contents, 23, 25 and 30% water in the three samples tested, the kneading was not adequate to give a homogenous mass from which truly duplicate samples could be taken. This was an unexpected turn of events that could possibly have been avoided if analytical procedures had been readily available locally.

Fermentation product analyses from two animals, however, followed consistent patterns which may be reliable. They are given in Table 5.1.

Table 5.1: Volatile Fatty Acid Production in Hindgut Content from Two 102 cm Male Iguanas

Animal No.	Body Weight (g)	. . . Acid Production, μM/hr. . . . Acetic	Propionic	Butyric	Energy Equivalent (cal/hr)
17	1,365	184	68.4	231	185
18	802	168	11.6	218	153

It is assumed that the majority of these fermentation products were absorbed since most of the hindgut pH values were approximately 6.8 to 7.0 at the time the contents were removed from the animal, indicating that acids were being absorbed as rapidly as they were produced. The drop in pH to 6.0 or less in 15 to 30 minutes was evidence for an active fermentation. There were some deviations from this pattern, however; animal 12 had an initial hindgut pH of 5.5. In no case was peristalsis of the hindgut observed, but it is not common in anaesthetized animals.

Analyses of blood samples showed measurable amounts of VFAs in both the efferent veinous blood, about 10 μM/ml, and the arterial blood supply, less than 4 μM/ml. Blood flow through the hindgut vascular system was 1.5 to 3 ml/min as measured in 10 animals by severing the main vein leaving the hindgut.

Energy Considerations: The observation of a hindgut fermentation is only meaningful in light of whether a significant contribution to the energy requirement of the animal is made. Nagy (this volume) has determined that the field metabolic rate of iguanid lizards is proportional to the 0.80 power of the body weight. Thus the equation 2.23 x body weight$^{0.80}$ = calories per hour required by the animal to sustain itself under field conditions. The energy derived from the hindgut fermentation in animals 17 and 18 is presented in Table 5.2.

Table 5.2: Percentage of Energy Requirement Derived from the Hindgut Fermentation

Animal No.	Body Weight (g)	. . Energy, cal/hr . . Required	Yield	Percent of Requirement
17	1,365	612	185	30
18	802	400	153	38

When one calculates the energy provided to the body by absorption into the blood stream (on the basis of average figures since useable blood samples were not collected for animals 17 and 18), it is found that 6 μM VFA/ml are taken up by 2.25 ml blood/minute for a total of 810 μM VFA/hr. This compares with 483 and 398 μM VFA/hr produced in animals 17 and 18 respectively. These results would indicate that several of the animals had fermentation rates considerably higher than those measured in animals 17 and 18.

Microorganisms: Direct microscopic counts of hindgut bacterial populations showed numbers in the range of 30 x 10^9 clumps of bacterial cells per gram. Colony counts in anaerobic cultures ranged from 3.3 to 23.5 x 10^9 per gram of content. Both sets of results are comparable with bacterial populations found in the hindguts of other animals. No protozoa were observed, but this may have been due to using formalin as a preservative. Nematodes were not present in remarkable numbers.

The dominant bacterial species in three iguanas were of the genus *Clostridium* whereas in the other eight cultured iguanas, the dominant bacteria were of the genus *Leuconostoc*. Several of the cultures may be of previously undescribed species. Both of these genera could be derived from soil and natural contamination of leafy material in the diet. There was no correlation between the kinds of dominant bacteria and the collection area or the size of the animal. Among the 140 pure cultures obtained, there were five of the genus *Lachnospira* but none from the genera *Bacteroides, Fusobacterium* or *Ruminococcus.*

Urea: Analyses of hindgut material from three animals for urea showed no indication of this compound. It would appear that the iguana does not secrete urea into the hindgut to provide a nitrogen source for the bacteria there as is found in some animals (Houpt, 1963).

DISCUSSION

The combination of anatomical and biochemical examinations leaves no doubt regarding the presence of a bacterial fermentation in the hindgut of *Iguana iguana* that contributes significantly to the energy requirements of the animal. Using Nagy's (this volume) equation for calculating the field metabolic rate, the two animals for which consistent fermentation rate data were available received energy in the form of VFAs from the hindgut at the rate of 30 and

38% of their requirements. An average of data obtained from eight animals from which adequate blood samples and blood flow rates were obtained, indicated that absorbed VFAs could contribute in excess of 50% of their energy requirements. These figures are higher than any obtained from hindgut fermentations in birds or rodents (McBee, 1977). Since the blood data are not from the same animals as the hindgut fermentation data, the relative rates of absorption of the three most abundant VFAs, acetic, propionic and butyric, cannot be calculated from these data.

The ratios of acetic to butyric acid produced in the hindgut fermentation of the iguana are of interest because acetic is the more abundant acid in homeothermal animal digestive system fermentations. The reason for this high production of butyric acid in the iguana is not apparent.

If one interprets a nearly neutral hindgut content which shows a rapid drop in pH upon removal from the animal as an indication of an active fermentation, then all animals collected between 9 A.M. and nearly midnight had such fermentations in progress and were absorbing the products. It is assumed, therefore, even in the absence of data for the midnight to 9 A.M. period, that the fermentation may supply energy for the entire 24 hour day, although the rate may not be uniform.

Factors that may affect the fermentation rate other than amount of food intake, would be (1) the body temperature, which ranged from 27° to 36°C at the time of examination and would be the highest when iguanas were basking in the sun; (2) moisture content, which in the rainy season would probably be considerably higher than the 23 to 30% found during the dry season; (3) the presence of fruits during certain seasons; the largest iguanas collected all contained fruits in the hindgut and they were assumed to have a higher concentration of fermentable carbohydrates than leaves; (4) the variety and state of the leaves consumed; and (5) the rate of absorption of fermentation acids. If these acids were not absorbed as rapidly as produced, a low pH might cut down on the fermentation rate, conserving energy until it was needed as was hypothesized for the willow ptarmigan (McBee and West, 1969). One iguana had a hindgut pH of 5.5.

Bacterial populations were about what one would expect with respect to numbers. The bacterial types, predominantly *Clostridium* or *Leuconostoc*, could be interpreted as those fermentative bacteria that would be enriched from the environment. The leaf fermentation would be similar to a beginning sauerkraut fermentation and the constant absorption of acids would prevent the development of the more acidophilic population found in the later stages of sauerkraut. In such a situation either *Leuconostoc* or *Clostridium* would be a reasonable genus to expect as the dominant microflora. A recognized method of transmitting hindgut flora from adult iguana to off-spring has not been found, and since the two dominant genera are both found abundantly in nature it is hypothesized that each iguana establishes its own fermentation hindgut flora from what it takes in for food. Motion pictures of young iguanas emerging from the nesting chamber show them repeatedly touching their tongues to the surrounding soil (see Burghardt, et al., 1977). It could also, therefore, be postulated that the young iguanas pick up fecal organisms left in the soil of the nesting areas by the females several months previously. Since the second hypothesis is only a modification of the first and since the necessary study to substantiate the second has not been made, the simpler of the two hypotheses will be advocated (but see Troyer, 1972).

The green iguana is thus another example of an herbivorous animal that has capitalized upon a hindgut fermentation to obtain more energy from a given amount of food than would be available through animal digestive enzymes. Such fermentations occur in all major animal groups (McBee, 1977), but in general, have been inadequately studied from both the microbiological and energy contribution view points.

Using what data are available, however, it appears that the green iguana may obtain from 30 to 40% of its energy from the hindgut fermentation. This would significantly reduce the time required for feeding and the exposure to predation with less than a 10% increase in the body weight. It also allows the iguana to survive on types and amounts of vegetation which would not otherwise be adequate.

When one compares the hindgut fermentation of the green iguana with that of other animals, the iguana, with 30 to 40% of its energy requirement coming from the fermentation, rates high on the scale. The beaver, *Castor canadensis,* receives about 19% of its maintenance energy from the hindgut (Hoover and Clarke, 1972); the yellow-haired porcupine, *Erethizon dorsatum epixanthum,* from 5 to 33% (Johnson and McBee, 1967); the laboratory rabbit from 4 to 12% (Bailey and McBee, 1964) and 10 to 12% according to Hoover and Heitmann, 1972. Among the birds, adequate data are available only for the ptarmigan. McBee and West (1969) estimated that the hindgut fermentation could supply 6 to 30% of the basal energy requirement of the willow ptarmigan *(Lagopus lagopus)* whereas Gasaway (1976b) found the yield to be 11, 4.8 and 3.8% of the energy requirement of the birds living under maintenance, standard and free-living conditions respectively. A similar fermentation in the rock ptarmigan *(Lagopus mutus)* produced 18% of standard and 7% of free-living requirements (Gasaway, 1976a).

The only other reptile that has been studied is the green turtle *(Chelonia mydas)*. Bjorndal (1979) found that this turtle living on seagrass received at least 15.2% of its daily energy budget from the hindgut fermentation, but considered this to be a low figure because of an extensive colonic fermentation that was not measured.

The fermentations in the green iguana and the green turtle are assisted by environmental temperatures of about 30°C or higher, thus approximating the fermentation rates found in warm blooded animals. In all cases of these other animals the fermentation products were similar, being mostly short chained fatty acids with lesser amounts of lactic acid and ethanol sometimes present.

SUMMARY AND CONCLUSIONS

A study of the hindgut fermentation in the green iguana *(Iguana iguana)* revealed that in males of the weight range of 420 to 1365 g the hindgut contents were about 10% of the body weight. This material was actively fermenting with the production of acetic, propionic and butyric acids which were absorbed into the bloodstream. Approximately 30 to 40% of the iguana's energy requirement may be supplied by this fermentation. The iguana facilitates the fermentation by basking in the sun, keeping its body temperature above 30°C when possible.

The dominant bacterial types present in the hindgut were either *Clostridium* or *Leuconostoc,* both of which could have been derived from natural sources.

The energetic value of this fermentation to the iguana is as high as, or higher than that found in other animals having hindgut fermentations.

Acknowledgments

The personal assistance of Dr. A. Stanley Rand in helping to solve the collecting and logistical problems, as well as his original encouragement towards this project, is gratefully acknowledged. Dr. Beverly Dugan also helped with collecting iguanas and supplied a great deal of background information on iguanas that was invaluable to the study. Laboratory space and vehicles were provided by the Smithsonian Tropical Research Institute. Adequate collections of iguanas would not have been possible without the aid of several poachers and an eager group of Panamanian boys. Financial support was provided by the National Science Foundation, Grant No. DEB77-08440.

6

Feeding Strategy of the Caicos Ground Iguana, *Cyclura carinata*

Walter Auffenberg
Florida State Museum
University of Florida
Gainesville, Florida

INTRODUCTION

Relatively few lizards eat plants (Iverson, this volume; Szarski, 1962; Ostrom, 1963; Sokol, 1967; Pough, 1973), and of these many are presumed to be omnivorous, especially as juveniles (Sokol, 1967). Only iguanines, a few agamids, and a few scincids have significant morphological specializations for an herbivorous diet (Sokol, 1967; Iverson, this volume) although behavioral modifications are common. Of the studies in which these behaviors have been reported few have viewed the problem of herbivory from the standpoint of food resource availability in respect to foraging efficiency and character. It was to obtain data on these and related variables that the current study was undertaken.

Cyclura carinata is a relatively small iguanine ($\overline{X}$ total length and weight for males 605 mm, 905 g; females 500 mm, 426 g, Iverson, 1979), restricted to the most southern Bahamas and the Turks and Caicos Islands, British West Indies. Density in undisturbed populations is as high as 36.8/ha. Iverson (1979) provided detailed information on home range, activity patterns, and refuging habit of *Cyclura carinata*.

Study Area: The investigation was conducted on Pine Cay, located on the northern edge of the Caicos Bank, 22°N, 72°W, about 180 km N Hispaniola. The small island (348 ha) is formed largely of indurated coralline sand and shell storm ridges separated by shallow swales, some of which contain brackish to fresh water. The highest ridge (7.9 m) lies near the windward coast and is composed of hard, sandy limestone, with solution holes and fissures filled with lateritic sandy clay. On the northern leeward coast a broad white sandy beach, continuously fed by longshore currents, is backed with moderate sand dunes.

Solar radiation is about 160 Kcal/cm^2/yr, with about 2600 total hr of sunshine (*fide* Lansberg, 1965). In general, edaphic and climatic conditions, as well as dispersal barriers, limit floral communities to relatively few species. Important wetlands are mangrove forest and inland marsh. Mangrove forest (restricted to small areas along the southwestern and eastern periphery) is almost exclusively comprised of *Rhizophora mangle.* Dominant plants of the inland marshes are *Mariscus jamaicensis* and *Conocarpus erectus.* The terrestrial habitats fall into four main categories, representing two major successional seres. The hydric sequence is probably salt marsh *(Juncus)* to freshwater marsh (*Mariscus* or equivalent) to pine flatwoods (dominated by *Pinus caribaea*). The terrestrial sere is apparently represented by direct succession of active dunes (dominated by *Scaevola*) to open (beach) scrub on the indurated storm ridges (dominated by *Coccoloba uvifera* and *Strumpfia mahogani*), to coppice with more mesic-adapted species (such as *Swietenia maritima* and *Coccoloba krugii*). The scrub vegetation grades into that of the active dunes near the northern and western shores, and into the coppice toward the high eastern ridge. The coppice communities are divisible into a more mesic northeastern portion and a more xeric one on the rocky eastern storm ridge.

In this study seven major plant communities were recognized. These and the percent of the total each occupies on Pine Cay are: Beach and Rocky Coast (12.05), Open Scrub (13.55), Thick Scrub (34.41), Rocky Coppice (4.39), Mesic Coppice (3.02), and Mixed Woodland (19.90). The remaining area is comprised of wetlands and developed zones. Maps and additional details are in Iverson (1979). Similar West Indian floras have been described in detail by Beard (1949) and D'Arcy (1975). *Cyclura carinata* inhabits all natural Pine Cay plant communities except Marsh and portions of Mixed Woodland, reaching maximum abundance in Rocky Coppice.

METHODS AND MATERIALS

The work was conducted intermittently from July, 1973, through August, 1975 (total 297 man field days). Data were collected as follows.

Plant Distribution, Density, and Diversity: All analyses were based on 30 sampling units in the form of belt transects so located that all major plant communities were included. Although local floral and topographic features caused transect lengths to vary (50-400 m), the width was consistently 1 m. Frequency of occurrence of woody plants and the dominant grasses and forbs was recorded.

Food Plant Biomass and Geometry: Analyses were conducted on individual plants representing the major food species of *Cyclura carinata.* Data were obtained on all parts normally eaten (number, size, and weight of the leaves, fruits, buds, and/or flowers), as well as major plant geometric features (height, crown width, and trunk inclination). Samples for all studies were randomly selected from plants located along the belt transects.

Caloric Content: Plant and animal caloric determinations were made on a Parr model 1411 adiabatic standard bomb calorimeter, corrected for nitric acid formation. In all instances the weighing error was less than 1%.

Food Plant Utilization: The food resource data presented below are based on 308 droppings, 126 stomachs, and 28 colons. The most complete record of food eaten was obtained by study of stomach contents. Because leaves were often quite digested and difficult to identify, and since some soft foods, such as mushrooms, were completely digested, study of food remains (mainly seeds) in the droppings was useful only to determine the kind and proportion of fruits eaten. Food plants sometimes showed unmistakable evidence of browsing, such as tooth marks.

Droppings were collected during summer (July-August) and winter (November-December) from eight stations that, together, represented a cross section of those habitats in which the lizards occurred (beach, scrub, and rocky coppice). Most stomachs and colons examined were obtained largely from individuals that had been killed by feral dogs and cats (see Iverson, 1978). The stomachs represented both sexes (males 55, females 71) and a range of both size (75-350 mm SVL) and season (investigations conducted during 10 months of the year).

RESULTS

General Utilization Patterns

Quantity and Volume: One hundred and twenty-five food species (7714 items) were recognized in the 126 excised stomachs, almost all of which were identifiable. The number of species/individual varied from 0 to 13 ($\overline{X}$ = 2.1, mode 3) and the volumes/stomach from 0 to 250 ml ($\overline{X}$ = 41), suggesting that the stomachs are typically at most one-sixth full. Approximately 5% were empty. The mean wet weight of food in the stomachs of adults (adult weight ($\overline{X}$ = 1234 g, N = 16) is 91 g. If the stomach content is considered a meal, then the mean weight to body weight ratio in nature is 7.4%–very close to determinations based on captives (see below).

Fifty-three food species (2153 items) were found in the colons (N = 28). The number of food species/individual varied from 6 to 33 ($\overline{X}$ = 18.5, mode 15).

Taxa and Parts Utilized: Of the total ingested material, plants constitute 95.6% of the number of items and 94.8% of the volume.

Animal prey (4.4% of total) are tabulated and the percent utilization shown in Table 6.1. A single slug (sp.?) was undoubtedly ingested when it was alive. Almost all the insect prey are nonflying types. The termite *Nasutitermes costalis* is represented by only the wingless soldier caste, and the cicada *Ollanta caicosensis* by wingless nymphal stage. Larval beetles are few and found burrowed within decomposed and swallowed mushrooms, so that they are undoubtedly accidentally ingested. Several nematodes were associated with fish carrion in one stomach and also very likely fortuitous prey. The most common species of the adult Coleptera found in the stomachs is typically found on both flowers and fruits of *Erithalis fruticosa;* its ingestion may also be accidental. In his study of the colon of *C. carinata,* Iverson (1979) also found some larvae of the moth *Pseudosphinx tetrio,* a honeybee, a few weevils, odonate wing fragments, and a solpugid.

Besides bird, fish, and rodent carrion in the stomachs we have seen lizards eat several types of canned meats, especially sardines. Recently killed animals are also eaten, for birds shot on the wing are immediately seized and devoured by adult lizards if they fall nearby. Hermit and land crabs *(Clibanarius* sp. and *Cardisoma guanhumi)* found in the stomachs are only represented by limbs, especially chelae. These were probably pulled from the living crabs and then swallowed. Wiewandt (1977) reported similar data for *C. cornuta* on Mona Island.

Table 6.1: Frequency and Volume of Animal Prey in *C. carinata* Stomachs and Feces, in Percent of Total Plant and Animal Matter

Taxa	Stomach (N=126) Frequency	Stomach (N=126) Volume (in ml)	Feces (N=308) Frequency
Cyclura shed skin	0.0018	0.0069	0.0037
Cyclura carinata	0.0001	0.0218	0.0000
Crabs *(Clibanarius* and *Cardisoma)*	0.0005	0.0160	0.0010
Termites	0.0340	0.0037	0.0153
Cicada nymph	0.0057	0.0173	0.0091
Beetle larvae and adults (sev. sp.)	0.0010	0.0043	0.0050
Slug (sp.?)	0.0001	0.0002	0.0000
Dipteran larvae (sp.?)	0.0001	0.0009	0.0000
Carrion	0.0004	0.0069	0.0003
Total	0.0437	0.0780	0.0344

Two lines of evidence suggest that *Cyclura carinata* may employ a unique method of obtaining termites. First, the termites are represented in the stomachs by only the soldier caste, rarely accompanied with any debris. Second, some termitaria are deeply scratched in a manner that could only have resulted from the claws of adult *C. carinata* (Figure 6.1). The same scratch pattern has been found on termitaria of Andros Island, where the stomachs of *C. cychlura* also contain termites. I believe that adults of both these species deliberately damage the termitarium surface and lap up the soldiers as they defend the broken wall.

Table 6.1 shows that (1) the proportion of animal prey in the droppings and stomach are more or less equivalent, (2) the animal prey makes up a very small proportion of the total prey spectrum, and (3) that in general there is an inverse relation between number of items and volume per species. Native plant species utilized are indicated by an asterisk in Table 6.2[1] These comprise 57.5% of the total macroflora of Pine Cay. Excluding mixed woodland, in which *C. carinata* is not common, percent utilization per community varies from 62.6 to 91.3. There is an inverse relation ($r = 0.40$; $p < 0.05$) between community floral diversity and utilization (Figure 6.2), so that a greater proportion of plants are eaten in simple habitats than in complex ones. The mean utilization of plants for all habitats combined is 70.8% ($SD = 5.8$).

Figure 6.1: Termitarium of *Nasutitermes costalis*, showing surface scratches made by a foraging *Cyclura carinata*.

Table 6.2: Common Native Plants of Pine Cay and Their Distribution and Utilization*

Taxa	Habitat	Beach	Rocky Coast	Open Scrub	Thick Scrub	Rocky Coppice	Mesic Coppice	Mixed Woodland	Parts Eaten**
*Acacia acuifera**	Shrub			X	X				L+
*Ambrosia hispida**	Forb	X							L–
*Amyris elemifera**	Shrub				X	X	X	X	Fr–
*Andropogon glomeratus**	Grass			X				X	L, Fr–
*Antirrhea myrtifolia**	Shrub			X	X	X			L–, Fr+
*Argythamnia seriacea**	Shrub				X	X			L–
Borrichia cf *arborescens*	Forb			X	X				
Bourreria cf *ovata**	Shrub					X			L–
*Bucida buceras**	Tree							X	L–
*Bumelia americana**	Shrub				X	X			L–
*Byrsonema cuneata**	Shrub				X	X	X		L–. Fr+
Caesalpinia bahamensis	Shrub			X	X	X	X		
Calyptranthes cf *paliens**	Shrub					X			Fr–
*Casasia clusiaefolia**	Tree			X	X				Fr–
*Cassia biflora**	Shrub			X	X	X	X		L+
Cassia lineata	Shrub			X	X	X			
Catesbaea foliosa	Shrub						X		
Cenchris sp.	Grass	X		X				X	
*Chamaecyce buxifolia**	Shrub	X		X					L–, Fl
*Chamaecyce vaginulata**	Shrub	X		X					L–, Fl–
Chloris cf *petraea**	Grass			X				X	L–, Fr+
Citharxylim cf *fruti-sans*	Shrub				X	X	X		
*Coccoloba krugii**	Tree				X	X	X		Fr+
*Coccoloba uvifera**	Tree	X	X	X	X	X	X	X	L–, Fr+
*Conocarpus erectus**	Tree	(Swales and edges fresh water)				X	X		L–
Cordia cf *bahamensis*	Tree		X			X	X		
*Crossopetalum rhacoma**	Tree				X	X			L–, Fr+
Cynanchum cf *eggersii**	Shrub				X	X			L–
Cyperus fuliginosa	Grass	(Swales)						X	L–, Fr–
Cuscuta americana	Vine				X	X			
Dichromena colorata	Grass	(Swales and disturbed areas)						X	
*Digitaria filiformis**	Grass	(Swales and disturbed areas)						X	L–, Fr+

(continued)

Table 6.2: (continued)

Taxa	Habitat	Beach	Rocky Coast	Open Scrub	Thick Scrub	Rocky Coppice	Mesic Coppice	Mixed Woodland	Parts Eaten**
Dodonaea viscosa	Shrub			X	X			X	
Drypetes cf *lateriflora*	Shrub					X	X		
Echites umbellata	Grass				X	X	X	X	
*Erithalia fruticosa**	Shrub		X	X	X	X	X	X	L+, Fr+, Fl+, B+
*Ernodia millspaughii**	Shrub	X	X	X	X	X	X	X	L–, Fr+, Fl+
Erythroxylum rotundifolium	Shrub			X	X				
*Eugenia foetida**	Shrub				X	X	X		L+, Fr+
Evolvulus sp.	Forb			X	X	X			L+
Fimbristylis inaguensis	Shrub	(Swales, etc.)							
Fimbristylis ovata	Shrub	(Swales, etc.)							
Forestiera cf *segregata*	Tree					X	X		
Guaiacum sanctum	Tree					X	X		L+
*Guapira obtusata**	Tree		X		X	X			L–, Fr+
*Gundlachia corymbosa**	Shrub	(Disturbed mesic areas)					X	X	Fl+
Guettarda cf *krugii*	Tree					X	X		L+
Gyminda cf *latifolia*	Shrub				X	X	X		
*Hippomane mancinella**	Tree				X	X	X		Fr–
Hypelate cf *trifoliata**	Shrub				X	X	X		L+
*Jaquinia keyensis**	Shrub			X	X	X	X		L–, Fr–
Lantana cf *involucrata*	Shrub			X	X				
*Manilkara bahamensis**	Tree		X		X	X	X		L–, Fr+, B–
Mariscus jamaiciensis	Grass	(Swales, etc.)						X	
*Maytenus buxifolia**	Shrub				X	X	X		L+, B+
*Metopium toxiferum**	Tree				X	X		X	Fr–
Myrcianthes fragrans	Shrub				X	X	X		
*Paspalum laxum**	Grass	(Disturbed areas)							L+, Fr+
Phialanthus cf *myrtilloides*	Shrub		X			X			
*Phyllanthus epiphyllanthus**	Shrub		X		X	X			L–
Pinus caribaea	Tree							X	L–
*Pithecellobium guadelupense**	Tree				X	X	X		L–, B–

(continued)

Table 6.2: (continued)

Taxa	Habitat	Beach	Rocky Coast	Open Scrub	Thick Scrub	Rocky Coppice	Mesic Coppice	Mixed Woodland	Parts Eaten**
*Plumeria obtusa**	Tree				X	X			
Psidium longipes	Shrub		X	X	X	X	X	X	L+, Fl+
Psychotria cf *ligustrifolia*	Shrub					X	X		
*Rachiallis americana**	Shrub	X	X	X	X	X			L+, Fl+
*Reynosa septentrionalis**	Tree				X	X	X	X	L+, Fr+
*Rhizophora mangle**	Tree	(Coastal)							L–
Sabal palmetto	Tree						X	X	
Savia cf *bahamensis*	Tree					X	X		
*Scaevola plumieri**	Shrub	X							L–, Fr–
Scleria lithosperma	Grass	(Swales, etc.)						X	
Setaria sp.	Grass	(Swales, etc.)						X	
Solanum bahamense	Forb	(Disturbed areas)							
Sophora tormentosa	Shrub	(Disturbed areas)						X	
*Strumpfia maritima**	Shrub	X	X	X	X	X		X	L–, Fr+
Suriana maritima	Shrub	X							
Swietenia mahagoni	Tree					X	X		
Tabebuia cf *bahamensis**	Shrub			X	X	X			Fl+, B+
Thouinia cf *discolor*	Shrub			X	X	X	X		
*Thrinax microcarpa**	Tree		X	X	X	X		X	Fr+
Thyralis sp.*	Shrub				X	X	X		L–
Tournefortia guaphaloides	Shrub			X					
*Tournefortia volubilis**	Shrub	X							Fl+
Uniola paniculata	Grass	X							
Verona cf *arbuscula*	Vine			X	X	X	X		
Xanthoxylem cf *coriaceum*	Shrub			X	X				
*Zizyphus taylori**	Tree		X			X	X		L+
Mushrooms (at least 2 sp.)*				X	X		X	X	(entire)

*Plants eaten by *Cyclura carinata.*

**Fr = fruits, Fl = flowers, L = leaves, B = buds; + = commonly, – = rarely.

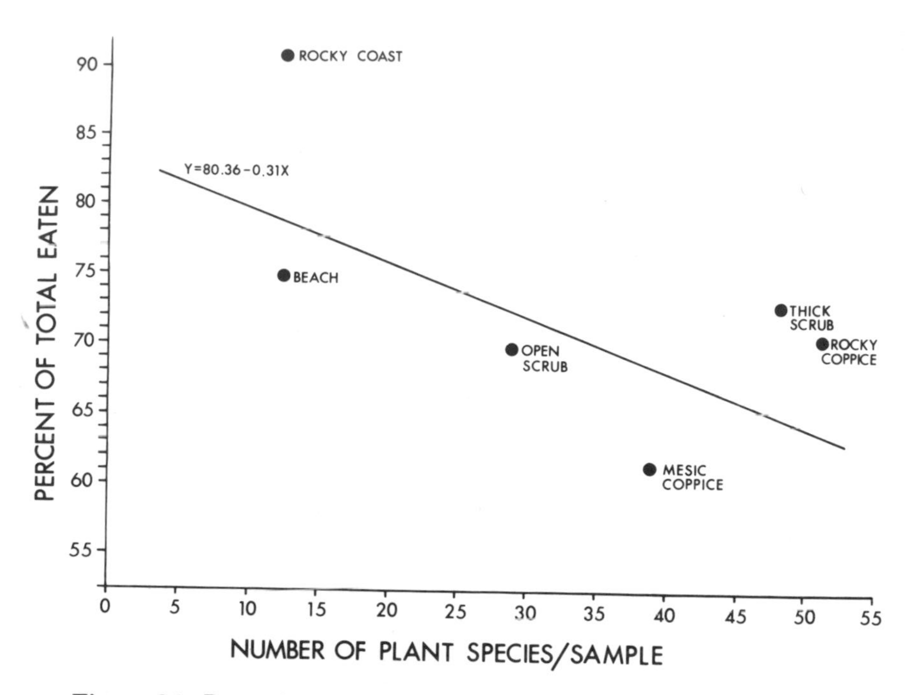

Figure 6.2: Proportionate numbers of plants eaten in habitats of increasing complexity.

Of the 58 species utilized, the leaves of 78% are eaten, fruits 46%, flowers of 12%, and buds (flowers or leaves) of 8% (Table 6.2). Though there are many more leaves eaten than fruits, the total volume of each is about the same and, though common, flowers comprise a small amount of the total volume (Table 6.3). The leaves of some species are so small (especially *Chamaecyce* species) that large numbers are eaten with each bite, often including the tips of the branches. Mushrooms provide the highest volume per item ratio because they are often eaten entire. Table 6.2 also shows that only half of the plant species eaten provide more than one food category (leaves, fruits, etc.); only one *(Erithalis)* provides all four; two *(Ernodia* and *Manilkara)* provide three; 22 provide two, and the remaining 25 only one. Almost all leaves are eaten fresh–the few withered, dry ones are probably accidentally ingested.

Table 6.3: Comparison of Total Items and Volume for Major Plant Foods

Plant Foods	Total No. Items	Total Volume (in cc)	$\overline{X}$ Volume (in cc)/Item
Leaves	4197	171.3	0.041
Flowers	1492	8.2	0.005
Fruit	995	168.9	0.171
Buds	680	58.2	0.086
Other (mushrooms)	16	32.4	2.025
Totals	7380	439.0	

Table 6.4 shows the proportional abundance of the common fruit and leaf food in the stomachs of *C. carinata* (buds and flowers are difficult to determine because they usually fall apart). The range in percent of the total volumes for each species is nearly the same in both food categories; leaves are represented by from 0.9 to 35.5% and fruits by from 0.4 to 23.1%.

Table 6.4: Percent of Total Volume of Leaves and Fruits in 126 Stomachs of *Cyclura carinata*

Leaves	%	Fruits	%
Acacia acuifera	2.4	*Antirhea myrtifolia*	0.6
Ambrosia hispida	3.0	*Brysonema cuneata*	1.2
Cassia lineata	4.9	*Coccoloba krugii*	23.1
Chamaecyce buxifolia	7.1	*Coccoloba uvifera*	2.3
Chamaecyce vaginulata	35.5	*Erithalis fruticosa*	9.0
Conocarpus erectus	2.9	*Ernodia millspaughii*	0.4
Crossopetalum rhacoma	5.7	*Eugenia foetida*	10.3
Grasses (combined species)	3.7	*Manilkara bahamensis*	1.0
Guettarda krugii	0.9	*Metopium toxiferum*	0.7
Gundlachia corymbosa	2.8	*Psidium longipes*	2.5
Psidium longipes	13.4	*Reynosa septentrionalis*	11.7
Rachiallis americanus	9.7	*Scaevola plumieri*	1.8
Thyralis sp.	1.5	*Strumpfia maritina*	21.1
Zizyphus taylori	6.5	*Thrinax microcarpa*	12.2

Seasonal Factors: Being located in the subtropics, seasonal changes in insolation, rain, and sea-salt nutrients (for details see Iverson, 1979) lead to a strong leaf and fruit pulse (Figure 6.3). Though the fruit (and flowering) pulse is more species-variable than that for leaves, the generally sharper peaks of the former result in more significant seasonal modification in the feeding patterns of *C. carinata* (see below).

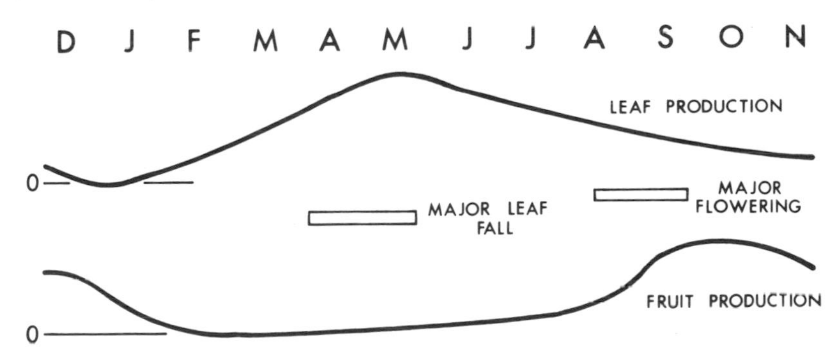

Figure 6.3: Diagrammatic representation of seasonal changes in availability in major food classes.

Table 6.5 shows that while fruits of most plant food species are available throughout the fruiting season, the proportional utilization changes from the early to the late parts of the season. Thus, seasonal availability leads to periodic switching of plant foods. The fruits of *Coccoloba krugii* and *Strumpfia maritima* are heavily utilized during both parts of the year. A few fruits are characteristically eaten only in the early and late periods. Of the 12 species regularly cropped during both periods, only 25% show no significant difference in utilization. Seasonal differences in leaf utilization are less marked. Flowers are never commonly eaten, but show marked seasonal consumption patterns based on availability.

Table 6.5: Proportion (in %) of Fruits Eaten in Early (July-August) and Late (November-December) Parts of the Fruiting Season on Pine Cay, Based on Seeds in Droppings

Arrows show shifts of statistical significance.

Early (N=177)	Early		Both		Late	Late (N=191)
26.82	*Reynosa*					
2.67	*Ryrsonema*					
0.13	*Thyralis*					
9.01	*Coccoloba uvifera*					
4.20		←	*Scaevola*			0.13
14.68			*Coccoloba krugii*	→		23.80
22.82		←	*Strumpfia*			15.41
0.47			*Crossopetalum*	→		2.22
0.17			*Metopium*			0.91
0.33			*Ernodia*			0.40
0.67			*Eugenia*	→		13.83
1.70		←	*Manilkara*			0.36
4.30		←	*Psidium*			0.16
0.63			*Antirhea*			0.48
0.23			*Thrianx*			18.74
20.18		←	*Erithalis*			0.48
					Amyris	14.07

Iverson (1979) has shown that in *C. carinata* the diel feeding pattern shifts from a bimodal one in spring, with morning and afternoon peaks, to a strongly unimodal one in winter, with peak feeding at 15-16 hrs. Both John Iverson and I conclude, on the basis of different data, that feeding is reduced in both summer and winter when compared to spring. This is somewhat corroborated by the fact that the abdominal fat bodies are significantly smaller in winter (Iverson, 1979). Reduced summer feeding may be related to reproductive behavior.

Sexual and Ontogenetic Factors: There is no significant difference in the numbers of species or quantity of plant and animal food of males versus females. There are, however, some important differences related to lizard size.

Size is not only related to the amount of food eaten, but to the number of species ingested. Thus, hatchlings have from 0 to 4 species in their stomachs, large adults as many as 13. The overall mean is 2.1 species/stomach (for specimens < 100 mm SLV $\bar{X} = 1.6$, N = 13; 101-150 mm SVL $\bar{X} = 1.5$, N = 63; 151-200 mm SVL $\bar{X} = 2.1$, N = 34; > 200 mm SVL $\bar{X} = 2.5$, N = 16). There is also, as expected, a highly significant positive relationship between the number of items per stomach (0-112) and lizard size (SVL < 100 m $\bar{X} = 8.5$ items, 101-150 mm $\bar{X} = 12.2$, 151-200 mm $\bar{X} = 11.5$, > 200 mm $\bar{X} = 66.8$).

In general, the proportion of plant to animal foods is similar in both juveniles and adults (hatchling: number items 97.7% plant, 2.3% animal; adults 93.9% plant, 6.1% animal; volume 97.2% and 2.8% plant and animal respectively). These data show that young *C. carinata* are primarily herbivorous (as in *C. cornuta*, Wiewandt, 1977). Suggestions that young of other *Cyclura* species are carnivorous (Carey, 1975) are based on observations of captives. While the absolute volume of animal food in the individual stomach increases greatly from smallest to largest lizard size classes (0.016 cc/individual in near hatchlings to 0.437 cc/individual in the largest size class), when related to iguana body weight it is clear that animal prey is consumed at the same proportion throughout life.

Cannibalism is recorded only once (a mature female SVL 203 mm having eaten a hatchling), but Iverson (1979) reported observing a probably fatal attack by an adult male on a hatchling.

Ingestion and Assimilation: All *Cyclura* species have teeth adapted for browsing. Parts of larger, particularly rigid leaves (*Guapira, Rhizophora,* etc.) are not snipped off, but linearly perforated with the teeth. The piece is then torn off along the punctures. Large mushrooms are also sometimes bitten into pieces. Smaller leaves and all buds are pulled off and swallowed entire. The only fruit not swallowed entire is that of *Casasia.* These large hard-skinned fruits are only eaten after they fall, overripe, to the ground and become decayed at one end. Individuals then push their heads inside the capsule, feeding on the decomposing flesh and seed. Individuals that feed at the exact same part of a plant day after day become particularly efficient at harvesting, for they become intimately familar with the growth characteristics of certain plants. The advantage of such repetitive browsing is discussed further below.

Data on approximate food consumption and rate were obtained from five captive adult *C. carinata* ($\overline{X}$ wt = 1225 g) to which grapes ($\overline{X}$ wt = 3.1 g, $\overline{X}$ vol = 3.3 ml, N = 80) were available all day (N = 28). Variation in number of times each individual fed per day was from one to five times, each feeding period lasting from 1 to 33 min ($\overline{X}$ = 9.5), total consumption per feeding from 2 to 40 grapes ($\overline{X}$ = 10.7), and a mean total weight per feeding of 33.2 g and mean volume of 35.3 ml. Consumption rate varied from 0.75 to 8.0 grapes per min ($\overline{X}$ = 2.5), for weight from 2.7-28.5 g/min ($\overline{X}$ = 8.9 g/min), and for volume from 2.5-26.4 ml/min ($\overline{X}$ = 8.3 ml/min). Total daily consumption ranged from 9 to 40 items ($\overline{X}$ = 24.8), and 71.1-117.5 g/day ($\overline{X}$ = 78.9 g/day), being 6.4% of the mean adult weight. This is very close to conclusions obtained from field data (see above).

In his study of thermoregulatory behavior, Auth (1979) has shown that mean body temperature during feeding on the ground is 36.7°C (30.0-40.4, SD 2.42, N = 14), while body temperature during arboreal foraging is significantly higher ($\overline{X}$ = 38.6°C). Though body temperature range during feeding is very broad (23.6-40.4°C), the rate and amount eaten is correlated with it.

To establish gut passage rates, three adult *C. carinata* (1440-1510 g) were fed small (10 x 16 mm) telemetric devices in the field (N = 4) during studies by David Auth. Passage was completed in from 93 to 253 hr ($\overline{X}$ = 135). A passage rate of about 96 hours was reported in *Ctenosaura pectinata* (Throckmorton, 1973). If one unusually long passage time in *Cyclura carinata* is excluded (it is over 100% greater than the next longest period), the mean passage times for both species are almost identical.

No data are available on digestibility coefficients (see Nagy, 1977) of any *Cyclura* species. However, the work by Iverson (1979, in press, this volume) suggests that in this species characteristic segmentation, microflora, and microfauna assure high caloric and nutrient assimilation.

Floral Composition and Utilization

Food Species Density and Diversity: Table 6.6 shows the density of major food plants in each habitat. Some major food species are distributed through all or most of the habitats in which *Cyclura* is found *(Coccoloba uvifera, Strumpfia, Psidium, Ernodia),* while others are very restricted *(Chamaecyce, Bourreria, Coccoloba krugii, Conocarpus, Gundlachia).* However, Figure 6.4 shows that there is no correlation between food plant density and utilization as percent of total number of items. Both fruits and leaves are selected irrespective of density.

Table 6.6: Mean Number Major Food Plants/ha in Each of the Major Pine Cay Habitats

Food Species	Beach	Rocky Coast	Open Scrub	Thick Scrub	Rocky Coppice	Mesic Coppice
Acacia	–	–	117	33	–	–
Ambrosia	23	–	–	–	–	–
Antirhea	–	–	384	546	657	–
Byrsonema	–	–	–	8	37	–
Bourreria	–	–	–	–	44	–
Bucida	–	–	–	–	–	–
Bumelia	–	–	–	13	50	–
Calyptranthes	–	–	–	–	2	–
Casasia	–	–	12	74	–	–
Cassia	–	–	7	37	112	–
Chamaecyce (combined)	97	–	37	–	–	–
Coccoloba krugii	–	–	–	6	483	–
Coccoloba uvifera	89	284	446	645	198	82
Conocarpus	–	10	–	–	32	20
Crossopetalum	–	–	–	8	10	–
Cuscuta	–	–	–	2	3	–
Erithalis	–	281	310	1,327	731	1,020
Ernodia	542	49	1,283	769	12	690
Eugenia	–	–	6	37	223	–
Evolvulus	–	–	11	27	20	–
Guapira	–	333	–	9	235	–
Gundlachia	–	–	–	–	–	30
Jacquina	–	–	40	60	80	102
Manilkara	–	568	–	285	855	–
Maytensis	–	–	–	20	80	73
Metopium	–	–	–	74	112	82
Phyllanthus	–	7	–	186	1,090	–
Psidium	–	1,568	1,290	2,108	1,326	2,554
Rachiallis	267	1,774	397	99	198	–
Reynosa	–	–	–	87	371	144
Rhizophora	. . . . . .	. . . . .	(out from, and at shore)	. . . . .	. . . . .	. . . . .
Scaevola	443	–	–	–	–	–
Strumpfia	812	2,097	1,271	818	124	1,329
Thrinax	–	88	99	868	359	155
Thyralis	–	–	–	14	86	76
Zizyphus	–	333	–	–	446	–

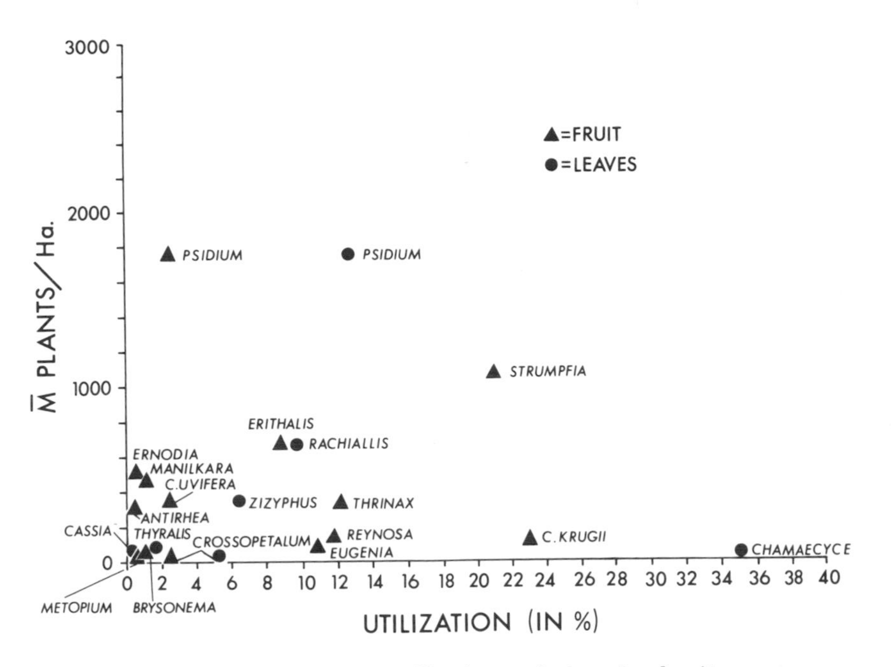

Figure 6.4: Percent utilization vs food species density.

Lizard densities (see Iverson, 1979, for details, based on 547 transects) are very significantly correlated with number of individuals of food plants. Thus, Rocky Coppice has a mean of 10,750 food plants per ha and has a lizard density of 27.2 individuals/ha; Thick Scrub has a mean of 8,020 food plants/ha and 17.7 lizards/ha; Open Scrub 5,540 food plants/ha, lizards 1.7/ha. In general, individual lizard activity ranges are larger in poorer habitats (see Iverson, 1979). The mean ranges of subadults ($\overline{X}$ wt = 300 g) is 700 m² and for adults ($\overline{X}$ wt = 1200 g), 1800 m².

Figure 6.5 expresses the floral diversity of three habitats by comparing the accumulated numbers of food species/sample and the number of individuals of each plant/sample (convenient because areas can be compared without having to standardize for area and sampling intensity). It shows that of these communities the greatest number of species per number of individual plants is found in Rocky Coppice, lowest in Mesic Coppice. In the Beach community there are generally only six common food plants, and the illustration shows that this level diversity can be expected to occur within a sample of 140 plants. There are 18 common foods in Rocky Coppice and these can be expected in a sample of 210 plants. In Mesic Coppice only eight common food species occur, but they can be expected only in a sample greater than 210 plants. The data thus show that of these three communities, the greatest food plant diversity is found in Rocky Coppice–undoubtedly an important factor in the energetic costs of searching for foods (see below). Calculated Coefficients of Community (Whittaker, 1976) show that the least similarity among the habitats' floras is between Mesic and Rocky Coppice (= 0.76) and the highest of all between Thick Scrub and Rocky Coppice (= 0.80).

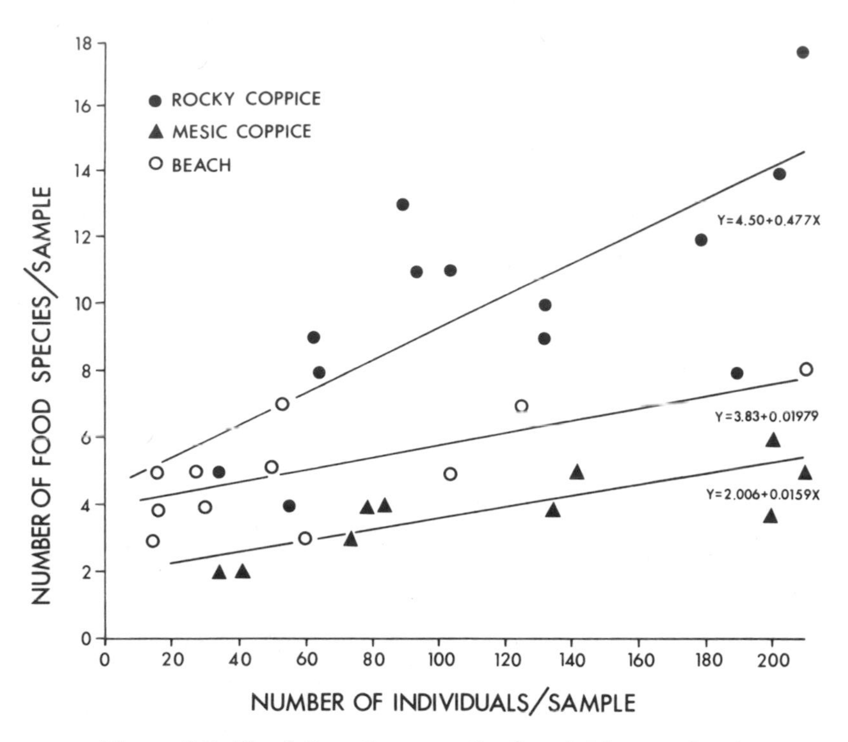

Figure 6.5: Floral diversity among the three habitats analyzed.

In Figure 6.6 it is shown that the number of food species/sample is related to area in such a way that Rocky Coppice has a much greater food plant diversity/area than the other two habitats. When the mean single feeding foray search area of adult *C. carinata* is superimposed on these data, it is seen that in Mesic Coppice this area can be expected to contain only 2 food species, whereas about 6 are expected in the same size area of Rocky Coppice. Thus, the average foraging area in Mesic Coppice would have to be almost three times larger to include the same food plant variety.

Calculation of food preference indices for each of the plant food species eaten on Pine Cay (Table 6.7) indicates that only about half (about one-quarter of the total island macroflora) are eaten in greater proportion than they occur in nature, and hence may be considered preferred. However, because of the total time that plant foods are available throughout the year, the index is skewed in favor of leaves, rather than short-lived fruits and flowers. The index is also skewed to those plant species with small, numerous leaves, since several are taken in one bite (i.e., *Chamaecyce*).

Table 6.8 provides the expected search times for encountering a sample of food plants during fall in each of the major habitats. This table shows that, given the mean time of 107.8 min/daily food search at this time of year (see below), all the food plants in the Beach and Mesic Coppice habitats can be expected to be encountered in the daily foray: 78% of those on the Rocky Coast; 64% of those in Open Scrub; 60% of those in Thick Scrub; 89% of those in Rocky Coppice. Thus Rocky Coppice, which tends to have the highest lizard densities, not only has one of the highest floral diversity levels, but the greatest density of food plants. The mean expected time to contact at least one of all the food plant species in Rocky Coppice is 67.7 minutes–well within the mean forage period of adult *C. carinata*. The mean expected times to contact all food species are considerably greater (143.4 to 287.2 min) in Rocky Coast and Open and Thick Scrubs and significantly lower (30.4 to 36.4 min) for Beach and Mesic Coppice.

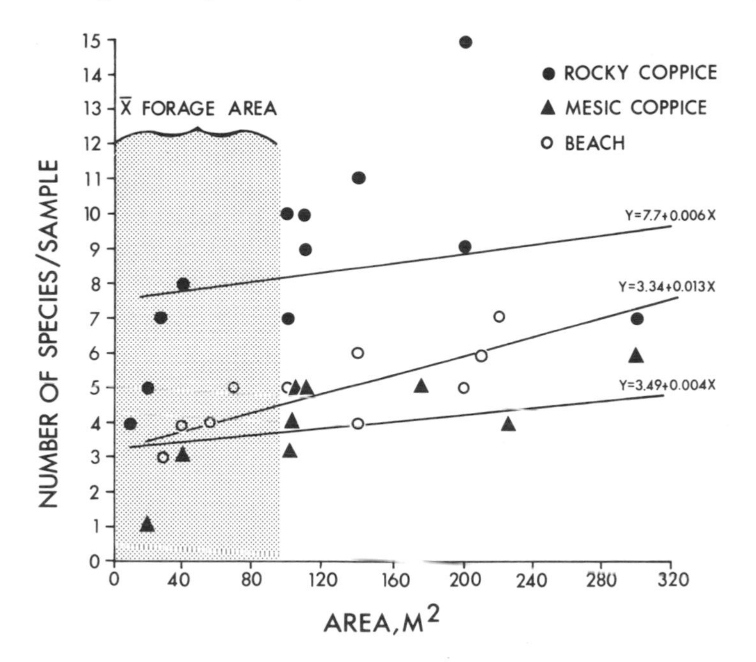

Figure 6.6: Rocky Coppice has a significantly higher diversity of plants for equal area than either Beach or Mesic Coppice. The mean forage area of adult *C. carinata* (shaded) would have to be almost twice as large to provide the same food species diversity in Mesic Coppice as in Rocky Coppice.

Food Resource Yield: Table 6.9 shows the yield of leaves and fruits in number of items and weight per plant and per island. There is no relation between utilization and number, or between utilization and weight of either leaves or fruits for sample areas of 800 cm^2, for total plant, for overall plant size, or for the entire island. The overall range in each of these categories is considerable, and there is no significant correlation between food species selectivity and any yield measurement. Furthermore, there is no significant relationship between any of these factors and the extent to which each plant species is utilized. There are no common fruits that are not eaten, though this is not true of leaves. Thus these data suggest that within each category of fruits and leaves some factor other than quantity determines the actual species harvested.

Table 6.7: Number of Stomachs in Which Plant Species Occurred, Food Plant Density (% of Total) in Stomachs, and Food Preference Indices*

Food Plant	Number Stomachs**	Percent in Stomach (by item)	P.I.
Acacia	12	3.43	4.90
Ambrosia	3	0.01	0.99
Antirhea	25	7.03	1.27
Byrsonema	2	0.06	0.40
Bourreria	3	0.04	0.22
Bucida	2	0.02	0.40
Bumelia	1	0.38	1.00
Calyptranthes	4	0.02	1.00
Casasia	3	0.04	0.13
Cassia	23	6.19	41.27
Chamaecyce virginulata	25	5.40	67.50
Chamaecyce buxifolia	10	2.54	63.50
Coccoloba krugii	46	12.15	7.11
Coccoloba uvifera	9	2.89	0.57
Conocarpus	3	0.25	0.83
Crossopetalum	6	0.93	13.29
Cuscuta	1	0.54	18.00
Erithalis	27	5.71	0.61
Ernodia	38	7.51	1.01
Eugenia	18	4.03	1.20
Evolvulus	1	0.04	0.80
Guapira	5	0.53	0.26
Grasses (combined)	13	3.36	16.00
Gundlachia	1	0.17	3.40
Jacquini	1	0.09	0.10
Manilkara	17	0.54	0.09
Maytensis	3	0.35	5.00
Metopium	6	0.28	0.42
Phyllanthus	2	0.02	0.01
Psidium	28	4.10	0.21
Rachiallis	54	12.35	1.43
Reynosa	8	1.57	0.98
Rhizophora	2	0.01	0.50
Scaevola	9	1.60	1.54
Strumpfia	28	7.74	0.51
Thrinax	31	6.66	1.12
Thyralis	2	0.12	0.17

*P.I. = percent number of items in stomach per percent number of individuals in wild; values greater than 1.0 indicate preferential selection.

**Total stomachs examined 124; most with several plant species included, see text.

Table 6.8: Expected Search Time, in Minutes, for Encountering Food Plants During the Fall in Major Habitats on the Basis of Food Plant Density and Foraging Patterns*

Food Species	Beach	Rocky Coast	Open Scrub	Thick Scrub	Rocky Coppice	Mesic Coppice
Antirhea	∞	∞	18.2	13.4	11.1	∞
Byrsonema	∞	∞	∞	912.4	197.3	∞
Casasia	∞	∞	608.3	98.6	∞	∞
Cassia	∞	∞	1042.8	192.3	65.0	∞
Chamaecyce (both sp. combined)	79.3	∞	197.3	∞	∞	∞
Coccoloba krugii	∞	∞	∞	1216.6	15.1	∞
Coccoloba uvifera	82.0	25.7	16.4	11.3	36.9	89.0
Crossopetalum	∞	∞	∞	912.4	∞	∞
Erithalis	∞	26.0	23.6	5.5	10.0	7.2
Ernodia	13.5	149.0	5.7	9.5	608.3	10.6
Eugenia	∞	∞	1216.6	192.3	32.7	∞
Guapira	∞	21.9	∞	811.0	31.1	∞
Manilkara		12.9		25.6	8.6	
Metopium				98.6	65.2	89.0
Phigalanthus		1042.8		39.2	6.7	
Pithecellobium				251.7	41.9	
Psidium		4.7	5.7	3.5	5.5	2.9
Rachiallis	27.3	4.1	18.4	73.7	36.9	
Reynosa				83.9	19.7	50.7
Scaevola	16.5					
Strumpfia	9.0	3.5	5.7	8.9	58.9	5.5
Thrinax				912.4		
$\overline{X}$	30.4	143.4	287.2	248.6	67.7	36.4

*See Morrison 1978 for computation methcd, etc.

Table 6.9: Biomass of Common Food Plants

Species	$\bar{X}$ No. Fruits per 800 cm^3	$\bar{X}$ No. Leaves per 800 cm^3	$\bar{X}$ Fruit Weight per 800 cm^3 (g)	$\bar{X}$ Leaf Weight per 800 cm^3 (g)	$\bar{X}$ Plant Size (cm^3) x 10^4	$\bar{X}$ Total Fruit Biomass per Plant (g)	$\bar{X}$ Total Leaf Biomass per Plant (g)	Fruit Biomass per Island (g)	Leaf Biomass per Island (g)
Antirhea myrtifolia	7.5 (N=32)	N.A.	1.74	N.A.	2.8 (N=58)	61	–	9.68×10^4	–
Byrsonema lucida	1.6 (N=12)	63.2 (N=41)	0.42	1.80	20.8 (N=8)	109	16,432	1.71×10^4	7.39×10^{-5}
Casasia clusiaefolia	–	N.A.	42.53	N.A.	14.0 (N=11)	744	–	6.4×10^4	–
Coccoloba uvifera	204.3 (N=58)	–	23.20	–	52.2 (N=61)	15,138	–	2.64×10^7	–
Erithalis fruitcosa	22.4 (N=48)	18.3 (N=76)	1.51	3.13	21.0 (N=43)	396	4,804	1.45×10^6	1.76×10^7
Guapira obtusa	–	41.9 (N=57)	–	2.47	34.4 (N=12)	–	18,017	–	1.04×10^7
Manilkara bahamensis	1.3 (N=49)	24.2 (N=41)	11.87	1.21	30.2 (N=38)	4,496	917	7.68×10^6	1.56×10^7
Phyllanthus eryphyllanthus	–	78.0 (N=35)	–	36.16	0.2 (N=17)	–	195	–	2.50×10^5
Pithecellobium guadelupense	–	81.4 (N=52)	–	16.26	15.9 (N=11)	–	16,178	–	3.28×10^6
Psidium longipes	2.3 (N=50)	292.3 (N=38)	0.91	25.02	48.7 (N=25)	554	177,938	4.92×10^6	1.57×10^9
Rachiallis americana	–	4,883.0 (N=115)	–	20.90	1.6 (N=32)	–	977,660	–	2.67×10^9
Reynosa septentrionalis	1.2 (N=49)	–	228.0	–	30.3 (N=16)	864	–	5.20×10^5	–
Scaevola plumierii	1.7 (N=47)	76.3 (N=49)	4.22	109.11	458.5 (N=47)	24,181	437,199	1.07×10^7	1.9×10^8
Strumpfia maritima	25.1 (N=58)	1,125.0 (N=21)	1.26	15.16	1.2 (N=56)	19	16,875	1.22×10^5	7.9×10^8
Thrinax microcarpa	74.3 (N=46)	–	0.37	–	0.9 (N=17)	4	–	6.3×10^3	–
Zizyphus taylori	–	474.4 (N=44)	–	41.82	29.8 (N=32)	–	176,565	–	1.38×10^8

Caloric yield is a potentially important preference factor and was investigated in this study. However, it is important to point out that bomb calorimetry makes no distinction between unavailable calories (presumably as in cellulose) and those available (as in fats, sugars, etc.). Unfortunately the following data do not reflect this distinction.

Table 6.10 shows that variation in total caloric values of major foods of *C. carinata* is from 3906 cal/g dry weight of certain leaves to 6707 for hatchling *Cyclura*. In general, cal/g dry weight are highest in vertebrate prey (5890-8708, $\bar{X}$ = 6724), followed by termites (5700), fruits (4219-5707, $\bar{X}$ = 4594), all other insects (2516-5663, $\bar{X}$ = 4523), leaves (3908-5797, $\bar{X}$ = 4412), mixed grasses (4161-4700, $\bar{X}$ = 4272), fungi ($\bar{X}$ = 3856, *fide* Cummins, 1967), and hermit crabs ($\bar{X}$ = 2516, *fide* Cummins, 1967). There is no significant relationship (r = 0.2) between caloric content of either those leaves or fruits preferred as food (see below) and those not as highly preferred ($\bar{X}$ calories nonpreferred plant species = 4341.2 ± 130.79, $\bar{X}$ calories of preferred = 4615.0 ± 441.41). However, there is a significant positive relationship (r = 0.7) between caloric content and volume so that at least the larger fruits tend to have proportionately higher caloric content/g dry weight than the smaller ones.

Annual caloric yields ($\bar{X}$ leaves or fruits per plant x $\bar{X}$ wt/g x $\bar{X}$ cal/g per food type for each species) of a random sample of leaf and fruit foods is shown in Table 6.11. Not surprisingly, it shows that leaves produce higher total calories per plant ($\bar{X}$ = 358,059 Kcal/yr) than fruits ($\bar{X}$ = 26,950 Kcal/yr/plant), though the energetic cost of harvesting leaves may be greater because caloric content per item is much lower. Once found, fruits offer the highest single item caloric reward, but they are more seasonal and far less common on a per-item basis than leaves. Further, there is no significant correlation between utilization by ground iguanas and either total annual yield in both leaves and fruits per individual food plant or all the plants of a species on the entire island (r = 0.28 and 0.39 respectively). But if leaves and fruits are considered separately, there is a significant tendency for the lizards to utilize plant species producing fewer total calories per plant in either fruits or leaves (r = 0.65 and 0.61 respectively, see Figure 6.7) irrespective of plant size. However, when total yield/island is considered (Figure 6.8), utilization appears positively correlated with leaf yield, but negatively with total island fruit yield. The reason for this difference is not apparent but may be related to encounter rate, or the lizards may be selecting fruit crops of low abundance and leaf crops of high abundance.

Strategies of Food Location

Cyclura carinata exhibits several strategies to cope with (1) the irregular distribution of food plants and their edible parts, (2) the competition known to exist for some plant products, and (3) the anticonsumption strategies of at least some of the plants upon which they feed. The major question to which the remaining sections of this paper are directed is, "How does this ground iguana maximize its net energy gain per unit time?"

It has been shown that the food of this species is dispersed in an environment exhibiting great variation in the abundance of edible species and/or their parts. Such factors have great importance in determining how long an individual continues to search in a given area, where to search, and what species to look for. As discussed below, the foraging patterns and persistence of at least the adults make it clear that they have learned the food distribution patterns within their activity ranges and gauge the length of time they should spend searching in a particular place in such a way as to maximize net energy gain in accordance to these.

Table 6.10: Caloric Value of Common Food Resources of *Cyclura carinata* on Pine Cay

Food Resource	N	$\overline{X}$ Caloric Value	Remarks
Plants:			
Antirhea	5	4480.66±12.6	fruits
Cassia lineata	3	4470.11±32.0	leaves
Coccoloba uvifera	6	4328.34±14.8	fruits
Erithalis	4	4674.19±16.5	leaves
Erithalis	8	4760.72±13.0	fruits
Guapira obtusa	8	4359.23±11.8	leaves
Manilkara	8	5694.89±63.2	fruits
Phyllanthus	5	4459.75±33.8	leaves
Pithecellobium guadelupensis	7	4469.84±31.3	leaves
Psidium	5	4337.44±10.3	fruits
Rachiallis	8	4210.17±13.8	leaves
Reynosa septentrionalis	6	4377.79±10.6	fruits
Reynosa septentrionalis	6	4422.40±10.5	leaves
Rhizopora	3	4298.23±105.0	leaves
Scaevola	6	3908.85±13.8	leaves
Scaevola	5	4470.49±15.3	fruits
Thrinax	9	4218.84±21.3	fruits
Zizyphus taylori	6	4782.65±16.4	leaves
Fungi	16	3856	Cummins 1967
Mixed grasses	260	4178	Cummins 1967
All leaves	53	4412.24±33.8	
All fruits	46	4593.71±41.4	
Animals:			
Coleoptera	5	5668	Cummins 1967
Cyclura carinata hatchling	1	6707.34	entire
Decapoda	10	2516	Cummins 1967
Nasutitermes (soldiers)	5	5700.45	Weigert 1974
Orthoptera	12	5386	Cummins 1967
All vertebrates	40	6274	Cummins 1967
Feces *(Cyclura)*	6	3613.8	

Table 6.11: Annual Yield in cal/plant for Food Plants

Fruit Species	Fruit Annual Kcal/Plant	Leaves Species	Leaves Annual Kcal/Plant
Scaevola	94,520	*Scaevola*	1,763,669
Coccoloba uvifera	65,500	*Zizyphus taylori*	844,449
Manilkara	25,604	*Guapira*	78,540
Psidium	24,000	*Pithecellobium*	72,313
Reynosa	3,782	*Erithalis fruticosa*	22,455
Erithalis fruticosa	1,888	*Rachiallis*	4,116
Antirhea	273	*Phyllanthus*	869
Thrinax	17		

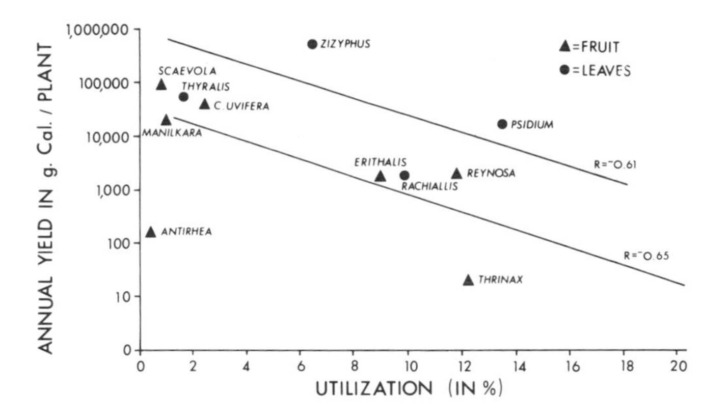

Figure 6.7: Relationship between utilization and annual caloric yield per plant when both leaves and fruits are considered.

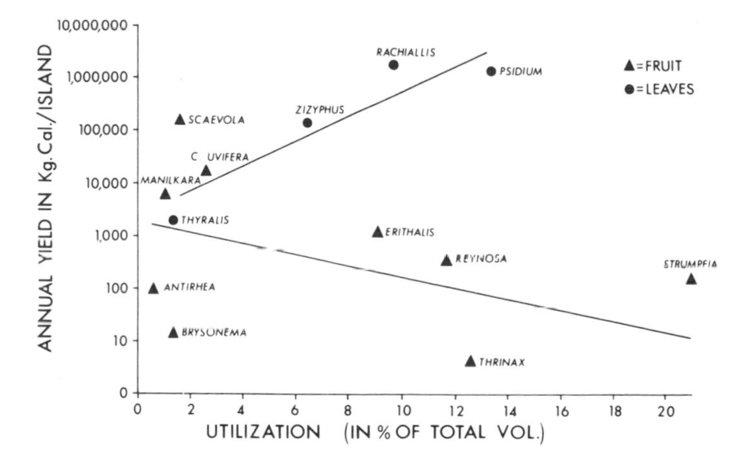

Figure 6.8: Relationship between total annual caloric yield and utilization of certain leaves and fruits on the entire island.

Temporal and Areal Components of the Search Pattern: The feeding patterns of four free-ranging adults ($\bar{X}$ SVL = 344 mm) telemetered on Water Cay (adjacent and connected to Pine Cay; same habitats) in October (N = 27 forays) and February (N = 25 forays) show that (1) there are up to five daily feeding forays ($\bar{X}$ = 1.93) with very little overlap between them, and (2) the mean distance covered in each foray is 149 m (108 m in October, 190 m in February). Observation of our captives suggests that the mean search path width is about 0.7 m. Thus on the basis of the October and February data the mean daily adult foraging area is estimated as 201 m^2 (146 m^2 in October, 256 m^2 in February; difference undoubtedly due to seasonal variation in food density). Mean elapsed total daily searching and feeding time is 107.8 min, or about 18% of the total diel activity phase and 15.6% of the total day length for this time of year. Mean foraging rate is 1.25 m^2/min. The mean size of the adult activity range is 0.16 ha (Iverson, 1979), variations depending mainly on body size and habitat characteristics. Therefore, it appears that approximately 1.0% of the average home range is covered in a single foray, and 2.0% of the average home range in a day.

While it is unknown to what extent food density and distribution determine size of activity range, it is assumed that there is an inverse relation between them (Pitelka, 1942; McNab, 1963; Schoener, 1968; Simon, 1975; and others). Auth (1980) has suggested that foray length and total time depend on temperature.

Most resident adult *Cyclura carinata* seem to know the exact location of food resources within their activity ranges (see below). They approach these directly, obviously with some expectation of success. For at least part of the year these resources include ripe fruits that have fallen from the trees *(Reynosa, Scaevola, Coccoloba, Byrsonema, Casasia,* and *Crossopetalum).* The fallen fruits are not only concentrated under certain trees, but even more frequently cluster in large numbers due to surface irregularities. As an example, the mean of the fallen large fruits of *Reynosa* in 400 cm^2 beneath the trees was 176. Smaller fruits, such as those of *Strumpfia* and *Crossopetalum,* often fall into the leaf litter, where they have to be scratched out of the surface debris by the lizards. Fruits rarely eaten on the ground, but picked from the parent plant, are *Antirhea, Coccoloba krugii, Cordia, Erithalis, Eugenia, Ernodia, Manilkara,* and *Thrinax.* There is no significant difference in caloric content, size, or weight between those fruits eaten from the ground and from the trees. Only ripe fruits are eaten.

Vertical Components of the Search Pattern: Many iguanine lizards are known to climb into shrubs and trees to obtain food (see Iverson, 1979, for review), so that *C. carinata* is not unusual in this regard. At least one large *Cyclura* species does not climb for food (Wiewandt, 1977), and the smaller size of *C. carinata* may be an important adaptation for harvesting a greater variety of potential foods on an island with low floral diversity. However, large *C. nubila* introduced on Magueyes Island, Puerto Rico, often climb for fruits and some leaves.

Many of the food plants of *C. carinata* are harvested from above the substrate. The fruits and flowers of *Ernodia* and the leaves of *Rhachiallis, Chamaecyce,* and *Phigalanthus* are usually found in sprawling shrubs within 10 cm of the ground, requiring simply lifting the head. The shapes and sizes of other plants demand only a bipedal position (*Ernodia, Strumpfia, Scaevola*, sprawling types of *Conocarpus, Cassia,* and both *Chamaecyce* species). The maximum distance to which adults can browse in a bipedal stance is about 250 cm. All sizes of *C. carinata* feed bipedally, though adults use the technique more commonly than juveniles. Bipedal and climbing behaviors are usually preceded by considerable visual investigation. The lizard repeatedly tilts its head laterally as well as turning it to the side, peering upward while walking past or circling the plant.

Cyclura carinata climbs to at least 7 m to feed. Plants regularly browsed to this height include *Eugenia* (fruits) and *Bucida* (buds). However, most browsing occurs between 1 and 2 m.

Plant crown diameter is an unimportant barrier to browsing, since almost every conceivable shape is included among the more common food plants. In general, food plants with stiff horizontal branching or many small terminal twigs are usually harvested to the branch tips, but rarely beyond 2 m from the plant central axis ($\overline{X}$ = 0.8 m). There is no significant correlation between browsing extent and plant height, branch tip height, or crown width, or crown shape. However, browsing is more frequent in plants with (1) trunk diameters greater than 5 cm (X^2 = 9.28, df = 2, P = 0.01) and (2) angles to ground less than 60° (X^2 = 9.50, df = 3, P = 0.089). There is no significant correlation with position or type of trunk furcation. Thus, given a sufficient stimulus, individuals will climb to almost any part of almost any shape of food plant. The major exceptions are slender, willowy plants (*Tabebuia, Casasia,* and others). Climbing for food is evidently less common on windy days (Iverson, 1979).

Food Orientation Cues: Very little is known about how lizards find their food. As in many other vertebrates, vision is obviously an important cue in *C. carinata,* for captives will investigate new objects thrown nearby; they are attracted to bright colors. Wild, foraging individuals also constantly tilt their heads to peer into the foliage above. Observations by Iverson (1979), Auth (MS in prep.), and myself indicate that *C. carinata* also can orient accurately toward an odoriferous upwind food source from distances as great as 200 m. During foraging, olfactory cues (indicated by frequent substrate tonguing), as well as visual ones, are evidently extensively utilized. Norris (1953) and Carey (1975) observed similar behavior in *Dipsosaurus dorsalis* and *Cyclura pinguis,* respectively. Iverson (1979) has reported the importance of auditory or vibratory cues in food orientation, for he observed that individuals initiate search behavior in the correct direction when the heavy ripened fruits of *Casasia clusaefolia* struck the ground. In fact, an investigative charge could usually be initiated by throwing any small object through the canopy.

Adaptations Modifying Harvesting Rate

Repetitive Browsing: *Cyclura carinata* exhibits several important morphological and behavioral modifications that tend to maximize food procurement. Of these, one of the most important is the repetitiveness of some foraging patterns, in which the same area, the same bush, or even the same parts of a plant are repeatedly visited by a particular individual. Iverson (1979) noted individuals feeding on a particular branch of a plant over the course of several days, until all of the fruits or leaves were consumed, whereupon that tree was abandoned and the same pattern repeated elsewhere. However, I have noted that some individuals will feed on small-leafed or quick-growing plants for much longer periods of time, with different parts of the plant repeatedly harvested at different levels (Figure 6.9). For a foliovore, such behavior assures a reasonably secure food source, for repeated cropping tends to increase local production by encouraging additional shoot and foliage development (Figure 6.10). The nutrients of new plant growth are not only more readily available (Milton, 1979), but often contains less lignin than mature growth (McNaughton, 1976; Field, 1976; Moss, 1977), fewer secondary compounds (Rockwood, 1974; McKey, 1974; Freeland and Janzen, 1974), and higher protein/dry weight (Feeny, 1970; Hladik, 1978), though exceptions are apparently common (Milton, 1979).

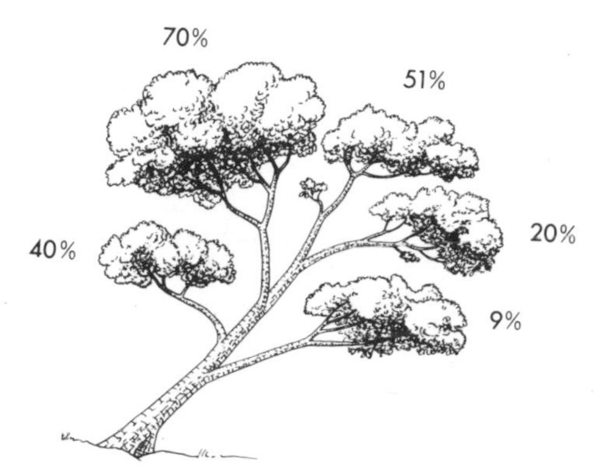

Figure 6.9: Proportion of leaves torn or nipped off *Zizyphus taylori* by long term repetitive browsing of a single adult *Cyclura carinata.*

In some cases these repetitive feeding patterns depend on complex factors. As an example, several individuals in mangrove areas could reach the particular *Rhizophora mangle* trees on which they regularly browsed only during low tide. Thus their feeding visits were timed such that they could regain the shore before the incoming tide made the journey impossible.

It is unknown why only certain individuals of some, but not all food resource plants are regularly visited, while others of the same species are ignored. Plant geometry may be an important, but not a decisive factor. It is not known whether repetitive feeding is based on habitual response patterns, taste, nutritional factors, or a combination of these. Plants in which it occurs most commonly as a long range characteristic are *Zizyphus, Rhizophora, Gaupira, Conocarpus, Scaevola,* and *Bucida*–all species in which only the leaves are eaten. Fruits are, of course, taken from below all individuals of some plant species, such as *Reynosa* and *Manilkara.* However, as far as is known, it does not occur in flower resource plants (*Tabebuia,* etc.)–perhaps because these parts are present for too short a period to make the association profitable.

The role of native herbivorous lizards in determining the composition and shape of vegetation has not received serious attention until very recently. The fact that plants are seldom seen suffering from defoliation in nature and that vegetation is commonly abundant are both poor reasons for discounting important interactions between herbivorous organisms and their food plants (Harper, 1961; Merton *et al.*, 1976).

Figure 6.10: (A and B) unbrowsed and browsed individual of *Rachiallis americana;* (C) *Maytensis buxifolia* is normally a small tree, but is here so heavily browsed it is only a few centimeters high; (D) heavily browsed (lower) as opposed to adjacent unbrowsed (upper) branch of *Zizyphus taylori;* (E) heavily browsed *Gaupira obtusata.*

Though food plants are generaly available on Pine Cay throughout the year, (1) some parts and some species are clearly eaten more often than others, and (2) the common habitual search pattern of this species means that some parts of the activity range are missed or ignored while others are heavily utilized. These factors are surely important in selective seed dispersal, for almost all seeds pass through the gut of *C. carinata* relatively unchanged, and at least some (probably all) are capable of germination (perhaps enhanced because of mechanical and chemical effects, but see Janzen, 1976). Thus, some Pine Cay plants may have their seeds dispersed by *Cyclura* more than others. However, the level and importance of this remains unknown.

In any event, the repetitive feeding patterns clearly modify the geometry of particular plants within the activity range; overbrowsing in some cases (particularly *Guapira,* Figure 6.10) causes plant death, while encouraging vegetative production in others (Figure 6.10).

Toxic Foods: Though there is no evidence suggesting that plant species maximize consumption of their fruits for seed dispersal, or that *C. carinata* plays any role in this, the toxicity of certain plants is undoubtedly an adaptation to minimize consumption. Interestingly, *Cyclura carinata* regularly feeds on some of these normally very toxic foods, while others, such as the milk weeds (Aescleipiadacae, containing a strong cardiac glycoside), are completely ignored. Stomach pH is an important factor in detoxification efficiency; the alkaline condition contributing to the degrading of a wide variety of toxic secondary compounds, permitting a diet high in essential oils and phenols (Nagy and Tengerdy, 1967). Unfortunately, pH of the stomach of *Cyclura carinata* is unknown, though captive *C. cornuta stejnegeri* have very acidic (pH 1-3) gastric fluids.

Toxic plants eaten are poisonwood *(Metopium toxiferum)* and manchineel *(Hippomane mancinella)*. Both fruits and leaves of these two plants are eaten, though fruits more commonly. In at least *Metopium* the fruits are less toxic than the leaves (Freeland and Janzen, 1974), so that the least poisonous part is more commonly consumed. To my knowledge, no vertebrates other than reptiles (*Geochelone* and *Cyclura*[2]) eat any part of the manchineel, suggesting the possibility of a very different detoxification system than in birds and mammals. However, the white-crowned pigeon *(Columba leucocephala)* regularly feeds on poisonwood fruit (pers. obs. and well known among Bahamians).

Competition: Interspecific food competitors are few. Table 6.12 lists the only birds on Pine Cay known to feed on fruits (based on stomach contents). First, the number of competitors is small (7 species = 7.9% of total Pine Cay bird fauna). Second, of the 88 plant species used by *C. carinata* on the island, only three were eaten by these seven bird species, suggesting that food competition from birds is low. Eight feral rat stomachs contained many seeds, stems, and only a few fruits of *Strumpfia,* so that they do not seem to be important food competitors. The insects in 20 *Anolis scriptus* stomachs showed only a 1.5% overlap with the numbers of suborders of insects eaten by *C. carinata* and 26 *Leiocephalus psammodromus* from the same island showed only a 5.6% subordinal overlap in insect food.

While there is no significant competition between hermit crabs (*Clibanarius* sp.) and *C. carinata* for food, there is an interesting interaction between them. Crabs abound on the island and apparently derive much nourishment from the

C. carinata droppings (cal/g dry wt $\bar{X}$ = 3613.8, N = 6). The same commensal behavior has been noted by my co-workers and me on several Bahamian Islands, and Wiewandt (1977) found it difficult to obtain large samples of *C. cornuta* droppings on Mona Island because of the local abundance of coprophagous hermit crabs.

Table 6.12: Bird Species Known to Compete with *Cyclura carinata* for Food on Pine Cay

Bird Species	Fruits Found in Stomachs			Remarks
	Strumpfia	*Erithalis*	*Scaevola*	
Zenaida Dove (N=3)				
Zenaida aurita	+*	–	–	Misc. seeds
White-winged Dove (N=2)				
Zenaida asiatica	–	++**	–	Misc. seeds
White-crowned Pigeon (N=3)				
Columba leucocephala	–	++	–	–
Cuban Crow (N=2)				
Mimus polyglottes	+	+		Cicadas, beetles
Pearly-eyed Thrasher (N-7)				
Margarops fuscatus	+	++	–	Spiders, grasshoppers, beetles, and lizards***
Least Tern				
Sterna albifrons	–	–	+	–

*+ = only a little in stomach
**++ = major food in stomachs
****Anolis scriptus*

Social Factors: Adult *C. carinata* are sometimes seen feeding in small groups (up to 7) in or under particularly heavily-fruited trees so that social facilitation (Greenberg, 1976) is possible. *Cyclura carinata* will sometimes steal food from one another (Iverson, 1979). These observations suggest that the conditions under which social learning occurs are sometimes met in *C. carinata* and that social learning in this species is a distinct possibility. In fact, absolute density is probably determined by the interplay of territorial behavior, food abundance, shelter availability, and social learning.

DISCUSSION

The evidence presented above clearly shows that there is no demonstrable shift from a carnivorous to herbivorous diet with age in *Cyclura carinata*. In fact, the proportion of animal to plant matter in the total diet in juveniles and adults is identical when mass is considered. Although individuals of all sizes (80-360 mm SVL, 15-1900 g) will consume animals opportunistically (dead or alive), the diet of all individuals of all ages and both sexes and during all seasons is largely composed of plants (see also *Iguana iguana*, Hirth, 1963, Henderson, 1965; *Ctenosaura similis*, Fitch and Henderson, 1978; and *Cyclura cornuta*,

Wiewandt, 1977; though Van Devender [this volume] reports a high proportion of insects in the diet of young *Ctenosaura similis*). The evidence accumulated so far suggests that, contrary to other views (Pough, 1973; Carey, 1975; and others), Iverson (1979) and Wiewandt (1977) are justified in questioning the frequency of an ontogenetic dietary shift from carnivory to herbivory in large iguanines. The only animal foods deliberately eaten by *C. carinata* are slow moving (slugs, nymphal cicadas, termites) or carrion.

Because it eats mainly plants, the food of *C. carinata* does not come in nutritionally balanced packages. It is then no surprise that it is not haphazard in selecting plant foods.

The native land macroflora of Pine Cay is composed of about 90 species. Of these 49% are regularly sampled by *C. carinata.* On Mona Island (approximately 40 times larger than Pine Cay) there are approximately 210 large, common, native land plant species (see Woodbury *et al.,* 1977), of which about 30% are eaten (data from Wiewandt, 1977). However, the absolute number of species utilized on the larger, florally more diverse island of Mona is 40% greater than on Pine Cay, suggesting that food diversity is directly related to floral diversity.[3] My current studies of feeding by *C. cychlura* should provide additional data on this, for the floral diversity on Andros Island is much greater than on even Mona.

While plants comprise most of the food, they exhibit considerable variety in their resistance to digestion by *Cyclura* species (Wiewandt, 1977; and others). The leaves of some species are only slightly altered (*Reynosa, Acacia, Rachiallis,* etc.), while those of others *(Ambrosia, Maytensis, Chamaecyce,* etc.). are almost completely digested, as are most fruits. There is no significant correlation between plant species eaten by *C. carinata* and caloric content–in spite of the general assumptions of caloric optimization in many mathematical analyses (see Pulliam, 1974, for review). Until available calories can be separated from unavailable calories correlations with leaves that have a large and variable amount of cellulose are difficult to interpret. However, there is an apparent positive relationship between utilization and total island caloric yield for a sample of leafy foods, but a negative one for fruits. The reasons for this are not clear. Nor is there any satisfactory explanation as to why some plant species are eaten and others are not; understood even less is why certain individuals of some species, or even only particular parts of them, are utilized. Hughes' model (1979) of the effect of learning as it relates to encounter rates may be important in maximizing intake.

Food choices based on caloric content do not necessarily lead to a balanced meal. Therefore it is probably not surprising that the food of *C. carinata* includes some uncommon and novel types. These are probably not sampled for their immediate nutritional advantage, but perhaps because of factors associated with the detoxification of secondary compounds. In general, most of the plants eaten contain no dangerous toxins. However, almost all certainly contain secondary compounds that must be chemically altered to avoid physiological stress (Freeland and Janzen, 1974).

Many workers have shown that the extent to which toxins are degraded depends on the physiological experience of the intestinal flora; that is, the individual only gradually acquires detoxification ability–probably through selection for bacterial species capable of degrading specific secondary compounds (Freeland and Janzen, 1974). A wide spectrum of plant foods is probably nutritionally advantageous, though conflicting selection pressures are best avodied through minimal sampling of certain foods. This may explain the apparent high level of opportunism suggested by the feeding behavior of *C. carinata*. Natural selection has probably served to develop a feeding strategy that *stresses selection of a variety of chemical food types, rather than maximizing the quantity eaten* of any one species, for it has been shown above that the amount eaten of any species at one time is far less than the total amount available within the activity range. *Cyclura carinata,* being a generalist herbivore, is probably forced to eat small amounts of a variety of foods in order to maintain a wide secondary compound detoxification potential. Thus, if food is a limiting factor in population size, as Iverson (1979) suggests, it is not the quantity or caloric quality that is most important, but the degree of food diversity and its assimilability by the lizard. This is strongly supported in the relationship between the lizard density and floral diversity in the several habitats compared.

The eating of highly toxic foods is a special category. It is undoubtedly significant that only the least toxic parts or stages of poisonous plant species are eaten. Nevertheless, plant species highly toxic to other vertebrates, such as poisonwood and manchineel, may be eaten because their nutritive value may exceed the costs of detoxification. Still, energetic costs of detoxification undoubtedly limit the amount of toxic compound ingested at one time and, at least in mammals, physiological damage cannot be avoided by simply eating less (Freeland and Janzen, 1974), if the appropriate microsomal detoxification enzymes are absent. Sex may also be a factor, for mammalian females (particularly if pregnant) are apparently less efficient in detoxifying certain plant secondary compounds (Mandel, 1972). It is not known whether the feeding on highly toxic plants is restricted to particular individuals.

In addition to novel foods, there are those that appear to represent staples, being preferred in a variety of habitats, irrespective of their densities (see Barnett, 1956, for similar patterns in rats). These are the foods that are probably sought and eaten because of their high nutritional level and/or easy assimilation (examples are *Manilkara, Strumpfia,* etc.). There is also that class of staple food that is abundant, but of low caloric value, often because of poor digestability (examples are *Cassia* and *Ernodia* leaves). These may serve more of a roughage function than anything else.

Though only a few plant species are generally consumed each day ($\overline{X} = 2$), the entire gut contents represent several days of selective ingestion, providing a mixture of nutritional staples, "roughage," and variously toxic species ($\overline{X} = 18.5$). Thus the foods selected during the cumulative search patterns of several days represent a compromise between the total utilizable calories available in the food plants recognized during the feeding foray and the variety of nutrients necessary for long range maintenance.

Of the many proposed feeding optimization models, those by Westoby (1974) and Hughes (1979) are particularly interesting for they are not only based on random food searching, but mediated by learning. Both models show that the chances of eating rare foods are not the same for each encounter, but that they probably change during each feeding period because the individual is selecting *meals,* rather than a series of non-related *foods.* Food selection in *C. carinata* is probably similarly adjusted constantly during each day on the basis of what foods have been eaten previously. These adjustments control the final "meal" composition (in this chapter considered the food restricted to the stomach). Carrion feeding and insectivory probably have a similar basis, for these related behaviors serve to increase a diet otherwise probably low in available carbohydrates and proteins (Field, 1967; and others), without which the intestinal flora of herbivores apparently becomes severly reduced (Williams Smith, 1967). Westoby's model, based on food species qualities rather than quantities and reflecting what he calls a "fallible nutritional wisdom," fits the *C. carinata* data quite well when viewed in the light of the requirements of both the energetics of the individual and long-term detoxification phenomena.

The most important factor in the total feeding strategy of *C. carinata* is probably food location learning, at least for staple foods within the activity range. Individuals that learn from social interaction or otherwise to search in only a few food-rich sites are undoubtedly better adapted than those searching randomly over a wide area with highly variable food densities. In this connection one of the most interesting behaviors is repetitive feeding at the same plant. While this pattern is obviously beneficial from several standpoints, the probable need for food variety and the impingement of seasonal factors (availability of fruits, etc.) suggest that rigid repetitiveness is not always desirable or possible. The location of new localized food sources (carrion, fruit, termites, etc.) depends on some randomization in search pattern. In some cases these discoveries may be fostered by social interaction, such as territorial boundary patrolling and social facilitation.

Cyclura carinata is a refuging herbivore (see Hamilton and Watt, 1970). That is, it returns to a burrow between feeding periods. Individuals are probably subject to a selection pressure resulting in overall minimization of distance from the refuge to various food sources. However, during repetitive return to a specific food patch, closer sources of the same food species were often passed enroute. Morrison (1978) discovered a similar "anomalous" behavior in fruit bats. His analysis showed that the search time component was more important than the commuting time component. That is, energetic costs were reduced if searching for a closer identical food patch was avoided, since the energy saved in commuting time was more than counterbalanced by the energy lost in increasing search time. The separation of commuting and actual searching times has not yet been made in *C. carinata.* However, optimal searching strategy ultimately depends on foraging path complexity (see Curio, 1972, and Morrison, 1978, for reviews), so that actual commuting time is minimized by search paths with relatively high turning rates. Furthermore, only a small, unknown fraction of an area's potential food plants are utilized by any individual, probably accounting for the partly overlapping foraging ranges often seen in these normally highly territorial animals.

Cyclura carinata is one of the smaller species of the genus. Future work on the feeding of other species may show that this smaller size is an adaptation to supplement an otherwise restricted diet on an island supporting low floral diversity by making the canopy available for feeding, for the larger species are apparently unable to climb into the trees, though *Iguana iguana* clearly does so. This and similar questions are being investigated in on-going studies of other *Cyclura* species on islands with more diverse floras.

SUMMARY AND CONCLUSIONS

The present paper reports the results of a study of the diet and feeding behavior of *Cyclura carinata* on Pine Cay, Caicos Islands, British West Indies. Ingestion, passage rates, foray patterns, caloric values, and annual yield of common food species are discussed.

Both juveniles and adults eat the same proportion of plant and animal foods, and there is thus no ontogenetic dietary change in this species. The primary foods are plants, and sixty percent of the macroplant species (including a few toxic forms) on the island are eaten. Complex food sampling patterns result in a varied diet of many plant species and parts, evidently differing in nutritional value and secondary compounds produced. All the plant species eaten fall into three food categories: (1) common staples of high caloric value, (2) common staples of low caloric value, and (3) foods frequently eaten in spite of the fact that they are uncommon.

The first category seems to form the energy base, providing needed calories for maintenance, growth and movement, etc. The second is perhaps important as roughage, for many leaves of species falling into this category pass through the gut relatively unchanged. The third category includes species of high preference level. Some of these are toxic and are undoubtedly sought out and eaten to maintain a system to detoxify certain secondary compounds. Unfortunately, protein content of none of the plant foods utilized was determined. The work of Milton (1979) and others on food selection in certain primate herbivores suggests that protein content is probably a major factor in food selection in the ground iguana as well, and that some of the plants in category three are very likely eaten for this reason.

It has also been shown that certain individual plants are repeatedly browsed, while others of the same species are ignored, perhaps due both to the energetic economy of harvesting known resources rather than searching for new ones and to the factors related to protein and fiber content in young sprouting leaves versus mature leaves. This repetitive feeding pattern has an effect on the geometry and density of some plants.

Thus, it seems clear that several factors interact to determine the food choice of this species of herbivorous lizard. Caloric content in the form of nonstructural carbohydrates does not seem to be an important factor in leaf and fruit choice. These conclusions do not support earlier models of optimum diets (MacArthur and Pianka, 1966; Emlen, 1966; Schoener, 1969, 1971), based on the supposition that the objective is to maximize energy yield/foraging time. Rather, the feeding of *Cyclura carinata* seems to better fit those optimal diets stressing a nutrient mix within a given amount of food (Westoby, 1974; Pulliam, 1975; Milton, 1979), often with dietary constraints induced by problems associated with toxic secondary compounds.

Notes

[1]Several introduced plants are also eaten, including horticultural and garden varieties.

[2]*Cyclura carinata* (this paper), *C. cornuta* (Wiewandt, 1977), and *Geochelone elephantopus* (Van Denburgh, 1905). Wiewandt (1977) also reported that *Cyclura cornuta* feeds on caterpillars whose toxicity is increased by feeding on *Plumeria*.

[3]The Mona Island food species list would probably have been higher if Wiewandt had been able to use stomach contents in place of only fecal pellets and visual observations.

Acknowledgments

Without the field work funds provided by the New York Zoological Society this study would have been impossible. This assistance is greatly appreciated and acknowledged.

Of the many people in the Caicos Islands who made this study possible, special thanks are due C.W. (Liam) Maquire, formerly of the Meridian Club, Pine Cay. He was extremely generous during my many visits to the island, in spite of the fact that our work often interferred with his busy schedule. Donald Correll and the staff of the Fairchild Tropical Gardens were very helpful in providing plant identifications. J.C. Dickinson, Jr., Florida State Museum, helped in important ways during the study of potential competitor species. Robert Woodruff, Florida Division of Plant Industries, generously supplied insect identifications. David Auth, Karen Bjorndahl, John B. Iverson, and Thomas Wiewandt, all formerly of the University of Florida, read the manuscript in several stages and helped remove many of my errors, for which I am very grateful. J. Bacon, San Diego Zoological Society, kindly arranged to determine stomach fluid pH values in captive *Cyclura* in his care. I also wish to thank my sons Garth and Troy for their help with many of the plant transects.

Section III

Demography and Life History Strategies

Life history strategies are the patterns of growth and reproduction of animals and how these relate to the environments in which they live. The evolution of these strategies has received a great deal of theoretical attention. In this section the applicability of life history strategy theory to field studies is deliberated and some detailed examples of strategies in different species presented.

Wiewandt (Chapter 7) considers at some length the wide variety of both environmental and social factors that influence the reproductive strategies of a species with an emphasis on female nesting behavior. With examples selected from a variety of iguanines he strongly makes the point that no single factor analysis can explain the differences we see between species but rather that each species has adapted uniquely to its environment and that each environment may have selective pressures unique to it. As a striking example he suggests that sea lion bulls cavorting on nesting beaches in the Galapagos may collapse large marine iguana nest burrows and so limit the upper size to which females grow.

Harris (Chapter 9) and Van Devender (Chapter 10) both address rates of growth in green iguanas in Colombia and Costa Rica respectively. They demonstrate different rates of growth in different localities. It seems plausible though certainly not proven, to relate these differences to differences in food, probably quality of food rather than quantity. Harris reports that in the Colombia population growth was much faster in early rainy season than later in the year. Van Devender found growth faster where iguanas were rare suggesting that intraspecific competition may occur, at least occasionally. Van Devender also shows that iguanas begin to reproduce while they are still growing rapidly; growth does not slow dramatically in either *Iguana iguana* or in *Ctenosaura similis* as it does in the large sympatric iguanid, the basilisk, *Basiliscus basiliscus*. Growth must obviously slow subsequently, but the shape of this curve cannot be determined until we have more data on the growth of adult iguanas. Interestingly, of the three large sympatric species that he compares, *Ctenosaura,* which begins life as an insectivore and later becomes herbivorous, grows most rapidly while the strictly herbivorous *Iguana* grows more rapidly than does the largely insectivorous *Basiliscus*.

Though growth rates of young iguanas vary significantly in different places the variation is small and Rand and Greene (Chapter 8) use an average growth rate in order to calculate from museum specimens as well as field data the approximate hatching and laying times for green iguanas across their range from Mexico to Brazil. It appears that hatching is usually synchronized with the beginning of the rains throughout the range and it is suggested that this is adaptive in a number of ways, perhaps the most important being that hatchlings appear when new vegetation is most abundant.

In the final chapter in the section, Case (Chapter 11) examines the life history patterns as well as other aspects of the ecology of the giant chuckawallas found on islands in the Gulf of California. He argues that reduced predation and erratic fluctuations in food availability have been the forces selecting for gigantism here. His observations on sociality and low levels of aggression should also be kept in mind as the more pugnacious species are chronicled in later sections.

In their patterns of reproduction and growth the iguanines show perhaps more diversity than in any other aspect of their biology. Some of the differences fit well with current life history strategy theory, others seem to require more *ad hoc* explanations and these latter emphasize the dangers in applying general theory to specific cases.

7

Evolution of Nesting Patterns in Iguanine Lizards

Thomas A. Wiewandt
Department of Ecology[1]
and Evolutionary Biology
University of Arizona
Tucson, Arizona

INTRODUCTION

The concept of reproductive effort, defined as the proportion of the total energy budget an organism invests in reproduction, is receiving increasing attention from biologists interested in population biology and life history evolution (Williams, 1966a, b; Tinkle, 1969; Gadgil & Bossert, 1970; Tinkle et al., 1970; Schaffer, 1974; Wilbur et al., 1974; Tinkle & Hadley, 1975; Hirshfield & Tinkle, 1975; Pianka, 1976). Studies of reproductive effort have spawned theoretical models, predictions, new models, and more predictions, all aimed at formulating general theory on the evolution of life history types. Tinkle (1969) and Tinkle & Hadley (1973, 1975) attempted to relate various measures of reproductive effort in lizards to demographic parameters such as age at maturity, body size, clutch size, and nesting frequency. The outcome was not terribly encouraging (Hirshfield & Tinkle, 1975, p. 2227):

> The difficulties in understanding the evolution of reproductive effort stem from the fact that predictions from theory are, in many cases, results of assumptions in the models which require careful examination before the predictions may be considered relevant to organisms in nature. Difficulties also arise because it is not clear what data constitute adequate measures of reproductive effort. It is impossible at present to decide whether failure of the data to be consistent with theory is due to inappropriate estimators or to inadequate theory.

In this chapter I examine social and ecological factors influencing reproductive effort in a small group of lizards, the Iguaninae, by taking a close look at the adaptive significance and evolution of nesting patterns. I hope to stimulate thinking about factors seldom considered in studies of lizard life history evolution. My approach is qualitative and somewhat speculative, because demographic data on iguanines are fragmentary (for a comprehensive, pre-1978 compilation, see Table 6 in Iverson, 1979), and there are few comparative data on social behavior and habitats. Following a generalized description of the "typical" reproductive pattern for iguanine lizards are five sections concerning different, though interrelated, aspects of an iguana's nesting strategy: (1) size and age at first nesting; (2) clutch, egg, and hatchling sizes; (3) nesting phenology; (4) nest-site selection and migrations; and (5) origins and consequences of communal nesting.

GENERALIZED REPRODUCTIVE PATTERN FOR IGUANINE LIZARDS

Most small insectivorous lizards attain sexual maturity 1-2 years after birth or hatching (Fitch, 1970), but iguanas normally require at least 2-3 years and then reproduce regularly for three or more years thereafter, until death. All iguanines are oviparous. Typically breeding is annual and seasonal, with mating taking place in male territories 3-7 weeks before nesting.

Gravid females select nesting sites on the ground in sunlit areas and excavate burrows large enough to accommodate a female's entire body. Each burrow terminates in an egg chamber wide enough to permit the female to turn around inside. She lays only one clutch, and then blocks the passageway to the eggs by filling and packing, leaving a pocket of air over the eggs (see Figure 7.1). Sites are vigorously defended against conspecifics during nest preparation.

Incubation commonly requires 10-14 weeks. Nest temperature remains nearly constant, with mean values of 28°-32°C for most species. Eggs from each clutch usually hatch in synchrony, and young dig to the surface without parental assistance. Hatchlings are highly vulnerable to predation and are cryptic. The young disperse from nest sites shortly after emerging.

This synthesis was drawn from a multitude of papers, which include the following key references: *Amblyrhynchus* (Carpenter, 1966; Boersma, 1979; Trillmich, 1979); *Conolophus* (Werner, this volume; Christian & Tracy, this volume); *Ctenosaura* (Fitch, 1970; Fitch, 1973a; Fitch & Henderson, 1977a; Hackforth-Jones & Harker, MS.); *Cyclura* (Carey, 1975; Wiewandt, 1977, 1979; Iverson, 1979); *Dipsosaurus* (Norris, 1953; Mayhew, 1971); *Iguana* (Rand, 1968a; Müller, 1968, 1972; Fitch & Henderson, 1977b; Dugan, this volume); *Sauromalus* (Johnson, 1965; Nagy, 1973; Berry, 1974).

While there are few known exceptions to this generalized pattern among New World iguanines, Fiji iguanas, *Brachylophus,* show several pronounced departures (see Gibbons and Watkins, this volume). Sexual activity appears to be relatively less seasonal (ca. 5 mos/yr) in *Brachylophus,* and females possibly lay more than one clutch annually. Mating takes place about six weeks before, and again about two weeks after, nesting. No air space is left over the egg clutch (this is evidently also true for *I. iguana*–A.S. Rand, pers. comm.). Furthermore, incubation time is extremely long, ranging from 18 to 35+ weeks; intermediate incubation periods of 14-17 weeks have been noted for *Conolophus*.

Figure 7.1: Reconstructed cut-away view of a Mona iguana nest. During nest-covering females refill only the passageway to the egg chamber, leaving an air space over the eggs, essential for proper development and emergence of the young. This clutch is near hatching–indentations in the shell are normal during the last week of incubation. Photo by T.A. Wiewandt.

SIZE AND AGE AT FIRST NESTING

At what point in an organism's lifetime should it first attempt reproduction? Theoretically, selection should favor individuals making the earliest possible investment in reproduction without sacrificing long-term reproductive success (see discussion by Pianka, 1976). For a long-lived species, this might mean delaying the onset of breeding to permit growth to a larger and more advantageous body size. Among iguanine lizards, for example, young, inexperienced males are at a severe disadvantage in combat over breeding territories. To avoid injury and probable loss of long-term competitive ability, members of this size class must forego breeding or adopt an alternative and less profitable mating strategy (see Dugan & Wiewandt, this volume).

For a female, egg production may result in an appreciable drain on her energy reserves, requiring growth to be slowed or suspended temporarily (e.g., Van Devender, 1978), and an abdominal cavity packed with eggs apparently limits food intake (Rand, 1968; Wiewandt, 1977). In some ecological settings, female iguanas have much to gain by trading-off an early reproductive effort for maximal growth. Fecundity typically increases with advancing age/size (e.g., see Berry, 1974; Fitch & Henderson, 1977b; Iverson, 1979; Hackforth-Jones & Harker, MS.), and in a review of reptilian reproductive cycles, Fitch (1970) noted that larger females of the same species not only produce larger broods, but also produce them with greater consistency or at shorter intervals. Susceptibility to predation may be significantly reduced (e.g., see Van Devender, 1978; Harris, this volume) or completely eliminated in adults of large body size (Wiewandt, 1977; Werner, this volume). Where suitable nest sites are scarce (e.g., on Mona I., P. R.) a female's size affects her chances of successfully defending a site from intruding conspecifics (Wiewandt, 1977). Moreover, larger lizards are endowed with added resistance to dehydration and starvation in harsh, unpredictable environments. Such benefits may be particularly meaningful to iguanas due to their predominantly herbivorous habits (Pough, 1973; Iverson, this volume; Van Devender, this volume).

Just as reproduction and growth are inseparable phenomena, both are closely tied to an organism's food environment. Pianka (1976) emphasized that interactions and constraints between foraging and reproduction are vital to understanding an animal's time and energy budget and that these factors have barely begun to be considered jointly in empirical studies of either growth or reproductive effort. There have been few rigorously designed comparative investigations of growth and reproduction in lizards (Van Devender, 1978), and none are available for the Iguaninae (but see Van Devender and Case, both in this volume). Unfortunately many iguana populations are now precariously small and can provide the large data base statistically desirable for such analyses only through unscrupulous sampling or a long-term commitment to studying marked individuals in the field. The latter approach is nicely exemplified by Case's on-going integrative research on mainland and insular *Sauromalus* populations and should serve as a model for others to follow. In spite of the status quo, enough information is available on the ecology of some iguana populations to allow a few preliminary comparisons (Table 7.1).

Three species categories can be distinguished in Table 7.1 using broad ecological criteria: (1) desert dwellers of the Temperate Zone–*Dipsosaurus* and *Sauromalus;* (2) species of dry subtropical islands–*Cyclura;* and (3) mainland forms from moist subtropical or tropical environments–*Ctenosaura* and *Iguana.* The same three groups are evident in demographic data comparing species by the degree of pre-reproductive growth in females relative to growth normally achieved through adulthood. *Dipsosaurus dorsalis* and *Sauromalus obesus* do

Table 7.1: Comparisons of Relative Size and Age at First Nesting and Relative Pre-Reproductive Growth Rates in Females from Ten Populations of Iguanine Lizards

Values in parentheses indicate odd samples, estimates taken from different populations; asterisks indicate data source.

		Body Size (SVL in mm)						
Species	Location	$\bar{X}$ Hatchling A	Smallest Breeding Female B	$\bar{X}$ Adult Female C	Percent Pre-Reproductive Growth Relative to Growth to $\bar{X}$ Adult Female Size (B-A) ÷ (C-A) x 100	Approximate Age at Maturity D	Growth/Year as Percent Female Size at Sexual Maturity (B-A) ÷ D ÷ B x 100	Sources (A, B, C, D) and Remarks
Dipsosaurus dorsalis	California	(49*)	110	120	86	5-6 yr	9	Mayhew, 1971; *Parker, 1972
Sauromalus obesus	California	54	150	~170	83	4-7 yr	11	**Berry, 1974
	California	(54**)	149	~163	87	—	—	Johnson, 1965
Cyclura carinata	Caicos Is.	80	190	225	76	6-7 yr	8	Iverson, 1979
Cyclura pinguis	Anegada, BVI	(~105)	~375	468	74	7-9 yr	9	Carey, 1975 small sample
Cyclura stejnegeri	Mona I., P.R.	119	~375	475	72	5-7 yr	11	Wiewandt, 1977 small sample
Ctenosaura similis	Nicaragua	57	200	276	65	21-33 mo	32	Fitch & Henderson, 1973a
Iguana iguana	Costa Rica	57	212	335	56	21-33 mo	32	Fitch, 1973a
	Nicaragua	74	250	327	70	21-33 mo	31	Fitch & Henderson, 1977b
	N.E. Colombia	68	201	280	63	21-33 mo	29	Müller, 1968, 1972 See note 2 in text.

comparatively more of their growing before the onset of nesting, while *Ctenosaura similis* and *Iguana iguana* do the least; *Cyclura* are intermediate. Age at maturity is clearly independent of species body size, with *C. similis* and *I. iguana* being outstanding in their relatively early start at egg production. These two iguanas are also distinctive in showing unusually rapid overall growth rates during their pre-reproductive years; females grow approximately three times faster than any of the other species under consideration. A look, albeit superficial, at the ecological settings of each species appears to provide at least a partial evolutionary explanation for the above life history differences.

Iguanas from North American deserts have short activity seasons (ca. 6 mos.) limited by both low winter temperatures and unpredictable dry periods which reduce food quality and abundance (Norris, 1953; Nagy, 1973; Berry, 1974). Consequently, annual growth tends to be low and variable, and reproduction risky. For example, female *Sauromalus* grow slowly and do not produce clutches every year (Nagy, 1973; Berry, 1974). They apparently have lower survivorship than males, possibly due to the high energetic cost of reproduction in an uncertain environment (Berry, 1974). A spent female chuckawalla may have difficulty surviving through winter hibernation following a dry desert summer because suitable forage is typically gone when oviposition occurs. *Dipsosaurus dorsalis* probably experiences less severe limitations because it is well adapted for activity in extreme heat (Norris, 1953) and may be less susceptible to desiccation. Desert iguanas also appear to be better suited for utilizing subterranean refugia offering more favorable microclimates. All else being equal, however, smaller size is a physiological handicap under stressful conditions. That females of both species do not begin reproducing at smaller sizes seems to reflect a climatically imposed need to postpone reproduction and channel surplus energy resources into growth, until reaching a size that will lower the risks of starvation and dehydration.

Iguanas native to islands of the Bahamas and Greater Antilles *(Cyclura)* are larger than Temperate continental desert iguanines, require as long or longer to reach sexual maturity, and tend to grow at proportionally similar rates (Table 7.1). Furthermore, a greater proportion of the average cycluran female's growth occurs after maturity, assuredly related to the fact that adults of at least the larger forms are extremely long-lived (Wiewandt, 1977) and growth continues, though slowly, throughout life. Major ecological differences which have accompanied the evolution of these attributes are lessened climatic extremes, low predator diversity, and limited nest site availability.

Cyclura habitats are influenced by maritime climates with high humidity year-round and appreciable thermal stability, especially at Antillean latitudes. While rainfall exceeds that of the Temperate deserts by three- to eleven-fold, little of this moisture is plant-available and productivity is curtailed due to highly permeable limestone (karst) substrata with shallow surface soils (Lugo et al., 1979). Unpredictable rainless periods and a pronounced annual dry season are also characteristic (see data for Mona I., P. R., in Wiewandt, 1977). Consequently, many, if not most, important iguana foods remain highly ephemeral in occurrence, and pre-reproductive growth rates are slow (Wiewandt, 1977; Iverson, 1979). The availability of seasonally transient foods of high nutritive value is, however, relatively more certain on these subtropical islands than in desert environments, allowing more consistent year-to-year production of off-spring (see Wiewandt, 1977; Iverson, 1979).

The smallest of the three cycluran iguanas represented in Table 7.1, *C. carinata*, lives on tiny islands with comparatively little structural and biotic diversity. Reduced feeding options may restrict growth significantly, and the small size of this species facilitates climbing and canopy-foraging (Auffenberg, this volume). Of added importance, *C. carinata* populations occur far enough north to experience sporadically cool and dry winter conditions unfavorable for activity year-round (Iverson, 1979).

Strongly skewed pressure from island predators and competitive interference between nesting females would select against early maturity in cycluran iguanas. Because the only iguana predators native to the Greater Antilles and Bahamas are birds, snakes (which tend to be small), conspecific iguanas, and possibly crabs (Carey, 1975; Wiewandt, 1977; Iverson, 1979), reaching a large size early in life confers an enormous adaptive advantage. Iguanas exceeding approximately 600 g in weight and 25 cm in snout-vent length surpass the expected size limits of prey for any of these predators (Wiewandt, 1977). Large females are also better equipped to defend their interests while nesting in a highly competitive social environment arising from a natural scarcity of suitable nesting areas (see Communal Nesting, Origins & Consequences). Both factors favor attaining a large body size before being burdened with egg-carrying, laying, and nest-defense.

The subtropical/tropical mainland iguanines *Ctenosaura similis* and *Iguana iguana* exhibit the fastest growth, earliest maturity, and smallest proportion of growth between hatching and first-nesting relative to mean adult female size (Table 7.1). These attributes, taken together with data on clutch size and nesting behavior (discussed in upcoming sections), hatchling behavior (Burghardt et al., 1977), and tail-break frequencies (Harris, pers. comm.), clearly indicate that predation is the key ecological force operating on the reproductive patterns of these populations. Apart from varying degrees of seasonality imposed by rainfall, climates are generally equable and favor year-round activity.

Of special interest is the relatively slow growth of *I. iguana* studied by Müller (1972). This atypical population represents a distinct ecological race of the semi-arid coastal region at Santa Marta in northeastern Colombia. Food is seasonally scarce, and iguana growth is stunted relative to that in other populations examined in Central and South America. Compared with the population studied by Harris (this volume), Santa Marta juveniles grow only half as fast (see Harris' Figure 9.5) and females mature at a size approximately 25% smaller.[1] Where drought-induced stress on vegetation is greatest, productivity and availability of preferred food resources can be expected to be lowest (see Pielou, 1975). Much drought-resistant vegetation is unpalatable and difficult to digest (see Auffenberg, this volume). Iguanines characteristic of subtropical-dry and subtropical-moist mainland ecosystems, i.e., most *Ctenosaura*, probably experience similar limitations in drier parts of their ranges.

While information on longevity of other *I. iguana* populations is unavailable, Müller calculated that females in his study area survive for a maximum of only 5-8 reproductive seasons. This figure is, however, difficult to interpret from an evolutionary perspective since females are intensively hunted today by humans in NE Colombia, as elsewhere in Latin America. In contrast, reproductive life expectancy for tropical and subtropical insular iguanas and Temperate desert iguanines is believed to be at least twice as long: 10-15 years (Johnson, 1965; Berry, 1974; Iverson, 1979) and upwards (Wiewandt, 1977; Werner, this volume).

CLUTCH, EGG, AND HATCHLING SIZES

Few measurements of lizard offspring have been systematically collected or reported. Reasonably complete records exist for only 11 New World species of iguanas (Table 7.2), half of which are data compiled from two or more sources. Interspecific diversity in sizes of adult females, clutches, eggs, and neonates is extreme. Adult females of the smallest species *(Dipsosaurus dorsalis)* are nearly identical in body length to hatchlings of the largest *(Cyclura stejnegeri)*[3]; their eggs approximate the size and shape of seedless grapes and goose eggs, respectively. In general, hatchling size increases with increasing species size (Figure 7.2). Larger species also tend to lay larger clutches, but the correlation is not close across generic lines. The largest clutches, for example, ($\overline{X}$ = 43.4 in *Ctenosaura similis*) are produced by iguanas roughly half the adult weight of those with the smallest clutches ($\overline{X}$ = 2.3 in *Amblyrhynchus*).

How each female apportions her expenditure in progeny is, like growth vs. reproduction, a give-and-take proposition. The cost of larger hatchlings is a smaller clutch and that of a larger clutch is smaller hatchlings. This relationship is evident in iguanas if clutch size is compared with relative female investment in individual offspring, termed Expenditure Per Progeny or EPP by Pianka (1976) and others (Figure 7.3). EPP is frequently offered as a measure of reproductive effort, a synonymy that I believe is unsatisfactory and one that is not implied here.

The 11 species examined invest between 0.8 and 5.7% of their body weight in each hatchling (Figure 7.3). Iguanas investing the most live on subtropical islands or in Temperate continental desert environments. All have small clutches. The five species laying an average of more than 15 eggs per clutch have uniformly low EPP values, near 1% (0.8-1.4%). They comprise an assortment of island and tropical/subtropical mainland forms.

Ecological factors that encourage a stepped-up investment in individual progeny are not often considered. They are undoubtedly diverse, varying with respect to both local conditions and the lizard's habits. One rather obvious benefit of large hatchling size in the marine iguana, *Amblyrhynchus,* is decreased vulnerability to physical dangers associated with life at the surf edge (see Boersma, 1982). Large egg and hatchling size in Mona iguanas *(Cyclura stejnegeri)* probably offers several advantages, some of which may be shared by other iguanine lizards. First, by virtue of their lower surface-to-volume ratio, large eggs and hatchlings are less susceptible to desiccation than are small ones. Moreover, an iguana's soft-shelled eggs dehydrate much faster than hard-shelled crocodile and bird eggs (Rand, 1968b). In semi-arid habitats with uncertain rainfall, as on Mona, even a slight reduction in desiccation rate might be critical in some years. Second, large hatchlings are less likely to be trapped un-

Table 7.2: Egg and Hatchling Data for the Eleven Best-Studied Species of New World Iguanine Lizards

Where known, sample sizes given in parentheses and ranges beneath sample means.

Species Location	$\bar{X}$ Adult Female Weight (g)	$\bar{X}$ Clutch Size	Eggs: $\bar{X}$ Weight (g)	Eggs: Dimensions (mm)	Hatchlings: $\bar{X}$ Weight (g)	Hatchlings: $\bar{X}$ Body Size (SVL in mm)	Sources and Remarks
Dipsosaurus dorsalis CA; AZ; NV	70	– 3–8	– –	24 x 17 (5) 22–76 x 15–20	4* –	47 (9) 44–52	Norris, 1953; Mayhew, 1971; Parker, 1972.
Sauromalus obesus California	175 –	8.6 (27) 6–13	8 (1 clutch) –	20 x 15 –	8 –	54 –	Berry, 1974; Iverson, 1979; Case, this vol.
S. hispidus Angel de la Guarda I., Baja	900* –	22.0 (5) 14–29	10 (2 clutches) –	25 x 24 –	10 –	70 –	Case, this vol. (hatchling size estimated).
S. varius San Estebán I., Baja	1,200* –	23.4 (5) 16–32	18 (1 clutch) –	40 x 28 –	14 –	75 –	Case, this vol. (hatchling size estimated).
Ctenosaura similis NW Costa Rica; W. Nicaragua	651 (263) –	43.4 (69) 12–88	– –	28 x 19 (117) 20–31 x 16–28	5.1 (13) 4–8	58 (13) 54–60	Fitch & Henderson, 1977a, 1978; Hackforth-Jones & Harker, MS.; Van Devender, pers. comm.
Iguana iguana Costa Rica; Nicaragua; Colombia; Panama	1,195 (169) 400–2,150	35–43 14–76	– 9–14 (86)	39 x 26 (41) –	11.6 (23) 8–15	76 (23) 72–79	Fitch & Henderson, 1977b; Harris, this vol.; Licht & Moberly, 1965; Müller, 1972; Van Devender, pers. comm.; Rand, pers. comm.

(continued)

Table 7.2: (continued)

Species Location	$\bar{X}$ Adult Female Weight (g)	$\bar{X}$ Clutch Size	Eggs: $\bar{X}$ Weight (g)	Eggs: Dimensions (mm)	Hatchlings: $\bar{X}$ Weight (g)	Hatchlings: $\bar{X}$ Body Size (SVL in mm)	Sources and Remarks
Cyclura carinata	476 (32)	4.3 (11)	26 (20)	52 x 31 (32)	15 (9)	80 (20)	Iverson, 1979.
Caicos Is.	205–1,135	2–9	19–30	46–58 x 29–54	13–15	76–83	
C. ricordi	1,285 (8)	8.6 (5)	–	–	30 (11)	87 (11)	Castro & Duval, 1979 (zoo data); Carey, 1975; Wiewandt & Gicca, unpubl. zoo & field data.
Dominican Republic	908–1,634	4–18	–	–	26–35	82–92	
C. cornuta	3,519 (6)**	15.3 (15)	–	–	51 (82)	105 (82)	Castro & Duval, 1979 (zoo data); Wiewandt, 1977 (zoo & field data); Wiewandt & Gicca, unpubl. notes.
Dominican Republic	2,951–4,767	8–24	–	–	38–68	96–112	
C. stejnegeri	4,750 (9)	12.0 (37)	104 (68)	80 x 51 (100)	74 (66)	119 (65)	Wiewandt, 1977.
Mona I., P.R.	3,400–5,400	5–19	82–159	70–95 x 45–61	60–92	105–127	
Amblyrhynchus cristatus	1,370 (26)	2.3 (18)	96 (38)	93 x 42 (38)	72.2 (6)	–	Carpenter, 1966; Bartholomew et al., 1976; Boersma, 1982.
Narborough I. & Santa Cruz I., Galapagos Is.	–	2–3	79–121	78–100 x 40–52	–	105–130	

*Estimate made by author.

**Weights from three unhealthy captives and three healthy field animals.

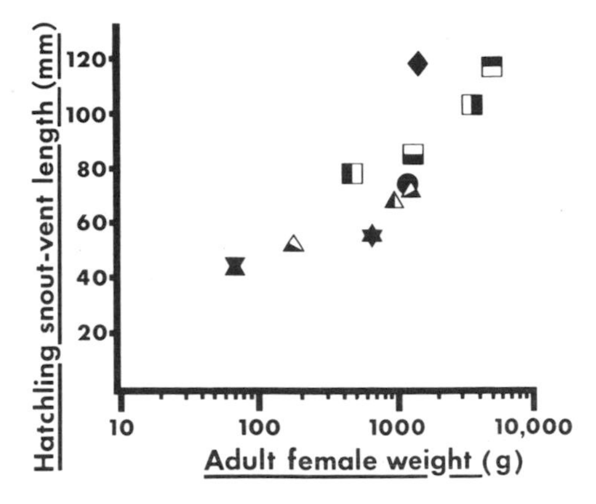

Figure 7.2: Scatter diagram relating mean adult female size to mean hatchling size for 11 species of iguanine lizards.

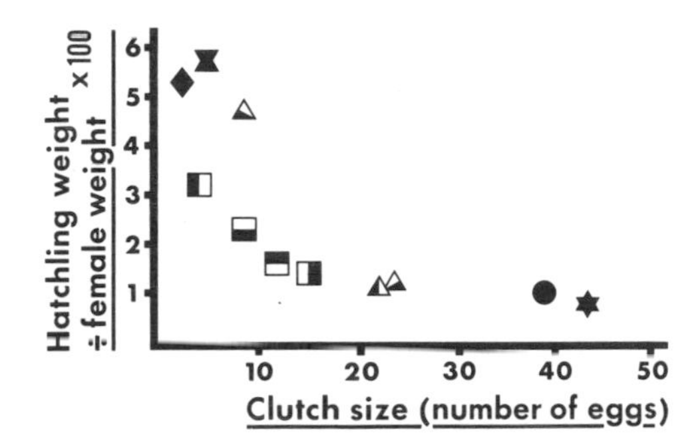

Figure 7.3: Scatter diagram comparing clutch size to a female's "Expenditure Per Progeny" (EPP), the percentage of mean hatchling weight in mean adult female weight, for 11 species of iguanine lizards.

Key

- ● *Iguana iguana*
- ✱ *Ctenosaura similis*
- ◬ *Sauromalus obesus*
- ◭ *S. hispidus*
- ◮ *S. varius*
- ◧ *Cyclura carinata*
- ◨ *C. cornuta*
- ⬓ *C. ricordi*
- ⬒ *C. stejnegeri*
- ⧗ *Dipsosaurus dorsalis*
- ◆ *Amblyrhynchus cristatus*

derground, for they are stronger and have metabolic advantages associated with larger body size. Assuming that all hatchlings carry an equivalent proportion of yolk per gram body weight, those of larger body size would be capable of sustaining activity longer without additional food. Third, because juveniles, like adults, are predominantly herbivorous, smaller hatchlings would be at a physiological disadvantage in their ability to utilize plant foods (see Pough, 1973; Wilson & Lee, 1974; Van Devender, this volume). Fourth, large hatchlings would have a greater variety of food types available. This is probably important on Mona because the fruits of several common trees are near the upper size limit for consumption by hatchlings. Finally, any increase in body size helps to shorten the interval during which juveniles are exposed to predation, particularly important on Mona since hatchlings are near the upper prey size limit for two species of snakes, the only known non-avian predators native to the island (Wiewandt, 1977). Clearly all such factors are interrelated. Perhaps the remarkable uniformity in body size (SVL) among Mona hatchlings (coefficient of variation = 4.2 for a sample taken from nine clutches) reflects the force with which smaller hatchlings are selected against in this environment.

Possibly the most peculiar reproductive pattern known for the subfamily is that of the marine iguana *(A. cristatus)*. Its clutch size is consistently small, never exceeding three eggs (mode = 2), and each egg is relatively enormous. In contrast, young female Mona iguanas, which attain sexual maturity at a body size near the maximum recorded by Carpenter (1966) for adult female marine iguanas, lay five or six eggs, and older females lay as many as 18 or 19. Eggs of both species are similar in size (Table 7.2).

Why has selection not favored slightly larger adult female marine iguanas capable of (1) avoiding all risk of predation by Galápagos Hawks, (2) laying more, but not smaller, eggs, and (3) gaining a competitive edge during disputes over nest sites? A number of observations suggest that limits on clutch and body sizes are imposed by special problems associated with nesting in loose sand. Carpenter (1966) noted that stamping heavily on the sand above a known *Amblyrhynchus* nest had no effect on the underground air space in the egg chamber, vital for proper development and emergence of the young. This outcome is to be expected since sea-lions frequent the same beaches during nesting (Carpenter, 1966; Eibl-Eibesfeldt, 1966; Boersma, 1982). Clearly, the larger nest chambers required for larger clutch volumes are more likely to collapse under the weight of a heavy animal on the surface. Since the diameter of a nest tunnel should reflect the size of the iguana digging it, selection probably also opposes large body size in female *Amblyrhynchus*. Carpenter and Boersma each found females fatally trapped in burrows that had collapsed around them, and wider burrows probably amplify such a risk. Nesting Mona iguanas select stabilized soil or sand deposits, the upper 20 cm of which are typically supported by a mat of plant roots. Consequently, burrows seldom collapse as females dig, but the weight of a person walking over a nest can readily destroy its air space. As no large mammals are native to Mona, nest-trampling by people and introduced mammals is a serious threat to the iguana's future (Wiewandt, 1977).

The two species laying the largest clutches by far and showing the lowest expenditure per progeny (Figure 7.3), *I. Iguana* and *Ctenosaura similis,* both inhabit tropical and subtropical mainland environments where predation is assuredly most intense. High fecundity in these forms appears to be essential for keeping ahead of losses to predators, well supported by lines of evidence already presented. Noteworthy in this context is Fitch and Henderson's (1977b) observation that *I. iguana* nesting for their first time produce eggs that are only three-fourths the weights of those laid by large adults. Since these females often begin laying small clutches at the end of their second year, at a very small body size, possibly reproductive success at this age depends more upon offspring number than upon the size of each. I know of no other instance where egg size of iguanines varies between age or size classes. Unlike *I. iguana, Ctenosaura similis* exhibits an ontogenetic shift in diet (from insectivory to herbivory), enabling its numerous tiny hatchlings to experience accelerated early growth, thereby bypassing physiological handicaps associated with herbivory at extremely small sizes (Van Devender, this volume).

Iguanas making relatively low per progeny investments accompanied by moderately large clutch sizes are *Cyclura cornuta* and the two insular *Sauromalus* populations (Figure 7.3). Among West Indian ground iguanas, those populations on or near large islands produce the bigger clutches with smaller eggs, presumably a demographic response to relatively greater diversity and abundance of predators (Wiewandt, 1977; Iverson, 1977). Comparison of the species pair *Cyclura cornuta* and *Cyclura stejnegeri,* for example, shows that clutches laid by occupants of the Dominican Republic are 28% larger than those of Mona Island, and hatchlings are 31% smaller by weight; snout-vent lengths of adult females are similar in both (Wiewandt, pers. observ.). The two insular *Sauromalus* studied by Case (this volume) occupy inhospitable desert environments and experience mass mortality of all age classes during prolonged rainless periods. By mainland standards (i.e., compared with *S. obesus*), recruitment is low in spite of their larger clutch sizes (accompanying larger body sizes) and lower susceptibility to predation. Reproduction in the insular populations is relatively infrequent and follows an "all-or-nothing" pattern (Case, this volume). Being able to produce young in quantity during years when food and moisture conditions are favorable is evidently critical to the fitness of these iguanas.

FREQUENCY AND TIMING OF NESTING

Most tropical or subtropical lizard species do not have short, well defined breeding and nesting seasons (Fitch, 1970), but iguanines tend to concentrate nesting activity into a 1-3 month period. Females lay single clutches each reproductive season, regardless of latitude. The tropical Fiji iguanas, *Brachylophus,* may be an exception (see Gibbons, this volume).

Nesting phenology is broadly defined by abiotic forces, such as climate, which impose limits on the time and duration of a female's growing and reproductive season. Iguanines inhabiting North American deserts nest near the middle of their activity seasons, during or after peak abundance of high quality foods (Norris, 1953; Nagy, 1973). Since rainfall and plant foods vary from month to month and from year to year, earlier nesting might jeopardize a female's survival through the following winter, whereas later nesting would probably take a heavy toll on hatchlings (see Parker, 1972). Whether reproduction is even possible in any given year appears to depend on the foraging conditions before the nesting season, and the availability of energy reserves stored from the previous year (see Johnson, 1965; Nagy, 1973; Berry, 1974). The eggs of both *Dipsosaurus* and *Sauromalus* evidently require little moisture for successful incubation and hatching; and nesting normally occurs during the hot, dry summer months throughout the major portions of the ranges of these species.

Amblyrhynchus cristatus is another species whose reproductive phenology appears to be closely tied to seasonally fluctuating food resources. Boersma (1979) noted that the breeding cycle is coordinated with tidal rhythm. Females eat algae on exposed reefs. Egg laying occurs after the lowest tides of the year, when food is most accessible, and occurs first in the western, more productive, waters of the Galápagos archipelago.

Carpenter (1966) may also be correct in suggesting that the annual nesting season of the marine iguana is timed to avoid the cool and dry "garua" season which might inhibit hatching. Licht and Moberly (1965) showed that eggs of *I. iguana* have remarkably stringent thermal requirements, with high mortality at temperatures only a few degrees above and below 30°C; this might be true for *Amblyrhynchus* as well. Boersma (1979) discounts Carpenter's hypothesis on the grounds that occurrence of the garua season varies from year to year, and the iguanas show a rigid rather than a flexible breeding schedule. However, a female laying eggs in February or March cannot possibly forecast weather conditions two or three months later. I believe selection would favor females that lay during the period producing the best long-term results, a strategy that should result in a relatively fixed breeding schedule. The same line of reasoning can be applied to any situation in which female iguanas lack the cues needed to evaluate the fate of their progeny in an environment that is unpredictable from year to year.

In most iguanine lizards, reproduction is timed to match seasonal conditions most favorable for survival of eggs and hatchlings. The tropical and subtropical iguanines *I. iguana* and *Ctenosaura similis* typically oviposit during the driest part of the year. This pattern provides insolation and thus high nest temperatures during incubation, minimizes the probability of nest flooding, and places hatching near the onset of the rainy season, when food is most plentiful. This permits emergent young to benefit from rapid early growth (Rand, 1968a; Fitch, 1973a; Rand and Greene, this volume).

In drier, subtropical environments, such as *Cyclura* habitats, the pattern is similar, though somewhat different. Nesting begins after the spring dry season and hatching coincides with the late summer or fall wet season. This places incubation at a time when the eggs are not likely to desiccate, and hatching occurs during the best period for emerging and foraging on fruits and ephemeral herbs (Wiewandt, 1977; Iverson, 1979). *Conolophus pallidus* follows this pattern on Isla Santa Fe (Christian & Tracy, this volume). Such breeding cycles are probably cued by photoperiod; Wiewandt (1977) found extraordinary year-to-year precision in the timing of nesting by Mona iguanas, regardless of pronounced variations in rainfall.

Many species of iguanas place their eggs deep in the ground, which requires a concerted digging effort from hatchlings attempting to reach the surface.

Consequently, emergence from the nest may typically be a team effort. In Mona iguanas, eggs of each clutch hatch in synchrony and the young dig to the surface as a group through packed sandy soil and a superficial mat of roots. Those last to hatch are sometimes buried alive by their siblings, accounting for about 1% of the in-nest mortality. Most emergence occurs after heavy rains. Without cooperative digging and periodic rain to soften the ground, many hatchlings would probably die underground. Presumably the large yolk reserves carried by each hatchling is an adaptation to ensure the animal's survival during this critical stage. Although hatchlings can apparently survive without food for at least 2-3 weeks underground, delayed emergence similar to that documented for some turtles (see Gibbons and Nelson, 1978) is unknown for iguanine lizards.

NEST SITES AND MIGRATIONS

Where an iguana nests is closely related to the physical structure of its habitat. Conditions permitting, the lizard's dwelling place may double as her nesting place. In *Dipsosaurus dorsalis* and *Cyclura carinata* females prepare their nests within sandy burrows normally used for shelter (Norris, 1953; Iverson, 1979). In other species, females must seek special places for nesting, and long-range movements may be necessary when suitable nest sites are scarce.

Karst terrain, the principal habitat type of cycluran iguanas, has little surface soil (LeGrand, 1973) suitable for nesting. On Mona Island, less than 1% of the island's surface area provides conditions that nesting *Cyclura stejnegeri* prefer, and gravid females must migrate distances up to 6.5 km to reach nesting areas (Wiewandt, 1977). Apparently three other West Indian iguanas, *Cyclura collei* of Jamaica, *Cyclura nubila lewisi* of Grand Cayman, and *Cyclura n. caymanensis* of Cayman Brac and Little Cayman, face similar problems (Lewis, 1944, p. 97):

> This early French naturalist [deTertre] relates that *Cyclura* come down from the mountains during May to lay 13 to 25 eggs in a heap in the sand on the seashore. *C. caymanensis* was found to conform to those observations.... On Grand Cayman, *Cyclura macleayi lewisi* [= *C. nubila lewisi*] is said to have been common in former years along the north coast...where there is a wide sandy coastal shelf. The people hunted the species for food, and in one way or another the population has been greatly reduced. According to the evidence of our searches in 1938, the species no longer frequents the coast,–not even during the breeding season–but digs nests in the earth [red phosphatic clay] in the more or less secure uninhabited east central portion of Grand Cayman.

Lewis also stated that the only evidence of past nesting by *C. collei* on Great Goat Island was found in the "red earth" near the island's coastal mangrove swamp. How many, if any, of the iguanas surviving in the rugged interiors of these islands continue to nest successfully is unknown. Migration in itself indicates the important role such soil-rich coastal areas have played in the evolutionary history of these West Indian populations. Females on Mona commonly attempt nests in soil-filled potholes on the plateau, but most of those efforts are abandoned. The only completed nest I uncovered in this type of a situation suffered total mortality apparently from overheating or desiccation. On a portion of Andros I., Bahamas, where iguanas *(C. cychlura)* have no soil available, eggs are laid in large termite nests (Auffenberg, pers. comm.).

In the Pacific Ocean, on the island of Fernandina, Galápagos land iguanas *(Conolophus subcristatus)* sometimes migrate 15 kilometers or more to reach suitable nesting areas (Werner, this volume). An unusual combination of lava terrain and little sunshine over much of Fernandina during the nesting season evidently precludes nesting everywhere except within the island's volcanic cone. Nesting and incubation coincide with the cool season. Gravid females climb 1400 m to the crater rim and then descend to the crater floor in search of soil deposits. The sky normally remains clear over the crater, and nest holes are dug in places warmed by direct sunlight and/or fumaroles (Werner, this volume). Why these iguanas favor the cool, wet season for nesting is unclear.

Apart from the data on the Mona Iguana and Galápagos land iguanas (including *Conolophus pallidus*–see Christian & Tracy, this volume), the only other well documented evidence of long-distance nesting migration in lizards is for the green iguana, *Iguana iguana,* on Barro Colorado Island, Panama. Montgomery *et al.* (1973) followed the postnesting movements of four *I. iguana* that swam to nest on Slothia, an islet adjacent to Barro Colorado, and found that upon returning to Barro Colorado, the four females traveled 300 m, 950 m, 2.2 km, and 3.0 km into the depths of the rainforest. Here the availability of favorable nest sites appears limited by the presence of nest predators native to Barro Colorado and by "shading-out" from the dense tropical understory and canopy vegetation (Rand, 1968a). Even on Mona, where the canopy is comparatively thin, otherwise acceptable sites that receive shade much of the day are not used for nesting.

How and when do female iguanas learn where suitable nest sites are located? During the nesting season, there is conspicuous and widespread trial-and-error digging on Mona's plateau, suggesting that females acquire a migration pattern through experience. Once a female has encountered an appropriate site, the best strategy to adopt would be to return to that spot the next year and continue doing so as long as conditions remain favorable there. This interpretation appears to hold for the iguanas of Barro Colorado Island, for although Slothia is a young islet created within the past 50 years, it is now a major nesting area. If such finds occur largely by chance, a female would not necessarily encounter equally attractive sites nearer to her normal activity area. My observations agree with those of Montgomery *et al.* (1973) who were puzzled that some females travel considerably greater distances to nest than appeared necessary, assuming each had a single home area. We cannot assume that female iguanas know all parts of their home islands, and choice of a migration route is probably limited by individual experience. The possibility that neonates learn through imprinting to return to nest at their birthplace merits investigation.

A recent study of the migratory movements of the snake *Coluber constrictor* to and from winter hibernacula revealed that a formative period may exist during which dispersal distance and direction, as well as the home range occupied, become "fixed" behavioral attributes in many members of the population (Brown and Parker, 1976). I suspect that such an ontogenetic process may operate in iguanas. Extended radio-telemetry work on the movements of newly matured females would shed light on this point.

COMMUNAL NESTING, ORIGINS AND CONSEQUENCES

Origins of Communal Nesting: The natural scarcity of suitable nest sites on Mona has been a selective force of paramount importance to that iguana population. Presumably seasonal migrations and communal nesting were once a necessity, and in most parts of the island iguanas still are forced to compete for limited available nest space (Wiewandt, 1977). Similarly, in the Galápagos, competition for nest sites is intense because accumulations of soil are sparse (see Carpenter, 1966; Eibl-Eibesfeldt, 1966; Christian & Tracy, this volume), a problem compounded in one case by a shortage of thermally favorable areas resulting from great intra-island differences in climate (Werner, this volume). Iguana nesting aggregations found in the wet tropical lowlands of Central America occur where well drained sites warmed by sunlight are in short supply (see Rand, 1968a; Fitch, 1973).

Natural selection may favor communal nesting for other reasons as well. A female choosing a site already worked by other iguanas will find digging easier, may be able to usurp another female's hole, or may lower the risk of predation to herself or her emerging hatchlings by nesting in synchrony with others. On the other hand, costs associated with aggressive interactions and the risk of losing a clutch to iguanas nesting later may be considerable. An assemblage of nests might also be more attractive to potential egg-predators than a solitary nest. In Guanacaste Province, Costa Rica, *Ctenosaura similis* utilize a network of subterranean passages for nesting. Several females oviposit in the same connecting burrow system, but individual clutches are laid in separate chambers (Hackforth-Jones and Harker, MS). In this case, there was no evidence of a shortage of potential nest sites, so communal nesting probably evolved in response to other selective pressures, such as predation or energy constraints (Hackforth-Jones and Harker, MS). Whether *Ctenosaura* utilize the same sites in successive years is unknown.

Traditional use of the same localities year after year may actually improve conditions for nesting, enhancing the attractiveness of such sites. Rand (1968a) noted that group nest-digging by *I. iguana* over the years effectively maintained a clearing in the forest. While excavating nests in a study area on Mona, I discovered that approximately half of the nest tunnels were forked, one passage leading to the fresh clutch from that season and the other going to an old clutch of empty shells. Such old passages, though sand-filled, were still identifiable by their looser soil. With the low density of iguanas on Mona today, many apparently favorable nesting sites on the island's southwestern coastal plain are not used and a locally clumped nesting pattern predominates. Gravid females exploring for suitable sites at the onset of nesting are strongly attracted to shallow depressions left from the previous season (Wiewandt, 1977).

Nest Guarding: Guarding behavior (against conspecifics) *after* nest covering has been reported for eight iguana populations: *I. iguana* in Chiapas, Mexico (Alvarez del Toro, 1972); *Cyclura stejnegeri* on Mona I. (Wiewandt, 1977, 1979); *C. carinata* on Pine Cay, Caicos Is. (Iverson, 1979); *Amblyrhynchus cristatus* on Hood I. (Eibl-Eibesfeldt, 1966), on Fernandina I. (Trillmich, 1979), and on Caamano Islet (Trillmich, 1979), all in the Galápagos Is.; *Conolophus pallidus* on Santa Fe I. (Christian & Tracy, this volume); and *C. subcristatus* on Fernandina I. (Werner, this volume). Observations in Gibbons and Watkins (this volume) suggest that guarding completed nests may also occur in the two South Pacific iguanines *(Brachylophus)*. The absence of such behavior has been specifically reported in *A. cristatus* on Fernandina I. (Carpenter, 1966; Eibl-Eibesfeldt, 1966). While female *I. iguana* studied in a Panamanian nesting aggregation may return to add surface fill to nests for up to 4 days after laying (Rand, pers. comm.), parental care is not exhibited to the degree found by Alvarez del Toro. Some Mexican females visit their nests daily for 15 days to pile on more sand and debris, making repairs if for some reason they appear open or disarranged. These iguanas apparently were not assembled in aggregations and may have nested within their usual activity areas. No mention was made of the duration of such visits or where the females passed intervening hours.

The fact that the Panamanian *Iguana* depart from the nesting area soon after laying and presumably migrate back to their normal home ranges (see Montgomery *et al.,* 1973; Rand, 1968a) suggests *not* that they would have little to gain by tending their nest sites for a few extra days, but rather that they cannot afford the additional energy expenditures. During nesting, *I. iguana* typically have shrunken muscle masses at the base of the tail and the animals' pelvic bones show, suggesting that they have metabolized both fat and protein (Rand and Rand, 1976). In contrast, shrunken tail musculature was seen in only four females nesting on Mona (Wiewandt, 1977) and has not been observed in *C. carinata* (Iverson, pers. comm.).

Owing to the geological and ecological similarities in *Cyclura* habitats throughout the range of this group, I would expect nest-tending behaviors to be common to most, if not all, West Indian ground iguanas. Mona iguanas, *Cyclura stejnegeri,* nest communally, away from normal activity areas. Spent females normally guard their covered nests for 1-10 days, a significant portion of the two-week nesting season. Even so, roughly 10-15% of egg-filled nests are dug into by intruding gravid females (Figure 7.4). Holes left by feral pigs that have rooted up and plundered iguana nests are diligently refilled the following day by nest-tending females. Iguanas do not attempt to challenge a hungry pig and are evidently unaware of their egg losses (Wiewandt, 1977).

Caicos Island iguanas, *Cyclura carinata,* defend completed nests for several days to a month or more after oviposition (Iverson, 1979). Because eggs are laid in a terminal portion of burrows normally used as retreats, they need protection not only from other females seeking nest burrows, but also from subadults and males seeking shelter after the nesting season. Guarding females are consequently aggressive towards all conspecifics. By reducing their daily home range, spent females can resume foraging and guard their nest sites simultaneously, typically watching for intruders while perched in nearby food plants (Iverson, 1979).

Figure 7.4: After having been displaced from her nest by an intruding gravid female, a smaller, spent female Mona iguana attempts to defend the site by attacking the intruder and by kicking loose soil into the burrow. Photo by T.A. Wiewandt.

Descriptions of nesting activities in two similar races of Galápagos marine iguanas (*A. c. cristatus* on Fernandina I. and *A. c. venustissimus* on Hood I.) show how local differences in ecology can affect nesting behavior. From positions on nearby rocks, female iguanas on Hood I. keep watch over their egg-laying sites for a few days after filling the nest burrow, occasionally descending to check the spot with tongue flicks and to scrape more dirt over the egg cache (Eibl-Eibesfeldt, 1966). Neither Eibl-Eibesfeldt nor Carpenter (1966) found such guarding behavior on Fernandina and females there are less prone to fighting over burrows than are the Hood iguanas (Eibl-Eibesfeldt, 1966). These authors emphasize that both populations are faced with a natural scarcity of suitable digging sites and are forced to nest in aggregations. Nevertheless, one important ecological difference between the two islands is evident: In contrast to the open nesting beaches on Fernandina, the area used by *Amblyrhynchus* on Hood I. offers very little loose sand above the high tide line; females must dig their burrows in small patches of hard, gravelly soil upon the plateau, and many burrows are abandoned because of lava blocking the way to further digging (Carpenter, 1966).

With respect to nesting opportunities, the situation on Hood I. closely parallels that on Mona. Presumably difficult digging encourages females to exploit the efforts of their neighbors, before and after nest covering. In both cases this pressure has been countered by extended nest-guarding behavior and may have led to the reduction in male/female sexual dimorphism. Eibl-Eibesfeldt (1966) noted that, unlike the other races of the marine iguana, the Hood females assume a bright, male-like coloration during the egg-laying season. Individuals so ornamented may hold a competitive advantage by being better equipped to intimidate other females by means of an impressive challenge display. Similarly, female Galápagos land iguanas, *Conolophus subcristatus,* on Isla Fernandina take on male-like colors shortly after breeding and throughout nesting (Werner, pers. comm.), a condition that accompanies intrasexual strife over nest sites and nest guarding (Werner, this volume).

Communal Nesting as a Modifier of Nesting Phenology: The nesting season of *Cyclura stejnegeri* on Mona Island is extremely short and nearly identical in duration and timing year after year. The entire season lasts 18 days (±4 days), with over 80% of the oviposition falling into an 8-day period islandwide. Yet, Mona's subtropical, maritime climate does not seem sufficiently limiting to explain such concentrated nesting.

I suggest that competition between communally nesting females for favorable nest sites can act as a nesting synchronizer. In Figure 7.5, three hypothetical stages are outlined for the evolution of an abbreviated nesting pattern from a less restrictive seasonal pattern with limits imposed only by climate and associated variables, such as food availability. Two assumptions have been made: (1) extended nest guarding is an advanced form of parental care in iguanine lizards and is derived from nest-hole defense that characteristically accompanies burrow digging and filling operations, and (2) the island's climate limits successful nesting to a period of two months, mid-June to mid-August, an intentional simplification.

In Stage 1, accompanying the expansion of the founder population the genetic contributions of two classes of females would be strongly selected against, those nesting outside the June-August season and those nesting early in the season. Early females run a high risk of losing their eggs to disturbance by the latecomers. Clearly, the outcome would be pronounced directional selection.

As directional selection proceeds, in Stage 2, a greater proportion of the population would nest synchronously later in the season resulting in intensified competition. This would be accompanied by increased nesting in places with more difficult digging and less favorable incubation conditions. Late nesting would probably take a heavy toll on hatchlings that miss optimum periods for emergence, dispersal, feeding, and getting established at a retreat before the onset of the dry season in January. Furthermore, tendencies towards extended nest guarding would be favored in females that nest relatively early.

Stability is attained, in Stage 3, when selection against late nesting balances that against early nesting. In addition to the limitations imposed by climate, females arriving relatively late in the season may have difficulty finding suitable unoccupied sites once nest-guarding behavior is established in the population. Nesting early is advantageous only if a female is able to protect her eggs until most other females have departed from the nesting area. How long she can profitably afford to stay to fight off intruders must vary with the costs involved. Under conditions of intense competition, nest defense is energetically taxing and the risk of physical injury is considerable.

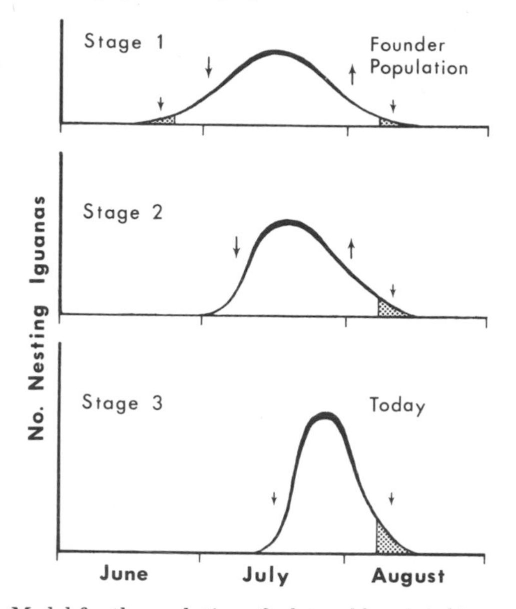

Figure 7.5: Model for the evolution of a late, abbreviated iguana nesting season where competition among females for suitable nest sites is keen. Stippled areas represent nesting efforts under environmental conditions marginal for incubation, hatching, and/or survival of first year young. Arrows indicate the action of selection. The curve is driven to the right by egg-loss in females that nest early to those nesting later and is kept to the left by limitations from the physical environment.

In the context of intense female/female competition for nest sites, large body size and delayed maturity should also be favored by selection if mature females have no natural predators, as on Mona, and can benefit from long life expectancy. Clearly, large females can exploit the efforts of other nesting females and defend their interests more effectively than small individuals. The highest pay-off, in terms of adult survivorship and reproductive success, should therefore go to females that postpone breeding until reaching an age and stature at which the costs of competition are lowered. I suspect that the evolutionary consequences of male/male competition for mates and female/female competition for nest sites are sufficiently similar to explain much of the male/female convergence in appearance found in Mona's iguana population (see Wiewandt, 1977, 1979).

SUMMARY AND CONCLUSIONS

All lizards of the subfamily Iguaninae are oviparous, and most have short, predictable, annual nesting seasons. Populations differ with respect to age at maturity, nesting phenology, nesting place, gregariousness, parental care, and size of clutches, eggs, and hatchlings. Primary determinants of nesting patterns include climate, predation pressure, competition, and the quality and quantity of nest sites. Nesting phenology is broadly defined by climate, particularly as temperature and moisture conditions affect length of the activity season, incubation, hatchling emergence, and food availability. Decreasing predation intensity permits longer life expectancy, slower growth, later maturity, and smaller clutches, with increased investment in individual progeny. Scarcity of favorable nest sites encourages nesting migration and aggregation. Intense competition among females probably selects for seasonally late, synchronous nesting, nest-guarding, delayed maturity, and reduced sexual dimorphism.

The degree of sexual dimorphism evident in iguana body size, form, and color must be cautiously interpreted. There is no evidence for sexual dimorphism favored by food resource partitioning in iguanine lizards. Where studied, diets of both sexes are essentially identical. Even in the marine iguana, separation of foraging niches appears to be a consequence of, rather than an explanation for, size dimorphism (Boersma, 1982; Trillmich, 1979). Just as sexual selection may be strongly operative in male iguanas (see Dugan and Wiewandt, this volume), promoting intersexual character divergence, other selective pressures acting specifically on the female sex may promote character convergence. For example, keen female-female competition for nest sites apparently favors large body size and other male-like attributes that would enhance a female's competitive ability. Other factors, such as ecologically imposed upper limits to nest volume (as suggested here for *Amblyrhynchus*), may have the opposite effect by selecting against large females, thereby increasing size dimorphism. Furthermore, because the cost of reproduction (with respect to energy budgets and exposure to predation) varies between the sexes in different ecological settings, optima for adult male and female body size can be expected to differ.

Understanding life history patterns and identifying factors responsible for their evolution require much more than a tally of life history attributes. Failure to recognize and investigate the complexities of an animal's ecological and social position in its community has been an interpretive stumbling block in most treatments of reproductive effort. Valid comparisons between and within species must be selectively drawn from broad, integrative studies conducted over extended periods of time; few such studies are presently available.

Notes

[1]Current address: 3436 N.E. 2nd Ave. Ft. Lauderdale, FL 33334

[2]Müller (1972, p. 121) says "Clutch size is dependent upon female body size, varying between 14 (in one or two year-old *I. iguana*, SV: 16-21 cm) and 70 in larger females (SV: 35-40 cm)." This statement, which suggests that females 16-21 cm SVL lay clutches of 14 eggs, is evidently in error. Data in Figure 1 and Table 1 of Müller (1968) indicate that females do not begin nesting before an age of 21 mos., at 20.1 cm SVL. Instead of selecting the 15.8-21.1 cm size class, measured in April/May, he should have cited his Dec./Jan. data of 20.1-23.3 cm SVL (probably the same cohort) taken just prior to nesting.

[3]In this chapter, I have chosen to retain the name *Cyclura stejnegeri* Barbour and Noble for the Mona iguana, even though Schwartz and Carey (1977) favor the trinomen *C. cornuta stejnegeri,* which emphasizes its close relationship to the Hispaniolan rhinoceros iguana, *C. cornuta.* Schwartz and Carey's evaluation was severely hampered by necessarily small sample sizes and their choice of only two taxonomic criteria–body coloration and scalation. Because Mona is thought to have always been well removed geographically from neighboring islands, on theoretical grounds one would expect the Mona iguana to be far less similar to Hispaniolan populations than they are to each other. Until further data become available to permit comparisons within this population complex, no meaningful species/subspecies designations can be made. The limited comparative information that I have been able to collect about the life history of Hispaniolan *C. cornuta* suggests that Mona females differ significantly in reproductive tactics (see Table 7.2 and text).

Acknowledgments

I wish to thank several persons who helped make this synthesis possible by generously sharing their field experiences, opinions, and in some cases, unpublished data: Walter Auffenberg, Dee Boersma, Michael Carey, Jude Duval, Donald Harker, John Iverson, A. Stanley Rand, R. Wayne Van Devender, and Dagmar Werner. I am also grateful to the Florida State Museum for permitting me free access to their reprint collections. Colleagues who offered valuable suggestions for improving the content and readability of an early version of this paper are Gordon M. Burghardt, Martha Crump, Dennis Harris, John Iverson, A. Stanley Rand, and especially Kentwood Wells.

8

Latitude and Climate in the Phenology of Reproduction in the Green Iguana, *Iguana iguana*

A. Stanley Rand
Smithsonian Tropical Research Institute
Balboa, Republic of Panama

and

Harry W. Greene
Museum of Vertebrate Zoology
University of California
Berkeley, California

INTRODUCTION

Slothia is a tiny islet in Lake Gatun, Panama, just off Barro Colorado Island. The yearly nesting migrations of female green iguanas *(Iguana iguana)* and the hatching of the young have been observed here for over a decade (Rand, 1968; Burghardt et al., 1977). One result of these studies and others at nearby localities (Dugan, this volume) is that we have gained some understanding of the relation of the iguana's breeding cycle to seasonal climatic changes in central Panama (Figure 8.1; Dugan, this volume). Females require two to three years to reach maturity and thereafter breed every year. Courtship and mating occur in December and January, coinciding with the beginning of the dry season and arrival of the trade winds. Eggs are laid in February and early March, the early part of the dry season and hatch in late April through May, at the beginning of the rainy season. Females visit Slothia only to nest. The nest clearing is visible on a year-round basis from the front porch of the dining hall on Barro Colorado Island. Observations of marked animals on Slothia show that females nest only once per migration (Rand, 1968, and subsequent observations), and nine years of records demonstrate that migrations are confined to a few weeks in late January, February and early March every year (Figure 8.2).

Literature records confirm that this general pattern occurs at other localities (Fitch and Henderson, 1977; Beebe, 1944), although nesting occurs at somewhat different times in different localities. Tropical lizards typically have extended breeding seasons and some species may even breed continuously (Fitch, 1970; Licht, in press); the situation in *Iguana iguana* is therefore highly unusual (although at least some other iguanines are also seasonal breeders; Wiewandt, this volume) and invites explanation. Does reproduction in iguanas vary geographically, such that it tracks latitudinal variation in the timing of rainfall or some other environmental variable? If so, what is the adaptive significance of this pattern? Our purposes here are to answer the first question and to offer some informed speculation on the second.

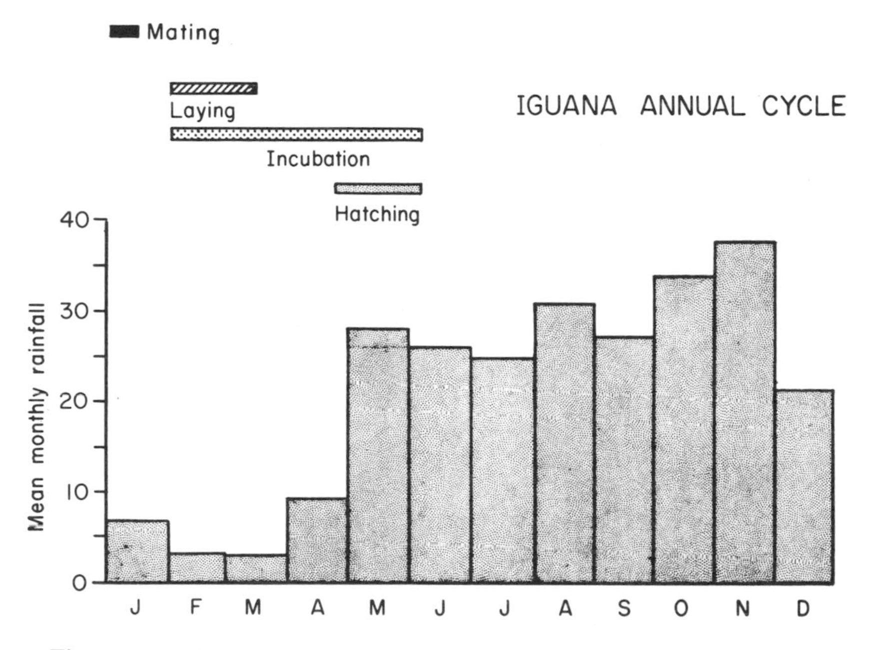

Figure 8.1: The annual reproductive cycle of *Iguana iguana* (after Dugan, 1980) and the mean monthly rainfall in inches for Barro Colorado Island, Republic of Panama.

METHODS

To determine seasonal changes in gonadal conditions (for reviews, see Fitch, 1970, and Licht, in press), herpetologists typically autopsy samples throughout the activity cycle. This is usually done by collecting fresh material on a regular basis or by examining existing preserved material in museums. Neither of these options was available to us, because female iguanas are so large that they are rarely preserved in series with viscera intact, and we lacked time and funds to do the necessary collecting ourselves. Instead, we examined the specimens of juvenile iguanas which are available in museums to estimate dates of egg-laying and hatching throughout the species range. We obtained snout-vent lengths, collecting dates, and localities for juvenile iguanas in the

Harvard Museum of Comparative Zoology; University of Kansas Museum of Natural History; Museu de Zoologia, Universidade de São Paulo; University of Texas at Arlington; Texas A & M University; and the United States National Museum. As a basis for interpreting these measurements, we used regression analysis to compare sizes and capture dates for ca. 400 young iguanas captured on BCI over a period of several years. We compared these estimated hatching dates with rainfall records where these were available to us (Wernstedt, 1972; Azevedo, 1974).

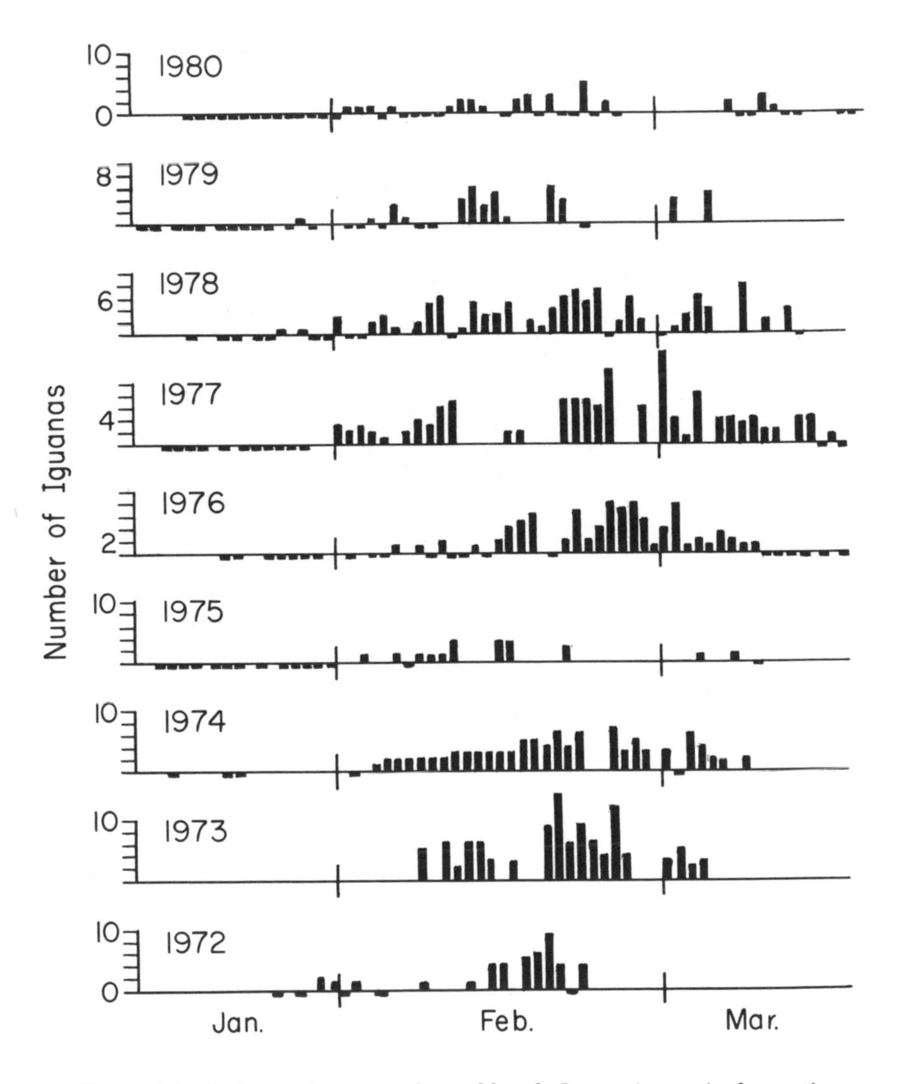

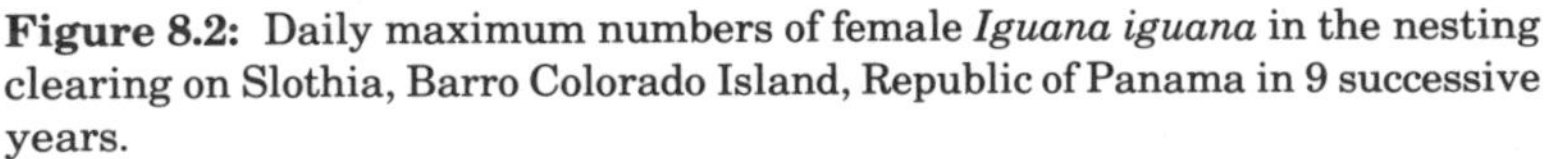

Figure 8.2: Daily maximum numbers of female *Iguana iguana* in the nesting clearing on Slothia, Barro Colorado Island, Republic of Panama in 9 successive years.

RESULTS AND DISCUSSION

Regression analysis gives a mean growth rate of 0.232 mm/day for Barro Colorado Island, and measurements of over 300 hatchlings at the nest clearing on Slothia in three different years (1975, 1976 and 1978) gives a mean snout-vent length of 73.5 mm (Figure 8.3). When these results are used to predict a hatching date for the total sample, there is a definite trend to earlier nesting in southern localities and it continues across the equator. If we assume a 90-day incubation period (cf. Fitch and Henderson, 1977; Burghardt et al., 1977), these hatching dates can be used to estimate laying dates and thereby display the entire breeding cycle at each locality (Figure 8.4).

It is important to note that our results are dependent on several assumptions. Growth rates vary among individuals in a population (Rand et al., unpublished), among nearby populations (Müller, 1968; Van Devender, this volume), and among widely separated populations. However, our figure of approximately 0.23 mm/day agrees well with data from Colombia (0.24 mm/day, Mullër, 1968), and Belize (0.22 mm/day, Henderson, 1968). All of these are a bit lower than data from Costa Rica (0.31 and 0.37 mm/day; R.W. Van Devender, this volume) and Colombia (0.58 and 0.36 mm/day; Harris this volume). Furthermore, egg size and consequent hatchling size may vary geographically. Finally, collectors probably favor smaller specimens, because they are more easily caught and more easily prepared and stored. Our assumptions about growth rates, hatching sizes, and representativeness of samples seem reasonable but the hatching dates estimated from them can only be approximate.

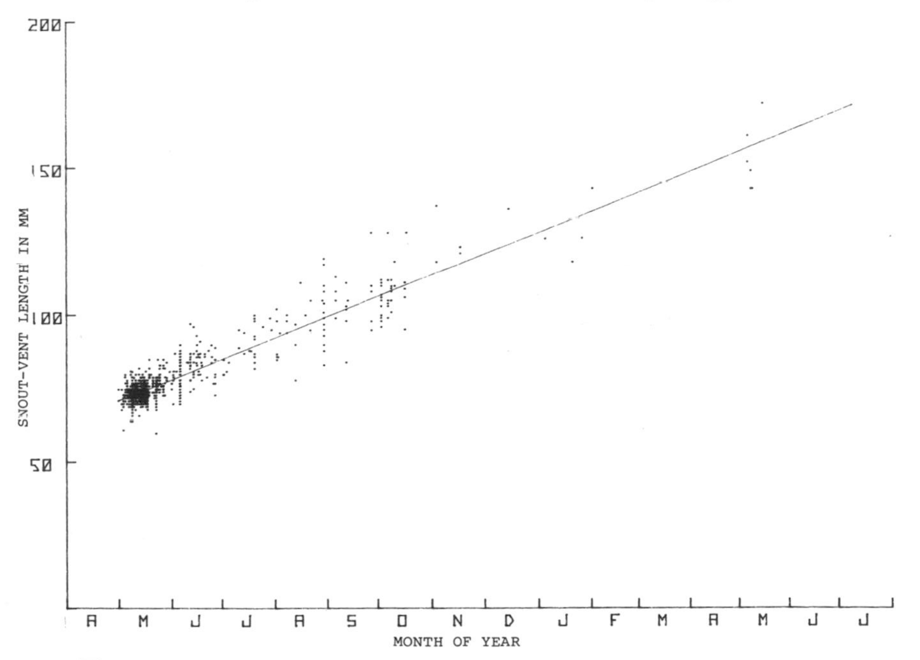

Figure 8.3: Date of capture and snout vent length in mm of young *Iguana iguana* captured on Barro Colorado Island, between 1965 and 1979. (Y = 43.2 + 0.232X).

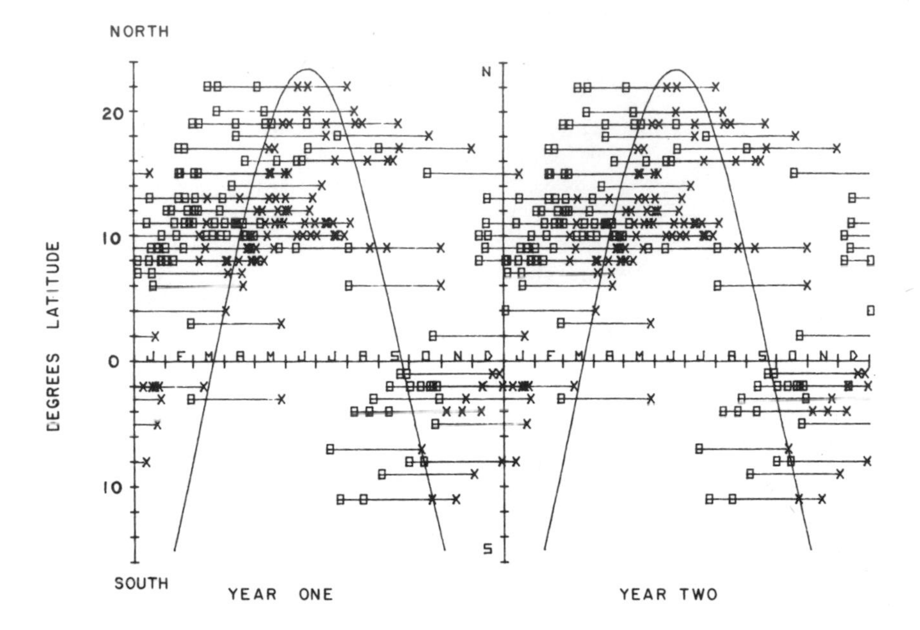

Figure 8.4: Dates of laying □ and hatching X, estimated from sizes of museum specimens of young *Iguana iguana*, plotted against latitude. To allow easy visualization of a complete year, data is plotted twice as if in 2 successive years. Curves give date and latitude when the sun is directly overhead.

The latitudinal trend in iguana breeding cycles parallels the progression of the sun (Figure 8.4), but not precisely enough to suggest that it is cued to specific day length throughout the species range. Using the available information on precipitation (Wernstedt, 1972; Azevedo, 1974), it appears that hatching usually takes place very early in the rainy season, mating and laying usually occur in the dry season (Figure 8.5).

In 36 of the 40 localities for which we have both rainfall and iguana data (Figure 8.5), hatching is estimated to occur during the first four months of the rainy season. Obviously this concentration of hatching during the beginning of the rains is more than coincidental. The exceptions to this pattern may be errors due to local variation in growth rates, in hatching size or to our small sample sizes.

However, the outlying points invite further consideration. Particularly interesting are the late rainy season hatching dates for Gulfo Dulce (12) and Bocas del Toro (16). Both localities have a heavy annual rainfall and a poorly developed dry season. The Bocas data suggest a double nesting season, and a double season is also suggested for Paramaribo in Surinam. For both these localities, the rainfall records suggest two dry seasons each year. However, we have been unable to confirm the existence of a second nesting season in Bocas del Toro and Hoogmoed (1973) reports only a single nesting season for Surinam. Nesting appears to occur very late in Guadaloupe (28) and indeed is somewhat

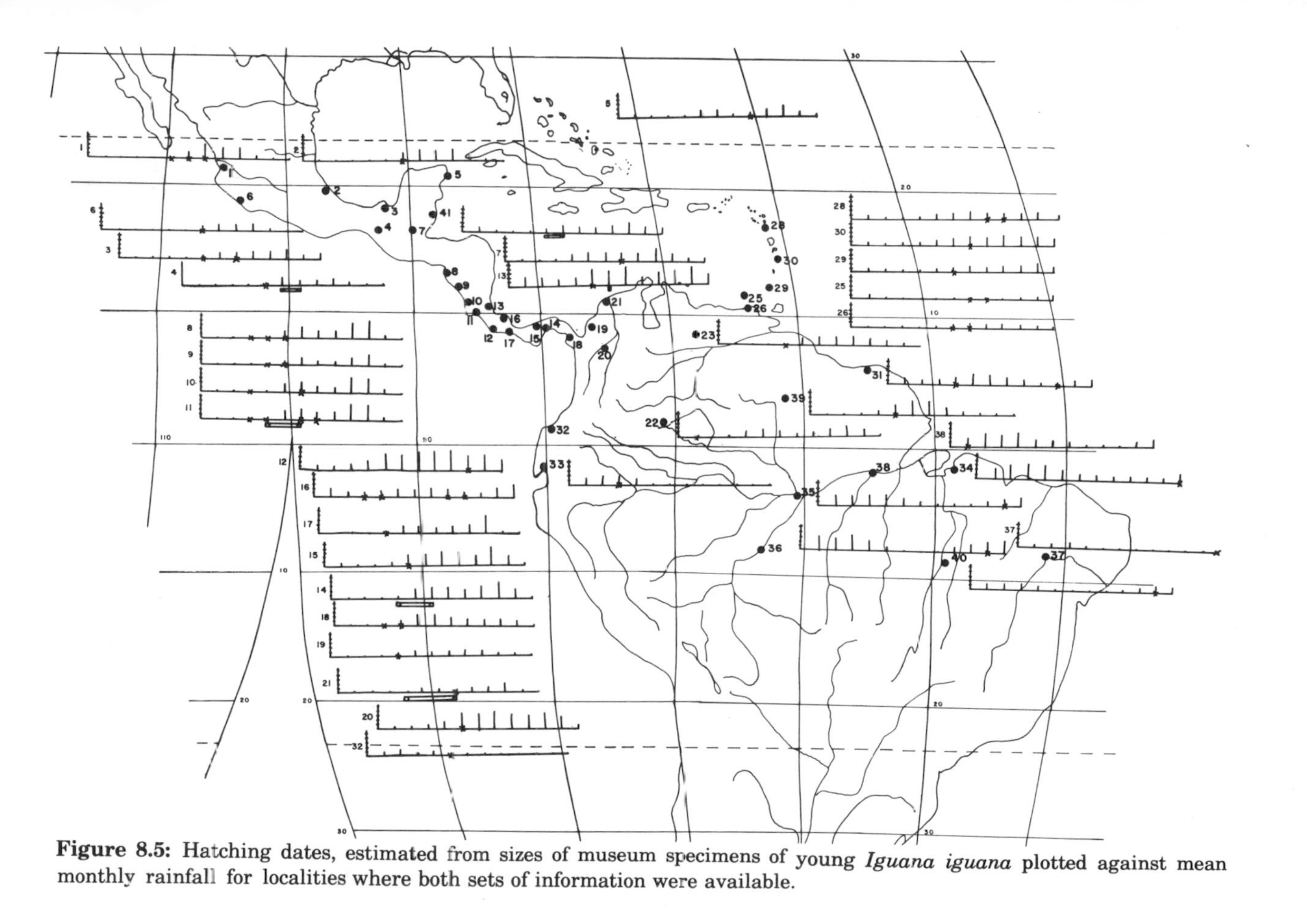

Figure 8.5: Hatching dates, estimated from sizes of museum specimens of young *Iguana iguana* plotted against mean monthly rainfall for localities where both sets of information were available.

late in two of the other four Caribbean islands represented, Cozumel (5) and St. Lucia (30); in contrast, Grenada (29) and Margarita (95) follow the more usual pattern. The Vaupes (22) locality in Amazonian Colombia, is a bit early, but this may be a sampling error and we have no other nearby data to check against. The Ecuadorian coastal localities, Esmeralda (32) and Guayaquil (33), are both later than would be expected from the reported rainfall and more similar in timing to the Panamanian populations to the north. It is tempting to suggest ad hoc explanations for these anomalies but there seems no point in doing so until they have been verified and local conditions investigated. They are certainly not numerous enough to question the general relationship.

Our data implicate the seasonality of rainfall as an important factor influencing reproductive events in *Iguana iguana,* a relationship that has been extensively documented in certain other iguanids (Licht, in press). As Licht pointed out, conclusions based on latitudinal correlations with single environmental factors must assume that other environmental features are not important and that all populations of the species make identical physiological responses to such cues. These assumptions remain to be tested for *Iguana,* and the studies of yearly variation at a single locality called for by Licht would be useful (Figure 8.2 suggests that at BCI such variation is slight but present).

Finally, we want to speculate on why iguanas nest when they do. First, Dugan (this volume) noticed that males were able to spend longer periods on exposed display perches during windy periods, even when it was sunny, than they could during still periods. These windy periods occur most frequently during the dry season; they thus coincide with the mating season, when the dominant males spend long hours displaying, patrolling, and mating in open, conspicuous display areas. A second reason is that dry season conditions are favorable for incubation. Licht and Moberly (1965) showed that iguana eggs incubate best at ca. 30°C, and Rand (1972) demonstrated that soils in a Panamanian nest clearing reach this temperature only during the dry season. Furthermore, heavy rains during the rainy season could cause nest flooding and fungal infections with resultant egg mortality. Third, hatching occurs early in the rainy season, when new flush foliage is available to the young (Wolda, 1979). Young leaves are probably a higher quality food resource than are old leaves. Young iguanas are folivores and, because they are small and growing rapidly, probably need better quality food than do adults. Preliminary observations by Troyer (pers. comm.) indicate that young iguanas do seek out young leaves, in preference to the old ones that are more easily available to them. Fourth, the rains also soak the soil around the nest, making it easier for the young to dig their way out.

Certainly there need be no single reason for the reproductive pattern in these animals. The timing of mating, laying, incubation, and hatching have evolved together, and the total response might represent an adaptive compromise. The available data do suggest that hatching during the first half of the wet season is more important than concordance of incubation with the dry season, because, for a number of localities, much of the incubation occurs during the rainy season. Green iguanas clearly offer exciting prospects for future studies of proximate and ultimate factors controlling reproduction.

SUMMARY AND CONCLUSIONS

Latitudinal variation in the breeding cycle of *Iguana iguana* is not clearly correlated with calendar date. It does relate to position of the sun and day length. Our studies of juvenile lizard size suggest that hatching date varies most closely with the onset of rains over a transect from Mexico to Brazil. As a result, mating usually occurs during the windy beginning of the dry season, when animals can spend more time on exposed display perches. Laying occurs such that eggs incubate at a time when the soil is appropriately warm and there is little risk of water-related mortality. Hatching occurs with the onset of rains, when soil is easy to dig through and new leaves are available as food for the young iguanas.

Acknowledgments

We thank R. Thomas, P.E. Vanzolini, and L. Maleret for supplying measurements on some lizards; G. Zug, E.E. Williams, and W.F. Pyburn for permission to examine specimens; R.W. Van Devender and K. Troyer for unpublished observations; all those helping observe, collect, measure, mark, and recapture iguanas on BCI; and the Smithsonian Tropical Research Institute and NSF Grant BNS 78-14196 awarded to G.M. Burghardt for financial support.

9

The Phenology, Growth, and Survival of the Green Iguana, *Iguana iguana,* in Northern Colombia

Dennis M. Harris
Museum of Zoology
University of Michigan
Ann Arbor, Michigan

INTRODUCTION

In Colombia, as elsewhere in Latin America (see Fitch et al., this volume), there is concern that the green iguana, *Iguana iguana,* is being overexploited. The egg is the principal product from the iguana in Colombia (Figure 9.1). Eggs are excised from the living female which is then stuffed with leaves (Friedmann Köster, pers. comm.) or ashes (Jorge Hernández, pers. comm.), sewn up, and released with the belief that she will reproduce again. The eggs are boiled in salt water and sold as a novelty food at more than twice the price of chicken eggs by weight. Iguana meat is less commonly eaten as it is considered a low class food, although Indians and other poor Colombians may depend heavily on it for protein. Iguanas are exported for the pet trade and are also a favorite bait for catching crocodilians (Alan Lieberman, pers. comm.). The degree of exploitation of iguanas is certainly cause for concern about their status, and is a major reason for the present study.

The iguana, by inhabiting major portions of tropical America, is exposed to a wide spectrum of climates. It should then be expected to show high diversity in various aspects of its life history. To test this expectation we must know details of the natural history of several different iguana populations. This chapter describes phenology, growth and survival of the green iguana from a site in northern Colombia.

THE STUDY AREA

This study was conducted at Los Cocos, the headquarters of Parque Nacional

Figure 9.1: Iguana eggs for sale in Ciénaga, Colombia.

Natural Isla de Salamanca, lying 9 km east of Barranquilla, Colombia (11°01′N, 74°41′W) from 20 November 1977 to 22 July 1979. The study area was approximately 10 hectares and included a 900 x 100 m swath of land between the road that crosses Salamanca and a canal to the south, a 400 x 10 m swath on the north edge of the road, and a 100 m swath along the edge of a brackish watercourse, also north of the road.

The Island of Salamanca is a geologic spit 60 km long which forms the northern, coastal boundary of an estuarine system lying between the Sierra Nevada de Santa Marta and the Magdalena River. Salamanca is characterized by a dry tropical climate, though because of its extreme lowland situation, moisture from runoff is also present. Because of the influence of the Sierra Nevada de Santa Marta, Salamanca is drier at its eastern end. The airport 15 km southwest of Los Cocos, the reporting station furthest from the mountains, receives 753.7 mm of rain annually (30 year average), the town of Tasajera, 40 km east of Los Cocos, receives 441 mm (13 years), and the Santa Marta airport, 15 km northeast of Tasajera, receives 307.9 mm (22 years) (HIMAT, 1979; HIMAT data). The latter two stations lie within the arid coastal zone in which Müller (1968, 1972) studied a population of iguanas. In the region there is a long annual dry season from December through April (Figure 9.2) during which defoliation of the vegetation occurs. During this study, however, the region was exceptionally moist with each weather station reporting above average rainfall in 1978. During 1978, Los Cocos itself received 927 mm of rain. The elevated rainfall resulted in less intense defoliation during this dry season than the previous one.

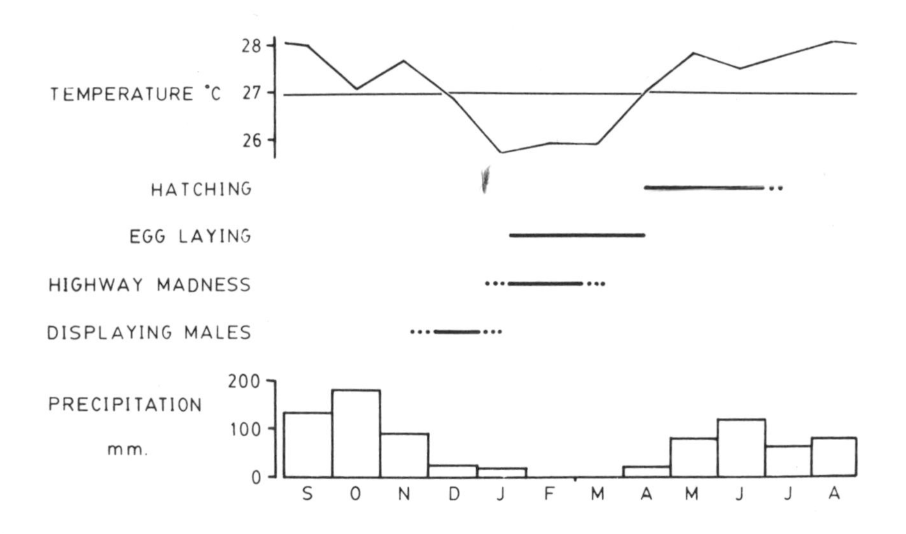

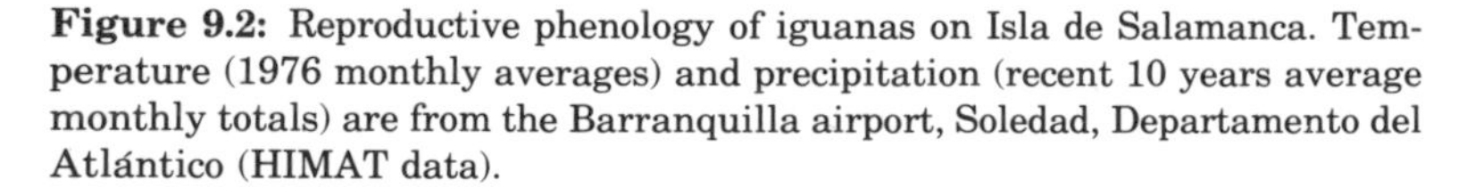

Figure 9.2: Reproductive phenology of iguanas on Isla de Salamanca. Temperature (1976 monthly averages) and precipitation (recent 10 years average monthly totals) are from the Barranquilla airport, Soledad, Departamento del Atlántico (HIMAT data).

The study area contains several plant associations including a stand of introduced coconut palms *(Cocos nucifera)* in a field of the grass, *Paspalum vaginatum,* emergent associations along the canal and in a small freshwater swamp, and a saline flat of *Batis maritima* and *Sesuvium portulacrastrum.* Iguanas were most frequently encountered either in the roadside thornscrub association of *Poponax tortuosa, Prosopis juliflora* (Mimosaceae), and *Capparis flexuosa* (Capparidaceae), or in a field in early succession dominated by *Cassia bicapsularis* (Caesalpiniaceae) and the vinaceous species *Cayaponia metensis* (Curcurbitaceae) and *Odontocarya paupera* (Menispermaceae). Similar to Henderson's (1974) study site in Belize, the latter plant association contains a species of *Mimosa. Mimosa pigra* was visually the most abundant species in that association at the beginning of the study, but the elevated rainfall apparently encouraged a very lush growth of vines with the resultant extirpation of many of the *Mimosa* plants which had not died off the year before. The study area is bordered on the north and east by mangrove swamp, on the south by grassy marsh, and on the west by freshwater swamp forest. Botanical sources are Romero Castañeda (1971), and the Los Cocos and Universidad del Magdalena (Santa Marta) herbaria.

The study site contains five buildings and some crocodile enclosures. Iguanas commonly lived in the buildings and in associated woodpiles, as well as in with the crocodiles. During the study, two adult iguanas were observed being eaten by the crocodiles.

MATERIALS AND METHODS

Iguanas were collected quarterly (March, June, September, and December; see Table 9.1 for dates). They were taken mainly at night, often with the aid of a noose on a 5 m long telescoping fiber glass pole. Snout-vent length (SVL) was measured to the nearest millimeter. Marking was accomplished by clipping a unique combination of toes for each individual and by painting with acrylic paint. Paint markings (numbers drawn on the body, or a series of four small color-coded dots on the base of the tail) were lost when the iguanas shed, but some marks lasted nine months. The animals were then usually released within a day of capture, however some females were retained in order to obtain information on reproduction. The site was visited frequently during times in which intensive collecting was not being done in 1979 to supplement phenological observations. During these additional visits, a few iguanas were collected.

Habitat alteration by humans during the study probably had little effect on the iguanas. A minimal amount of cutting of the vegetation was done to maintain paths within thickets. Also, a small amount of the area was altered for construction purposes by the park staff. One of these constructions, a mesh-walled caiman pen, became a rather effective trap for adult iguanas. The vegetation within the pen was frequently removed, an activity which encouraged the growth of weeds to which the iguanas were lured. They could then be run down as they tried to escape through, rather than over the mesh.

Statistical significance througout the paper is judged at the alpha level of 0.05.

Table 9.1: Size Increase and Survival of Hatchling Iguanas of 1978 in Their First Year

Mean Date	30 June	6 Sept.	13 Dec.	21 Mar.	2 July
Initial Date	7 June	30 Aug.	1 Dec.	9 Mar.	21 June
Final Date	11 July	20 Sept.	24 Dec.	2 Apr.	22 July
Mean SVL, mm	97.51	136.55	169.53	202.47	238.12
Standard Deviation	15.16	15.84	18.64	19.05	18.35
Range	70-136	101-172	135-219	160-255	202-270
Sample Size	117	96	64	195	66
Cumulative Marked	117	196	245	413	459
. . . . . Captures and Recaptures					
30 June 1978	117 (100%)	19+10* (24.8%)	10+8 (15.4%)	9 (7.7%)	3 (2.6%)
6 Sept. 1978		96+12 (100%)	15+16 (28.7%)	17+1 (16.7%)	4 (3.7%)
13 Dec. 1978			64+14 (100%)	30+4 (43.6%)	5 (6.4%)
21 Mar. 1979				193+2 (100%)	23 (11.8%)
2 July 1979					66

*Recaptures plus animals inferred to be present because they were recaptured at a later date followed by the percentage of the initial sample that they represent.

RESULTS AND DISCUSSION

Reproductive Phenology

Breeding occurs once a year in the dry season. As the amount of rainfall decreases, toward the end of November (Figure 9.2), the large males at Los Cocos, of which there were six in each year of the study, may be seen displaying from prominent perches in palm trees and in the low bushy trees. This behavior ceases early in January. Elsewhere (Dugan, this volume) it has been shown that high rates of display are associated with the process of mating.

At the end of the displaying males period, large adult iguanas become more active on the road (Highway Madness, Figure 9.2). At this time, from January to March, iguanas are casually more visible than during the rest of the year. Road frequenting is most striking along a part of the Salamanca highway which passes through a swamp of black mangrove trees *(Avicenna nitida)* lying 10-15 km east of Los Cocos. There, hundreds of iguanas may be killed by automobiles each breeding season, particularly during early February.

The sandy road embankment, but 30-40 years in existence, is a major breeding site on Salamanca, particularly for the purpose of egg laying, although, because large numbers of both males and females may be found on the road, it therefore must serve more than just the functions of the females. Because the increase in abundance of males on the highway occurs just after the time when males are seen displaying at Los Cocos, it seems plausible that some quality about the highway enhances recuperation from breeding. For example, thermoregulation may be facilitated by the open conditions and by the warming of the pavement. I however do not discount the possibility that many of the road killed males could still be actively breeding. Because the periods of egg laying and hatching are relatively longer than the period over which I saw males displaying at Los Cocos, I suspect that the mating season is actually of longer duration than the estimate indicates. For the reproductive male, the roadside area may be attractive because of there being relatively more suitable perches. Road killed males may have been moving between perches. For example, one male at Los Cocos would shuttle between a palm tree used for sleeping and escape from the investigator, and a two meter high perch in a *Poponax* tree from which it did most of its displaying–it would bask and forage in the open area between the perches. The road itself may serve for a display site with the advantages that it is elevated several meters above the swamp, a displaying animal would be very visible, and there are lots of females there. My observations of live animals on the road, however, are few and I did not observe them either head bobbing or mating.

Egg laying and hatching each encompass about three months, mid-January to mid-April and mid-April to mid-July, respectively. Hatchlings of 1979 were first seen and captured on 16 April and lastly on 21 July. The last hatchling measured 77 mm and possessed an egg tooth which in most of 86 captive hatched iguanas was lost within two days after emergence. Sand slides, presumably from iguana digging, may be found on the road bank around the end of January and the last of three captive females oviposited on 16 April 1979, but actual nesting in the wild was not observed. These observations correspond to the given hatching schedule with an approximately three month long incubation period (Ricklefs and Cullen, 1973; this study).

Young from three clutches of iguana eggs, artificially incubated at ca. 31°C, emerged in 74, 78, and 81 days with first emergence to last hatching in one clutch lasting three days. These incubation times are longer than the previous record of 73 days for Colombian iguana eggs (Licht and Moberly, 1965),but still must be regarded as conservative because of possible egg retention by the females that, respective to the above incubation times, were held for 14, 40, and 10 days prior to their ovipositing in a plastic drum half filled with moist packed sand.

Breeding in the dry season is advantageous to the iguana for reasons based upon nutrition and nest site availability. Normally, food may be scarce when the adults are reproducing, but simultaneously they may not need to eat as much, or at all then since a full stomach may hinder reproductive function by displacing the eggs and there is less time for foraging if a territory must be maintained. Fasting in the reproductive epoch is possible by drawing upon bodily fat reserves accrued during the rest of the year. Nutrition is probably most critical to the hatchlings, which appear with the first rains in mid-April when new foliage is also appearing (providing food and also concealment for the green neonates–though some are beige in color). Bare ground for nesting is more abundant in the dry season and it is also less likely that flooding of nest sites will occur, in fact, many potential nesting grounds in the area are annually inundated.

Growth

Growth rates from 114 recaptures of free-roaming iguanas are summarized by the method of least squares linear regression in Figure 9.3. The intervals between captures for animals used in this analysis ranged between 33 and 581 days, with over half between 60 and 120 days. There were only eight points for which the interval between first and last capture exceeded one year and these were all from large individuals. Growth over time intervals of less than a month were excluded so that actual growth could be distinguished above mensuration error, which becomes more important in measurements taken over relatively short periods (Van Devender, 1978).

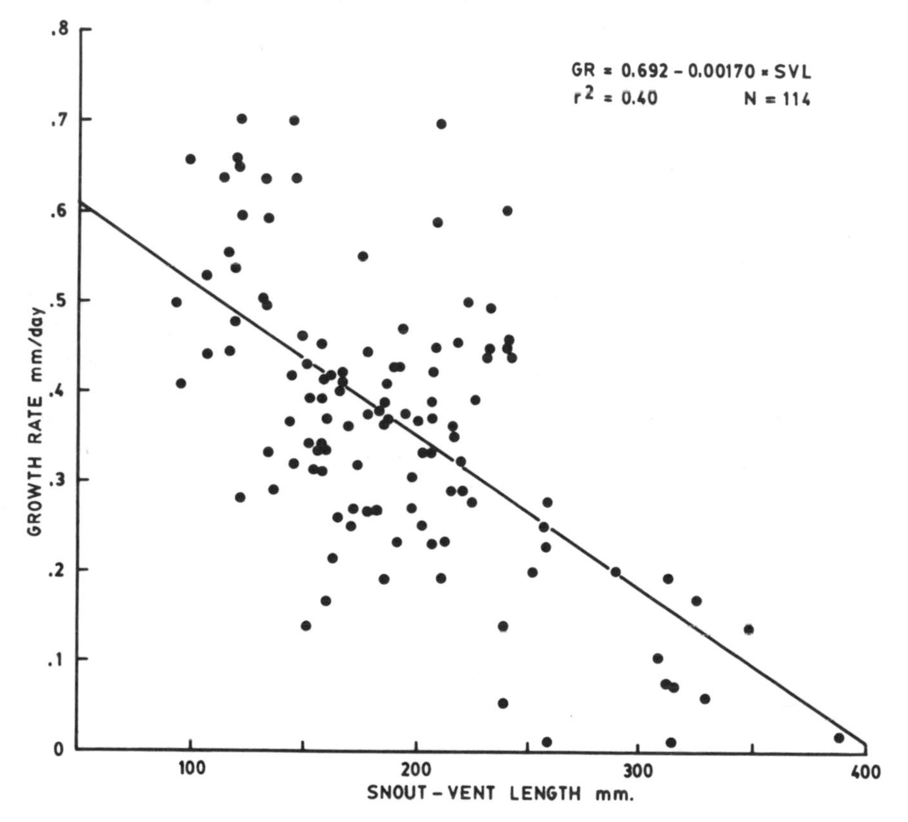

Figure 9.3: The relation between growth rate and snout-vent length of Salamanca iguanas.

The rate of growth (GR), which ranged from 0.01-0.70 mm/day, shows a significant inverse relationship to body size (Figure 9.3), the equation being GR = 0.692 − 0.00170 x SVL. However, a large amount of residual variation in growth rate is evident from the low value of the coefficient of determination ($r^2 = 0.40$). Part of this variation may occur because of the influence of season on growth. The relative effects of size versus season in this study are inseparable due to the iguanas' high degree of reproductive synchrony.

In Figure 9.4, individual growth rates from collection to collection are summarized for the juveniles of 1978. Growth rates are initially very high, averaging 0.58 (1 standard deviation = s.d. = 0.10) mm/day over a 2-3 month interval.[1] Then they drop off significantly at the height of the rainy season to 0.36 (s.d. = 0.09) mm/day and continue at 0.35 (s.d. = 0.10) mm/day through the dry season. At the beginning of the next rainy season, average growth rate increases to 0.42 (s.d. = 0.12) mm/day, but not significantly (t = 0.073, d.f. = 39). Growth rates determined from sample means of SVL repeat this same pattern until March-July, where the growth estimate shows little increase.

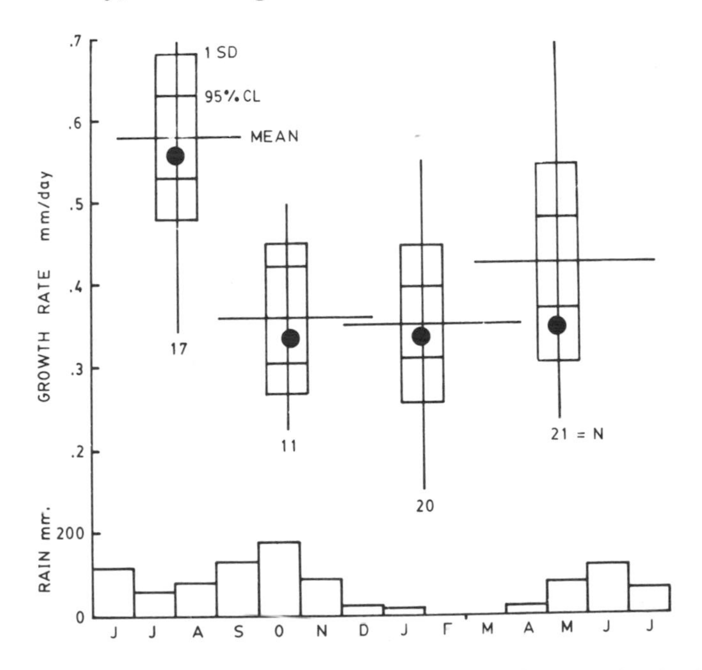

Figure 9.4: Juvenile growth with season. The growth rates of individual iguanas, taken over 2-3 month intervals, are presented with respect to time and rainfall pattern. The horizontal lines at the growth rate means denote the maximum lengths of the intervals between samples. The vertical lines are ranges. The boxes indicate plus and minus one standard deviation (SD) and 95% confidence limits (95% CL). Solid circles are values calculated using the mean SVL of each sample from Table 9.1.

The decrease in juvenile growth rate between June-September to September-December occurred despite an abundance of vegetation being present, and so, appears not to be caused by a depletion of food resources. Juvenile growth rate was sustained during the December-April dry season. Defoliation then was considerable, but foliage was present throughout and ripened fleshy fruits were at their peak abundance. The growth of these iguanas, then, does not seem to be food limited as it is in Santa Marta, substantiating Müller's (1968) expectation. He hypothesized that growth is more rapid in "banana zone" iguanas (akin to those of Los Cocos) than it is in those of the dry coastal zone of Santa

Marta allowing "banana zone" iguanas to attain larger adult size. He proposed that this was due to nutritional stress. Remarkably, Salamanca iguanas on the average, grow as much in one year as Santa Marta iguanas do in two (Figure 9.5). Note also that the distribution of SVL of Salamanca juveniles becomes increasingly dispersed to the extent that it may bridge three of Müller's age classes (Figure 9.5, Table 9.1). The increase in variation is great enough that the sample variances can be shown to be nonhomogeneous (X^2 = 9.49, d.f. = 4, Bartlett Test). In Belize (Henderson, 1974), juvenile growth rate (0.22 mm/day) was similar to that in Santa Marta (ca. 0.25 mm/day, calculated from data in Müller, 1968). Growth rate is considerably slower in adult iguanas than it is in juveniles. Ten iguanas, 250 mm SVL or greater at first capture grew at an average rate of 0.085 (s.d. = 0.065) mm/day.

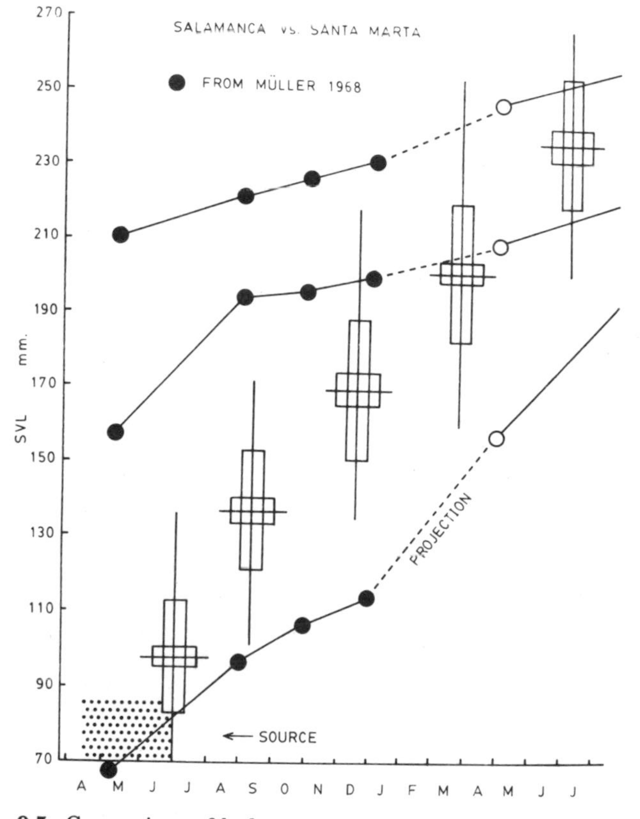

Figure 9.5: Comparison of body size increase between iguanas from Santa Marta (based on data from Müller, 1968) and Los Cocos (Table 9.1). SVL is plotted against months. "SOURCE" indicates the range of size and time-of-emergence of the hatchlings. Statistical diagrams are of the same form as in Figure 9.4.

Iguanas measure 70 to 86 mm SVL at hatching and in a year may grow to about 220 mm SVL. The smallest gravid female found at Los Cocos measured 265 mm SVL, though it would be reasonable to expect that sexual maturity is attained at 250 mm SVL (Fitch and Henderson, 1977). Average female size is 292 mm SVL (s.d. = 34.84, N = 56, range = 250-394). The Salamanca females may grow to minimum sexually mature size in 1.5 years. If size is the determining factor for maturation, with the large degree of variation in growth rate considered, it is expected that the major proportion of second year females are reproductively mature.

Survival

Four hundred and fifty-nine iguanas that hatched in 1978 were captured a total of 564 times in their first year of life. The recapture data are summarized in Table 9.1. The first sample was composed of 117 recently hatched iguanas of which 24.8, 15.4, 7.7, and 2.6% were known to be alive after one, two, three, and four successive quarters, respectively. An individual was known to be alive if it was actually recaptured within a sample or if it was captured at some later date. The recovery success of individuals from three successive samples is treated in the same way as the first to see if they repeat the trend of the first sample. In each of these three, the initial sample size has been augmented by a number of animals which were known to exist in the population because they had been previously captured, then recaptured at a later date. Individuals which were captured outside of the defined sampling periods have been included in this analysis.

Three factors affect recapture success. First, the vegetation was structurally complex and in places impenetrable, especially during the September and December collections, making the neonates difficult to find, thus, many marked ones could have easily been overlooked. This problem is partially compensated for by including animals that were known to be alive, but not captured. Second, in the final sampling period, about the date of 2 July 1979, the iguanas were more elusive than before. For example, some animals would drop from their sleeping perches and run with no more than a light shining on them as a stimulus. Some that took flight were known to be marked because they had missing toes, but they were not assignable to a particular prior sample, and thus not included in Table 9.1. The recapture values in the last period must also conservatively represent survival because there is no contribution of individuals known to be alive but not actually captured. Third, movement was not controlled, therefore the survival figures are influenced by migration.

The recovery trends may be effectively compared by graphical means by plotting the natural logarithm of numbers, with the initial sample size adjusted to 1000 individuals, against time (Figure 9.6). The single estimate of survival in the first quarter shows that it is poor with only one fourth of the individuals being recovered. The two estimates in the second quarter do not agree with one another, the first recapture of the second sample being poor (28.7%), reminiscent of first quarter survival, and the second recapture of the first sample being relatively good (62.1% surviving of those known to be alive in the second period). I suspect that the former value may be more strongly influenced by movement (e.g., the relocation of activity range as a result of flight from a predator, or an investigator) than the latter, and considering the structural complexity of the habitat mentioned above, an iguana would not have to move very far to become overlooked. Some animals however appear to be rather sedentary, continually frequenting the same areas and being repeatedly recaptured.

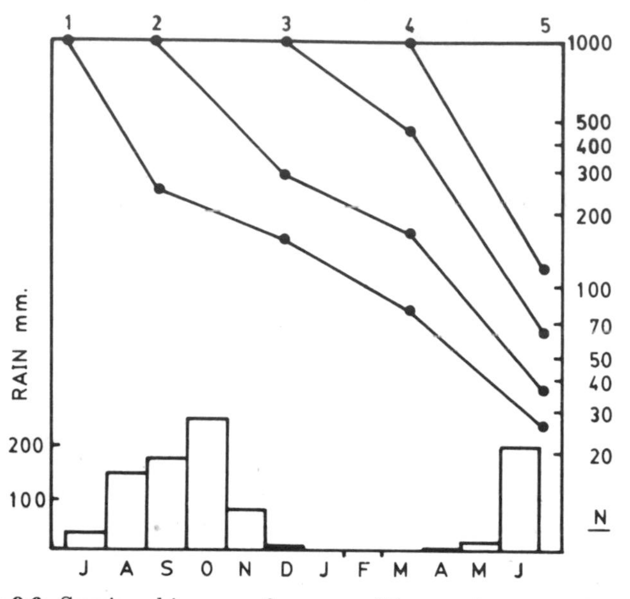

Figure 9.6: Survivorship curves for young-of-the-year iguanas and rainfall at Los Cocos, 1978-1979. Survival data are from Table 9.1, with sample sizes adjusted to 1000. Unpublished rainfall data are courtesy of the Instituto Colombiano de Hidrología, Meteorología, y Adecuación de Tierras (HIMAT).

The three survival estimates of the third quarter are quite consistent with 50%, 58.1%, and 43.6% of the individuals known to be alive at the beginning of the period surviving through it. This consistency was a result of very thorough sampling during the March collection period. At that time most of the low thicket vegetation was defoliating whereupon the iguanas became more concentrated in the thorn scrub vegetation along the roadsides where they could be more effectively sampled.

Recovery success at the end of the fourth quarter of those animals known to be alive at its beginning was relatively poorer than for the third quarter, being 33.3%, 22.2%, and 14.7% for the animals from the first, second, and third samples, respectively, and from the fourth sample only 11.8% were recovered. Immigration is a likely explanation of this, along with the factor of wariness discussed above. Also, a trend exists such that animals most recently marked are least likely to be recaptured, thus, animals long established in the area seem most likely to remain there at a time when the tendency to migrate may be strong because of nutritional or social stress caused by there being a high density of animals congregated in the thorn scrub habitat. The last sample coincided with elevated rainfall and improvement of the vegetation, thus, iguanas may be expected to move away from areas where they are congregated. Most iguanas that breed at Los Cocos do not live there permanently. For this reason also, emigration of the young is expected to occur.

The fourth sample of March 1979 included nearly all of the resident young;

for convenience, there were about 200. If this number is increased according to the survivorship curve of sample one, it may be calculated that the Los Cocos area produced about 2600 hatchlings in 1978.

SUMMARY AND CONCLUSIONS

Green iguanas in Parque Nacional Natural Isla de Salamanca near Barranquilla, Colombia mate during December and possibly later, lay eggs from mid-January to mid-April, and hatch from mid-April to mid-July. Mating and nesting occur in the dry season and hatching occurs in the wet season, a schedule which seems beneficial to the nutrition of the hatchlings and to survival of the nest, by avoiding periods of flooding.

Growth rate is inversely proportional to body size and it is also related to season, the two factors being inseparable because of the high degree of synchrony in reproduction. Juvenile growth rate is high in the first quarter (0.58 mm/day), but lower and reasonably constant through the remainder of the first year at ca. 0.38 mm/day. The juvenile growth rate in the moist environment of western Salamanca is about twice that reported for dryer climates (Henderson, 1974; Müller, 1968). Based on growth rate, females may reach the size at which they are sexually mature in 1.5 years, so it is expected that most of the females lay eggs in their second season.

Three-fourths of the hatchling population disappears in its first quarter year of existence, after which about half of the hatchlings are lost per quarter for the rest of the first year, with no less than 2.6% surviving after one year. Relatively poor recovery success in the final quarter of the first year may be strongly influenced by emigration of the yearlings from the Los Cocos "breeding ground" to surrounding mangrove and fresh water swamp forests where most of the breeding population lives. Based upon survivorship and an estimate of density in March, it is calculated that roughly 2600 iguanas were produced in the vicinity of Los Cocos in 1978.

Note

[1]This very high initial rate of growth was also observed among hatchlings of 1979 whose mean was 0.602 (s.d. = 0.179) mm/day over an average of 26.3 (17-64) days among 26 juveniles averaging 104 mm SVL.

Acknowledgments

I thank the contributors to this symposium, especially Drs. A.S. Rand, B.A. Dugan, and R.W. Van Devender for their comments and suggestions. I also thank Señora Neovis de López for her aid in identifying plants, and Drs. Stephen C. Ayala, Richard D. Worthington, and Friedmann Köster, and my parents for their support. This work was funded and otherwise aided by the United States Peace Corps and El Instituto de Recursos Naturales Renovables y del Ambiente (INDERENA); they are gratefully acknowledged.

10

Growth and Ecology of Spiny-Tailed and Green Iguanas in Costa Rica, with Comments on the Evolution of Herbivory and Large Body Size

Robert Wayne Van Devender
Department of Biology
Appalachian State University
Boone, North Carolina

INTRODUCTION

Green iguanas *(Iguana iguana)* and ctenosaurs *(Ctenosaura similis)* are the largest and most conspicuous lizards in Central America, yet few details of their natural history were available until recently. This paucity of data is only one example of the general lack of such data for tropical lizards (Tinkle 1969; Tinkle, Wilbur and Tilley 1970; Fitch 1970). Few tropical species have been subjects of intensive field work, especially in the Neotropics. A second difficulty limiting data on these species is their rather long life spans. Monitoring cohorts from birth to death requires many years of field work–time intervals not available to most workers. Another complication is that these lizards are not easy field subjects; they are wary, hard to capture and relatively difficult to handle. Nevertheless, recent interest in several fields has stimulated work on many aspects of iguanine lizard biology.

The major aspects leading to the recent popularity of these species are the following: (1) Their size and visibility make them superior subjects for studies of social interactions and the relationship between behavior and evolution (Evans 1951, Carpenter 1979, Dugan this volume, Dugan and Wiewandt this volume). (2) The unusual nesting situation for *Iguana iguana* on Slothia off Barro Colorado Island, Canal Zone, allows a number of unique opportunities for studies of oviposition, movements of females and young, female-female interaction, juvenile interactions and nest predation (Rand 1968, 1969, 1972; Montgomery, Rand and Sundquist 1973; Burghardt et al. 1977; Drummond and Burghardt this volume). (3) The unusual herbivorous habits of these species also attract attention (Lönnberg 1902, Swanson 1950, Hotton 1955, Montanucci 1968, Pough 1973, Rand 1978, Iverson this volume, McBee and McBee this volume). (4) Interest in collecting tropical data to test general ecological hypotheses (e.g., Levins 1968; MacArthur and Wilson 1967; Tinkle, et al. 1970; Pianka 1970, 1972) has generated work on the general ecology and life history of these species (Hirth 1963; Müller 1968, 1972; Fitch 1970, 1973a, 1973b; Henderson 1973, 1974; Fitch and Henderson 1977a, 1977b; Henderson and Fitch 1979; Harris this volume; Rand and Greene this volume). (5) Conservation measures and the role of large lizards as human food reserves are additional stimuli (Fitch and Henderson 1977a, 1977b; Fitch et al. this volume). The holistic evaluations of the life histories of these lizards should be possible in the near future.

One weak link in our present knowledge is the transition from hatching to maturity. I hope that this chapter will help to fill that gap. Here I report the results of a four-year mark-release-recapture study of iguanas and ctenosaurs living in two Costa Rican streamside habitats. Most of the data are for juvenile lizards and allow me to discuss habitat choice, food habits, one form of survivorship, early growth and several other aspects of their ecology.

MATERIALS AND METHODS

Study Areas

The two study sites were in the Rio Corobici drainage, NW of Cañas, Guanacaste Province, Costa Rica. The sites differed in several respects, but the small distance (2 km) between the sites and the similar elevations (40-50 m) limited any climatic differences. Both were in the transition belt between the Tropical Moist and Tropical Dry life zones (Holdridge 1967, Tosi 1969). An official weather station was located on Finca La Pacifica within 0.5 km N of one site. The mean annual rainfall was about 1700 mm, but little or no rain fell between mid-November and early May (Fleming 1974, Glander 1975, 1978, Van Devender 1975, 1978, Instituto Meteorologico Nacional 1974). By the end of the dry season, essentially all non-riparian trees lose their leaves. Temperature varied little through the year with monthly averages between 26° and 31°C. Both sites were subject to periodic flooding during the wet season.

The Rio Corobici site (RC) was a roughly-rectangular plot (200 x 125 m) on Finca La Pacifica, 4.5 km NW of Cañas. It included a section of the river, several large islands and the riparian zone on each bank. The area was bounded by pasture to the south, mango orchard to the north and the Pan American Highway to the west, but there were no real barriers upstream and downstream. Most lizards used the riparian zone but some adult ctenosaurs lived in pastures and some iguanas ventured into the mango orchard. One hectare of the 1.89 ha total area was riparian habitat usable by the lizards. The stream was large and swift with a rocky bed and clear water. The vegetation included 171 trees with an average canopy height of 11.4 m (Glander 1975). In 9.9 ha of the same riparian zone, including most of the site, Glander (1975) reported 96 species of trees. The most common trees were *Anacardium excelsum, Hymenaea courbarill, Luhea candida, Lysiloma seemannii, Andira inermis,* and *Licania arborea*. Important shrubs were *Mimosa pigra, Inga vera* and *Ardisia revoluta*. Grass thickets were present in several parts of the area. Several families lived adjacent to the site and moved through it regularly. The residents captured or killed iguanas on several occasions.

The Rio Sandillal site (RS) was along a small tributary of the Rio Corobici 2.5 km NW Cañas. The plot was long and narrow (40 x 400 m) and included 0.79 ha. Of this area, 0.5 ha was riparian zone habitat. The stream was narrow and shallow with a muddy bottom. The area was surrounded by pasture to the N and S and abutted the Pan American Highway at one end. The canopy was essentially closed along the stream, but there were several open areas inside the plot and the pasture edges were open and sunny. The 330 trees in the area included at least 58 species with the following species dominating: *Luhea candida* (25 trees), *Anacardium excelsum* (21), *Inga vera* (18), *Licania arborea* (18), *Sloanea terniflora* (17) and *Tabebuia ochracea* (13). In December 1973 approximately 30% of the trees were removed to make way for a power line. The one family that lived adjacent to the site probably had little effect on the lizards.

Capture and Marking

Iguanas and ctenosaurs were captured at night and permanently marked by clipping unique combinations of up to three toes. At the time of capture the following data were recorded: snout to vent length (SVL), length of tail, regenerated part of tail, species, sex (when apparent), site, weight, date, reproductive status and location within the site. Most lizards were processed immediately and released at the site of capture. Some larger individuals and clusters of hatchlings were brought into the lab for processing. These were released at the capture site within 24 hours of capture. Field work was divided into monthly intervals, and each site was visited 3-7 times per month. Some data were gathered in each of the following periods: June to August 1971, May to November 1972, June 1973 to May 1974 and May 1975. The 1975 sample was destructive and all lizards were killed and preserved.

Survivorship

Survivorship curves were prepared from the recapture data by determining the number of monthly samples between the first and last capture for each lizard. This was the survival time for that lizard. Survival time and the log of number of lizards surviving *at least* that long were plotted. The resulting curves were compared with those of Deevey (1947). It must be noted that this approach equates emigration with death.

Habitat Choice

The location data for lizards of different ages were used to determine ontogenetic changes and species differences in habitat preference. These observations were reinforced with numerous observations on lizards outside the study areas.

Food Habits and Reproduction

Thirty iguanas and forty-five ctenosaurs, including all 1975 specimens, were sacrificed and dissected. Stomachs were removed for subsequent diet analyses. Food items were categorized as insects, leaves, flowers, fruit (including seeds), or "other." Food items were identified at least to order, and the relative contribution to the stomach content volume of each was recorded. Gonads were removed from males and preserved in 10% formalin. They were later squashed and searched microscopically for mature spermatozoa. Four gravid female iguanas captured in the study areas provided a little data on reproductive seasons. The appearance of hatchlings in the sites in different years also provided data on seasonality of reproduction.

Growth

The growth analysis had two primary parts: to describe the size-to-age relationship in these species by relating weight, SVL and age. Two models were used to compare SVL with age: one that assumed that linear growth conformed to the von Bertalanffy growth curve (Fabens 1965, Van Devender 1978, Dunham 1978, Smith 1977, Schoener and Schoener 1978) as shown in Appendix I equations 1 and 2; one that assumed that growth in length was constant, regardless of SVL, as in Appendix I equations 3 and 4. Both models were compared with data for lizards of known age. The short reproductive season and rapid growth allowed accurate age determination of all lizards in their first two years. Older lizards were called known-age animals only if they had been marked in the first two years. The constant growth model was also used to estimate ages of larger, older lizards. The weight-to-length relationship was treated as an exponential relationship. The exponent was estimated directly by linear regression of natural log transformed SVL and weight data (Appendix I, equation 5) and approximated by cubic relationship (equation 6). The weight-age relationship was determined by substitution of the values for SVL in equations 2 and 4 into equation 6 and by the log-log regression of weight and age. Separate regressions were prepared for each site and compared by means of Analysis of Covariance (ANCOVA). Principal statistical source was Draper and Smith (1966).

Biases

This study contained three important biases: (1) Restriction of the study to the riparian zone certainly influenced the outcome. Some iguanas and many ctenosaurs used the riparian zone for a while then left it. Ctenosaurs usually moved into the surrounding pastures, while iguanas moved into orchards and other stands of trees. Little effort was made to estimate the magnitude of lateral migration; but casual observation suggested that many of the lizards, particularly the ctenosaurs, did leave the study areas. (2) Capturing the lizards on night perches biased the study in favor of juveniles. Adult ctenosaurs were usually seen in tree holes or in burrows, where they were never adequately sampled. Adult iguanas perched high in trees and were loath to move from their perches, so they were often not captured even when located. The overall effect of the first two biases was that the study became one of juvenile lizards, rather than the complete study originally envisioned. (3) In each species juvenile males and females were indistinguishable. This reduced the number of meaningful comparisons of growth performance.

RESULTS

Individuals, Captures and Survivorship

A total of 275 iguanas was captured at the sites and 262 recaptures were made (Table 10.1). Iguanas were more common at the RC, but individuals at the RS were captured more often (Table 10.2). While most lizards did not remain in the population for long ($\overline{X}$ = 1.4 mo.), one iguana was monitored for twenty-two months (Table 10.3). The smallest iguana handled was 70 mm in SVL, and the largest was 453 mm. Almost 90% of all captures (471 of 537) were of lizards less than 170 mm SVL and in their first year (Table 10.4). Despite the facts that many juveniles emigrated and that adults were sampled less than were juveniles, the survivorship or extinction curve for iguanas was interesting (Figure 10.1). Most of the lizards were marked as hatchlings and failed to remain in the populations. The initial rapid drop in numbers of individuals with increasing time resulted partly from movements and partly from mortality.

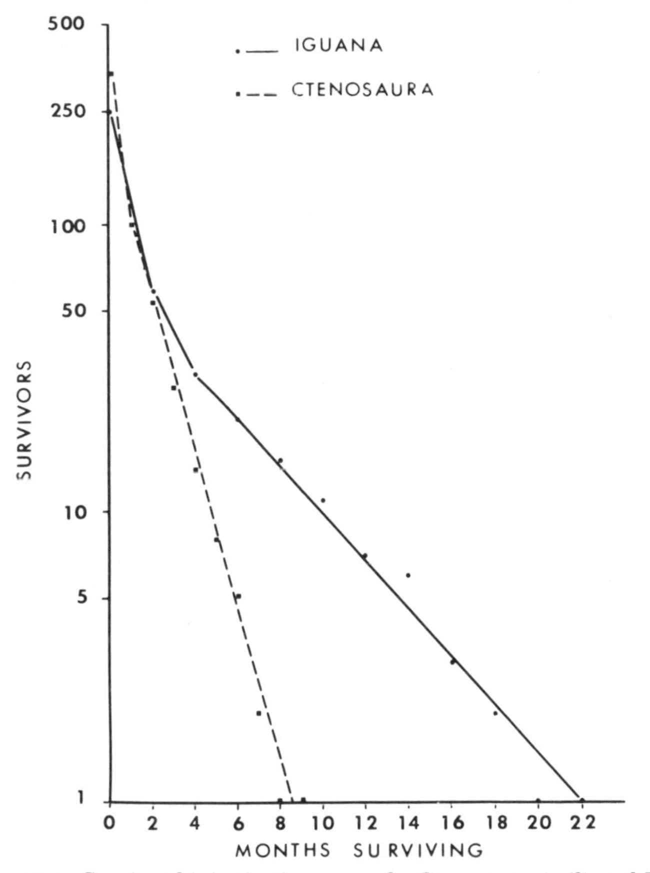

Figure 10.1: Survivorship/extinction curves for *Ctenosaura similis* and *Iguana iguana*. Data from both sites and all years were pooled. Straight lines on the semi-logarithmic plot indicate constant, per capita, loss rate per unit time. Losses were due to mortality and emigration.

After about three months in the population, the loss rate became constant (modified Type II curve of Deevey 1947), as indicated by a linear relationship on a semi-log plot. During the later stage of the survivorship curve, about half the lizards left the population or died every four months. The mere presence of adults at the sites indicated that the mortality rates shown in Figure 10.1 were too high.

Table 10.1: Annual Distribution of Effort, Animals Captured and Recaptures

			. . . . Iguanas		. . . Ctenosaurs . . .	
Site	Year	Months	Lizards	Recaptures	Lizards	Recaptures
RS	1971*	0	0	0	0	0
	1972	5	5	21	48	35
	1973-4	12	14	31	110	80
	1975**	1	0	0	7	0
	Total	18	18	52	165	125
RC	1971*	2	9	14	18	14
	1972	7	42	57	69	72
	1973-4	12	190	139	92	33
	1975**	1	26	0	8	0
	Total	22	257	210	187	119
	Both Total	–	275	262	332	254

*Little emphasis was placed on these species in 1971.
**All 1975 lizards were sacrificed and dissected.

Table 10.2: Capture Frequency for Lizards Marked in the Two Areas. Data for 1975 were Excluded

Captures	. . . . . . Iguana			. . . . .Ctenosaura.		
per Lizard	RS	RC	Both	RS	RC	Both
1	7	132	139	102	111	213
2	4	42	46	26	35	61
3	0	29	29	19	13	32
4	1	13	14	6	13	19
5	0	9	9	0	5	5
6	1	2	3	2	0	2
7	1	3	4	0	0	0
8	1	0	1	1	2	3
9	0	0	0	1	0	1
10	3	0	3	1	0	1
11	0	0	0	0	0	0
12	0	1	1	0	0	0
Total	18	231	249	158	179	337
Average	3.9	1.9	2.1	1.7	1.7	1.7

Table 10.3: Interval Between First and Last Capture for pre-1975 Lizards

	...... Iguana			.... Ctenosaura		
Months	RS	RC	Both	RS	RC	Both
0	18	231	249	109	129	238
1	0	23	23	20	26	46
2	3	14	17	15	11	26
3	4	8	12	5	8	13
4	1	5	6	3	3	6
5	0	3	3	2	1	3
6	0	3	3	2	1	3
7	0	3	3	1	0	1
8	0	2	2	0	0	0
9	0	2	2	1	0	0
10	0	2	2	0	0	0
11	0	2	2	0	0	0
12	0	1	1	0	0	0
14	0	1	1	0	0	0
15	0	2	2	0	0	0
17	0	1	1	0	0	0
18	0	1	1	0	0	0
22	0	1	1	0	0	0
Mean			1.45			1.18

Table 10.4: Size Distribution of all Captures

Size interval corresponding to the beginning of the second year indicated by*. Numbers in parentheses are animals dissected.

	...... Iguana		..... Ctenosaura	
SVL Interval	Frequency	Cumulative	Frequency	Cumulative
50-69	0	0	33 (1)	33 (1)
70-89	136 (19)	136 (19)	137 (5)	170 (6)
90-109	138 (5)	274 (24)	236 (14)	406 (20)
110-129	129 (3)	403 (27)	117 (7)	523 (27)
130-149	41	444 (27)	36 (4)	559 (31)
150-169	27 (1)	471 (28)	28 (7)	587 (38)
170-189*	15 (1)*	486 (29)	16 (3)*	603 (41)
190-209	9 (1)	495 (30)	6 (1)	609 (42)
210-229	8	503 (30)	1	610 (42)
230-249	11	514 (30)	2 (1)	612 (43)
250-269	3	517 (30)	1 (1)	613 (44)
270-289	2	519 (30)	3 (1)	616 (45)
290-309	8	527 (30)		
310-329	5	532 (30)		
330-349	1	533 (30)		
350 & up	4	537 (30)		
Maximum	453 mm		282 mm	

A total of 332 ctenosaurs were marked and 254 recaptures made (Table 10.1). Equivalent numbers of lizards were handled at each site despite the smaller area of the RS site (0.5 ha vs 1.0 ha). Recapture frequencies were rather low at both sites (Table 10.2) with an average of only 0.7 recaptures per lizard marked. The average interval spent in the areas was only 1.18 months (Table 10.3). The smallest ctenosaur handled was 54 mm in SVL while the largest was 282 mm. Over 95% of all captures (587 of 616) were of first year lizards less than 175 mm SVL (Table 10.4). The survivorship or extinction curve for ctenosaurs (Figure 10.1) was more nearly a Type II curve with a continuous loss of about half the lizards per month. Many of the losses were probably due to emigration and changes in habitat choice (see below). No ctenosaur was monitored for more than nine months.

Reproduction and Hatchling Emergence

Only four gravid iguanas were captured: two in January and two in February. Other observations outside the study sites indicated a short, mid-dry season reproductive bout as described by Fitch (1973a) for other areas in Guanacaste. Oviposition probably peaked in January and February.

The first hatchlings of each species appeared in April (Table 10.5), and very small individuals were found until June for ctenosaurs and until August for iguanas. For both species, essentially all hatchlings emerged between April and June, even with annual differences in time of emergence included. The smallest lizards marked were 54 mm and 70 mm in SVL for ctenosaurs and iguanas, respectively, but most hatchlings were considerably larger than these minima. Some iguanas up to 90 mm SVL had open umbilical scars, indicating recent hatching. The largest ctenosaur with an open umbilicus was 70 mm SVL. Hatchlings of the two species differed in behavior. Young ctenosaurs chose night perches on low branches and were rarely clumped. Iguanas also used low branches, but they were often in groups. It was commonplace to find 10-20 young iguanas in a space of only several m^2. Lizards were often in physical contact. Aggregations were most common at the time of emergence and were located in grass thickets. Most hatchling ctenosaurs were brilliant green in color, but two specimens that emerged very early, before the onset of the wet season, were the mottled brown color mentioned by Fitch (1973a).

Habitat Choice

Iguanas were primarily arboreal and strongly heliothermic. Their habitat requirements included trees and basking sites, which were more common in the riparian zone than outside it. Young iguanas usually spent the day in one of four situations: basking on the ground, in grass thickets, in *Mimosa pigra* thickets and in clumps of *Lindinia rivularis*, a common semiaquatic plant in both areas. Iguanas were usually well camouflaged and hard to see in any of these habitats. In the evening juvenile iguanas climbed into the available vegetation, which was usually less than 2-3 meters high. Adult iguanas spent most of their time in trees. Favorite areas were trees with thick foliage and direct sun exposure. Liana tangles were also used extensively. Night and day perches were essentially the same in adults. Adults were occasionally observed basking on the ground, especially in areas little affected by humans.

Table 10.5: Temporal Occurrence of Hatchlings and Small Lizards

Ctenosaura							
	Month	Size Interval in mm SVL					
Month	Rank	50-59	60-69	70-79	80-89	90-99	Total
April	4	2	0	0	0	0	2
May	5	8	13	15	0	0	36
June	6	1	7	16	22	3	49
July	7	0	0	8	29	25	62
August	8	0	0	1	21	48	70
September	9	0	0	0	1	24	25
October	10	0	0	0	0	4	4
Total	—	11	20	40	73	104	248
Mean rank	—	4.9	5.4	5.9	7.0	8.0	—
Mean month		April	May	May	July	August	—

Iguana							
Month	Rank	70-79	80-89	90-99	100-109	110-119	Total
April	4	23	39	0	0	0	62
May	5	1	49	18	2	1	71
June	6	2	13	12	3	1	31
July	7	0	4	13	33	19	69
August	8	0	1	2	8	15	26
September	9	0	0	2	10	15	27
October	10	0	0	1	0	4	5
November	11	0	0	0	0	1	1
Total	—	26	106	48	56	56	292
Mean rank	—	4.2	4.9	6.2	7.4	8.0	—
Mean month	—	April	April	June	July	August	—

The Table contains the number of lizards in each size category captured throughout the study. Since both hatching and growth are involved the months are ranked and average rank (month) for each size interval calculated.

Ctenosaurs also chose different areas as they grew older. Juveniles were usually bright green and spent considerable time in the understory of the riparian zone, most commonly on the ground in direct sunshine. Night perches were similar to those of young iguanas, low vegetation. Adults were dark brown, reddish or black and less restricted to riparian habitat. The most important element in their habitat was a retreat or refuge such as a burrow, hollow tree or rock pile. Appropriate refuges were not common, and each was usually occupied by a ctenosaur. Such retreats were widely distributed within and out of the riparian zone, as were adult ctenosaurs. Larger ctenosaurs used the retreats to escape predators and at night. Finding a safe retreat was probably the most important factor determining whether a lizard would survive.

Food Habits

Stomach contents of 20 iguanas and 19 ctenosaurs were analyzed for food (Table 10.6). The smallest ctenosaurs fed primarily on insects while larger individuals took progressively more plant matter. One interesting aspect of the herbivory in ctenosaurs was that leaves were not the primary food of any size class. Observations outside the study areas indicated that ctenosaurs were opportunistic feeders. They grazed on lawns, fed in fruit trees and even stole birds from mist nets.

Table 10.6: Food Habits of Juvenile *Ctenosaura similis* and *Iguana iguana*

Tabulated values are mean percent of stomach volume in each food category.

Species	SVL	n	Fruit	Leaves	Flowers	Insects	Others
C. similis	50-99	4	5	7.5	4	82.5	1*
	100-149	6	7	41	8	44	0
	150-199	8	40	47.5	6	4	2.5**
	235	1	73	25	0	2	–
Total		19					
I. iguana	79-99	15***	9	85	33	0	0
	100-203	5	15	85	0	0	0
Total		20					

*One millipede.
**One juvenile iguana and one bird.
***Two had empty stomachs.

Iguanas in Guanacaste were not insectivorous as juveniles. None of the stomachs contained any insect parts. Even hatchlings with some yolk left in the intestine fed on vegetable matter. Undoubtedly, these lizards would feed on readily-available insects and they will often even eat small mice in captivity, but animal matter was not utilized to any appreciable degree. The smallest iguanas fed heavily on flowers, which were not major food items of larger lizards. Iguanas of all sizes fed primarily on leaves.

Size and Age at Maturity

Relatively little data on attainment of maturity were gathered. However, capture and dissection of four yearling male ctenosaurs (114 to 163 mm SVL) in January and March 1974 did indicate that these animals were sexually mature at an age of less than one year. They all had enlarged testes and convoluted vasa deferentia and the two examined microscopically had mature spermatazoa. Under favorable conditions, these yearling males could presumably have fertilized females. Size and growth data are consistent with maturity in the second full year for both ctenosaurs and iguanas as suggested earlier (Fitch 1973a, Fitch and Henderson 1977a, 1977b).

Growth

There were 104 recaptures of marked ctenosaurs at liberty between 25 and 400 days. Growth rates (GR) ranged between −0.1 and 0.8 mm SVL per day with an average of 0.372. Regression of GR and mean SVL for the growth interval revealed a significant decrease in GR with increasing SVL (Figure 10.2), but the decrease accounted for only four percent of the variance in GR. This regression, a form of the von Bertalanffy growth model, was used to estimate SVL at all ages. Linear regression of known age and SVL data was another estimate of the SVL–age relationship. Both models were compared with the known age data (Figure 10.3). The linear model fit the data much better, but the von Bertalanffy model could have been improved slightly by using date of oviposition as day 1 rather than date of hatching. SVL at one year was 145-185 mm, which was consistent with both models and with constant growth at average GR. The linear model gave a better average SVL estimate for lizards one year old. ANCOVA revealed no significant differences in growth between the sites.

There were 119 recaptures of iguanas at risk in the populations for 25-400 days with GRs between −0.1 and 0.6 mm SVL per day and an average of 0.273 mm/day. Regression of GR and mean SVL again detected a significant negative correlation, but r^2 was only 0.04 (Figure 10.4). This von Bertalanffy curve and the linear regression of known age and SVL were again used to describe the age-SVL relationship (Figure 10.5). As with ctenosaurs, the von Bertalanffy model was a poor fit which underestimates SVL for older lizards.

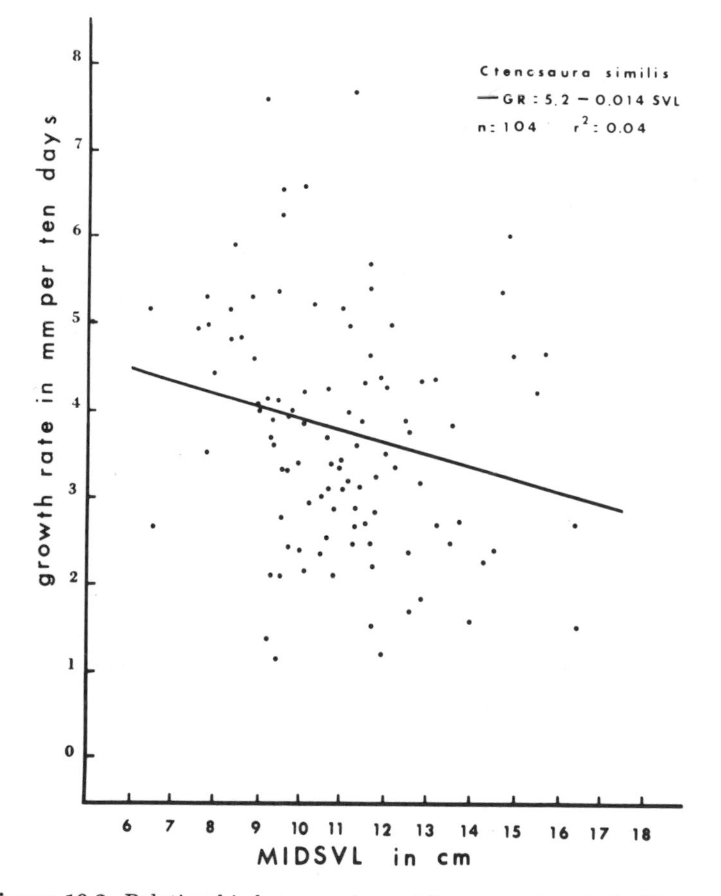

Figure 10.2: Relationship between size and linear growth rate in *Ctenosaura similis*. Each point represents a lizard at liberty 25-400 days between captures.

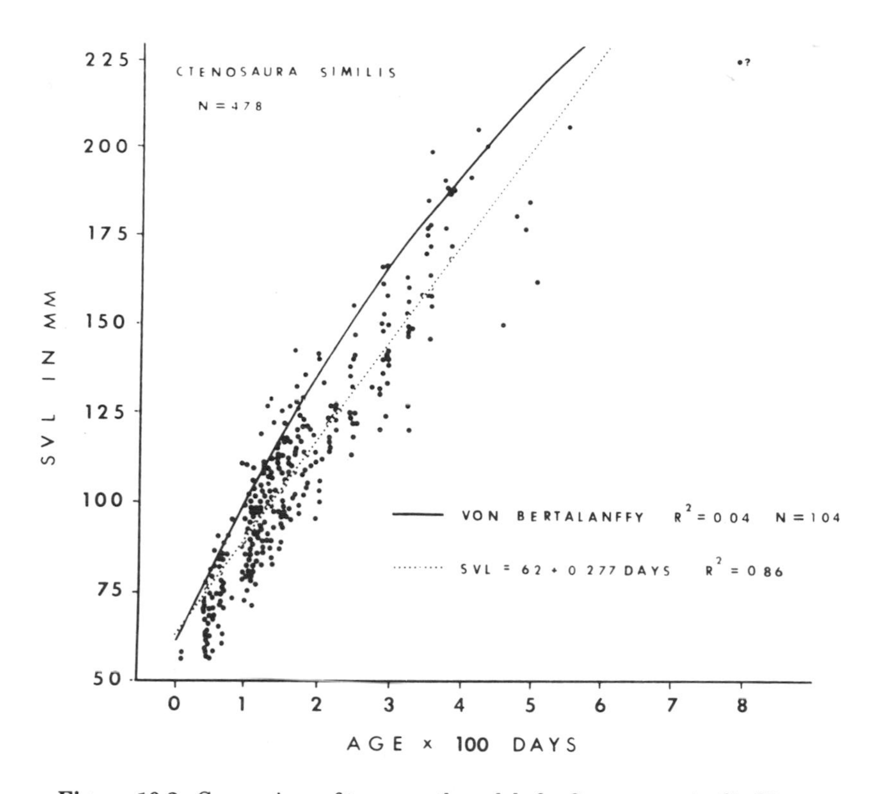

Figure 10.3: Comparison of two growth models for *Ctenosaura similis*. Linear regression of age estimates and SVL (N = 478) are compared with the integral form of the von Bertalanffy growth model for this species. The von Bertalanffy model would conform with the observed data better if an age of 90 days were assumed for hatchlings. Mean growth rate (0.372 mm SVL per day) was somewhat higher than the slope of the linear regression.

Examination of long-term growth trajectories (Figure 10.6) revealed another factor affecting GR, tail autotomy. Most of the low GRs for largest iguanas were associated with three lizards which had recently lost part of their tails. Iguana 1-16, a female, was marked on 12 XI 1972, when she had SVL = 239 mm and Tail = 627 mm. On 12 VIII 1973 she was 314 mm in SVL and had lost all but 338 mm of her tail. Three captures in the next six months showed no growth in SVL or in the basal segment of the tail, but 211 mm of tail were regenerated. This lizard also produced a clutch of eggs while replacing the lost tail. Iguana 2-18, a male of 252 mm SVL, showed minimal growth in SVL during three months when it lost most of its tail in two different autotomic events. A third lizard, Iguana 2-11, a female of 295 mm SVL, had regenerated most of a small loss when first captured. Even though she appeared to be in good condition, she exhibited minimal growth in either SVL or tail length in six months. If the large, autotomized lizards are removed, the von Bertalanffy regression is not significant at the p = 0.05 level.

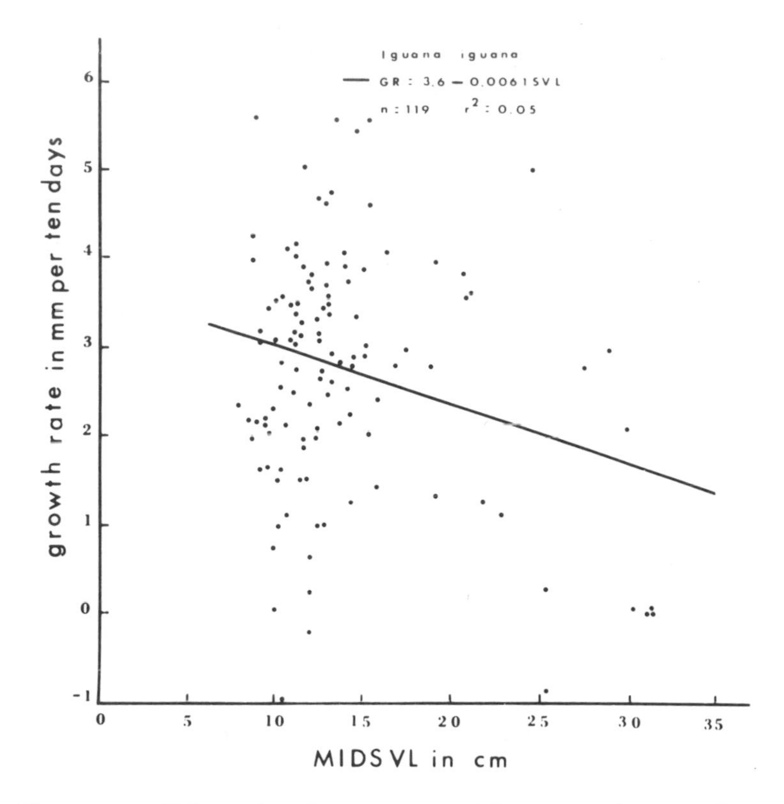

Figure 10.4: Relationships between size and linear growth rate in *Iguana iguana*. Each point represents a lizard at liberty 25-400 days between captures. The four points to the lower right are for lizards regenerating tails and may not be typical. Their removal renders the regression not significant.

ANCOVA of SVL and age data for iguanas less than two years old demonstrated highly significant differences in slope at the two sites. Regression equations were:

$$\text{RS SVL} = 28.8 + 0.379 \text{ mm/day (Age) } n = 33 \; r^2 = 0.87$$

$$\text{RC SVL} = 38.6 + 0.310 \text{ mm/day (Age) } n = 335 \; r^2 = 0.88$$

Iguanas grew more rapidly at the RS, where they were relatively uncommon.

As expected, weight was an exponential function of SVL in both species. Log-log regressions estimated the "true" exponents as 3.07 for ctenosaurs and 3.15 for iguanas (Table 10.7). These values were close to 3.0 or a cubic relationship, and cubic functions were found to be very good estimators of weight. Substitution for SVL in the linear growth model indicated that both species were adding mass in an exponential manner for at least the first two years. Log-log regressions of age and weight had slopes of 2.15 for ctenosaurs and 2.00 for iguanas. Weight increased with age squared.

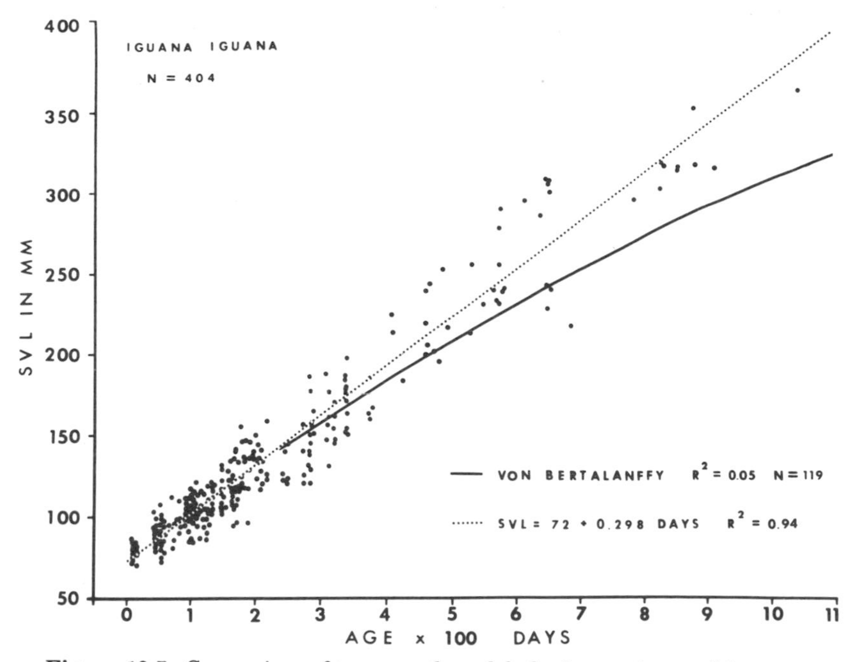

Figure 10.5: Comparison of two growth models for *Iguana iguana*. Linear regression of age estimates and SVL and the von Bertalanffy model are presented. The poor fit of the von Bertalanffy model may be due to the inclusion of data for animals regenerating tails. Slope of the linear regression is somewhat higher than the 0.274 mm SVL per day mean observed.

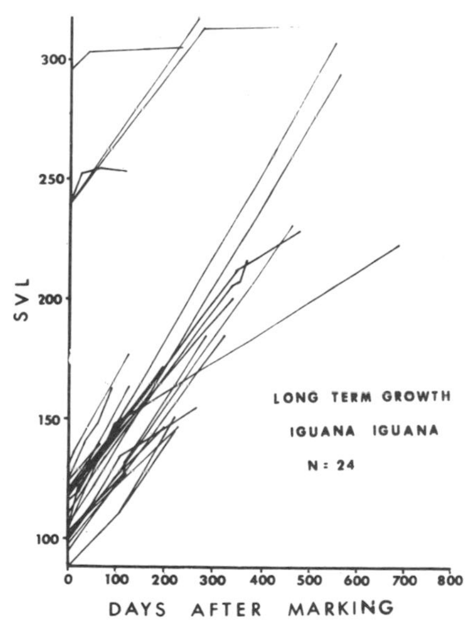

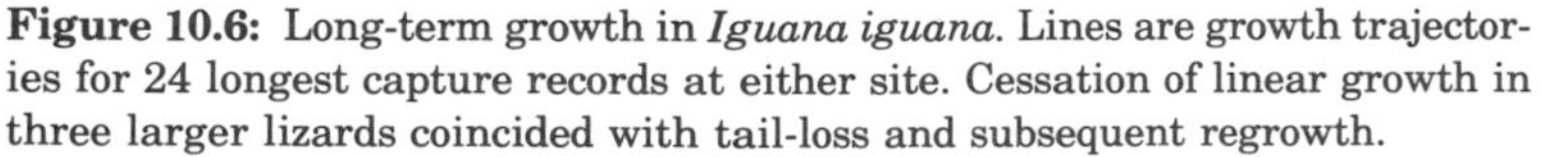
Figure 10.6: Long-term growth in *Iguana iguana*. Lines are growth trajectories for 24 longest capture records at either site. Cessation of linear growth in three larger lizards coincided with tail-loss and subsequent regrowth.

Table 10.7: Regression Data for Analyses of Weight*

Y	X	Species	a	b	n	r^2	Shape
Weight	SVL	*C. similis*	-144	1.796	402	0.80	Curved
		I. iguana	-390	4.135	441	0.60	Curved
ln(Wt)	ln(SVL)	*C. similis*	-10.7	3.07	402	0.983	Straight
		I. iguana	-11.0	3.15	441	0.972	Straight
Weight	SVL^3	*C. similis*	-1.00	0.000032	402	0.989	Straight
		I. iguana	-13.33	0.000040	441	0.962	Straight
Weight	Age	*C. similis*	-86.2	0.514	402	0.804	Curved
		I. iguana	-208.9	1.169	441	0.471	Curved
ln(Wt)	ln(Age)	*C. similis*	-8.32	2.15	402	0.806	–
		I. iguana	-7.06	2.00	441	0.838	–

*Regressions are of the general form: Y = a + bX, where Y is the dependent variable and X the independent. Shape refers to the overall appearance of the scatter plot. These regressions are valid only if the relationship between variables is a straight line.

Predators

Several species were observed to prey on iguanas and ctenosaurs (Table 10.8), but each species had one major, non-human predator in the riparian habitat. Lyre snakes, *Trimorphodon biscutatus,* were common in the habitat and seemed to specialize on eating ctenosaurs. At least ten separate snakes ate Ctenosaurs, while only one ate a green iguana. This species of snake achieves a length of over 2 m and can readily consume ctenosaurs in excess of 250 mm SVL. The major predator on juvenile green iguanas is the common basilisk, *Basiliscus basiliscus.* These lizards were common in the sites, and essentially every adult fed on hatchling iguanas when available. The iguanas are most susceptible to the basilisks in their first two to three months and soon become too large for most basilisks.

Table 10.8: Known Predators on *Ctenosaura similis* and *Iguana iguana.* Important Predators are Indicated with an Asterisk.

Class	Species	Ctenosaura	Iguana
Reptilia	*Basiliscus basiliscus*	+	+*
	Ctenosaura similis		+
	Oxybelis aeneus	+	
	Leptodeira annulata	+	
	Trimorphodon biscutatus	+*	+
	Loxocemus bicolor	+	
Mammalia	*Homo sapiens*	+*	+*
	Canis familiaris	+	+
	Philander opossum		+
Aves	*Quiscalus sp.*		+

Discussion

The riparian or gallery forest is a particularly important habitat in Guanacaste. Local land clearing practice leaves these the only relatively natural habitats in much of the province. The area has a prolonged dry season of 4-6 months with practically no rainfall and an almost constant, strong, dry wind from the Continental Divide. These factors produce severe drought and leave the streamsides as the only islands of green, succulant vegetation for most of the dry season. Riparian sites are the only ones favorable for herbivores throughout the year (Glander 1978) and they serve as short-term refugia for insects during the dry season (Janzen 1973a, b). It is, therefore, not surprising that gallery forests harbor large populations of ctenosaurs and iguanas, at least seasonally.

Guanacaste ctenosaurs apparently have one short reproductive season in mid-dry season, December to February (Fitch 1973a, b, Henderson 1973, Fitch and Henderson 1977a), somewhat earlier than Henderson (1973) reported for Belize. Oviposition and hatching take place in burrows prepared by the females, usually away from the riparian zone (Hackforth-Jones, personal communication). Hatchlings weigh about 5 g and emerge between April and June, a period bracketing the onset of the rainy season. Early hatchlings of the year are probably attracted to the riparian zone by the verdant vegetation and local insect abundance. Hatchlings move often and seem not to form strong site attachments. With the onset of the wet season, surrounding habitats become more favorable and many hatchlings leave the riparian zone while some new individuals move into the habitat. In their first year ctenosaurs grow from about 55 mm SVL and 5 g to 145-185 mm SVL and 100-200 g. They change from insectivory to herbivory and from green bodies to brown. By the end of the first year most ctenosaurs either leave the riparian zone or find "safe" refuges such as burrows or hollow trees. Females probably reach minimum reproductive size by their second dry season at an age of 20 months.

Iguanas use the riparian zone throughout life. Reproduction is concentrated in mid-dry season, and oviposition probably occurs in the riparian zone (Fitch 1973a). Hatchlings emerge between April and June and begin almost immediately to feed on vegetation–leaves, flowers and fruit. They have much greater site fidelity than do ctenosaurs and may remain in essentially the same place for over a year. In their first year iguanas grow from about 70 mm SVL and 12 g to between 160 and 190 mm SVL and 150-250 g. Females should reach minimum reproductive size when they are about 20 months old, but some probably do not reproduce until their third year.

Young ctenosaurs and iguanas have rather anomalous growth patterns. In most lizards smaller individuals grow more rapidly than do larger, older ones, and adults grow relatively slowly (Van Devender 1978, Dunham 1978, Schoener and Schoener 1978, Andrews 1976, Smith 1976). Over the age span studied here, the first two years, the tendency for older, larger lizards to grow more slowly is slight in ctenosaurs and insignificant in iguanas. For at least the first two years these lizards grow at essentially constant rates of 0.362 mm/day for ctenosaurs and 0.273 mm/day for iguanas. The ctenosaur GR is similar to that reported by Fitch (1973a), but iguana GRs differ from those reported by others. Müller (1968) and Henderson (1974) reported slower growth for iguanas in western Colombia and Belize, respectively; and Harris (present volume) recorded more rapid growth in another Colombia site. At present, the growth differences do not seem to correlate with any obvious habitat differences like rainfall or habitat disturbance.

Linear GRs for small ctenosaurs and iguanas are lower than initial GRs for the sympatric, carnivorous basilisk, *Basiliscus basiliscus* (Van Devender 1978), and higher than GRs reported for other, primarily herbivorous iguanids. The general pattern seen for the two iguana species is one of extended rapid growth, resulting in large size. This pattern is similar to that reported for *Cyclura carinata* (Iverson 1977), *Dipsosaurus dorsalis* (Mayhew 1971, Parker 1972) and *Sauromalus obesus* (Johnson 1965, Nagy 1973, Berry 1974). Perhaps this growth pattern will be found to be general for the larger, herbivorous lizards and different from that seen in carnivorous or insectivorous ones.

The constant, or slowly declining, linear growth in the ctenosaurs and iguanas results in exponential growth in mass. This somewhat surprising conclusion is borne out by the relationship between weight and age where weight is essentially a cubic function of age. This logarithmic growth in mass is particularly interesting in herbivorous lizards since plant matter is thought to be a relatively poor food (e.g., Pough 1973, Iverson this volume, McBee and McBee this volume). These results suggest a new and somewhat different interpretation of the evolution of herbivory in these species.

The most recent analysis of herbivory in lizards is that of Pough (1973). He observed the correlation between large body size and degree of herbivory in several lizard families and suggested that herbivory was causally related to underlying physiological processes. The main points in his argument were the following:

A. Most species of lizards occasionally eat plant matter, but very few feed predominantly on plants.

B. Plant parts provide fewer calories and less protein per gram than do animals.

C. Smaller lizard species and, often, smaller individuals of larger species are primarily carnivorous.

D. The largest species in several families are all herbivorous.

E. Small lizards have low energy needs, but high weight-specific needs; thus they utilize "better" food items.

F. Larger lizards require more total calories but fewer calories per gram of body weight.

G. The energetic cost of chasing animal food increases dramatically with lizard size, but the cost of obtaining plant food remains essentially constant.

His conclusion from these arguments was that large body size allows lizards to take advantage of the lower weight-specific metabolic rate and the lower cost of foraging to become herbivorous. His final statement was "...evolution of a large lizard both requires and permits a switch from carnivory to herbivory."

Pough (1973) argues further that only ecological or physiological specializations allow some large lizards to retain carnivory and some small species to be herbivorous. It seems just as reasonable to look for specializations that allow herbivory. Such specializations are now apparent in iguanine lizards (Lönnberg 1902, Iverson this volume, McBee and McBee this volume). They have greatly enlarged gastro-intestinal tracts with numerous modifications associated with fermentating vegetable matter. These adaptations presumably allow iguanines to process the large volume of plant matter necessary for maintenance and rapid growth. The low energetic cost of obtaining plant food also favors herbivory. With these modifications, even small iguanines can be herbivores, as witnessed by the iguanas in this study, *Dipsosaurus dorsalis* (Norris 1953, Mayhew 1971, Parker 1972), *Sauromalus obesus* juveniles (Johnson 1965, Nagy 1973, Berry 1974) and *Ctenosaura clarki* (Duellman and Duellman 1959).

Pough's hypothesis is certainly an interesting possibility; but I think it was premature and, at least in this case, essentially backwards. He was forced to make several assumptions about lizards before performing the analysis, including four that now seem unwarranted. He did not consider physiological adaptations increasing efficiency of utilizing plant matter (see Lönnberg 1902 and chapters by Iverson and McBee and McBee for refutation of this position). He oversimplified matters by pooling all lizards that fed on *any* part of a plant as "herbivores," even though the energy content and availability of fruits and leaves are certainly different. He placed unwarranted weight on the supposed ontogenetic shift from carnivory to herbivory in several species. In *Iguana iguana,* the sum total of the data for ontogenetic shift was a report of one juvenile eating one grasshopper (Hirth 1963). In Guanacaste even hatchlings fed exclusively on plants. Pough minimized the importance of small herbivores and large carnivores in other families. The result was a less than general model for the evolution of herbivory. Comparison of the behavior, growth and food habits of the three large iguanids in Guanacaste provides the basis for a different hypothesis about the origin of both herbivory and large body size.

The three large lizards are the common basilisk, *Basiliscus basiliscus,* which reaches about 600 g; the ctenosaur, *Ctenosaura similis,* which reaches over 2 kg; and the green iguana, which reaches over 4 kg. The basilisk is the most active species, and the most carnivorous. Juveniles are insectivorous, but adults also eat vertebrates, flowers and fruit. Feeding forays and social interactions result in rather long movements, even for small lizards. It is not uncommon for a 10 g lizard to move as much as 20 m across a stream to catch an insect or chase another basilisk. Ctenosaurs are also rather active, especially as juveniles. Hatchlings are insectivorous and chase food items as much as several meters. When disturbed, they will run across open areas of up to 20 m before retreating into thickets. Older ctenosaurs are more herbivorous and less active. They have specific retreats and move out from them to forage. Usually they have to run only a few meters to reach the retreats if danger threatens. Behaviorally, iguanas are quite different from the other two species. They are completely herbivorous throughout life and rather sedentary. Some hatchlings moved less than 20-30 m in their first two months of life, but a more usual pattern is to restrict their activities to a small area and to change such areas from time to time. Some of these movements were as large as 200 m. Hatchlings are very deliberate in their movements and spend most of their time in green vegetation, where they can escape predators passively by camouflage and immobility.

A comparison of first year growth of these species is presented in Table 10.9.

Basilisks have the smallest absolute growth and the lowest rate of increase. Iguanas have the highest absolute growth in the period but add mass at about the same relative rate as the basilisks. Ctenosaurs start smaller than iguanas but add mass more rapidly. It is curious that the species specializing on the "better" food has the slowest growth. Perhaps the energetic cost of catching the insects is indeed relatively high as suggested by Pough (1973) and Rand (1978).

Table 10.9: Food Habits and Growth of Three Large Iguanid Lizards at the Same Sites in Guanacaste Province, Costa Rica

Weights are in grams and weight at one year of age is near maximal.

Species	Diet	Hatchling Weight	Yearling Weight	. . Increase . . g	%
Basiliscus basiliscus	Carnivore	2	40	38	1900
Ctenosaura similis	Omnivore	5	200	195	3900
Iguana iguana	Herbivore	12	250	238	1980

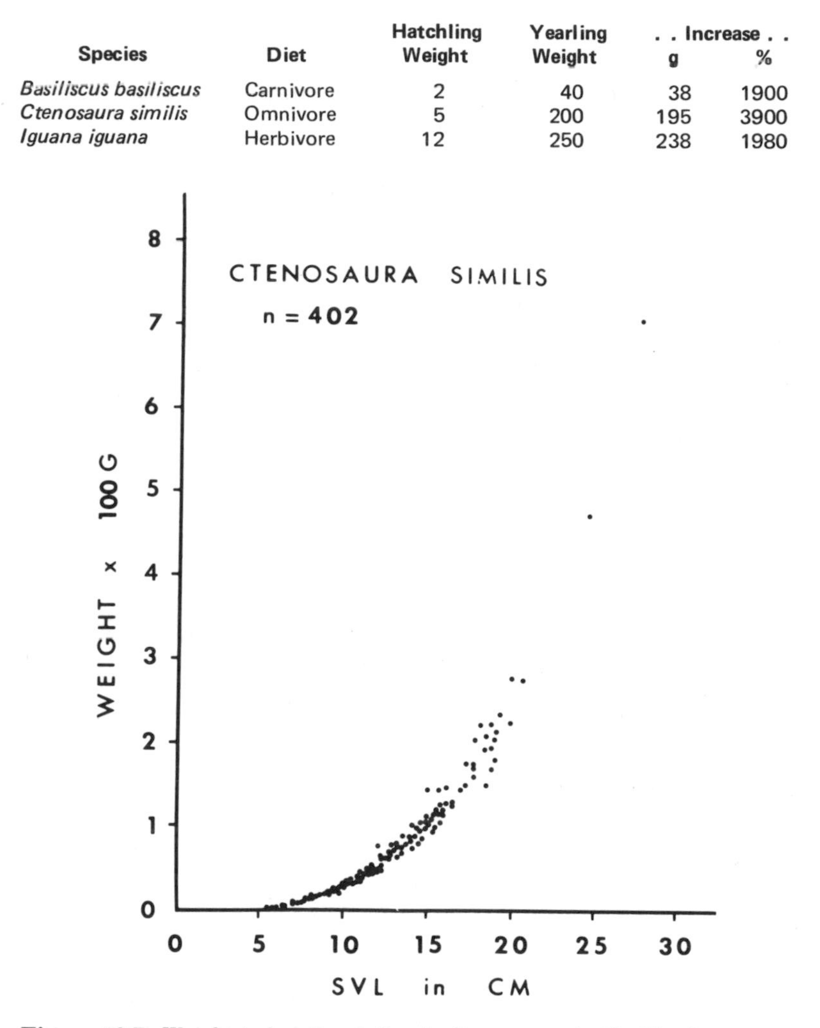

Figure 10.7: Weight-to-length relation in *Ctenosaura similis*. The data agree well with the cubic equation:

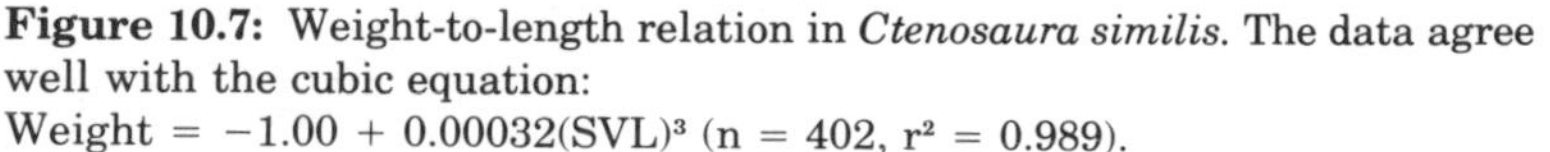

Weight $= -1.00 + 0.00032(\text{SVL})^3$ ($n = 402$, $r^2 = 0.989$).

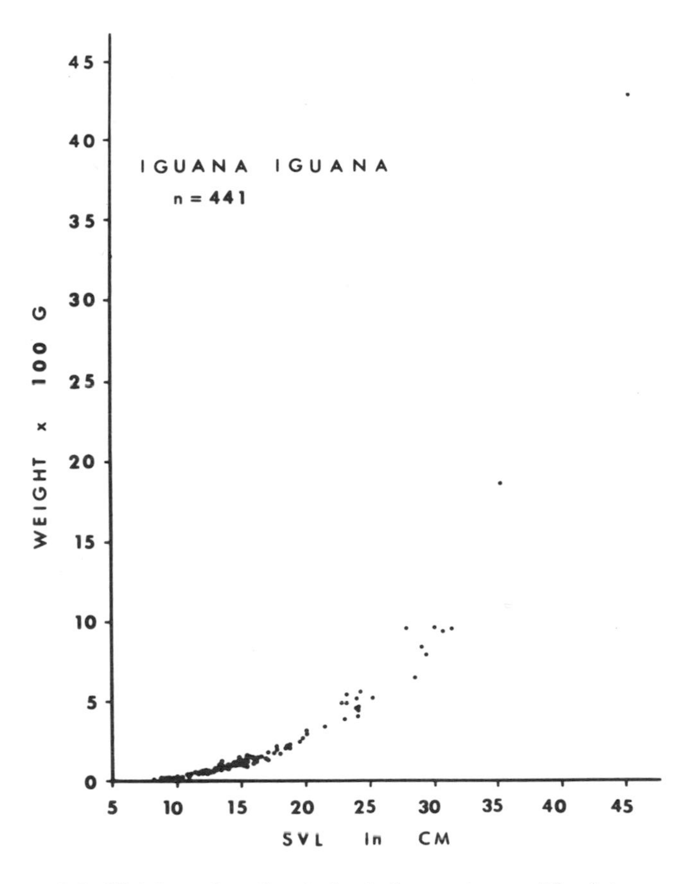

Figure 10.8: Weight-to-length relation in *Iguana iguana*. The data agree well with the cubic equation:
Weight = 13.33 ± 0.000040(SVL)^3 (n = 441, r^2 = 0.962).

These results relate to Pough's model in the following manner. Weight-specific metabolic rates for lizards of 10 g, 100 g and 1000 g are 0.28 ml O_2/g/hr, 0.19 and 0.12, respectively (Bennet and Dawson 1976). To maintain comparable levels of activity 10 g lizards must process 1.5 times as much food per gram of body weight as does a 100 g lizard and 2.3 times as much as a 1000 g lizard, assuming all use similar foods. Since plant foods provide less energy per gram, small herbivores must process more food per time unit than do carnivores. If the food flux rate is not sufficient to supply the necessary energy on plant foods, the small lizard must shift to higher quality animal foods.

The observation that iguanas never change diets is particularly significant. If the smallest lizard is capable of processing "enough" plant matter, the whole problem takes on a different aspect. Rather than the 10 g lizard "needing" 2.3 times as many calories per gram of body weight as a 1000 g lizard, it is the 1000 g lizard that "needs" only 43% as much food per gram as the 10 g lizard. In other words, what works for the 10 g lizard works consistently better as it gets larger. The difference in weight-specific metabolic rate may provide the energetic advantage that allows the herbivores to have exponential growth over such a large size and age range. The interesting problem becomes how does the small lizard manage herbivory.

This hypothesis is also consistent with a different scenario for the evolution of herbivory and large body size from Pough's. It seems most reasonable to suppose that the adaptations allowing herbivory are the original ones and that they occur in relatively small lizards. As soon as the lizard is capable of utilizing plant foods, it ceases to be energy limited in the sense that the carnivores are (i.e., they do not have to keep increasing the energetic costs of obtaining food and have more energy available for growth). Since the problems of dealing with the "poor" plant matter decrease relatively with larger size, selection could easily favor an increase in size, but only after the dietary change. Thus, herbivory does allow lizards to become larger than they could by remaining carnivores; but it is only one of several factors that can have this effect. Many different ecological "breakthroughs" could have the evolutionary "effect" of large body size (e.g., higher metabolic rates in varanids).

Growth in ctenosaurs is also consistent with this hypothesis, with one addition. If the herbivorous species can be an efficient predator when small, early growth can be accelerated by feeding on higher energy content foods. Ctenosaurs do change diets with size and age and seemingly profit from the change. They have higher relative growth rates than either of the species with a specialized diet.

SUMMARY AND CONCLUSIONS

Green *(iguana iguana)* and spiny-tailed *(Ctenosaura similis)* iguanas were subjects of a mark-release-recapture study between 1971 and 1975 at two sites in Guanacaste Province, Costa Rica. The 262 recaptures of 275 marked iguanas and 254 recaptures of 332 ctenosaurs provided information on reproduction, behavior, survivorship, food habits and growth during the first two years of life. Both species have reproductive periods in mid dry season, resulting in hatching near onset of the rainy season.

Iguanas hatch at about 12 g and 75 mm in SVL. They remain in the riparian zone, feed on vegetation, and grow rapidly. Most females probably mature by the end of their second year, but few hatchlings live this long. Linear growth is essentially constant at 0.273 mm in SVL/day for the first two years, while weight increases as the square of age.

Hatchling ctenosaurs are about 5 g and 60 mm in SVL. They utilize the riparian zone mainly in their first year. They are insectivorous initially but gradually change to herbivory. Sexual maturity is reached by the second or third dry season (1½-2½ years old). Growth is rapid with linear GR of 0.362 mm in SVL/day for the first two years. GR declines slightly with increasing size and age. Weight increases as the square of age.

Comparison of growth, behavior and food habits of these species and *Basiliscus basiliscus* suggests a novel scenario for the evolution of herbivory and large body size. Adaptations allowing herbivory permit subsequent developments of large body size.

Acknowledgments

I am pleased to have this opportunity to thank the people and institutions that made this work possible. Financial support for various stages of the research was provided by an N.D.E.A. Title IV Fellowship, Horace H. Rackham Fellowship and Edward C. Walker Scholarship, all at the University of Michigan, by timely employment by Dr. Jay M. Savage of the University of Southern California and by a Faculty Research Grant at Appalachian State University. Computer time was supplied by the Department of Ecology and Evolutionary Biology at the University of Michigan and by Appalachian State University. Logistic support during the early parts of the work was provided by the Organization for Tropical Studies, Mr. and Mrs. Werner Hagnauer, the Ministerio de Agricultura y Ganaderia de Costa Rica and Dr. Douglas Robinson of the Universidad de Costa Rica. Ms. Deborah Holle of Oklahoma State University assisted in analyses of stomach contents. Dr. Roy W. McDiarmid, Dr. Stanley Rand, Dr. Henry S. Fitch and my wife, Amy, provided intellectual stimuli and encouragement. My wife, Mr. Dennis Harris and the editors reviewed the manuscript and made many beneficial comments.

Appendix I. Equations used in growth analyses.

A. The von Bertalanffy Model:

eq. 1 $GR = a + b\overline{SVL}$

eq. 2 $SVL_T = (e^{b(T + C)} - a)/b$

where GR is linear growth rate, a is initial growth rate or GR at hatching, b is a damping function on GR, $\overline{SVL}$ is average SVL during the growth interval, T is lizard age and C is a constant of integration.

B. The linear model:

eq. 3 $GR = d$

eq. 4 $SVL_T = \mu + dT$

where μ is initial SVL and d is the constant GR.

C. Length to weight relationship:

eq. 5 $\ln(Wt) = j + f(\ln(SVL))$

where j and f are fitted constants estimating initial weight and the exponential term relating Wt and SVL

eq. 6 $Wt = g + h(SVL)^3$

where g and h are fitted constants.

D. Weight to age relationship:

eq. 7 $Wt_T = g + h(e^{b(T + c)} - a)/b$

eq. 8 $Wt_T = g + h(\mu + dT)^3$

11

Ecology and Evolution of the Insular Gigantic Chuckawallas, *Sauromalus hispidus* and *Sauromalus varius*

Ted J. Case
Department of Biology
University of California at San Diego
La Jolla, California

INTRODUCTION

The selective forces that influence an animal's body size are multifarious. I know of no cases, at least for terrestrial animals, where populational body size differences can be unambiguously assigned to single specific selective factors. Islands offer a favorable setting for untangling the knot of variables affecting body size and other life history characters. Insular endemics, species found only on islands, are notorious for having noticeable body size and behavioral differences from their mainland relatives (Case 1978). But, more importantly, because of the vagaries of colonization and extinction, a set of islands offers an array of different environments against which an investigator may assess the importance of suspected selective variables.

There are roughly 22 islands inhabited by *Sauromalus* in the Gulf of California, although no more than one species occurs on any single island (Soulé and Sloan, 1966). Two endemic species, *S. hispidus* and *S. varius,* inhabit the midriff islands (Figure 11.1) and have inordinately large body sizes for the genus (Mertens 1934, Shaw 1945). There are other islands in the Gulf which have endemic species and subspecies whose body sizes are indistinguishable from mainland relatives.

In 1970 I began to study this system with the hope of uncovering the selective agents responsible for body size divergence. Initially, this project's orientation was physiological and these results will be reported elsewhere. Only in the last two years have demographic studies been incorporated. This work is still in progress and the results and conclusions offered in this chapter should be viewed as provisional.

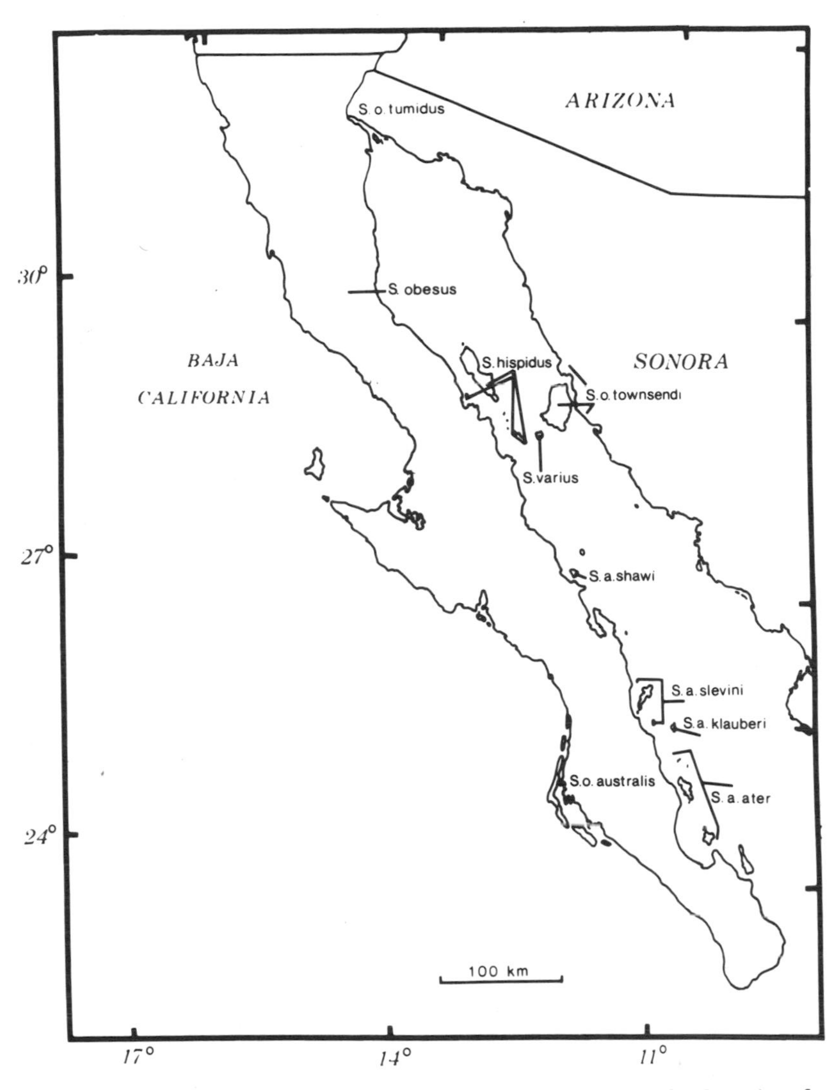

Figure 11.1: A distributional map of the various species and subspecies of *Sauromalus* in and around the Gulf of California. See Table 11.4 and Case (1976) for island names. Nomenclature follows Robinson (1972).

MODELS FOR INSULAR GIGANTISM

Three broad categories of factors are often invoked to explain insular body size differences.

(1) Optimal body size models (e.g., Case 1979, Schoener 1969) predict (among other things), that increases in the mean amount of available food (energy) for an individual should favor evolutionary increases in body size.

Islands are expected to have fewer competitors and the insular physical environment is often more benign. All else being equal, these factors might increase food availability. Yet all else is rarely equal, and the loss of predators on islands allows prey densities to increase so that food may become more limiting for these prey. Whether the supply to demand ratio (S/D) of prey for their own food is ultimately greater (due to greater food productivity) or lower (due to higher densities caused by reduced predation) will depend on the relative magnitude of these conflicting changes. A necessary condition for an animal's island S/D to be greater than that on the mainland is that it maintain individual feeding territories. In such cases, intraspecific squabbling will keep population densities from fully (linearly) increasing in response to any increases in food productivity (Case 1978).

For animals whose body sizes are not tightly bound within narrow limits by physical or competitive constraints, an increase in S/D should lead to (among other things) an evolutionary increase in body size. Case (1979) also found that optimal body size should increase with increases in the temporal variability of food availability. To see this latter relationship, recall that without any food the rate of decrease of an animal's relative body weight is inversely proportional to its initial body size. Therefore, a small animal will starve to death sooner than a large animal. If both a large and small animal are given *ad libitum* access to food, the large animal is also at an advantage because it has a larger gut volume and more internal space for fat storage. Hence, when food levels fluctuate severely, large animals will be at an advantage.

For future reference, I label this hypothesis (i.e., that insular gigantism is due to increases in the mean and variance in the food supply:demand ratio) as the food availability hypothesis.

(2) Predation itself may have a more direct impact on body size via two mechanisms. First, because predation intensities are reduced on islands, animals may forage, mate and conduct other vital activities in times and places which on the mainland were much too risky. Such a niche expansion may favor larger-sized individuals who may take advantage of a generally broader resource base (via the food availability argument above) and/or any unique resources which might demand large appendages or mouth parts for effective utilization. Secondly, and more directly, the size-preference of predators for prey may shift from mainland to island situations. Usually, the largest predators take the largest prey, although these larger predators are often the poorest island colonizers. Hence, the predators which do reach islands, are usually the smaller species who prefer smaller prey. On islands, there will be a selective premium for rapid growth and large size in prey organisms to escape this "window" of predator vulnerability. This premium is expected to be particularly great in animals like lizards which have indeterminate growth, because here clutch size increases with age and body size, yielding an extra payoff for older, larger females, once the risk of predation is removed. I will refer to this as the predation hypothesis.

(3) Empirically, many insular lizard populations have densities greater than that of relatives on the mainland (MacArthur 1972, Cagle 1946, Iverson 1979, Case 1975). Regardless of whether the root of this phenomenon lies in reduced predation or greater resource production, there are a number of potential

social consequences which might effect optimal body size. At least for polygamous/promiscuous species, intraspecific competition among males for females may become intensified. Since larger males are usually socially dominant, hold larger territories and mate with more females (Berry 1974, Brattstrom 1974, Ruby 1976), we might expect selection to lead to large male body sizes under high densities. I will call this the social-sexual hypothesis.

Of course, these three scenarios do not exhaust all possibilities and are not necessarily mutually exclusive. Moreover, each argument itself is multifaceted and depends on a host of assumptions regarding the insular setting; parts of each may apply. In what follows, I will review the current knowledge of the insular *Sauromalus* in an attempt to distinguish between these hypotheses. In particular I will show that the social-sexual hypothesis can effectively be eliminated and that the food availability hypothesis and the predator hypothesis remain viable options. Further studies should elucidate the relative importance of each and the validity of their respective subcomponents.

HISTORICAL EVOLUTIONARY BACKGROUND

Habitat diversity on the islands in the Gulf of California is limited by the uniform severity of the desert climate and by the absence of fresh-water habitats. The flora is typically Sonoran desert. The vegetation is more open on the north islands than it is on the more southern, which have a more impenetrable thorn-scrub or short-tree vegetation. The islands are ideal for evolutionary studies because of the near absence of permanent human settlements due to the lack of freshwater.

The most recent taxonomic account of the *Sauromalus* is that of Robinson (1972) who updated and revised Shaw's (1945) treatment (see also Etheridge, this volume). Physical descriptions of the various species and subspecies can be found in either source. Distributional ranges for reptiles are in Soulé and Sloan (1966) and Case and Cody (in prep.).

Sauromalus hispidus occurs on Angel de la Guarda (Angel) and all of its satellite islands (Mejia, Granito, Pond, Figure 11.1) which were connected to the main island by land bridges during the sea level minimum at the time of the last glacial maximum, 10 to 15 thousand years ago. It also occurs on San Lorenzo Norte and Sur which although probably never connected to Angel in the Pleistocene were certainly connected to each other less than 15 thousand years ago. Finally, *S. hispidus* is found on a number of small land bridge islands in Bahia de los Angeles including Smith, La Ventana, and Piojo. None of these islands presently are inhabited by mainland *S. obesus* which must have gone extinct after the land connection severed. None of these land bridge islands had late Pleistocene land connections with either Angel or the San Lorenzos, so the presence of the insular endemic *S. hispidus* argues for a very recent introduction perhaps by Seri Indians or Mexicans who occasionally eat these lizards. In keeping with this interpretation, no detectable morphological or chromosomal divergence has occurred between the various island populations of *S. hispidus* (Robinson 1972).

Similar distributional anomalies occur in the non-gigantic species. For example *S. ater slevini* (*S. slevini,* Shaw 1945) occurs on the land bridge island of Coronados as well as the deep water island of Monserrate. Because of the probable human influence here (Aschmann 1959), great caution must be exercised in interpreting the distributional data for *Sauromalus.* Simply because we now find a particular insular endemic species on a given island does not imply that the unique features of that species evolved in accord with that island's environment, or that an evolutionary equilibrium has in any way been reached. For *S. hispidus,* this may only be true for the population on the source island of Angel de la Guarda or San Lorenzo.

Sauromalus varius occurs naturally only on San Esteban (Figure 11.1) a small (43 km^2 deep-water island. On Isla Alcatraz, a land bridge island off Kino, in Sonora Mexico, recent human introductions have created a hybrid population involving introgression between *S. hispidus* and *S. varius* and *S. obesus* (Robinson 1972).

Although geologically diverse (Gastil, *et al.,* 1980) the deep-water islands inhabited by all the gigantic *Sauromalus* have probably existed as such for only about 1 or 2 million years (Moore 1973). There are no observable chromosomal differences between the insular and peninsula species, *S. obesus* or *S. ater* (Robinson 1972) and very little divergence at presumptive gene loci as electrophoretically discriminated (Murphy 1980). On the other hand, *S. ater slevini* on Monserrate is morphologically more similar to *S. obesus* but more different genetically from *S. obesus* than is *S. hispidus.* In general, there is no apparent relationship between the extent of morphological, chromosomal, and genetic divergence among the insular *Sauromalus* (Murphy, in prep.).

RESEARCH PROTOCOL

General reconnaissance visits and behavioral observations on *S. hispidus* were conducted on Angel de la Guarda in four main locations: the area at the south end near Isla Pond; Punta Diablo on the east side; Puerto Refugio at the north end; and Palm Canyon midway on the east side. As the name implies, Palm Canyon contains one of the most mesic habitats on the island and it is here that I laid out a 29.2 hectare grid with stations 20 meters apart for my demographic studies. A blind, for behavioral observations, was established in a mountain cave overlooking a large section of the grid. When a new animal is seen, the location, time of day, and ambient temperature at 1 m above ground are recorded. The animal is then captured by hand and its body temperature, sex, snout-vent length and tail condition are measured and recorded. The gut is felt for the presence of food and females are also palpated to detect eggs. The animals are marked by painting numbers on their backs and by toe clipping.

On San Esteban there is a single expansive arroyo on the southeast side of the island, where I established a 28 hectare grid to study *S. varius.* The characteristics of the grid and the capture protocol are very similar to that on Angel. Nearly 180 *S. hispidus* and 150 *S. varius* have been marked on these two grids. Each grid has been visited 6 times (1 to 3 weeks per trip) from spring 1978 to fall 1979. During each trip, an assistant and I spend about half the time marking and measuring animals and the other half observing animals from a blind.

BODY SIZE AND GROWTH

Table 11.1 provides an overview of the life history differences between *S. hispidus, S. varius* and *S. obesus;* this table incorporates data gathered from both museum collections, Berry (1974), and my own field studies. Hatchling lengths and weights were estimated from the smallest individuals found for each species and are consistent with estimates reached by extrapolating from mature egg sizes. The hatchlings of all 3 species are relatively similar in size (8 to 14 g) even though differences in adult body size are striking (380 to 1800 g).

An examination of Figures 11.2 and 11.3 reveals that juveniles are an obvious feature of the population in the small-sized island species of *S. ater,* in mainland *S. obesus* and in *S. hispidus,* but are notably absent in *S. varius.* (For over 200 specimens in California museums and zoos, collected over 80 years, only 3 juveniles exist. In my field work I have found only 3 juveniles in 10 years).

As yet, I have had too few recaptures of juvenile *S. hispidus* at Palm Canyon to estimate individual growth rates. However, a rough estimate of growth can be gained by comparing the shifts in the size distribution between 1978 and 1979 (Figure 11.3). The left-most peak in this distribution at about 98 mm snout-vent (SV) in 1978 represents an unusually large number of first year individuals that probably hatched the previous fall. The next spring (1979) there was very little new recruitment and the juvenile peak had shifted to 138 mm. This change of 40 mm over 320 days translates to an average growth rate of 0.25 g/day (after adjusting for the relationship between SV and body weight). For comparison, *S. obesus* of the same age class at China Lake in the California Mojave Desert had a growth rate of about 0.14 g/day for males and 0.06 g/day for females, or an average of about 0.10 g/day (Berry 1974), less than half the rate of *S. hispidus* young.

Table 11.1: Comparative Life History and Reproductive Features of *Sauromalus*

Feature	*S. hispidus*	*S. varius*	*S. obesus* (Locations)	Ref.*
Max. adult snout-vent length (mm)	♂298	324	180 Amboy	1
	♀304	314	178 Amboy	1
			223 Little Lake	1
			205 Little Lake	1
Max. adult body weight (g)	1,400	1,800	180 Amboy	1
			380 Little Lake	1
Avg. hatchling snout-vent (mm), weight (g)	70, 10	75, 14	54, 8 China Lake	2
Avg. egg size (mm) and weight (g) near laying	25 x 24, 10	40 x 28, 18	20 x 15, 8 Amboy	1
Hatchling growth rates (g/day)	0.25	–	0.10 China Lake	2
Max. clutch size	29	32	13 China Lake	2
% breeding ♀'s (N)	8 (82)	16 (70)	32 (47) all	2
			38 (32) all	1

*References for *S. obesus* only: 1, Case unpublished; 2, Berry (1974).

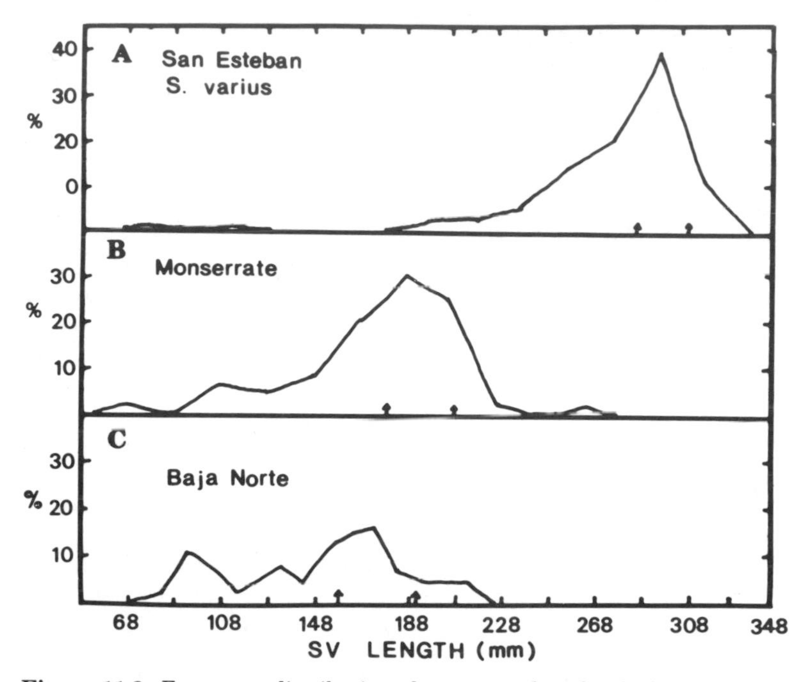

Figure 11.2: Frequency distribution of snout-vent lengths (for both sexes) in three populations of *Sauromalus*. The Monserrate population is *S. ater slevini*, and the Baja Norte population is *S. obesus obesus*. Sample sizes are listed in Table 11.2 and represent composites over all localities and collection dates. The arrows at the base of each figure denote the median and upper decile snout-vent lengths.

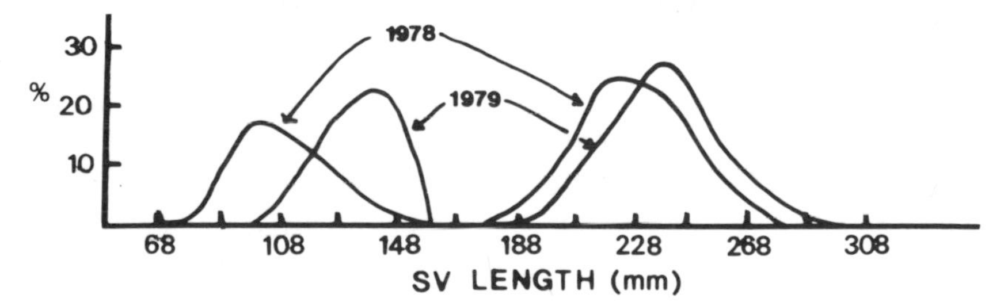

Figure 11.3: The frequency distribution of snout-vent lengths of *S. hispidus* at Palm Canyon on Angel de la Guarda for two subsequent years. N = 85 for 1978 and N = 60 for 1979. In the text an average juvenile growth rate is calculated on the basis of the shift in the left-most peak from 1978 to 1979.

The relative growth rate of *S. hispidus* adults can be compared to *S. obesus* by comparing growth rings in the dentary bone (Peabody 1961, Case 1976). Whenever carcasses were found in the field, their skulls were saved and the SV length estimated from regressions of dentary length versus SV (see Mortality). The dentary bones were sectioned, polished and examined under a drop of xylene with a compound microscope. Although there is surely not a one-to-one correspondence between these rings and age in *Sauromalus,* they can be used to compare relative growth rates between species or populations (Enlow 1969). Figure 11.4 contrasts the rate at which the number of rings increases with snout-vent length for 2 populations of *S. obesus* in California and for *S. hispidus* on Angel. Clearly *S. hispidus* grows faster than even the relatively gigantic *S. obesus* population at Little Lake. The maximum number of rings in all 3 populations is typically about 12 to 14. This, however, cannot be taken as indicating a similar maximum age for these populations since in some other reptilian species the older, larger individuals do not add growth rings, or if they do, the accretion of external rings may be accompanied by the dissolution of internal rings (Enlow 1969).

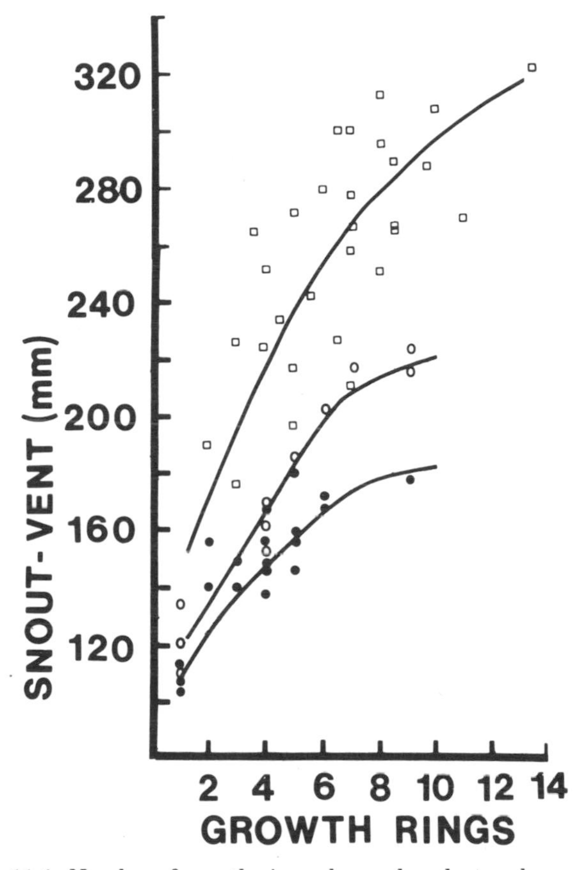

Figure 11.4: Number of growth rings observed on dentary bone sections (average of two counts for each lizard) as a function of chuckawalla snout-vent length. Open circles represent *S. obesus* from Little Lake, Calif.; closed circles represent those from Amboy; squares are for *S. hispidus* from Angel de la Guarda. Differences between these populations in adult body size are paralleled by differences in the number of rings for a given snout-vent length. Lines fitted by eye.

The size distribution in Figures 11.2 and 11.3 lump females with males. This seems justified since there is little difference in the maximum body size which the two sexes reach (Table 11.1). Interestingly, *S. hispidus* females average slightly larger than males, whereas in *S. varius* and *S. obesus* the reverse is true. This paucity of sexual dimorphism in the insular gigantics is also reflected in the total absence of coloration differences between the two sexes. Sexes can be distinguished on the basis of body shape (males have wider necks) and the presence of enlarged femoral pores in males.

REPRODUCTION

Lamentably, there is little information on breeding and reproduction in the gigantic *Sauromalus*, stemming from the infrequency of these events in natural populations. Moreover, I have sacrificed no animals during the last 5 years of my studies for conservation reasons. Examining females for clutches by palpation is unreliable unless the oviducal eggs are well-developed. For males, I have habitually noted the extent of femoral pore secretions, but whether these secretions are truly indicative of testes activity remains in doubt. Nevertheless, museum specimens are ample and the numerous animals which died (usually natural deaths) during my field studies were always autopsied. The extent of breeding in one year can also be assessed by determining the number of hatchlings recruited in the subsequent spring.

Clutch size is plotted as a function of snout-vent length for all *Sauromalus* in Figure 11.5. Unlike the case in many insular vertebrates (e.g., see Cody 1966), clutch size in the gigantic *Sauromalus* is not reduced below allometric predictions for their body size based on mainland *S. obesus*. In fact, clutch size in the gigantics seems higher than predicted, reaching a maximum of 32 eggs. The average size of full term oviducal eggs in *S. hispidus* (based on two clutches) is 25 x 24 mm, each weighing about 10 g. This is only slightly larger than eggs of *S. obesus* (25 x 15 mm; 8 g, based on 6 eggs in one clutch). *S. varius* has larger (18 g) and more oblong eggs (40 x 28 mm; data for one clutch). The reproductive mass (clutch weight/adult weight with eggs) is 37% for *S. varius* and 25% for *S. hispidus*. The typical reproductive mass is between 35 and 40% in *S. obesus*.

Perhaps the most unique reproductive feature in the insular gigantics is their reduced frequency of breeding. Berry (1974) examined museum specimens of adult female *S. obesus* which were collected during the months when ovarian follicles became enlarged, April, May and June. Out of 47 females only 32% were found to be gravid. Case (unpublished) examined a partially overlapping set of 32 museum specimens of *S. obesus* collected during the same spring months and found 38% gravid. On the other hand, a cohort of 82 *S. hispidus* females from the same months (museum specimens plus field animals collected or examined from all islands and years) contained only 8% gravid females. Only 16% of an equivalent group of 70 *S. varius* adult females were gravid (Table 11.1). Bull and Shine (1979) recently reviewed the incidence of low frequency reproduction among ectotherms and found that very few reptiles skipped more than 2 years between successive breeding attempts. Although it is still impossible to give an average value for the gigantic *Sauromalus,* my observaions indicate 4 years would be a conservative estimate for adult female *S. varius*. The near total lack of juvenile *S. varius* is problematic and suggests that eggs or hatchling predation may be severe. The introduced *Rattus rattus* on San Esteban could be responsible, yet if so, it is strange that the sympatric ctenosaur is not effected; juveniles are common in this species.

The incidence of gravid females in the gigantic specimens does not appear to be spread evenly over the years, although sample sizes are still too small for statistical comparisons. There are no weather stations on any of the study islands but annual rainfall can be estimated from records for Bahia de los Angeles and El Barril on Baja California and at Hermosillo, Sonora on the other side of the Gulf (Hastings and Humphrey 1964a,b, 1969a,b, and personal observations). During the dry years of 1972, 1975 and 1976 no detectable breeding (i.e., no recruitment or gravid females seen) was found for either *S. hispidus* or *S. varius*. In 1977, a moderately wet year following a dry year, considerable recruitment occurred in *S. hispidus* at the mesic Palm Canyon but not elsewhere on Angel or in *S. varius* on San Esteban. In the wetter years of 1978 and 1979, a few gravid females were found at both Palm Canyon and on San Esteban. In museum collections, gravid females have been found most prevalent in 1952 and 1968, years of heavy rainfall preceded by a year of at least moderate rainfall.

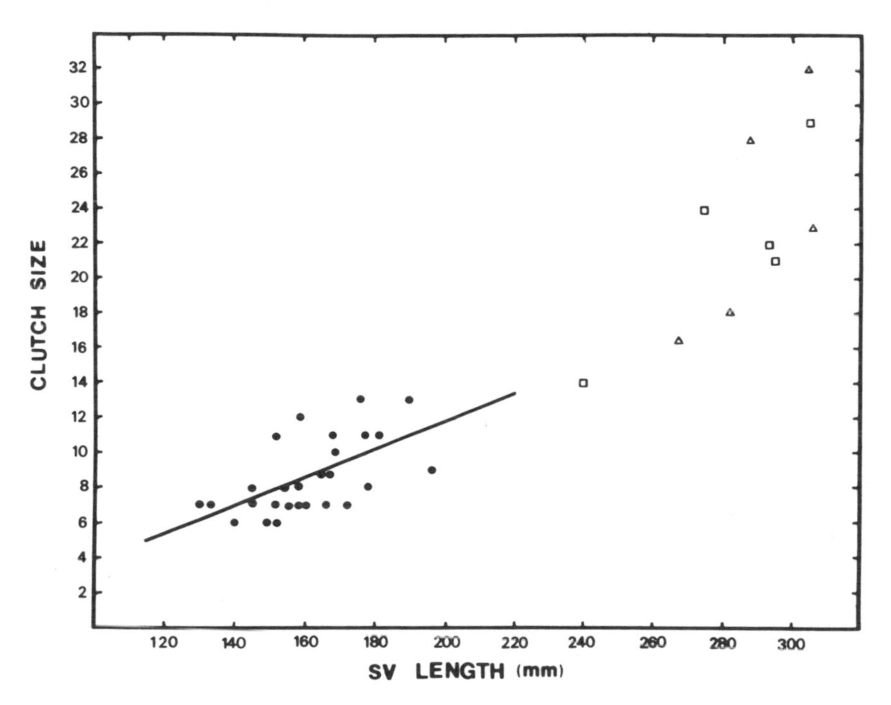

Figure 11.5: Clutch size versus snout-vent length in *Sauromalus*. Closed circles are for *S. obesus* (much of this data is from Berry 1974); squares are for *S. hispidus;* and triangles are for *S. varius*. The solid line is the least squares regression line for *S. obesus:* Clutch = 0.0775 SV – 3.81.

The picture that emerges is one of "boom or bust" iteroparity. An individual female either reproduces maximally or not at all. During most years a female will be reproductively inactive, but in wet years a significant number of adult females may produce clutches (although, perhaps no more than 30%; that is, the boom or bust iteroparity is not necessarily matched by a boom or bust population cycle). Based on limited data, the seasonal timing of reproduction (when it occurs) in both *S. hispidus* and *S. varius* does not seem to deviate from that of mainland *S. obesus*. Namely, mating occurs in the early spring, egg laying in early summer, and hatching in the fall. Further data are needed to substantiate this scenario: if hormonal titers could be measured in the blood or urine of field animals, a clearer picture of reproduction in both sexes should emerge.

As yet, I have not witnessed any courtship, mating or definitive nesting behavior in the field. One adult female *S. varius* was observed digging a burrow in the sandy arroyo floor on June 10, 1978. Sand and small pebbles were kicked out using both front and rear legs. The burrow was about 15 cm wide and 60 cm deep at an angle of 20° from horizontal. Later, this female was found to contain mature oviducal eggs so perhaps she was preparing to lay. Similar burrows are evident throughout the main arroyo on San Esteban and Angel wherever soil conditions permit. There seems to be no shortage of potential nesting sites. Although I excavated four of these burrows in October, I uncovered no egg shells.

MORTALITY

The frequency of individuals with broken or regenerated tails is often used as an index of predation intensity when it is known that intraspecific fighting is not a major source of these breaks (Pianka 1970, Rand 1954). This assumption appears to be fulfilled with *S. obesus* since out of hundreds of intraspecific aggressive encounters observed by Berry (1974) and myself, none resulted in tail loss. Furthermore, although females are rarely observed in aggressive interactions they usually display the same frequency of broken tails as males. Table 11.2 shows the upper decile snout-vent length of a number of island *Sauromalus* populations, along with their tail break frequencies. Nearly all of the insular populations have lower frequencies of tailbreaks than mainland populations of *S. obesus* where values of 25 to 40% have been found in populations as far apart as Utah to Mexico (Case 1976). (In both island and mainland populations males and females have statistically indistinguishable tail break frequencies so data for both sexes have been lumped.) The two gigantic species do not seem to have tail break frequencies which are consistently lower than nongigantic insular populations. Yet, it should be remembered that the average age of the gigantic species is probably much greater (based on their body size and their longevity at zoos; see below). A doubling in the median age of the population could compensate for a halving of predation rate (i.e., predatory encounters/time) yielding similar tail break frequencies. Hence, the fact that tail break frequencies in the gigantic *Sauromalus* are at least as low as the smaller species would indicate, if anything, a reduced predation rate were it not that gigantic tails are more difficult to break (see below).

Table 11.2: Body Sizes and Tail Break Frequencies for Insular *Sauromalus*

Location	Sample Period	Species	Median SV ... (mm)	Upper Decile SV ...	N	Tail Breaks (%)
Angel de la Guarda (entire island)	1968-1979	*S. hispidus*	222	263	191	12.0
Angel de la Guarda (Palm Canyon)	1978-1979	*S. hispidus*	221	253	145	13.5
Mejia*	all records	*S. hispidus*	214	280	51	2.0
San Lorenzo Norte plus San Lorenzo Sur	all records	*S. hispidus*	245	295	32	11.4
San Esteban	all records	*S. varius*	288	310	210	15.9
San Marcos**	all records	*S. ater shawi*	–	164	6	–
Carmen**	all records	*S. ater slevini*	148	181	22	19.1
Monserrate	all records	*S. ater slevini*	182	210	46	12.0
Santa Catalina	all records	*S. ater klauberi*	144	171	16	12.5
Santa Cruz	all records	*S. ater ater*	108	149	24	0.0
San Jose**	all records	*S. ater ater*	–	177	9	–
Espiritu Santo	all records	*S. ater ater*	144	172	18	31.0
Partida Sur**	all records	*S. ater ater*	144	172	18	31.0

*A satellite island off the north tip of Angel de la Guarda.
**Land-bridge islands.

The advantage of easily breaking tails in lizards as a predator escape mechanism is obvious, yet the partial loss of a tail carries some severe consequences. The animal will risk infection, and must devote extra energy to growing a new tail rather than increasing body size or allocating this energy to reproduction. An unexpected disadvantage was found by Berry (1974). Some *Sauromalus obesus* populations have a loose heirarchial social system dominated by a few large tyrant males whose territories encompass many of the smaller territories of females, juveniles, and subordinate adult males. A number of factors in concert determine a male's dominance but one factor of utmost importance appears to be the length of the tyrant's tail. Berry (1974) observed one tyrant in particular to immediately lose his social advantage after having lost his tail for unknown reasons. Iverson (1979) has observed essentially the same result in captive *Cyclura carinata.* Hence, the continued presence of tail breaking ability in *S. obesus* populations, in spite of these selective disadvantages, attests to its importance as a predator escape mechanism.

The actual means by which the tail break is accomplished is via a fracture plane across individual tail vertebra. It is at this plane that the break occurs, and the muscles, nerves, and blood vessels are easily separated. A typical *S. obesus* tail does not contain fracture planes in the first 8 or 9 vertebrae, but they are well defined in the remaining vertebrae. A similar sized *S. hispidus* individual usually does not have fracture planes in the first 14 to 18 vertebrae. Although the next 5 to 10 vertebrae do have partial fracture zones, they appear to require more pressure to separate than those in *S. obesus;* however, this difference was not quantified. The last few vertebrae lack planes altogether. I have not yet examined the vertebrae of *S. varius* but attempts to break their tails or those of *S. hispidus* were unsuccessful except with extreme pressure. *S. varius* has a somewhat higher tail break frequency than *S. hispidus* and in both insular gigantics, many broken tails do not regenerate, but simply heal over. Hence, the low frequency of broken and regenerated tails in these gigantic species may in large part be due to the loss of tail break facilitation. The ultimate evolutionary reason for the loss of this ability, however, must lie in the lower threat of predation on the islands where the adaptations of these species evolved.

A comparison of the potential lizard predators on various islands inhabited by *Sauromalus* is presented in Table 11.3. Islands suspected of being colonized by *Sauromalus* very recently are excluded from this table. All the islands with *Sauromalus* contain ravens (and turkey vultures). Neither Angel de la Guarda nor San Esteban have any mammalian carnivores, but then neither do most other islands in the Gulf. However, all the islands inhabited by *Sauromalus ater* with the exception of Monserrate and Santa Catalina, are land bridge islands and almost certainly had an ample diversity and numbers of carnivores only 10 to 15 thousand years ago. Moreover, while one could not call *Sauromalus ater slevini* on Monserrate gigantic, it is substantially larger than populations of the same subspecies on Carmen and Coronados. Hence, with the exception of Santa Catalina, there seems to be a loose pattern of increased body size in *Sauromalus* on islands which have had a long history of isolation from mammalian predators. Certain aspects of the gigantic species' behavior are very much in tune with this conclusion. These species have shorter approach distances (Schallenberg 1970) and they are often found in ludicrously careless hiding places.

Although I have been unable to find any thorough study of carnivore feeding habits in areas where *S. obesus* is present, some observations suggest that canids may be important predators. Berry (1974) observed that all the *S. obesus* in her study area fled into crevices at the sight of a coyote and did not emerge again that morning. At Amboy, California, a chuckawalla whose body temperature I was monitoring with a thermocouple probe, was pulled out of its crevice and carried away during the night by two kit foxes. Berry (1974) describes records of *Sauromalus* found in coyote scats. The literature is replete with instances of predation on island iguanids by feral dogs and cats (Iverson 1979, Wiewandt 1977, Schmidt and Inger 1957). The absence of these and native canids from Angel and San Esteban must reduce the risk of predation significantly.

Other potential predators on *S. hispidus* on Angel de la Guarda include the two rattlesnakes, *Crotalus ruber* and *C. mitchelli.* Only the latter is large enough to consume adult *S. hispidus,* but both snakes could probably eat juveniles. Twice, I have found individual *S. hispidus* adults inhabiting crevices or burrows with either *C. ruber* or *C. mitchelli.* Although they were actually touching one another, neither appeared interested in the other. Four stomachs of *C. mitchelli* and 2 of *C. ruber* contained only rodents, primarily *Perognathus.* On the other hand, J. Ottley (pers. comm.) has found 3 mummified *C. mitchelli* with large adult *S. hispidus* lodged in their mouth and throat. This suggests

Table 11.3: Potential Size-Specific Predators of *Sauromalus* on Islands in the Gulf of California

Island	*Sauromalus* Species	No. Small Snake Species[1]	No. Large Snake Species[1]	Avian Predators[2]	Native Mammalian Carnivores[4]	Feral Domestic Carnivores
Tiburon[3,5]	*S. obesus townsendi*	2	8	?	C,RC,GF	Cat, dog
Angel de la Guarda	*S. hispidus*	1	3	RT,B,SH,S	0	0
San Lorenzo Norte	*S. hispidus*	0	1	RT,S	0	0
San Lorenzo Sur	*S. hispidus*	0	1	RT,S,SH	0	0
San Esteban	*S. varius*	1	1	RT,SH	0	0
San Marcos[3,5]	*S. ater shawi*	2	2	RT,SH	0	Cat, dog
Coronados[3]	*S. ater slevini*	0	1	RT,B	0	Cat, dog
Carmen[3,5]	*S. ater slevini*	1	4	SH,S	0	Cat, dog
Monserrate	*S. ater slevini*	2	4	0	0	Cat
Santa Catalina	*S. ater klauberi*	1	2	0	0	Cat
Santa Cruz	*S. ater ater*	0	1	0	0	0
San Jose[3,5]	*S. ater ater*	4	5	SH,RT,S	RC	Dog
San Francisco[3]	*S. ater ater*	1	1	0	0	0
Espiritu Santo-Partida S.[3]	*S. ater ater*	2	4	SH,RT,S	RC	Dog

[1] Small snakes include potential lizard predators with a maximum total length less than 30 inches. Large snakes have a maximum length greater than 30 inches. Maximum sizes are taken from Stebbins (1966), and snake distributions from Soulé and Sloan (1966), plus personal observations.

[2] RT is the red-tailed hawk (*Buteo jamaicensis*).
B is the burrowing owl (*Speotyto cunicularia*).
S is the loggerhead shrike (*Lanius ludovicianus*).
SH is the sparrow hawk (*Falco sparverius*).
This data was partially supplied by Martin L. Cody.

[3] Land-bridge islands.

[4] C is the coyote *Canis latrans*.
RT is the ring-tailed cat *Bassariscus astutus*.
GF is the grey fox (*Urocyon cinereoargenteus*).

[5] Small human settlements occur on these islands in localized areas.

that the larger chuckawallas are typically inappropriate meals even for the large *C. mitchelli.* The only rattlesnake on San Esteban is an endemic subspecies of *C. molossus* which is much too small to consume anything but *S. varius* hatchlings. The stomachs of two individuals contained only rodents. In short, these snakes are probably not important predators at least of large adult *Sauromalus hispidus* or *varius.*

There are a few indications in the literature that predation by raptors or ravens may at times be severe. Van Denburgh (1922), Shaw (1945), and I have all found skeletons of *S. hispidus* and *S. varius* around osprey nests. Nevertheless, I am doubtful that these sea-feeding birds would attack chuckawallas. Although I have frequently seen osprey diving for fish I have never seen them hunt for terrestrial prey. More likely, the chuckawalla skeletons were gathered by the birds for nesting material. Mike Robinson (pers. comm.) has observed ravens feeding on *S. ater* on Monserrate and assumed they killed these lizards. He concluded that the eviscerated carcasses he observed on Angel too resulted from raven predation. My observations do not support this conjecture, but rather indicate that ravens are scavenging on dead or nearly dead animals.

Dead chuckawallas *(S. hispidus)* are particularly common on Angel (but not on its satellite islands). Carcasses *(S. varius)* are found in much fewer numbers on San Esteban and are virtually non-existent on the mainland. Carcasses appear more common in or after drought years than in or after wet years. For example, in 1972 Puerto Refugio (on Angel) bore large numbers of carcasses of *S. hispidus* representing probably several years of accumulation. A systematic collection was initiated and within four hours nearly 100 carcasses were found (and removed) by myself and an assistant. In spring 1977 (1976 was again a drought year), about 20 new carcasses were found at Puerto Refugio and many more on the xeric south end of Angel. As mentioned earlier, no measurable reproduction occurred at these sites in 1972 and 1976. Many live animals appeared emaciated and dehydrated; a few seemed close to death. I observed over 20 chuckawallas from a blind for five days in June and did not see any feeding activity. One adult died during this period within his rock crevice. Although he appeared emaciated, his body weight had not fallen to a level associated with starvation. His death must be attributed to disease which was, perhaps, provoked or encouraged by under-nourishment.

Carcasses were found in all stages of decomposition. Most specimens were found with fragments of skin surrounding a partial skeleton. Others were so fresh they still had full moist skin and muscle tissue which was being carried off by harvester ants. Significantly, these recently-dead adults of both *S. hispidus* and *S. varius* were not eviscerated and appeared intact, indicating a death by disease and/or starvation. One such intact and recently dead adult *S. varius* was marked with paint and left on the grid 5 months. Later I found it again but now it was delimbed and eviscerated. I did find two freshly dead juvenile *S. hispidus* which were already eviscerated and delimbed indicating that raptors may prey upon live juveniles.

The position of carcasses in the field also supports these conclusions. Most carcasses of *S. hispidus* on Angel were found below or within the nodes of large columnar cacti *Pachycereus pringlie,* (cardons): the larger the cardon, the more dead chuckawallas were found around it. As many as eight were found under a single cactus. A few other carcasses were found under elephant trees or on the peaks of hills. Of course, these are all areas that *S. hispidus* frequent but they are also the sites where ravens and red-tailed hawks perch to feed.

From a blind on Puerto Refugio in June 1972, I observed two ravens flying about 15 meters over a large adult *S. hispidus*. The ravens landed on a large cardon about 20 meters from the lizard. The chuckawalla turned its head upward in the raven's direction, head-bobbed a few times, and thereafter remained still for about five minutes. During this time the ravens fed on the cardon's fruit and seemed oblivious of the lizard's presence, although it was in plain sight. On three other occassions I have observed raven-adult *Sauromalus* interactions at Palm Canyon and San Esteban which followed this general pattern of mutual disinterest. However, on four occasions I have observed juvenile *S. hispidus* to run for cover at the sight of ravens. Ravens have been observed hunting for small lizards at Palm Canyon. They fly at a height of between 30 and 40 feet above the arroyo and swoop down on small lizards (e.g., *Callisaurus* and *Cnemidophorus*) as they move across open spaces. I have twice seen ravens with unidentified small lizards in their beaks. When juvenile *S. hispidus* cross open areas in the arroyo they run very rapidly; whereas, large adults increase their gait only slightly. Berry (1974) reported numerous incidents of *S. obesus* taking flight at the appearance of a hovering bird. These observations reinforce the conclusions reached earlier; namely, that large adult *Sauromalus hispidus* and *S. varius* are probably immune to attacks by ravens, although predation on juveniles may be severe.

Somewhat at odds with this conclusion is the scarcity with which juvenile carcasses are found. A large part of this rarity, particularly in *S. varius,* is simply due to the lack of juveniles in the population. This bias may be eliminated by comparing the size distributions of live and dead chuckawallas. Only for *S. hispidus* are sufficient data available for a meaningful comparison. Since many of the carcasses were found incomplete or distorted, it was necessary to extrapolate snout-vent length from the linear dimensions of either the dentary or femur bones. A sample of 25 freshly dead specimens of known snout-vent length were cleaned and their dentaries and femurs were removed and measured with calipers. Linear regressions were then made between these variables so as to predict snout-vent lengths of skeletal remains.

Figure 11.6 (bottom) displays the size frequency distribution for all the dead chuckawallas *(S. hispidus)* found on Angel de la Guarda between 1972 and 1979 (N = 160). Figure 11.6 (middle) shows the same distribution for live *S. hispidus* measured from 1968 to 1979 (N = 294). The two small arrows at the base of these distributions give the median and upper decile snout-vent length for the collection. The top section of Figure 11.6 plots the difference between these two curves as the percentage of live individuals of a given snout-vent length minus the percentage of dead individuals of the same length and reveals that there is a notable lack of dead one- and two-year old chuckawallas. I suspect that these small individuals are swallowed in whole by ravens or raptors, or at least torn apart so drastically that later recovery of their skeletons would be exceedingly unlikely. The lack of dead juveniles in my samples is probably due to this collecting bias rather than to an inherently low mortality for this cohort. In fact, my behavioral observations discussed earlier indicate that juveniles are in greater danger than adults from predation. Figure 11.6 also reveals an increase in mortality of the very largest individuals in the population which are presumably natural deaths due to senility and old age. For example, on San Esteban, a small proportion of adults have circular tumors about 1 or 2 cm in diameter in the head region (pers. obser. and Bostic, 1971). One recently dead individual that I found had two such tumors in its throat. Hence, I suspect that the ecological survival advantage of large size in very old individuals is confounded by the concommitant mortality simply due to old age and its associated diseases.

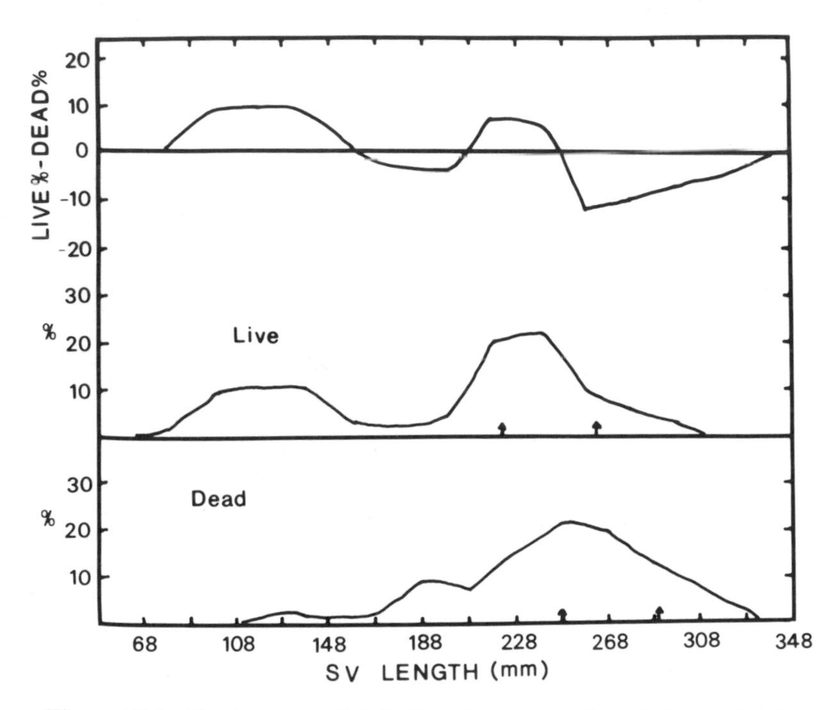

Figure 11.6: The frequency distribution of snout-vent lengths for dead *S. hispidus* (bottom) and live *S. hispidus* (middle). The dead sample (N = 160) is for all locations on Angel and for the period 1972 to 1979. The live sample (N = 294) is also for all locations on Angel and for the period 1968 to 1979. The top part of this figure plots the live percentage minus the dead percentage of each snout vent category. Values above the zero level indicate size classes which have relatively low mortaity and conversely values below the zero level indicate size classes with relatively high mortality. The apparent lack of small juvenile mortality may be an artifact of my inability to recover this size of carcasses. See text.

FEEDING AND HABITAT

A striking feature of Angel and San Esteban, is the conspicuous presence of the chuckawallas. Their densities appear greater than the smaller *Sauromalus* species on the mainland or on other islands in the Gulf. The lizards are more abundantly observed in a greater array of habitats and seen at more times of the day. In particular, the insular gigantics make extensive use of the arroyo sandy wash and beach dune habitats which are rarely utilized by the smaller *Sauromalus* elsewhere. *Sauromalus hispidus* is the most arboreal chuckawalla. At Palm Canyon, 45% of the juvenile *S. hispidus* were first observed in *Dalea, Hyptis, Acacia* or other perennials that were in bloom. Juveniles would climb to a height of 1.5 to 3.0 m and maneuver to the end of flowering stems where they would eat the blooms. Only 10% of the adults were first sighted in arboreal situations and then usually at about 1.0 m height in shorter, sturdier shrubs such as *Jatropha,* cholla or in cardon. At Amboy, I have never seen *S. obesus* climb shrubs although on Baja California and at other California populations (Berry, 1974; Nagy, 1973) a small portion of sightings are in low shrubs.

Sauromalus varius adults are also not inclined to climb vegetation; only 3% of my sightings were in such situations. Since juvenile *S. varius* are rarely encountered, their arboreal preferences are unknown but when the adults were seen arboreally they were usually in blooming and/or fruiting *Macrocereus, Cercidium, Prosopsis,* or *Olneya.* Compared to other chuckawallas, *S. varius* is found much more frequently under shrubs, particularly cholla and to a lesser extent *Jatropha, Bursera* and *Macrocereus.* In such situations, their mottled skin coloration makes them quite cryptic. Animals observed in these situations will remain still up to the moment that they are grabbed, presumably so as not to reveal their location. Interestingly, the arboreal niche on San Esteban is assumed by another herbivorous iguanine, *Ctenosaura hemilopha* (absent on Angel), which rivals the sympatric *S. varius* both in size and numbers. Like the sympatric chuckawalla, the ctenosaur feeds primarily on cactus fruits and assorted perennial flowers when they are available, but it is much more arboreal. For example, 53% of the individuals were first sighted in vegetation. Like *S. hispidus* on Angel, ctenosaurs occasionally perch on the tops of cardons to bask and feed.

The peculiar attraction that the large cactus, cardon, has for *S. hispidus* is unparalleled by any other species of *Sauromalus.* At Puerto Refugio on Angel about 25% of the *S. hispidus* individuals were found among the primary branches of cardon. This contrasts with only 0.2% for *S. varius* even though cardons are even more abundant on San Esteban. Intensive searches in comparable mainland habitats where cardons were abundant and on *S. ater* inhabited islands, failed to yield even a single cardon-inhabiting chuckawalla. The cardon flowers and fruits are typically near the very top of the cactus, at heights of over 6 m. Chuckawalla scats are occasionally found hanging from the topmost spines on cardons and more rarely, lizards may be seen perched on the very tops of the cardon. However, since red-tailed hawks, ospreys and ravens utilize the taller cardons for perches and roosting sites, the exposure of *Sauromalus* when climbing these branches represents a potential risk. Seemingly, the chuckawallas are usually content to seek refuge in the nodes at the branch bases. These situations offer protection similar to that of rock crevices with the added advantage that dried but still edible cardon fruits may accumulate in the forks.

The retreats utilized by *S. hispidus* and *S. varius* fall into two broad categories: *home-base retreats* which are large and secure, and utilized at night on a fairly permanent basis; and *temporary retreats* which are smaller, less secure, and not utilized every day but only during foraging forays into the surrounding area.

On Angel, the home-base retreats of *S. hispidus* are diverse and locally determined by the existing geological formations. At Puerto Refugio some animals occupy deep, extensive burrows in crossbedded sandstone. These burrows varied in height from 15 cm to more than 30 cm and most averaged about 30 to 60 cm in width. The depth was difficult to estimate because the burrows became narrower and changed direction past the entrance. Elsewhere on Angel and on the satellite islands of Mejia and Granito, the lizards use crevices between fragmented metamorphic rock or cracks among volcanic plates. On San Esteban, large burrows of *S. varius* are dug horizontally into the canyon banks. As on Angel, cracks among large volcanic rocks are also used as permanent home-site retreats. The temporary retreats on both San Esteban and Angel are usually among small, rocky outcrops in the arroyos; however, animals are also found in burrows (dug in sandy soil) and at the bases of thick brush. *Sauromalus hispidus* also uses the nodes of cardon branches. In many of these situations the animals are easily found and approached since these hiding places are often insufficient to hide the animals' entire bodies. If a lizard is spotted in the open, it will often freeze and only attempt to run to a retreat at the last possible moment. The behavior of fleeing animals suggests that they are very familiar with hiding places in their foraging areas. Even though many of these hiding places appear ridiculously inadequate, the potential avian and ophidian predators which they might encounter on the islands are probably unable to pry them loose from even the most superficial crevices.

For heliothermic reptiles, the maintenance of appropriate body temperatures is an activity of high priority. Although the mean preferred body temperatures of all the *Sauromalus* species are very similar in laboratory thermal gradients (Case, in prep.), the insular gigantics are more lax about regulating their body temperature in the field. Figure 11.7 plots the body temperature of active *S. hispidus* (Angel), *S. varius* (San Esteban), and *S. obesus* (Amboy, CA) versus the ambient temperature (1 m above ground in the shade) at the time of capture (see Research Protocol). Notice that the insular gigantics have more variable body temperatures and that they may be active at cooler ambient temperatures than *S. obesus*.

The expanded thermal niche of *S. hispidus* also affects their daily activity cycle. From a blind, I observed the behavior of *S. hispidus* at Puerto Refugio during 5 days in June 1972. Four of these large adults were then removed and placed on my *S. obesus* study site near Joshua Tree, in the California Mojave Desert. After a four-day adjustment period, their daily activity was observed from a blind along with the activity of the native *S. obesus*. The results of these studies are depicted in Figure 11.8. *Sauromalus hispidus* appears to emerge somewhat earlier in the day and recedes slightly later in the afternoon. Moreover, their activity is more evenly distributed throughout the day than that of *S. obesus* under identical conditions.

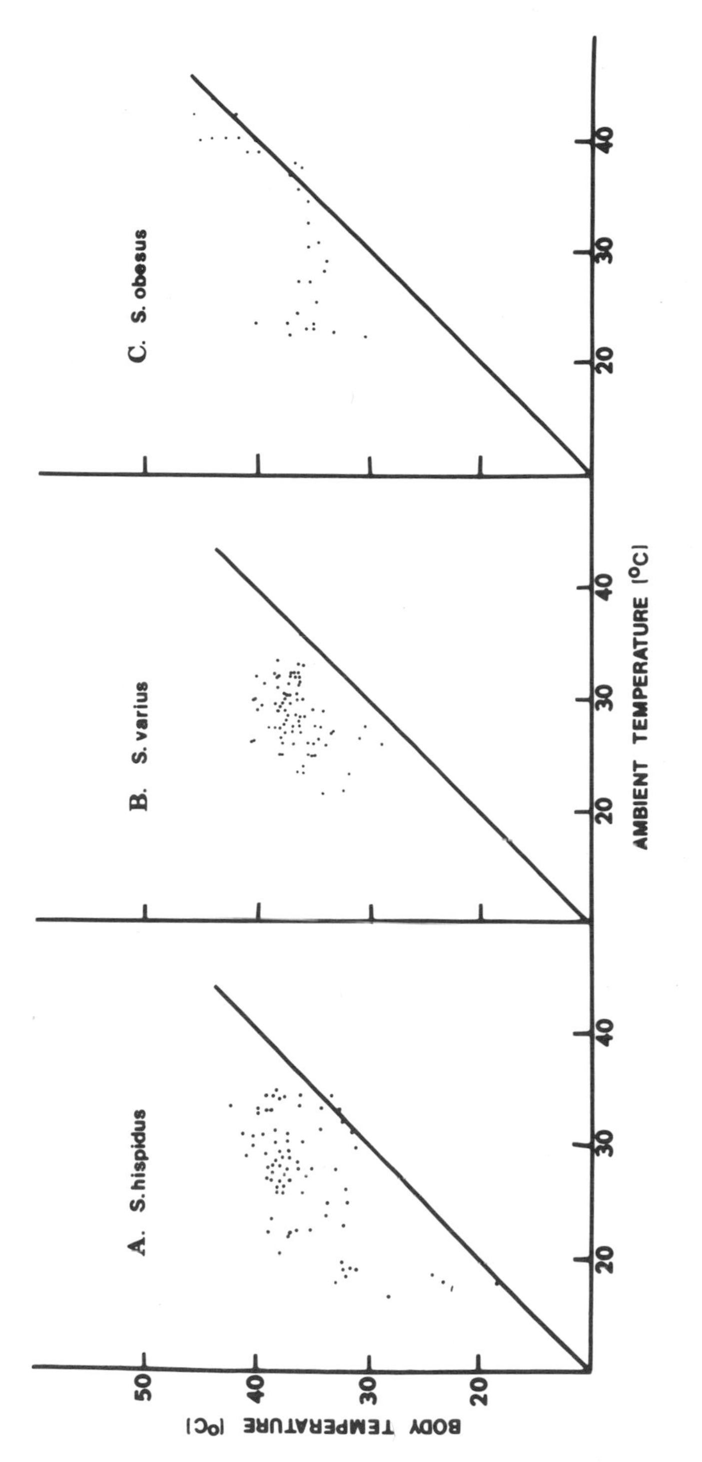

Figure 11.7: Body temperature as a function of ambient temperature for field-active *Sauromalus.* All temperatures were taken with a Schultheis thermometer and ambient temperatures were read in the shade 1 m above ground. The 45 degree line through each subfigure is drawn simply for reference.

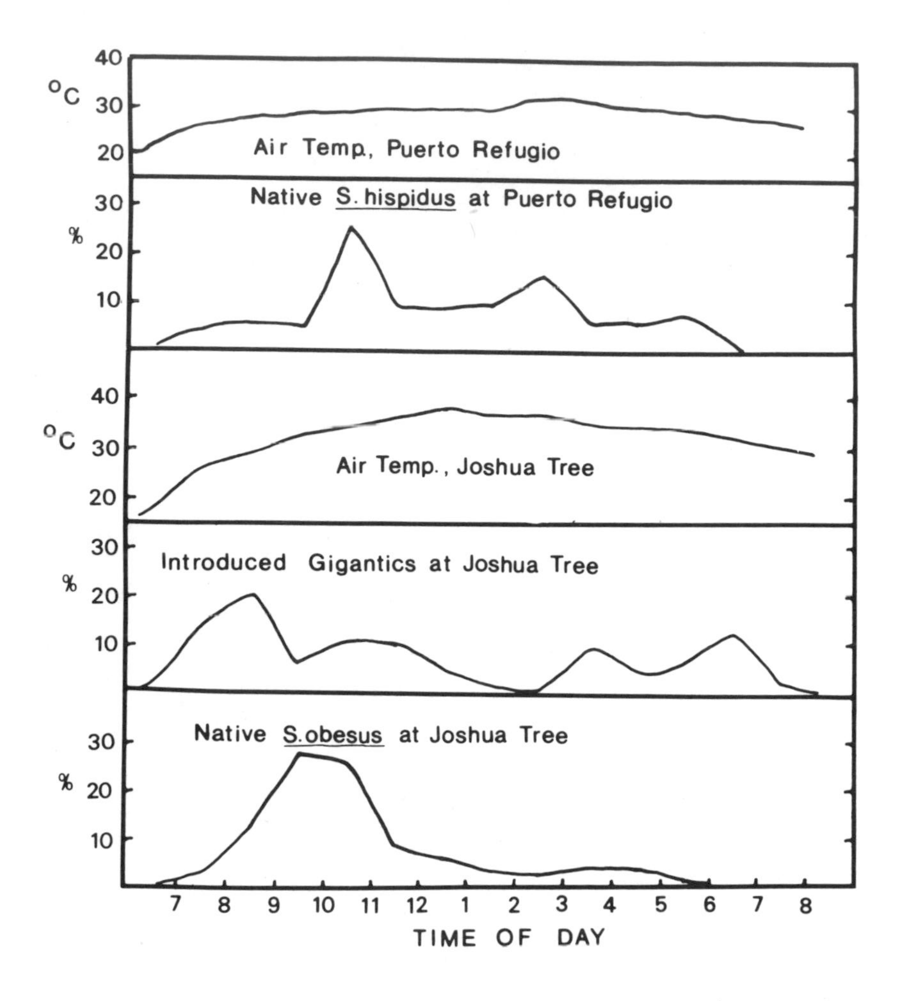

Figure 11.8: The distribution of *Sauromalus* activity during the day, plotted with corresponding ambient temperatures. The top 2 sub-figures are for normal activity at Puerto Refugio in June 1972. The bottom 3 sub-figures represent the activity cycle for 4 introduced, large *S. hispidus* adults near Joshua Tree, California, and (extreme bottom) all the native *S. obesus* at the Joshua Tree site. The large *S. hispidus* emerge earlier and retreat somewhat later in the day than *S. obesus* and their activity is more evenly distributed.

MOVEMENTS AND SOCIAL BEHAVIOR

Sauromalus hispidus: During the spring and summer, individuals typically emerge from their overnight retreats one or two hours after sunrise. Smaller animals seem to emerge earlier than larger ones, and the number of visible animals gradually increases until 10:00 to 11:00 hrs. About 30% of the animals are first seen in groups of 2 to 4 individuals. Initially, the animals remain near the burrows or retreat entrances, orienting their bodies to the sun and occasionally adjusting their positions with minor body shifts. After basking for about 10 to 40 minutes, the animals may move 1 to 20 m to new basking locations which are closer to the feeding sites located on plateau rims or in arroyos. Movements begin and end with a brief head-bob display. Before feeding actually begins, the animals may have moved to 3 or 4 separate basking sites. Around 10:00 hrs many animals defecate and afterwards begin feeding in earnest.

Juveniles typically climb the vegetation which they eat (*Hyptis, Cercidium, Dalea spinosa, Opuntia,* see Feeding and Habitat), reaching an average height of 1.8 m. Adults more often eat overhanging vegetation from positions on rocks or on the ground, but may climb short, sturdy shrubs (e.g., *Jatropha, Bursera* or *Opuntia*) or large cardons. On any single day, many individuals, perhaps the majority, do not feed at all but spend their entire time within retreats or shuffling between retreats and basking sites. Although many animals bask and feed in small groups, these groups do not move in concert, and the composition of basking and feeding groups does not appear stable. This is equally true for adults and juveniles. There may be 7 lizards in a single night-time retreat. Whether the membership of these groups is stable remains to be determined. As yet, the proportion of the population which is marked is too small for definitive answers.

There is a notable lack of aggressive behavior in both *Sauromalus hispidus* and in *S. varius.* Animals often come into physical contact with each other during the course of a day with encounters usually resulting in no more than head-bob displays.

During the day, ambient temperatures may rise above preferred levels, particularly in the summer, and the chuckawallas will seek temporary shelters, occasionally reemerging and making short movements to feeding sites or new retreats. Some adults move as far as 900 m from their initial emergence site in a single day. Later in the afternoon, many animals will begin making their way back to their more permanent retreats in rocky hills or at arroyo cliffs. Still others, however, remain in the arroyo in temporary retreats for the entire night. In the summer, activity diminishes severely in the afternoon but a secondary short burst of basking activity usually occurs at dusk.

As in mainland *S. obesus,* activity in *S. hispidus* is greatly diminished in the fall and winter. For example, during a three-day visit to Palm Canyon in November, I found only one animal active and could find only 11 additional animals by searching rock crevices, burrows, cardons and the like. Such capture success is at least one-fourth that typically experienced in the spring. Maximum air temperatures in October were around 25°C, similar to temperatures in early spring when chuckawalla activity is much greater.

Sauromalus varius: The pattern of activity on San Esteban differs mostly in degree from that already described for *S. hispidus.* Since juveniles are rarely found, nothing is known of their behavior. The adults seem to initiate feeding sooner after they emerge from night-time retreats than does *S. hispidus,* and are more frequently found under dense cholla patches which provide both food and protection from predators. Less often, *S. varius* individuals are seen under or in *Macrocereus, Bursera* or *Cercidium.* Nearly every lizard captured has cholla spines in its body, lips and tongue, and red fluid on their chins from the fruits of *Macrocereus*.

Animals typically feed into the early afternoon. However, as in *S. hispidus,* on any one day many animals never leave their retreats or simply bask near their entrances. Yet, compared to *S. obesus,* feeding activity is still much more commonly seen in both the gigantic species.

Sauromalus varius appears to be more philopatric than *S. hispidus.* Out of 23 recaptures, 8 of which were separated by at least a 5-month period, none has occurred at a distance greater than 40 m from the original capture site.

The entrance to some of the *S. varius* burrows are nearly large enough for a man to enter on his knees. The largest are found in cross-layered sandstone or in overhanging shelves of volcanic rock, and inevitably become narrower in their inner reaches providing a secure retreat for a number of individuals. Such burrows are usually occupied by two to four adults. Out of ten such groups which I have sexed, six were male-female pairs, two were triplets involving one male and two females, one was a triplet of two males and one female, and the last was a group of 4 including one male and three females. One particular male/female pair occupied the same burrow for at least six years, and another pair was known to be stable for one year. Whether group composition typically remains this stable still remains to be seen.

Because such dens are often only 5 to 20 m apart, the occupants of different burrows are visible to one another and frequently will cross paths on the way to foraging areas in the arroyo below or the mesa and slopes above. I have never seen any aggression accompanying these encounters nor have I ever observed animals to avoid moving into an area already occupied by adults of another group. On the contrary, it is not uncommon to find 3, 4 or as many as 8 adults in feeding areas within a 10 m radius of one another. Head-bobs and push-ups typically precede and punctuate each movement and are more common in *S. varius* than *S. hispidus. Ctenosaura* are sometimes found occupying permanent dens, temporary retreats, or feeding areas with individual *Sauromalus.* For one week in April, I watched the same pair of *Ctenosaura,* an adult male and a juvenile, and a large adult female *Sauromalus* use the same den and sunning spots. The three animals were at times in physical contact. On a number of occasions, the juvenile ctenosaur basked while lying on the chuckawalla's back. Other adult *Ctenosaura* that roamed into view were chased away by the adult *Ctenosaura* of this trio but the *Sauromalus* did not react. As yet, I have not seen any interspecific aggression between ctenosaurs and *Sauromalus* although ctenosaur-ctenosaur chases are not infrequent.

Of course, one may argue that this apparent lack of overt aggresson in the gigantic *Sauromalus* simply reflects the fact that the social structure has already settled into a stable order. Perhaps aggression in the past was important in shaping this order, but since the animals are so long-lived and little recruitment takes place, the social organization receives little if any perturbation. Aggressive displays in this situation are absent due to a lack of provocation rather than to a lack of any innate aggressive tendencies. To test this hypothesis, I performed some transplant experiments in late April, 1979. Two adult males and one adult female *S. varius* were removed from an area off the end of my grid and each was placed about one quarter mile away in three adjacent burrows.

Each of these burrows was already occupied by one or two resident *Sauromalus* (two contained male-female pairs, and one contained a lone female with two ctenosaur consorts). In two cases, the transplanted animals went inside the residents' burrows and were not seen again, although the residents emerged and carried on apparently normal activity, always returning to their same burrows. In the third case, the introduced animal (a male) and the burrow's resident male basked together for the next 2 days at the burrow entrance; occasionally the resident female would also emerge. All three gave head-bob displays prior to movements but no aggression was seen. Sometimes the two males basked on top of one another.

The experiment was discontinued after three days so the ultimate duration of the association is unknown. These brief results indicate that the lack of aggression is not simply due to a lack of provocation. The possibility remains that aggression is suppressed during the years that the animals forego reproduction (see Ryan, this volume). However, even if this is the case, such years appear to be the rule rather than the exception.

POPULATION DENSITY, SEX RATIO AND BIOMASS

The capture and marking procedure described earlier (see Research Protocol) is sufficiently traumatic for most animals that they retreat to shelters and are not seen again for the next few days. My mere presence in the study area frightens animals to such an extent that for successive days capture success on the grid usually declines markedly. To get around these handicaps, I rotate my visits to the islands on a weekly basis, thus giving the animals a respite to return to normal activity.

Utilizing an entire island visit as a single sampling interval and adopting a Lincoln index, the *S. hispidus* population density at Palm Canyon was around 52 animals/hectare (about ⅓ were juveniles) in 1978-79 and that on San Esteban was about 12.4 animals/hectare (with no juveniles; Table 11.4). Based on timed searches, chuckawallas are about one-fourth as numerous at the more xeric and less productive study areas on the south end of Angel and at Puerto Refugio as they are at Palm Canyon (e.g., 7 or 8 animals/hr compared to 30 to 40 animals/hr). Hence, chuckawalla density seems tuned to food productivity but as yet intra-island differences in predation intensity and refuge availability cannot be ruled out. Likewise, these census results are only for the relatively wet years of 1978 and 1979. The extent that chuckawalla numbers fluctuate from year to year is unknown, but I suspect that density fluctuations are not nearly as great as fluctuations in foods levels. One thing for certain is that maximum island densities can be many times greater than chuckawalla densities on the mainland or in the *S. ater* populations that I have studied. Presumably this density increase reflects the lack of island predators and the expanded chuckawalla niche and resource base (see Feeding and Habitat).

The male/female sex ratio for *S. hispidus* on Angel (lumping all locations and dates) is 0.86 (N = 245) and therefore slightly favors females (Table 11.4). This tendency also holds for each location considered separately and for populations on the satellite islands and the San Lorenzos. On the other hand, the sex ratio for *S. varius*, 1.05 (N = 154), is slightly biased towards males. However, neither sex ratio is significantly different from 1.0. Sex ratio in *S. obesus* can be highly variable and seems to favor males in populations with large body sizes and high densities (e.g., China Lake and Little Lake in the Mojave).

Table 11.4: Comparative Demographic Data for *Sauromalus* Populations

Feature	*S. hispidus*	*S. varius*	*S. obesus*	Location	Ref.*
Pop. Density (#/hectare)	52 (Angel, Palm Canyon)	12.4 (San Esteban)	13.8	China Lake	2
	15 (Angel, South end)		7.1	Red Rock	3
Sex Ratio (N)	0.86 (245)	1.05 (154)	1.6 (116)	China Lake	2
			0.91 (69)	Red Rock	3
			0.97 (114)	Amboy	1
Biomass (kg/hectare)	20.8 (Angel, Palm Canyon)	7.8	1.7	China Lake	2
	6.0 (Angel, South end)		0.8	Red Rock	3

*References for *S. obesus* work only; 1, Case unpublished, 2, Berry 1972; 3, Johnson 1965.

The biomass maintained by the insular endemic chuckawallas is large; 20.8 kg/hectare for *S. hispidus* at Palm Canyon and 7.8 kg/hectare for *S. varius* on San Esteban (Table 11.4). The *Ctenosaura* at the latter site probably adds at least another 5 kg/hectare. For comparison, Iverson (1979) found adult densities of the insular *Cyclura carinata* to be as high as 30 to 35 animals/ha and also found a good correlation between density and habitat productivity. He estimated the biomass of *Cyclura carinata* on the Turks and Caicos Islands to be 5.2 kg/ha for juveniles and 17 kg/ha for adults (Iverson 1979). Within an order of magnitude, these biomass values for large insular herbivorous lizards are comparable to those for large ungulates in African National Parks. For example, elephants and buffalo were the most common ungulates when Bourliere (1963) performed his Ugandan surveys. Their biomass ranged from 45 to 51 kg/hectare for elephants and 3.6 to 6.1 kg/hectare for buffalo. Various antelope maintained biomasses roughly 10 times less.

My impression has been that over the past 10 years chuckawalla population densities have declined at Puerto Refugio on Angel and on San Esteban. Both areas offer excellent anchorages and are visited frequently by tourists. On San Esteban, hundreds of *S. varius* have been collected over the past decade for the pet industry and by amateur and professional herpetologists. Because recruitment in these endemic species is normally very low, continued collecting may prove catastrophic.

FOOD LIMITATION

During the relatively moist years of 1978 and 1979, my impression was that food was not a limiting commodity for either *S. hispidus* at Palm Canyon on Angel or for *S. varius* on San Esteban. At Palm Canyon it is difficult to quantify this impression because the diet of *S. hispidus* is so varied (see Feeding and Habitat). Conversely, the diet of *S. varius* at San Esteban during the spring and summer months was composed primarily of cholla fruits and flowers. It is relatively easy to calculate the impact that these animals made on cholla, their favored food, by combining estimates of *Sauromalus* density, metabolic demands, and assimilation efficiencies, with measurements of cholla standing crop and fruit and flower production.

The median body weight at San Esteban is about 700 g. The energy requirements of a free living iguanid of this size would be 10.2 kcal/animal day (Nagy this volume). The population density is 12.4 animals/ha, so the energy demand is approximately 127 kcal/ha day for the spring-summer period. Since assimilation efficiencies are at best 77% (Case unpublished; Hansen and Sylber 1979), each chuckawalla must consume about 165 kcal/ha day.

The standing crop of cholla on my grid was 41 cholla/ha with an average of 15 flowers/cholla and 17 edible, ripe fruits/cholla. The average wet weight of flowers is 12 g and fruits is 18 g. Both contain about 70% water and I assume that they would yield around 1.8 kcal/gram dry weight (Cummins and Wuycheck 1971). Combining these figures gives a standing crop estimate of about 10,800 kcal/ha of edible cholla. Hence, the chuckawallas are only consuming about 1.5% of the standing crop per day.

To reach an estimate of the total demand on cholla fruit, it is necessary to consider not just the standing crop, but also fruit production during the growing season. The average total number of buds, flowers and fruits on individual cholla in the middle of the 1979 season yielded 205 potential fruits/cholla. The rates that buds turned into flowers, flowers into spiny, unripe fruits and the latter to ripe, edible fruits were estimated by marking these parts on 6 separate plants and observing their development over a two-week period. Combining these rates, the average number of ripe edible fruits produced per cholla in a 6-month period is estimated to be 308. Again, using 18 g/fruit and 1.8 kcal/g dry weight and 41 plants/ha, over a 6-month period 122,500 kcal of cholla fruit would be produced per ha. The amount of this energy, consumed by *Sauromalus* (assuming cholla fruits is all that they ate) is only about 23% of that produced. Since at least 30% of the chuckawalla diet consists of other plant species, they alone did not appear to be making a severe dent in the total food supply for 1979. Although no herbivorous mammals occur on San Esteban to compete for this food, the diet of the sympatric *Ctenosaura hemilopha* is nearly identical to that of *S. varius*. From my survey estimates *Ctenosaura* appears to be almost as numerous. Both iguanids together may have eaten, at most, 46% of the available energy tied up in cholla fruit during the moderately wet 1979 activity season. Hence, although food was not severely limiting in this wet year, it was probably not so superabundant that all individuals found food readily available. Perhaps, this explains why even in a relatively wet year like 1979, only a small portion of mature females became gravid.

Although I do not have similar quantitative estimates for food production during dry years or in more xeric sites, my impression is that food may often be critically limiting. During the 1972 drought period at Puerto Refugio on Angel, cholla flowers and fruits were scarce. A generous estimate would put fruit abundance at only one tenth that measured in 1979 on both San Esteban and at Palm Canyon on Angel. Many of the common perennial plants did not produce blooms and even lacked leaves (e.g., *Fourquiera, Bursera, Cercidium, Jatropha*). In still others, the vegetation was probably too dry to be edible (see Nagy 1973). Annuals were nonexistent. The vegetation which was still edible was noticeably cropped and stems were trimmed back to short nubs. As I mentioned earlier, many *S. hispidus* at Puerto Refugio appeared dehydrated and emaciated. Carcassas were common and there was no measureable reproduction.

In short, the population density of the gigantic *Sauromalus* may be set by the frequency and severity of drought periods, during which food and water will be in short supply. Yet because of their limited recruitment, long generation times, and delayed sexual maturity, the chuckawalla population cannot rebound rapidly when environmental conditions improve. Without such density increases, food is probably not limiting during periodic wet years.

THE EVOLUTION OF LARGE SIZE

The three evolutionary hypotheses for insular gigantism that were presented in the introduction are not equally consistent with the present findings (albeit, fragmentary) on the insular gigantic *Sauromalus.* While many elements of both the food limitation hypothesis and the predation hypothesis are highly compatible with present data, the social-sexual hypothesis is much less attractive. The sex ratios of *S. varius* and *S. hispidus* are very close to unity and sexual dimorphism is virtually absent in both body size and coloration. Moreover, these species are unique in the absence of overt aggressive interactions. Larger-sized males do not seem to accrue any selective advantage involving greater social dominance, larger territories, or more mates. Of course, it is possible that social interactions are suppressed except during wet years of substantial breeding effort (see Ryan, this volume), but even so, such years must be infrequent.

A provisional hypothesis which is consistent with present observations but still demands further study, is as follows. The loss of most predators and, to a lesser extent, competitors on these islands allowed chuckawallas to utilize feeding places and times which on the mainland were either already exploited or too risky because of exposure to predators. The lack of predators on the islands and the resulting expanded niche space, in turn, enabled chuckawallas to reach higher densities. Even given these density increases, the supply to demand ratio of the chuckawallas for their food is probably no lower than on the mainland, at least in moderately wet years. Why chuckawalla densities should be held below levels capable of more fully exploiting food is problematic. The reason is apparently *not* due to territorial squabbling as suggested by Case (1978). More likely, the frequency and severity of the ever recurring droughts, along with the limited breeding effort by these chuckawallas, causes densities to lag behind carrying capacity whenever environmental conditions improve. During periods of environmental deterioration, mortality increases dramatically, and reproduction may stop altogether. The net effect is a population whose density reflects longer the mark or "residue" of poor conditions than it does of the productive periods.

Because food is sometimes abundant but fluctuates rather severely from year to year, a larger average body size will be favored. Complementing this tendency is the relative advantage large chuckawallas have against the typical suite of predators found on Gulf islands. These feed primarily on smaller lizards because of their relatively small size. This scenario combines elements of both the food availability and the predator hypotheses. More data on size-specific predation and the extent of fluctuations in chuckawalla density and food productivity are sorely needed to judge the relative contribution of these two factors. The selective forces responsible for the unique "boom or bust" interoparity that individuals of the insular gigantics display, and their remarkable lack of aggression are also still baffling.

SUMMARY AND CONCLUSIONS

Giant chuckawallas, *Sauromalus hispidus* and *S. varius* are found only on a group of islands in the Gulf of California. These two species have exceptionally large body sizes compared to mainland relatives and to other endemic *Sauromalus* inhabiting different islands in the same area. This chapter presents results from an ongoing comparative study of the life histories and evolution of these lizards.

The insular species probably arose between one and two million years ago. Various lines of evidence suggests that these gigantic endemic *Sauromalus* have, by mainland standards, long lives and low recruitment. Moreover, population densities are high and biomass levels (7.8 to 20.8 kg/ha) can rival those typically found for mammalian ungulates.

S. varius individuals are extremely sedentary, and the same individuals are found year after year in the same locations. Overt aggression and defense of homesites are lacking, and individuals often cohabit preferred hiding places with specific other *S. varius* or with individual *Ctenosaura hemilopha*. Unlike other insular endemics the clutch size of *S. varius* and *S. hispidus* is not reduced below levels allometrically predicted based on mainland *S. obesus*. However, the frequency of reproduction is strongly reduced in *S. varius* and seems to coincide with years of heavy rainfall. Their diet consists largely of seasonally available flowering shoots of perennials such as ironwood, mesquite and palo verde; and cactus fruits–largely cholla and cardon. Food overlap with sympatric *Ctenosaura* is great but *Ctenosaura* are much more arboreal.

In *S. hispidus* reproduction is more frequent than in *S. varius,* but again is rarer than in mainland populations and is dependent on rainfall and moisture conditions. Juveniles are more arboreal than adults and overall *S. hispidus* is more arboreal than either the mainland *S. obesus* or *S. varius.* The diet of *S. hispidus* consists of seasonally available perennial flowering shoots (e.g., ironwood, *Hyptis, Dalea, Acacia*) and cactus fruits. Juveniles and small adults may fall prey to ravens and red-tailed hawks. Mortality of all age classes occurs in arid sites following a succession of drought years.

Compared to mainland *S. obesus,* both *S. hispidus* and *S. varius* spend a greater proportion of their time feeding and feed at a greater variety of places and times when food is available. During wet years food is probably not severely limiting but during drought years food limitation may be intense. Compared to mainland *S. obesus,* the island gigantics seem to devote a greater absolute and proportionate amount of energy to growth and maintenance at the expense of reproduction. Such a suite of life history characteristics is in keeping with theoretical expectations for evolution in a relatively predator-free environment. Because recruitment in these endemic species is normally very low, any additional mortality from human exploitation could prove catastrophic.

Acknowledgments

I am indebted to Ken Abbott, Martin Cody, Benita Epstein, Mike Gilpin, George and Molly Hunt, Bill Mautz, Don and John McMullen and Steven Ruth for assistance in the field. David Wake, Alan Leviton, Allan Sloan and John Wright allowed me access to museum specimens. Martin Cody, John Iverson, Bob Murphy, Ken Nagy, Ken Norris, J. Ottley, Mike Robinson, Norm Scott, and Mike Soulé shared assorted unpublished data and field notes. This work was supported by the Graduate Division at the University of California at Irvine, the Purdue Research Foundation, the National Geographic Society, and the Academic Senate at the University of California at San Diego. R.E. MacMillen provided guidance and criticism. I also thank Jack Bradbury, John Iverson and Bob Murphy for their comments on the manuscript.

Section IV

Adaptive Behavior and Communication

The behavior of iguanines, as of all animals, must be rather rigorously devoted to survival. We have already seen this in feeding behavior and female nesting. In this section we focus more directly on the mechanisms of behavior rather than the ultimate evolutionary reasons discussed in the following section on social organization.

The first two chapters, appropriately enough, deal with the head-bob displays of iguanine lizards. Herpetologists lagged far behind ichthyologists, ornithologists, and mammalogists in applying the concepts and methods of ethology to social behavior. The first, and probably still the most elegant, area studied was the analysis of bobbing in diurnal visually oriented lizards, especially in the family Iguanidae. Through the efforts of Charles Carpenter and his students a body of data on a variety of species was collected, quantified through film analysis, and interpreted in an evolutionary framework. Many lizards give head-bob displays in a variety of contexts, but male assertion displays seem the most ubiquitous and functionally clear. Iguanine lizards have not been studied with the detail of the smaller iguanid lizards. This is due to their large size and consequent difficulties in arranging suitable captive environments as well as the difficulties of field observation. In the first chapter in this section Carpenter (Chapter 12) gives an overview of the published literature on iguanines pointing out similarities, differences, and our general ignorance that hinders an evolutionary treatment of iguanine displays.

Greenberg and Jenssen (Chapter 13) provide a detailed display analysis of one of the most threatened and unknown iguanines, the Fiji banded iguana. They document the great amount of variability between individuals as well as the complexity of the displays. In addition, color changes occur rather quickly and presumably contribute to the information communicated. Other iguanines, including juvenile green iguanas, also change color, but to a reduced extent and for reasons largely unknown.

One of the problems in the study of iguanine behavior is the lack of a general ethogram that can be used in the analysis of behavior systems and in comparisons across species. Distel and Veasy (Chater 14) provide a catalogue of the basic movements of green iguanas based on extensive laboratory study. This work is important for viewing specific types of behavior, such as displays, in their appropriate context, and also served as a baseline in brain stimulation studies by these same workers. When more such data are in, including quantitative estimates and field observations, an ethogram in the classic mold will be possible.

The behavior of young animals is as important as that of the more commonly studied adults. Not only is juvenile behavior critical in surviving to adulthood but its study can address critical issues in the development of individual actions. Drummond and Burghardt (Chapter 15) have asked specific questions about how newly hatched iguanas migrate from an islet natal area, and the perceptual and social mechanisms employed. Their evidence for a socially mediated and endogenously motivated initial dispersal phase in young has implications for several areas of iguanine behavior and sociobiology as well as husbandry and conservation programs.

Marine iguanas on the Galapagos Islands are exotic and famous reptiles, figuring prominently in Darwin's writings and innumerable popular articles and television documentaries. And not without reason, as their aquatic specializations are truly unique. They live in a surprisingly harsh and limited environment. Boersma (Chapter 16) analyzes the nocturnal sleeping aggregations often found in this species and provides a persuasive case for heat conservation and resultant faster digestion as being the adaptive reason for this behavior. This chapter demonstrates again how useful reptiles are in answering questions on both the hows and the whys of behavior more difficult to deal with in the metabolically and behaviorally more active birds and mammals. Yet on the level of mechanism few qualitative differences between reptiles and endothermic vertebrates have ever been documented.

12

The Aggressive Displays of Iguanine Lizards

Charles C. Carpenter
Department of Zoology
University of Oklahoma
Norman, Oklahoma

INTRODUCTION

With possible minor exceptions, lizards of the family Iguanidae all exhibit aggressive displays which involve temporal movements of the head (head bobs and head nods), the head and anterior trunk or the entire lizard (pushups). These displays are performed primarily by males usually as a declaration of territory or in encounters between individuals (Carpenter, 1967a). The patterns of posturing and movements of these displays (display-action-patterns) are generally species-specific or species-typical. The display-action pattern (DAP) includes the sequence of movements through time and may occur in a fraction of a second or extend for a number of seconds (cadence). Movements (either head nods and bobs, or pushups) may vary in height (amplitude) and the length of time held in one position. The intervals between movements (pauses) may be a significant part of the display-action-pattern. These display movements through time can be depicted graphically (DAP graphs) and used for comparisons between species.

The objectives of this chapter are to describe the aggressive displays of the species of iguanine lizards for which information was available to me, limiting these descriptions to the display-action-patterns and associated behavior, to determine similarities and differences at the generic and specific levels, to relate these displays to lizard size and sympatry, and to compare with the aggressive displays of other groups of iguanid lizards.

METHODS AND PROCEDURES

My sources of information were varied (Appendix). I have used published data from my own and other papers, some of my unpublished data, as well as records, manuscripts, motion pictures, and videotapes generously loaned by others.

In order to make comparisons and present them in a simple manner, I have prepared all of the display-action-pattern graphs (Figures 12.1 through 12.3) to the same scale. Thus I have had to reconstruct some of the DAP graphs from published accounts, subject to my interpretation (Berry, 1974; Iverson, 1977; Lazell, 1973; Muller, 1972; Prieto and Ryan, 1978; Wiewandt, 1977). Analysis relied primarily on 16 mm motion pictures measured via a Vanguard Motion Analyzer, but Super-8 films and videotapes were analyzed with a stop watch. I believe both techniques produce comparable results. The sources of species recorded, where possible, are listed in the Appendix. I shall describe only briefly certain species. The table and figures can provide this information in a more succinct way.

As far as possible from the information available, the aggressive display of each species is described by eight characteristics (Carpenter, 1962, 1978): Site, Position, Posture, Parts Moved, Type of Movement, Units of Movement, Sequence, and Cadence.

It should be remembered that the data summarized in the table and figures have come from a variety of sources and have been subjected to my interpretation, that the sample sizes are quite variable, that the performing lizards may have been vigorously challenging with posturing, or only performing an assertion display (Carpenter, 1967a) which influences the amount of posturing, orientation and amplitude of movements.

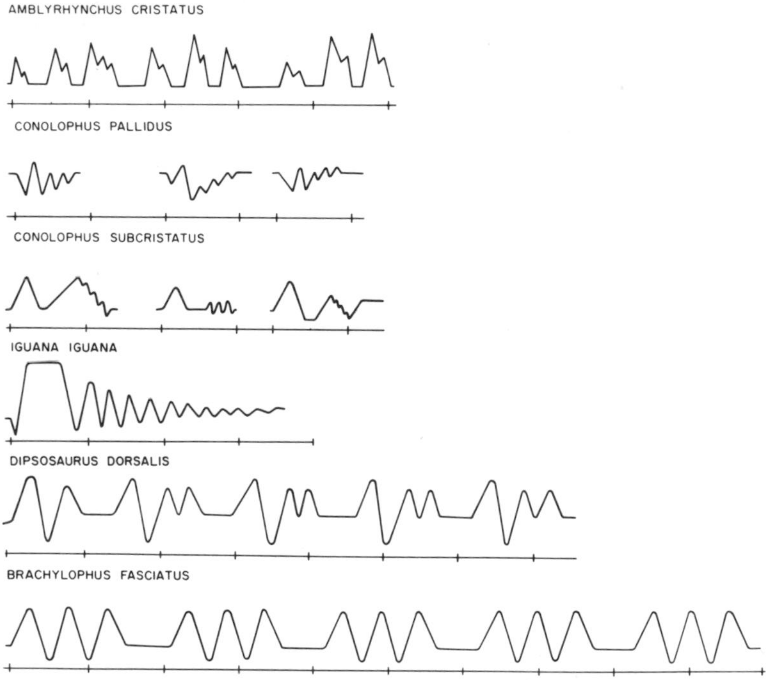

AMBLYRHYNCHUS CRISTATUS

CONOLOPHUS PALLIDUS

CONOLOPHUS SUBCRISTATUS

IGUANA IGUANA

DIPSOSAURUS DORSALIS

BRACHYLOPHUS FASCIATUS

Figure 12.1: Display-action-pattern graphs of species of genera included in the iguanine group of iguanid lizards, produced to a similar scale. Intervals indicated are 1 second. The vertical scale is relative to the type of movement, whether head bob or pushup.

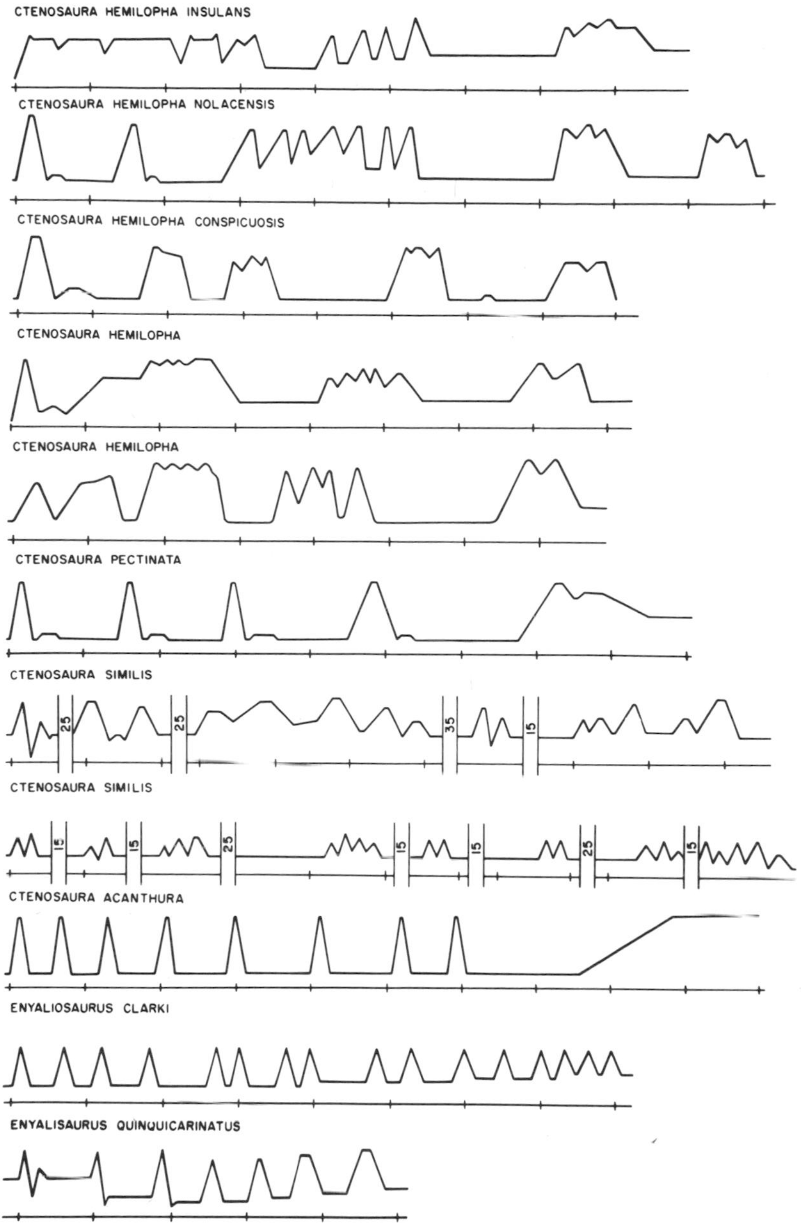
CTENOSAURA HEMILOPHA INSULANS
CTENOSAURA HEMILOPHA NOLACENSIS
CTENOSAURA HEMILOPHA CONSPICUOSIS
CTENOSAURA HEMILOPHA
CTENOSAURA HEMILOPHA
CTENOSAURA PECTINATA
CTENOSAURA SIMILIS
25
25
35
15
CTENOSAURA SIMILIS
15
15
25
15
15
25
15
CTENOSAURA ACANTHURA
ENYALIOSAURUS CLARKI
ENYALISAURUS QUINQUICARINATUS

Figure 12.2: The same as Figure 12.1. The vertical gaps with numbers for certain Spiny-tailed Iguanas indicate intervals in seconds. *(Enyaliosaurus = Ctenosaura,* see Etheridge, this volume.)

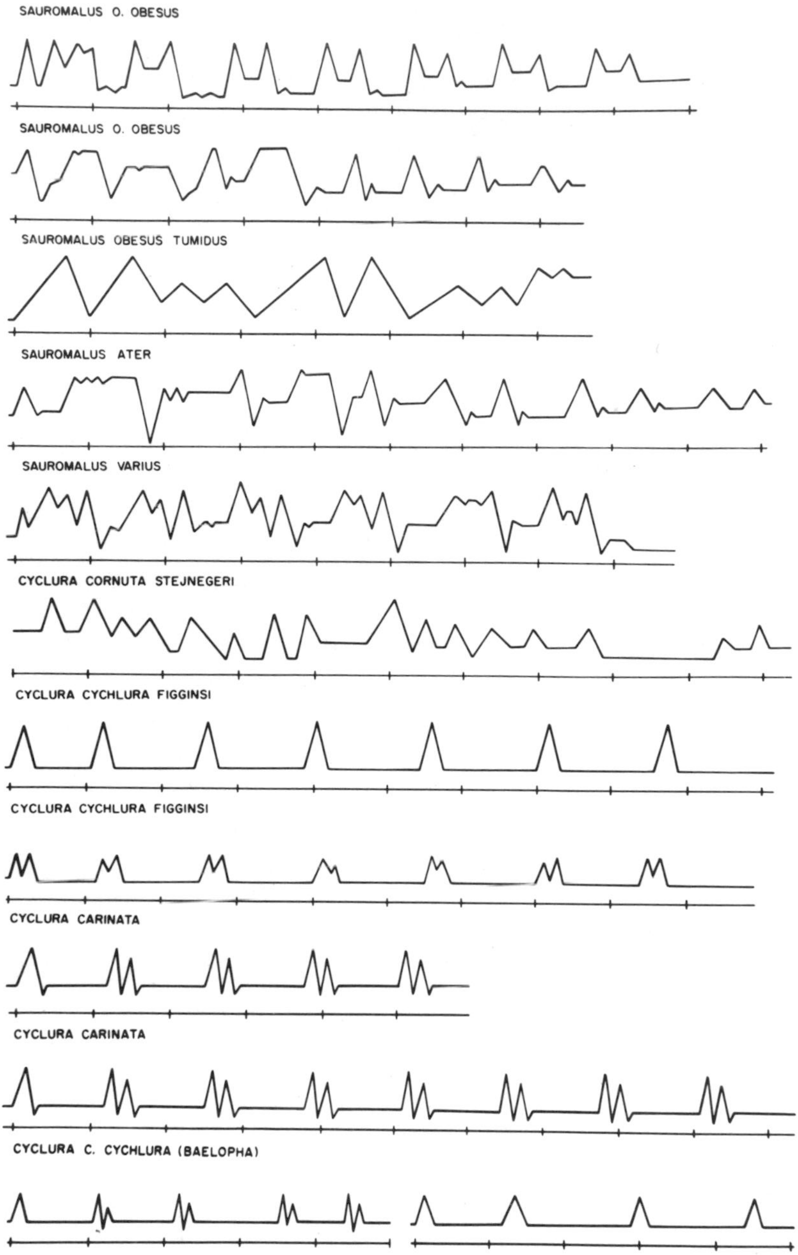

SAUROMALUS O. OBESUS

SAUROMALUS O. OBESUS

SAUROMALUS OBESUS TUMIDUS

SAUROMALUS ATER

SAUROMALUS VARIUS

CYCLURA CORNUTA STEJNEGERI

CYCLURA CYCHLURA FIGGINSI

CYCLURA CYCHLURA FIGGINSI

CYCLURA CARINATA

CYCLURA CARINATA

CYCLURA C. CYCHLURA (BAELOPHA)

Figure 12.3: The same as Figure 12.1.

RESULTS

The following brief discussion explains the categories and points out significant features for particular species. Table 12.1 presents the full analysis of all categories for all species.

Site: The site from which a lizard displays. Both raised sites with a broad view and sites on the ground or flat areas with restricted views are used by these lizards, though the raised site appears to be preferred by most forms. This may be related to the habitat niche of the species, i.e., the green iguana is quite arboreal and thus generally displays from a raised perch; the land iguana and the marine iguana display from flat areas but also will use boulder tops and the edges of cliffs as territorial display sites. The desert iguana lives on the flat or rolling desert so it displays from flat or prominent areas on the desert. All of these forms, except the green iguana, when in a close encounter with a conspecific, are usually on a flat area when displaying.

Position: This is how the lizard orients while displaying and usually depends on whether it is asserting or challenging and how far away the lizard being challenged may be. With assertion there is usually no particular orientation, but when encountering another male at close range, a few centimeters to a meter, the displaying male then presents himself laterally to his opponent. The other male responds with a similar orientation, heading in the opposite direction in the faceoff position.

Posture: Posturing involves relative changes in the parts of the lizard and functions to increase the size of the laterally viewed animal.

Most challenging iguanines lower the head (especially in those species that head bob), arch the back, inflate or bloat the trunk region and open the mouth, in some species exposing a fleshy tongue. The throat or gular region is expanded, but this varies. The green iguana has a large permanent gular flap or dewlap, while the spiny-tailed iguanas and the ground iguanas exhibit permanent, but proportionately smaller, gular enlargements. Lateral compression of the trunk is discernible, though only sightly, in the green iguana, desert iguana, and spiny-tailed iguanas, but apparently is not done by the marine iguana, land iguana, chuckawallas and ground iguanas. Lateral compression, where it occurs, does not reveal or accentuate color or marks as it does in many other iguanids.

Laterally viewed size is also increased by raising a roach of skin along the mid-dorsal region of the trunk and neck, which is usually topped by an erect crest of serrated scales. A roach is characteristic of displaying male marine iguanas, land iguanas, and ground iguanas, while a spiny-tailed iguana may raise its crest. Males of some species (i.e., green iguana) have larger crests than females. Apparent increase in size may also result when a male rises on four extended legs, especially when challenging. Aggressive male banded iguanas take on a brighter color which forms bands of light and dark green.

The tail in iguanines is not used as a part of a display, but may be used in fighting between two males, where each lashes or slaps the other with wide swings of his tail.

Parts Moved and Type of Movement: The parts that are moved while displaying and the type of movement involved are considered separately for analytical purposes. Some species move only the head (Figure 12.4) while other

Table 12.1: Characteristics of the Aggressive Displays of Iguanine Lizards (Refer to Display-Action-Pattern Graphs)

Species	Site	Position	Posture	Parts Used	Type of Movement	Units of Movement*	Sequence*	Cadence*	Aggression
Marine Iguana *Amblyrhynchus cristatus*	raised boulder or cliff	laterally presents faceoff	gape gular expansion inflates trunk lowers head erects roach and crest 4-leg rise exposed tongue	head	head bob	single or double or triple	repeated sets of 3 units	units = 0.125–0.5 sec pause = 0.1–0.3 sec seq. = 0.8–1.5 sec	head butt bite ride gape
Banded Iguana *Brachylophus fasciatus*	branch of tree	laterally presents faceoff	gular expansion	head and neck	head bob	single	triplets of 3 units	triplet = 1.2 sec pause = 0.57 sec	lunge bite color change
Land Iguanas *Conolophus pallidus*	raised or not	faceoff	lower head arched back inflates trunk gular expansion erects roach, crest	head	head bob	single and multibobs	single followed by descending multibob	seq. = 0.75 sec pause = 1.2–2.1 sec	lunge bite tail lash ride head butt (infreq.)

*Refers to symbols used in column appearing at the end of the table.

(continued)

Table 12.1: (continued)

Species	Site	Position	Posture	Parts Used	Type of Movement	Units of Movement*	Sequence*	Cadence*	Aggression
Conolophus subcristatus	variable raised or not		lower head gape crest and roach gular expansion inflates trunk	head	head nod or bob	multinod	series of 3 to 4 multinods of decreasing amplitude	seq. = 1.3 sec pause = 1.6 sec	lunge bite tail lash
Ground Iguanas *Cyclura carinata*	raised or level	laterally presents faceoff	gape gular expansion arch back roach and crest inflates trunk lateral compress 4-leg rise	head and neck	head bob	single or double	single and 3 doubles single, double and single	seq. = 0.28–0.35 sec D = 0.37–0.5 sec pause = 0.69–1.03 sec	tail lash gape dewlap fight
Cyclura cornutum stejnegeri		laterally presents	gape dewlap erects roach and crest	head	head bob and head roll (3-dimensional)	variable	irregular	S = 0.3–0.6 sec	
Cyclura cychlura (baelopha)	raised or level	faceoff		head	head bob	single or double	S-S-S or S-D-D or D-D-D	S = 0.19–0.3 sec D = 0.25 sec pause = 0.8–1.0 sec	chase gape

(continued)

Table 12.1: (continued)

Species	Site	Position	Posture	Parts Used	Type of Movement	Units of Movement*	Sequence*	Cadence*	Aggression
Cyclura cychlura figginsi	raised or level	laterally presents	lowers head gular extension erects roach and crest inflates trunk gapes	head	head bob	single or double	repeated S or D or both 2-5 up to 68	seq. = 0.3 sec D = 0.36 sec pause = 0.74–1.4 sec	lower head roach and crest
Cyclura nubila nubila *(macleayi)*		(faceoff)	gape crest	head	head bob or head shake		stereo-typed		gape lower head bite chase tail lash
Cyclura pinquis					sagittal plane		stereo-typed		
Cyclura ricordi					sagittal plane		stereo-typed		
Cyclura rileyi					head bob or pushup				
Spiny-Tailed Iguanas *Ctenosaura acanthura* (Vera Cruz, Mexico)	raised or level	(faceoff)	gape dewlap lateral compress crest arch back	head, neck anterior trunk	pushup	single	repeated singles	S = 0.25 sec pause = 0.5–0.75 sec	

(continued)

Table 12.1: (continued)

Species	Site	Position	Posture	Parts Used	Type of Movement	Units of Movement*	Sequence*	Cadence*	Aggression
Ctenosaura hemilopha insulans (Ceralvo Isl.)		(faceoff)	(lateral compress)	head, neck anterior trunk	pushup	LS MP M	LS-S-S-M-M	S = 0.3 sec LS = 2.2 sec M = 1.3 sec pause = 0.6–1.7 sec	
Ctenosaura h. hemilopha (Nolasco Isl.)		(faceoff)	dewlap arch back lateral compress crest gape	head, neck trunk	pushup and head bobs	S MP M	S-S-MP-M-M	S = 0.4 sec MP = 2.6 sec M = 1.0 sec pause = 0.9–1.8 sec	arch back bite chase
Ctenosaura hemilopha (San Estebán Isl.)		(faceoff)	dewlap arch back lateral compress crest gape	head, neck anterior trunk	pushup and head bobs	single pushups multibob pushups	S-S-M-M-M	S = 0.5–0.7 sec M = 0.8–0.9 sec pause = 0.5–1.5 sec	arch back bite
Ctenosaura hemilopha		(faceoff)	(lateral compress)	head, neck anterior trunk	pushup and head bobs	S LS M	S-LS-M-M	S = 0.6 sec LS = 0.9 sec M = 1.2–1.4 sec pause = 0.6–1.6 sec	
Ctenosaura pectinata	raised	(faceoff)	dewlap gape lateral compress	head, neck anterior trunk	pushup head bob	singles of increased length	repeated single	S = 0.3–0.6 sec pause = 1.1–1.6 sec	gape gular flap head high
Ctenosaura similis	raised or level		dewlap gape	head	head bob	single	2 to 5 low and fast 2 or 3 quick, vertical head jerks	D = 0.4–0.5 sec S = 1.25 sec long pauses	bite fight

(continued)

Table 12.1: (continued)

Species	Site	Position	Posture	Parts Used	Type of Movement	Units of Movement*	Sequence*	Cadence*	Aggression
Ctenosaura (Enyaliosaurus) clarki	raised		dewlap lateral compress	head, neck anterior trunk	pushup	single	groups of 4 singles up to 18	S = 0.25 sec 4S = 1.83 sec pause = 0.58 sec	
Ctenosaura (Enyaliosaurus) quinquicarinatus	raised			head, neck	head bob slight pushup	single	repeated singles	S = 0.25 sec increasing to 0.5 sec pause = 0.25–0.6 sec	
Desert Iguana *Dipsosaurus dorsalis*	raised or level	laterally present	lower head gular expansion lateral compress arch back	head, neck anterior trunk	pushup	single dougle	S D	S = 1.0 S + D = 1.2 pause = 0.4	lower head tail lash
Green Iguana *Iguana iguana*	raised		gape dewlap lateral compress crest and roach	head	head bob vibrate	single multibob decrease amplitude	S-M	S = 0.8 sec M = 2–2.5 sec	fight tail lash
Chuckawallas *Sauromalus ater*	raised	laterally present	lower head inflate trunk	head, neck anterior trunk	pushup and head bob	S M Dip	S-M-Dip-M-S-Dip-S-D-S-Dip	S = 0.3–0.6 sec M = 0.7–1.3 sec	

(continued)

Table 12.1: (continued)

Species	Site	Position	Posture	Parts Used	Type of Movement	Units of Movement*	Sequence*	Cadence*	Aggression
Sauromalus obesus	raised	laterally present	gape dewlap lower head arch back tilt inflate trunk	head, neck anterior trunk	pushup and head bob	S D M	S-M-D-D-D	S = 0.25 sec M = 0.7 sec D = 0.69–0.75 sec pause = 0.4–0.5 sec	fight bite tail lash chase
Sauromalus varius	level					M	M-M-M	M = 0.9–1.25 sec	

*S = single bob, nod or pushup
LS = long single
D = double bob, nod or pushup
M = multibob, multinod or multipushup (MP)
Seq. = time to complete sequence
Dip = lizard lowered to substrate rapidly

Note: Parentheses mean not observed but predicted for this form.

species move all parts from the head posterior to the anterior trunk. Thus the movement types vary from simple, shallow head bobs (marine iguana) where the head is moved rapidly up and down, to pushups where the head, neck and anterior trunk are raised and lowered by extension and flexion of the front legs (desert iguana).

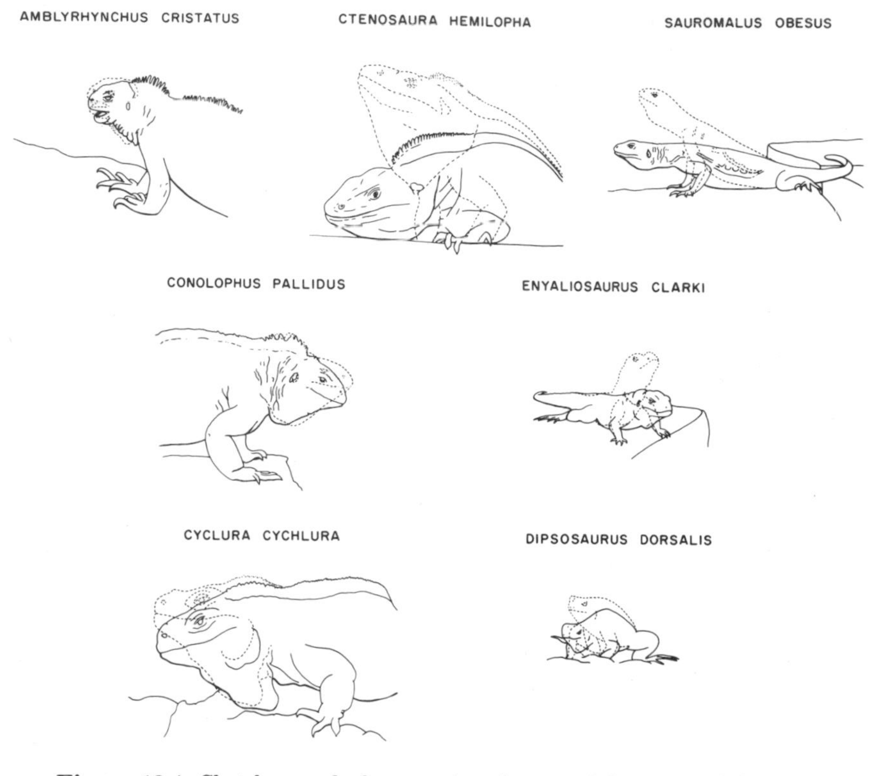

Figure 12.4: Sketches made from motion pictures of the type and degree of movement during the performance of the display-action-pattern for species of the iguanine genera of iguanid lizards. The genus *Iguana* (green iguana) is not shown. *(Enyaliosaurus = Ctenosaura.)*

Units of Movement: These units are arbitrary parameters selected for the convenience of the observer, to discern and analyze patterning. The units may be a single head bob or pushup, or double or triple head bob or a multibob (many bobs in rapid succession). The amplitude of the movement (the degree of movement up and down) may be high, low or held at a certain level, or it may involve a very low dip by the lizard. All of these units proceed through time and at a particular rate (cadence), which can be short (a fraction of a second) or long (over a second or more). The pauses between units are a significant part of the display pattern.

Examples of units are as follows (Figures 12.1 through 12.3): (1) simple head bob (ground iguana), (2) double and/or triple head bobs (marine iguana, ground iguana, spiny-tailed iguana), (3) single pushup (desert iguana, followed by a low dip) (spiny-tailed iguana), (4) multibobs (land iguana, spiny-tailed iguana), (5) pushups held at different amplitudes (spiny-tailed iguana, chuckawalla).

Sequence: A sequence consists of a series of units in succession forming the display-action-pattern, or the total display. Some sequences always consist of the same number of units. This type of sequence is termed Determinate. Other sequences consist of variable numbers of repeated units, i.e., from one or two up to 10 or 20 or more, and this type is termed Indeterminate. The best examples of Determinate and Indeterminate display-action-patterns are exhibited by other groups of iguanids. However, among the iguanines, Determinate display-action-patterns appear to be used by the desert iguana, marine iguana, and land iguana (Figure 12.1).

Cadence: The cadence is the duration of units, pauses and sequences, thus giving a temporal measure to these time-motion events. The cadence of a display-action-pattern is generally quite consistent. Thus a display-action-pattern is a series of movements through time producing a pattern which is typical for a species, a stereotyped or species-specific pattern (Figures 12.1 through 12.3).

Aggression: Aggression is mentioned here because some iguanines exhibit unique actions different from the display-action-pattern. Threat displays, without display movements, usually involve the same types of posturing as seen when displaying. It is during actual contact fights that peculiar behaviors are seen, other than lunging, biting, and tail slapping. For example, two male marine iguanas perform head butting bouts or contests, one attempting to butt the other from his territory (Carpenter, 1966b).

Comparisons of Display-Action-Patterns: When the display-action-patterns of congeners are compared it can be observed (Figures 12.1 through 12.3) that the species within each genus show similar characteristics. For those genera where I present data for more than one species, cadence and posturing are similar, as are the sequences of the display-action-pattern (ground iguana, land iguana, spiny-tailed iguana, chuckawalla), i.e., most species of ground iguanas perform repeated single or double head bobs, land iguanas, a single head bob followed by a multibob, chuckawallas, a series of repeated pushup patterns.

The forms of spiny-tailed iguanas show marked differences in their pushup-head bob patterns, with two species having simple repeated units *(C. acanthura, C. pectinata)* while others produce more complex display-action-patterns (Figure 12.1).

Intergeneric differences and similarities are apparent in both the type of movements and display-action-patterns. The marine iguana, land iguana, ground iguana and green iguana are similar in performing only head bobs, but differ in the display-action-patterns produced. The spiny-tailed iguanas, chuckawallas and desert iguana perform pushups, but have different display-action-patterns.

The larger species of iguanines (marine iguana, land iguana, ground iguana, green iguana) perform only head bobs, while the small species (desert iguana, chuckawalla, spiny-tailed iguana) perform pushups. I suggest this is related to size and potential increased energy costs for larger animals as proposed by Purdue and Carpenter (1972a, b) for sceloporine iguanids. The smaller iguanines also exhibit more elaborate postures.

The degree of species-specificity in the display-action-patterns in many iguanines may be related to the occurrence or non-occurrence of sympatry with similar sized iguanids. Those forms which are potentially sympatric have very distinct display-action-patterns: marine iguana and land iguana occur in close proximity on certain islands of the Galapagos Archipelago and have distinct display-action-patterns of head bobs. The desert iguana and species of chuckawallas occur in adjacent habitats, but have displays of pushups that are distinct. The green iguana may occur in sympatry with species of spiny-tailed iguana, and spiny-tailed iguana forms occur sympatrically with chuckawallas, but all have distinctly different display-action-patterns, as does the spiny-tailed iguana when compared with sympatric sceloporines in Mexico.

In the case of those iguanine forms which are allopatric with other large iguanids, such as many species of ground iguanas and certain insular forms of spiny-tailed iguanas, my data indicate display-action-patterns are more variable and differences have proably been affected both by phylogenetic and genetic drift factors.

When the aggressive displays of iguanines as a group are compared to other groups of the family (sceloporines, basiliscines, tropidurines, and anolines) certain features are remarkably similar. These similar family characteristics are that the aggressive display involves some form of display-action-pattern (head bobs or pushups), and that similar postural changes (gular expansion or dewlap, lateral compression of the trunk) and orientation (faceoff) are used. It is probably that these behaviors are homologs.

The non-iguanine groups are, for the most part, made up of smaller species. These size differences relate directly to differences between these groups and iguanines. The smaller forms generally have more exaggerated posturing (lateral compression of the trunk and extension of the dewlaps (notable exception is the green iguana) which may be an adaptation to displaying bright colors (usually by the males) or contrasting markings. As examples, many anolines have very elaborate dewlaps of contrasting colors and the dewlap pulsates during the aggressive display (Carpenter, 1967a), while sceloporines such as many fence lizards *(Sceloporus)* and tree lizards *(Urosaurus)* exhibit bright ventrolateral and gular colors brought out by their lateral compression and dewlap extension, and the sand lizards [earless lizards *(Holbrookia)*, zebra-tailed lizards *(Callisaurus)*, fringe-toed lizards *(Uma)*] exhibit marking or color and markings in a similar way. Basiliscines, which are intermediate in size, do not show bright colors.

Non-iguanine forms, generally, have more precise, stereotyped, rigid, species-specific display-action-patterns. Following my previous line of reasoning, this may be because they usually exist in sympatry with other iguanids of similar size.

SUMMARY AND CONCLUSIONS

The aggressive displays of the iguanine group of lizards exhibit the typical features known for the family Iguanidae. I have compared only the display-action-patterns and related posturings of 26 forms from the eight genera using eight categories (site, position, posture, movement type, parts moved, units of movement, sequence of movements and cadence) and presented these data in tabular and graphical form.

The site used for displaying relates to the habitat niche-ground and rock dwellers displaying there, and arboreal forms from their respective perches. Position (orientation) has no particular direction unless in a close encounter where some forms present laterally. The postures of the smaller species involve more body changes than seen in large species, but all use methods to increase size and indicate aggression, i.e., arching the back, inflating the trunk, extending the gular area, gaping the mouth, and some lateral compression of the trunk.

The types of movements vary from head bobs with little amplitude (marine iguana, land iguana, ground iguana), those with exaggerated head bobs (banded iguana, spiny-tailed iguana, green iguana) and sometimes involving slight front leg extension, to those performing pushups with marked front leg extension (desert iguana, chuckawalla). The parts moved by the lizard range from slight head bobs to exaggerated pushups involving the head, neck and anterior trunk. The amount of movement appears to be related to species size.

The units of movement vary from single head bobs or pushups, double and triple units in rapid succession, to multiple bobs and pushups of varying amplitude. Sequences of repeated patterns (Determinate) are more characteristic of the smaller species, whereas the sequences of the larger species are more irregular (Indeterminate). The cadence of units varies from 0.3 second for a single head bob to 1 second or more for head bob or pushup. Sequences may last only 1 to 2 seconds or may extend up to 10 seconds or more.

Those species belonging to the same genus exhibit similarities in their aggressive displays. The larger species performs primarily head bobs, the smaller species pushup movements. The degree of variability in display-action-patterns of a species seems to be related to sympatry with similar sized forms and to degrees of isolation, i.e., island forms.

When the aggressive displays of the species of the iguanine groups are compared to those displays of other iguanid groups (sceloporine, anoline, tropidurine, basiliscine), the differences can be related to size, for most of the small non-iguanine forms perform pushups or, as in the anolines, have transferred their main display signals to dewlap pulsing. The smaller forms generally have more pronounced posturing and sexual dimorphism. The non-iguanine groups generally have more precise (non-varying) display-action-patterns.

The display-action-patterns, as a part of the aggressive display behavior of iguanine lizards, have evolved as ritualized fixed action patterns. It is highly probable that these displays function as a visual communication system for species, sex, and possibly individual discrimination, and also serve other social functions relating to the contexts of territoriality and dominance. These functions will be further tested by future workers.

The rewards and insights I have obtained through my studies of the aggressive displays of the iguanines, as well as a very large number of other iguanid and agamid lizards, are many. For almost all species for which the display-action-patterns have been recorded, each form has a species-typical display–a signature–just as each species of bird has its own species-typical song. This is a remarkable achievement in evolution.

The insights have posed many questions for future research. Is there any relationship between types of displays and display-action-patterns and the ecological parameters of a species? Is there a type of display that best fits certain ecological conditions? Do similar sized iguanids and agamids, living on opposite sides of our planet, but under very similar ecological conditions, have similar aggressive behaviors, displays and social structure?–ecological and behavioral homologs? How did the display movements that produce the stereotyped display-action-patterns evolve–are they derived from a movement originally serving another function? Finally, how is such an innate time-motion mechanism contained and produced in the central nervous system of the lizard? These are but a few of the questions, but they indicate we are just opening the door on understanding what lizards can teach us through their aggressive displays.

Acknowledgments

I wish to thank the following for providing me with manuscripts, videotapes and motion pictures for my study: Robert F. Clarke, John B. Iverson and Walter Auffenberg, Thomas A. Jenssen, and Thomas A. Wiewandt. I also wish to acknowledge the assistance of grants from the National Science Foundation, the University of Oklahoma Research Institute, and the University of Oklahoma Biological Station.

Appendix

The sources of information for preparing the display-action-pattern graphs and approximate localities of the specimens, where known, in this paper are as follows: *Amblyrhynchus cristatus* (Carpenter, 1966b) – Galápagos Islands, Ecuador; Brachylophus fasciatus (Carpenter and Murphy, 1978) – Fiji Islands; *Conolophus pallidus* (Carpenter, 1969) – Barrington Island, Galápagos Islands; *Conolophus subcristatus* (Carpenter, 1969) – South Plazas Island, Galápagos Islands; *Ctenosaura acanthura* (Clarke and Robison, 1965) – Taumalipas, Mexico; *Ctenosaura hemilopha* (Clarke and Robison, 1965) – Mainland, Mexico; *Ctenosaura hemilopha conspicuosus* (Clarke and Robison, 1965) – San Estebán Island, Gulf of California, Mexico; *Ctenosaura hemilopha insulans* (Clarke and Robison, 1965) Ceralvo Island, Gulf of California, Mexico; *Ctenosaura hemilopha nolacensis* (Clarke and Robison, 1965) – Nolasco Island, Gulf of California, Mexico; *Ctenosaura pectinata* (Clarke and Robison, 1965; Evans, 1951) – Sinaloa, Mexico; *Ctenosaura similis* (motion picture film by T.A. Jenssen; Henderson,

1973) – Panama and Belize, repsectively; *Ctenosaura (Enyaliosaurus) clarki* (Carpenter, 1977) – Michoacan, Mexico; *Ctenosaura (Enyaliosaurus) quinquicarinatus* (Carpenter – motion picture film) – Oaxaca, Mexico; *Cyclura carinata* (Iverson, 1977) – Turks and Caicos Islands, British West Indies; *Cyclura cornuta stejnegeri* (Wiewandt, 1977) – Mona Island, *Cyclura cychlura cychlura (baelopha)* (Carpenter – motion picture film) – Andros Island, Bahamas; *Cyclura cychlura figginsi* (Carpenter – filmed in Dallas Zoo) – Exuma Island, Bahamas; *Cyclura n. nubila (macleayi)* (Wiewandt, 1977) – Cuba?; *Cyclura pinguis* (Wiewandt, 1977) – Anegada Island, British West Indies; *Cyclura ricordi* (Wiewandt, 1977) – Hispaniola; *Dipsosaurus dorsalis* (Carpenter, 1961) – S.W. Arizona; *Iguana iguana* (Müller, 1972)–N.E. Colombia; *Sauromalus ater* (Carpenter, motion picture film) – Santa Cruz Island, Gulf of California, Mexico; *Sauromalus o. obesus* (Carpenter, motion picture film; Berry, 1974); San Bernadino Co., California; *Sauromalus obesus tumidus* (Prieto and Ryan, 1978) – Pima Co., Arizona; *Sauromalus varius* (R.F. Clarke, motion picture film) – San Estebán Island, Gulf of California, Mexico.

13

Displays of Captive Banded Iguanas, *Brachylophus fasciatus*

Neil Greenberg
Department of Zoology
University of Tennessee
Knoxville, Tennessee

and

Thomas A. Jenssen
Department of Biology
Virginia Polytechnic Inst.
and State University
Blackburg, Virginia

INTRODUCTION

The social displays of many lizards have been described; however, there are no quantitative analyses of displays available for an iguanine species. The behavior of lizards of the genus *Brachylophus* are of particular interest because of their perplexing biogeography and phyletic affinities with other iguanines (Avery and Tanner, 1971). The genus appears to consist of at least two species, *B. fasciatus* and *B. brevicephalus,* isolated on the Pacific island groups of Fiji and Tonga, respectively (Avery and Tanner, 1970; but see Gibbons and Watkins, this volume). The natural history of *B. fasciatus* has been outlined by Cahill (1970) and Cogger (1974), and their ecology was presented by Gibbons and Watkins (this volume). Aspects of social behavior have been briefly described by Cogger (1974) and by Carpenter and Murphy (1978), and discussed in the context of related iguanines by Carpenter (this volume).

The work reported here was undertaken to provide additional information on the social behavior of individual *Brachylophus fasciatus* (Figure 13.1) and to quantitatively characterize their display behavior in a way that will be useful for interspecific comparisons.

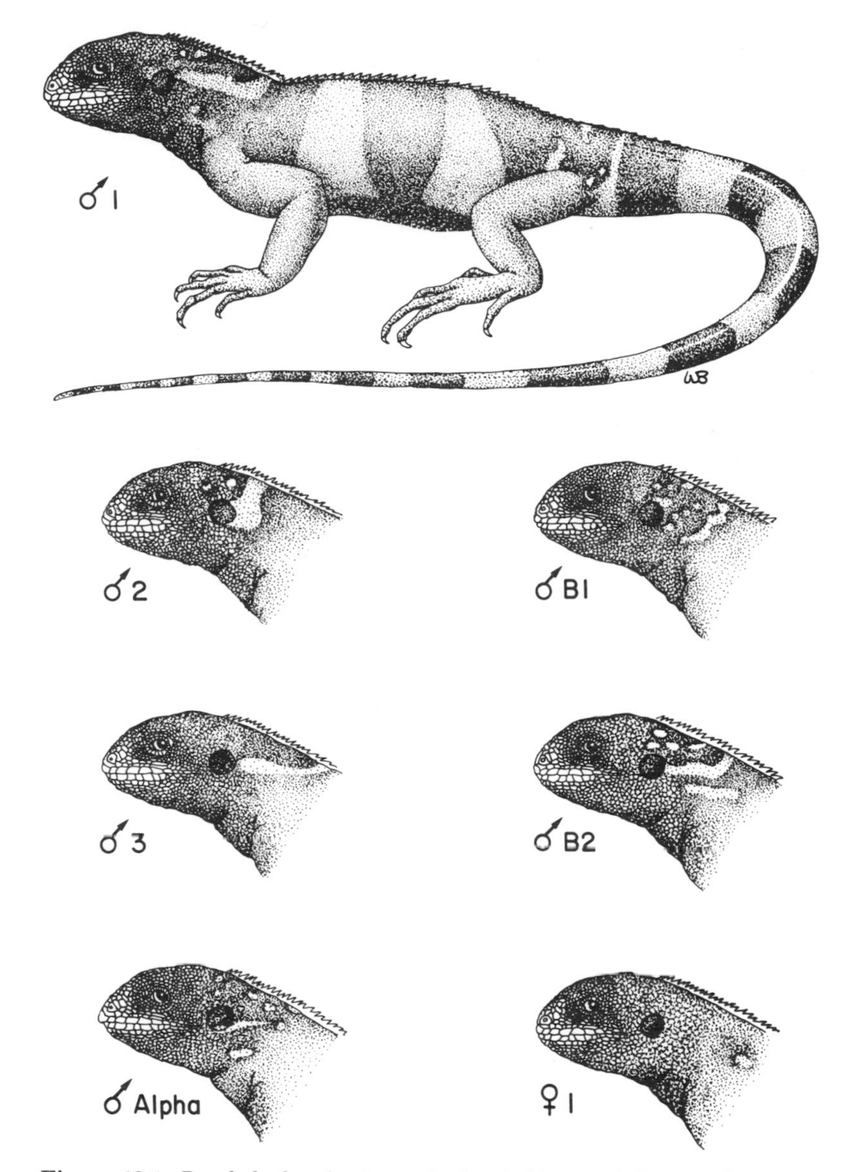

Figure 13.1: *Brachylophus fasciatus,* the banded iguana, indicating the variability of the bands or spots on the left profile of animals studied.

METHODS AND MATERIALS

Subjects and Maintenance Regimens

This analysis is based on the behavior of six adult male and two adult female *B. fasciatus.* Three males (numbers 1, 2, and 3) were captured as adults.

Males 1 and 3 have been kept at the Knoxville Zoological Park since February, 1975. At the time of this study, Male 2 was on loan to the Knoxville Zoo from the Memphis Zoo. A fourth male, "Alpha," was a young adult who was hatched at the Knoxville Zoo, January 17, 1977; its parents were a female which had died before our observations began and Male 1 or 3 (both had access to the mother). Two young males, "B1" and "B2" were also observed.

The two female subjects were obtained as adults along with males 1 and 3. Female 1 was on loan to the Knoxville Zoo through the auspices of the Memphis Zoo. Female 2 was not observed by us directly; her displays were recorded at the National Zoological Park by Dr. A. Stanley Rand who made his films available to us for analysis. Though the exact collection sites of the subjects are not known, it is believed that they came from the island of Kandavu, in the Fiji group (John Gibbons, pers. comm.).

Data for display analysis were collected at the Knoxville Zoo from July 23 to August 5, 1979. Our subjects were housed individually in home cages of various sizes. They were all healthy, and were being maintained on a diet of fruit, vegetables, and vitamin-dusted crickets. Cage temperatures were held between 25°-30°C under a 14:10 light:dark cycle.

Conditions of Observation

The study of these animals were greatly aided by their apparent insensitivity to observer effects. When the subjects were placed in a new habitat or allowed to make spontaneous forays from their cages, they showed little or no awareness of our presence. They responded rapidly to the sight of each other over relatively large distances, thus requiring precautions to prevent premature interactions.

When provoking social encounters, we could not allow contact between these valuable animals due to the possibility of injury. Therefore, interactions were staged in one of three ways. The first technique used two 75 x 30 x 40 cm terraria joined end to end by a fixed transparent partition and a sliding opaque panel. Each terrarium contained branches and artificial vegetation and was illuminated by overhead fluorescent lights. A male was placed in each compartment for several days prior to testing. During tests, the opaque panel was raised to allow males to see each other, while the transparent partition was left in place to prevent possible physical contact.

Because the appearance and expression of many display behaviors may covary with the distance between antagonists (Hover and Jenssen, 1976), a second technique utilized a 3 x 1 m arena to provide larger inter-subject distances. This testing site contained horizontal and vertical branches, artificial foliage, rocks and was lighted with heat filtered floodlights suspended from an overhead lattice. A small holding cage containing a male was placed at one end of the arena after its front glass was covered. The house lights were turned off and the floodlights on. A second male was released at the other end of the arena and his exploratory behavior and body color were recorded. The cover concealing the caged male was then removed and the resulting agonistic behavior was filmed.

A third technique employed the same display arena, but utilized two unrestrained subjects. In this situation the lizards were an adult male and female or two young adult males (Male Alpha and B1 or B2) who were less apt to engage in vigorous fighting than the older males.

During the paired interactions (male-male, male-female), the behavior of both subjects was simultaneously filmed (Kodak Ektachrome 160 type G film) using two super 8 mm cine cameras (18 frames/s). The staged interactions provided excellent documentation of complete display sequences. These displays were subjected to frame-by-frame analysis as described by Jenssen and Hover (1976).

Display Action Pattern Analysis

Head bob displays of *Brachylophus fasciatus* were found to be relatively long (some exceeding a minute), complex, and composed of several subdivisions which we labeled "elements." Because of considerable intra- and inter-individual variation, the displays were analyzed in terms of three characteristics: (1) durations between elements; (2) duration of each head bob within an element; and (3) duration of each inter-bob pause within an element. Descriptive statistics (mean, standard deviation of the mean, and coefficient of variation) were then calculated for the temporal characteristics of each unit within each individual's display patterns (Appendices 1-6). Since the magnitude of the standard deviation (S.D.) of a mean ($\overline{X}$) gives an indication of the amount of variability in a sample, one can use this measurement to indicate relative variability between unrelated samples. This is done by relating the S.D. to its $\overline{X}$ via the coefficient of variation (CV = S.D. x 100/$\overline{X}$) (Sokal and Rohlf, 1981). The CV has been regarded as an effective indicator of stereotypy (Barlow, 1977).

Color Change Analysis

Changes in body coloration are a striking feature of *B. fasciatus* social interactions. To document the extent of color change, we matched Munsell color patches (Munsell Color, 2441 N. Calvert St. Baltimore, MD 21218; see also Smithe, 1974) to the banded and unbanded portions of the torso of an isolated, behaviorally unaroused lizard and compared these patches with those matched to the corresponding body regions when the subject was socially aroused or subjected to handling. The latency to color change was determined with a stopwatch or by filming a stationary male at appoximately 1 frame/s from the time he was placed in the arena through the period when the subject was able to view another male. When the film was projected, the beginning and end points of the color transition were easily discerned and timed.

RESULTS AND DISCUSSION

From direct observation and our permanent record (600 m of film) we found several behavioral patterns consistently associated with social interactions. Most conspicuous of these were the vertical display movements of the forebody known as bobbing. Attending the bobbing movements were profile changes and color changes that function as display modifiers (sensu Jenssen, 1978), as well as other behavioral patterns of potential significance in the social organization of the species.

The principal social display in this species consists of coordinated sequences of head and/or limb movements that result in characteristic, more-or-less stereotyped patterns of vertical movements of the forebody (bobbing). Analysis of bobbing involves plotting these movements by time, yielding display action patterns (DAPs). Lizards had two basic head bobbing patterns–relatively brief bursts of rapid head bobs and prolonged sequences of slower head bobs. The slow bobs are quite complex and will be dealt with first.

Slow Head Bobs

The slow bob (SHB) had characteristics unlike those documented for most other iguanid lizards: The animals demonstrated distinctly individual patterns and the DAP was composed of several linked patterns (elements). They were performed at the beginning of male-female interactions and throughout male-male interactions.

Display Elements: The slow head bob displays were composed of relatively independent complex motor patterns, called "elements." Each lizard performed its display with a varying number of these elements. From an operational standpoint, the first element of a display was labelled *A* and the subsequent 1 to 16 elements as *B* (Figure 13.2). In each individual, *A* elements were qualitatively distinguishable from *B* elements, with the exception of Male 2. A lizard began and sometimes (<10%) concluded its display with a single *A* element; if the display continued beyond the initial *A* element, it was composed of repetitions of one or more *B* elements (Figure 13.2). A display was always concluded at the end of an element, never within one; thus, the elements were subject-determined.

Similar phenomenon has been described for *Anolis townsendi* (Jenssen and Rothblum, 1977). In *A. townsendi* the display was performed in "acts" (see Barlow, 1968:219-220, for precise definition). Though some acts were not always performed in each display, they always appeared in sequence (e.g., acts A, B, C, D or acts A, C, D, but never B, C, A). *B. fasciatus,* however, followed different rules with regard to the ordering of their display elements. An *A* element always began a display, but the ordering of the various *B* elements could be scrambled. For this reason we have designated the display subdivisions as "elements" rather than "acts."

Displays of Individuals: The subjects displayed striking individual differences in several respects. These differences involved bob number, bob amplitude, and bob cadence. Because of this an analysis of the displays of each individual is presented separately.

Male 1: This male had two possible *A* elements, A_1 and A_2 (Figure 13.2), which appeared with equal frequency. The two patterns differed in the presence or absence of Unit 4 (Appendix 1). Both *A* elements could open with an optional head bob. Approximately 60% (16/26) of the displays had this additional bob.

The *B* elements possessed the same basic pattern. The only variation was a progressive de-emphasis of the second spike-like bob (very rapid rise and fall of amplitude) in the *B* pattern (Unit 4, Figure 13.2); after several repetitions, only a single spike appeared. This source of variation in the duration of Unit 4 (second spike bob) is reflected by the unit's large coefficient of variation (Appendix 1). The number of *B* elements per display had a bimodal distribution (Figure 13.3). A review of the films provides some indication that *B* elements were more numerous when the male was interacting with a female or a distant male, and decreased in number when displays were performed close to another male.

Another trend observed in Male 1's displays was the progressive duration increase in successive inter-element pauses within a display (Appendix 6).

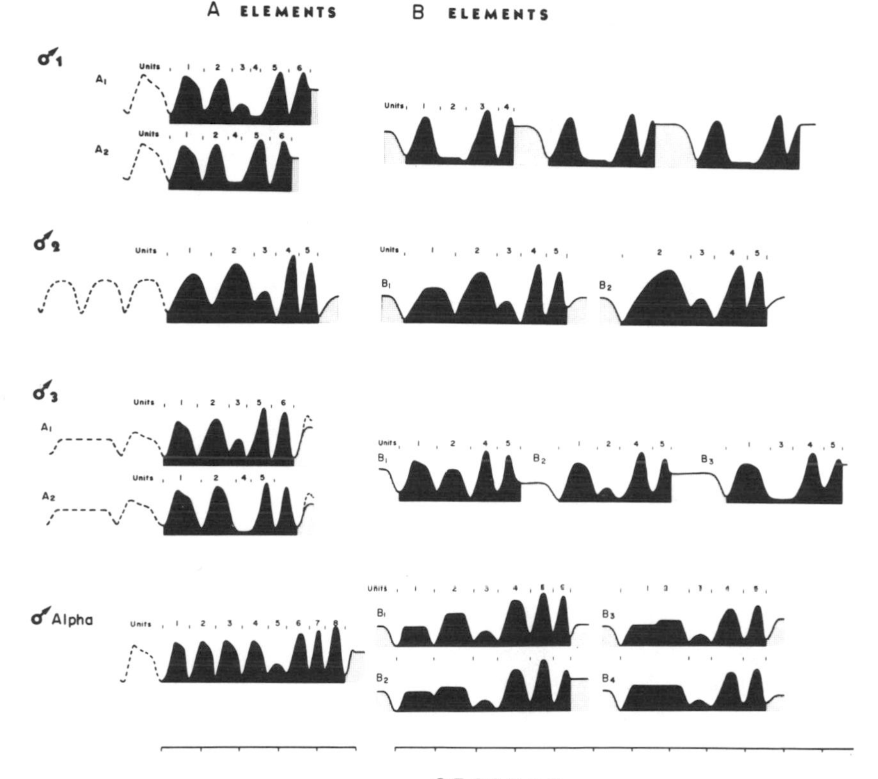

Figure 13.2: Display-Action-Pattern Graphs of the signature displays from four male *Brachylophus fasciatus*. Relative head amplitude (y-axis) of head bobs is plotted through time (x-axis). Solid black areas give the core pattern for each element, dashed lines give the patterns for optional introductory bobs, and stippled areas indicate inter-element head amplitude movement. Unbroken stippled areas between *B* Elements indicate a set temporal sequence to the *B* Elements.

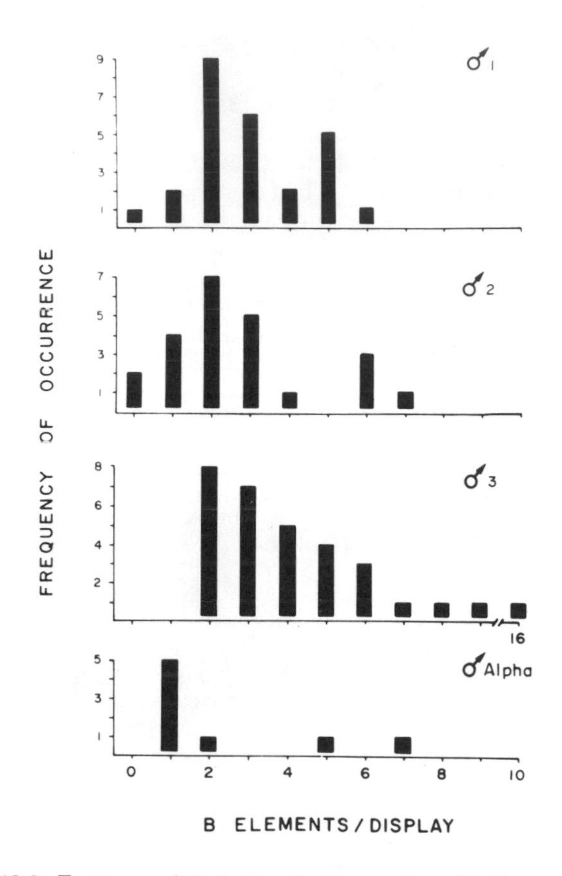

Figure 13.3: Frequency distribution for the number of *B* Elements appearing in each display performance by four male *Brachylophus fasciatus.*

Male 2: This male almost always (22/23) began his displays with a series of introductory bobs. This series averaged 5.8 bobs (range 0-10, S.D. 2.3). The introductory bobs were followed by his particular *A* element (Figure 13.2).

Usually the *A* element was accompanied by a series of *B* elements (Figure 13.3). Male 2 had two kinds of *B* elements (B_1 and B_2). The B_2 pattern was similar to the B_1 pattern, but lacked the first bob (Unit 1) while expanding the second bob (Unit 2, Figure 13.2). The B_1 elements appeared twice as frequently as the B_2 elements (33:17), with B_1 usually beginning the series of *B* elements. The sequencing of B_1 and B_2 elements, however, followed no predictable relationship.

Male 3: The *A* element was preceded by a prolonged, low amplitude bob in approximately 20% of the displays (Figure 13.2). Following this was another optional bob which occurred in greater frequency (33% of the displays). Next came the display's *A* element which could appear in two forms, A_1 and A_2. The A_1 pattern always showed Unit 3, a small bob, and A_2 replaced this with Unit 4, the inter-bob pause (Figure 13.2); the two forms were performed with similar frequency. There was a tendency for Male 3 to produce an occasional third spike-like bob at the end of both A_1 and A_2 patterns (Figure 13.2).

Male 3's *B* elements were performed in a method quite different from the other subjects: they comprised a series in which the pattern progressively transformed (Figure 13.2). The amplitude of the second bob (Unit 2) decreased with successive performances of the *B* element until it was replaced by an inter-bob pause (Unit 3).

To demonstrate this transformation, of the first performed *B* elements, the ratio of B_1 to B_2 variants (containing either a high or low amplitude (Unit 2) was 27:4 (31 displays), the B_1:B_2 ratio in the next *B* element of a series was 17:14 (31 displays), then 6:17 (23 displays), and finally 0:16 (16 displays). This shift in pattern is also seen in mean unit durations of Units 2 and 3 and their high coefficients of variation; Unit 2 durations shorten and Unit 3 durations increase as one inspects each successive *B* element given within a display (Appendix 3). One peculiarity of this male's *B* element performance was that all displays contained at least 2 *B* elements (Appendix 3). Two performances of the *B* element would be necessary to effect the predictable shift from the B_1 to the B_2 pattern.

Male Alpha: With only eight displays, we are less certain about the syntactics of this male's displays. However, his displays were clearly different from the others and were composed of a single individual-unique *A* element and four patterns of *B* elements. Male Alpha had the most bobs in this *A* element, no inter-bob pause, and three concluding spike-like bobs.

The *B* elements fell into four patterns (B_1-B_4). Because of the small sample size and intermediate forms to these patterns, we lumped the *B* elements into two classes (B_1-B_2, B_3-B_4) for the statistical analysis (Appendix 4). The B_1-B_2 elements contained six units and were the most frequently performed (10/17) of the *B* elements. When giving the B_3-B_4 elements, Male Alpha fused Units 1 and 2 into one elongated bob and dropped the third spike-like bob at the end of the element (Figure 13.2). In two of the recorded *B* elements, the small bob (Unit 3) of the B_1 pattern was omitted. Unlike Male 3, there was no apparent sequence to the use of the various *B* elements.

Females 1 and 2: Both females showed a single, individual-unique *A* element pattern. Female 1 used a total of seven bobs in her pattern and Female 2's *A* element always had at least four bobs (Figure 13.4; Appendix 5). There was no evidence that either female had a *B* element to their displays. It is possible that under the proper social context (e.g., female-female confrontations) they would exhibit displays with *A* and *B* elements. Our data, however, were only of females displaying at courting males.

Rapid Head Bobs

A second type of head bobbing behavior was recorded for the males. This is a weakly stereotyped burst of rapid head bobs (RHB) which frequently was repeated in quick succession (Figure 13.5). Several times we saw RHB superimposed upon a male's individual-unique display pattern. One such instance was filmed in which Male 2 gave RHB during his introductory bobs (Figure 13.5).

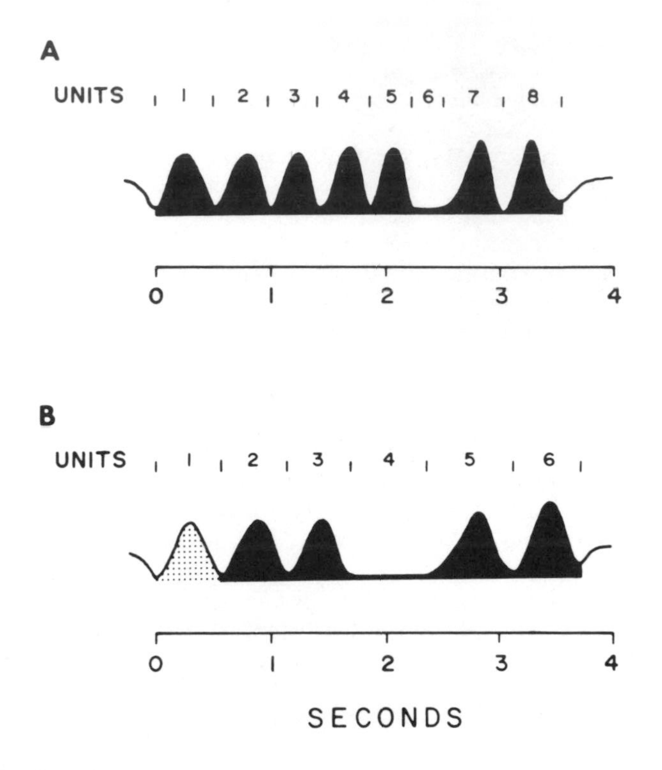

Figure 13.4: Display-Action-Pattern Graphs of the signature display from two female *Brachylophus fasciatus,* Female 1 (A) and Female 2 (B). Relative head amplitude (y-axis) of head bobs is plotted through time (x-axis). Solid black areas give the core pattern for the displays, and the stippled area gives the pattern of an optional introductory bob.

Rapid head bobbing is a common display type of many iguanid repertoires where it is often associated with courtship. In *B. fasciatus,* however, RHB were performed in other situations as well. The contexts in which we observed RHB were: (1) adult male performed RHB as he watched or approached an adult female; (2) large adult male performed RHB as he watched or approached another adult male; and (3) young adult male performed RHB while interacting at a distance with a large adult male. Considering the sexual dimorphism in aroused color state and display behavior, it is not likely that there was a mistaken identity of the recipient's sex. Similar situations have perplexed investigators observing RHB in other species as well (e.g., Rothblum and Jenssen, 1978; Ruby, 1977). Because of the variety of contexts in which RHB occurs, we have avoided a label that implies an exclusive courtship function for this display type. The contexts in which RHB occurs contrasts with those of SHB, which was observed at the beginning of both male-female and male-male encounters but continued throughout the interactions only of adult males.

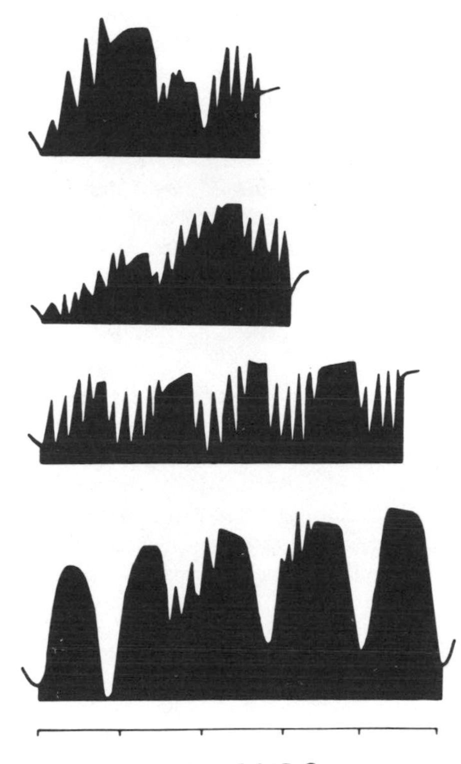

Figure 13.5: Four specific examples of Rapid Head Bobs (RHB) by male *Brachylophus fasciatus*. Relative head amplitude (y-axis) of head movement is plotted through time (x-axis). In the top three graphs the RHB are interspersed with several brief interruptions, and in the bottom example the RHB are superimposed on a portion of Male 2's introductory display.

Display Stereotypy

The coefficient of variation (CV, see Methods) is a good indicator of the stereotypy of a variable: CV values for the measured variables increases (increasing variability) as the relative stereotypy decreases. From his review of ethological literature, Barlow (1977:103) stated that most ritualized behaviors do not show great stereotypy (CV values range 15-35%), and suggests that highly stereotyped behavior should be characterized by CV values less than 15%.

Most of the *B. fasciatus* display units do not demonstrate strong intra-individual temporal stereotypy, most units falling within the 15-35% range for CV values (Appendices 1-6). Even at this intra-individual level the displays of this iguanine are only moderately stereotyped at best. This contrasts with the small (15%) CV values for all the *Sceloporus undulatus* units analyzed (Rothblum and Jenssen, 1978) and most of the *Anolis townsendi* units analyzed (Jenssen and Rothblum, 1977).

Besides relatively weak temporal stereotypy for units within displays having a consistent, recognizable bob pattern, there are also occasional instances of the pattern itself being altered ("pattern variability" sensu Jenssen, 1978:277). During such instances, a lizard deviates from one of its recognized display patterns to such a degree that the pattern is qualitatively disrupted. As examples, Male 3 would at times insert an extra spike-like bob at the end of his *A* element; during multiple display episodes, Males 1 and 3 both progressively de-emphasized the last spike-like bob of their *B* elements until it occasionally disappeared; and Male 2 showed intermediates to its *B* element patterns.

In summary, *B. fasciatus,* while performing recognizable individual-unique displays, showed weak temporal stereotypy, and the consistency with which they performed their display patterns was less than that reported for most iguanid species.

Inter-Individual Differences in Display Behavior: The magnitude of inter-individual display diversity found in *B. fasciatus* is apparently greater than that previously described for any iguanid. Population differences have been found (e.g., Ferguson, 1971; Gibbons, 1979; Jenssen, 1971), as well as some pattern variation between lizards of the same population (e.g., Berry, 1974; Dugan, 1979). One species *(Sceloporus undulatus)* possesses an individually specific display type in a multiple display repertoire (Rothblum and Jenssen, 1978).

The only common threads running through the fabric of male *B. fasciatus* displays were (1) the displays were performed as a variable series of elements and (2) each element ended in "spike-like" double or triple bobs (Figures 13.2 and 13.3).

Display Modifiers and Potential Social Signals

Behavioral patterns that accompany bobbing and can alter the stimulus configuration and possibly the message communicated are known as display modifiers (Jenssen, 1978). Two conspicuous behavioral patterns exhibited by *B. fasciatus* that may function in this way are profile changes and body color changes. Other behavioral patterns that occurred during encounters and may have potential social significance were cloacal discharge and tongue-flicking. Snorting also occurs during social interactions and will be briefly described.

Profile Changes: Changes in the profile of the torso or throat that resulted in an enlargement of the animal's apparent size consistently occurred during the aggressive interactions of male *B. fasciatus.* The changes were accomplished in varying degree by three mechanisms: extended throat, sagittal expansion, and crest erection. Extended throat was the first pattern utilized in an interaction and involved a change in the profile of the head effected by an extension of the basal elements of the hyoid apparatus. This occurred as quickly as 5 seconds after sighting a potential antagonist and often continued throughout an encounter. Planimetric measurements of profile change caused by extended throat indicate that it can be responsible for 35 to 40% increase in the apparent size of the head (see Figure 13.6).

Sagittal expansion of the profile of the torso is accomplished by lateral compression of the body. Like extended throat, this is also a rapidly effected phenomenon, usually occurring shortly after extended throat but sometimes apparently simultaneous with it. Apparent size of the torso alone can be increased between 15 and 20%.

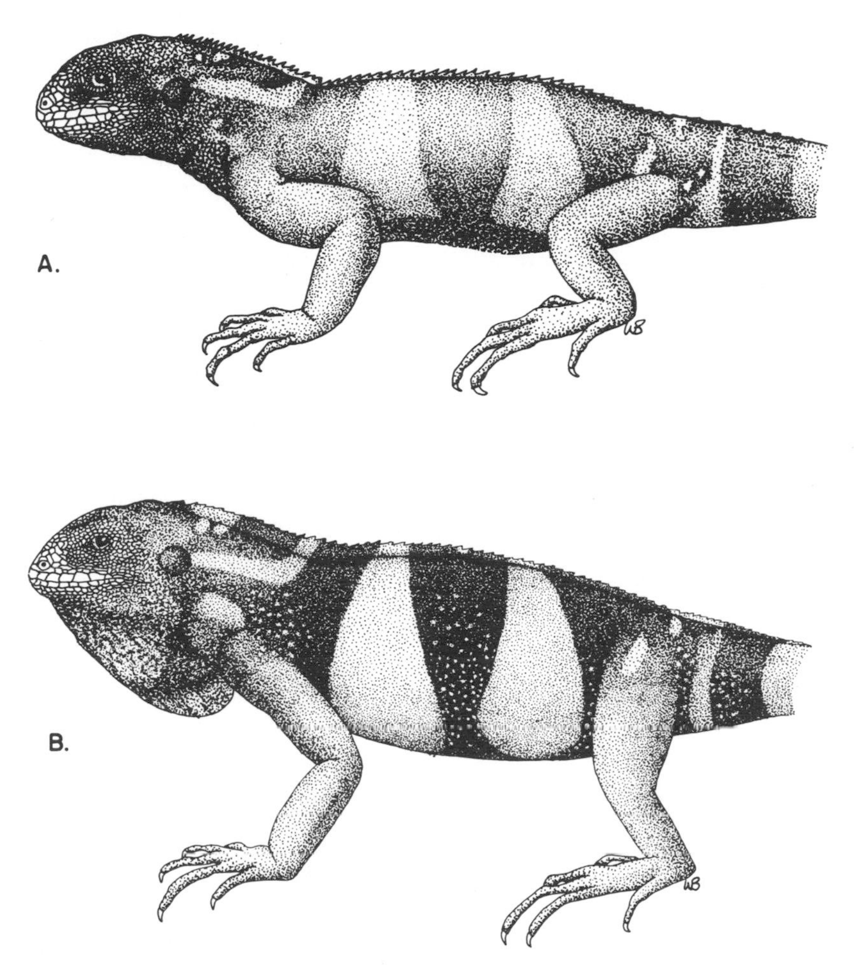

Figure 13.6: Profile changes in *B. fasciatus* that occur during aggressive social interactions. Changes are most obvious in the ventral throat (extended throat), torso (sagittal expansion), and dorsal crest. (B = aggressive stance.)

The dorsal crest in *B. fasciatus* contributes about 3% to the profile of a resting animal and its erection is slight and easily overlooked. It occurs relatively slowly, and is first apparent approximately 150 seconds after sighting an antagonist. The precise mechanism is not clear, but the contraction of slow muscle fibers along the dorsal midline from cervical to near sacral levels seems reasonable. Erection of the crest is greatest in the nuchal area and, while capable of an almost 80% increase in size, it contributes only about 5% to the profile of an aroused animal (Table 13.1).

Table 13.1: Changes in Apparent Body Size of Aroused *B. fasciatus**

	Contribution to Relaxed Profile	Contribution to Aroused Profile	Percent Increase of Component
Torso	76.5%	72.5%	18%
Head	20.2%	22.7%	40%
Crest	3.3%	4.8%	82%

*Planimetric measurements of Male 1, test number 8.79.

Changes in the apparent size of an animal are common during the aggressive interactions of many lizards that utilize visual displays. The most common of these changes involve altering the throat profile. Throat profile is typically increased by extending various elements of the cartilagenous hyoid apparatus. Hyoid changes have their effect by moving basal (basihyal) elements ventrally, resulting in an apparently "engorged" throat (extended throat), or by moving the longer medial retrobasal (ceratobranchial II) process, resulting in extension of a folded dewlap.

In *B. fasciatus* it is not clear that the retrobasal process participates in the profile change. The term "extended throat" was selected to emphasize this point because in some species the effects of basal and retrobasal hyoid elements on throat profile are utilized in two distinctive situations. For example, in *Anolis carolinensis,* retrobasal dewlap extension is relatively non-specific and occurs in many situations, while the basal hyoid "extended throat" occurs only as a component of the "throat" or "challenge" display (Greenberg, 1977a). When a potential adversary has been identified as a conspecific male, sagittal expansion is often added to extended throat. The constraints of the present study did not allow us to distinguish possible context differences between the use of extended throat or sagittal expansion by *B. fasciatus;* however, while extended throat occasionally occurred along with both rapid and slow head bobs, sagittal expansion was associated only with slow bobs. Several recent interpretations of the evolution of fighting behavior (Parker, 1974; Dawkins and Krebs, 1978) have identified changes in apparent size as a competitive strategy which could affect a conspecific's assessment of an antagonist's resource holding power.

Profile changes serve to enlarge the apparent size of an animal but in many instances they are coupled with displays of signal color such as the blue ventrum of *Sceloporus* (Greenberg, 1977b) or the red dewlap of *Anolis.* In *B. fasciatus,* profile changes are typically coupled with body color changes, but it is not clear that these color changes have signal value.

Body Color Changes: The common name of *B. fasciatus,* banded iguana, derives from two or three broad bands of relatively light blue-green vertically traversing a darker green background. In the females we observed that this pattern was less evident but still clearly present, contrary to most descriptions of this species (e.g., Avery and Tanner, 1970). In the dappled sunlight of its forest canopy habitat, it is likely that this coloration would make the animals difficult to detect (Cogger, 1974). Individual differences in the configuration of the bands allowed us to recognize each subject (Figure 13.1).

When sleeping or isolated and undisturbed, animals possessed banding and head markings of relatively low contrast with the body's green background color. This was termed the tonic color state. During staged encounters, however, aroused males manifested a dramatic color contrast as the green background color became darker, in some individuals to an almost black coloration, termed "dark color state" (Table 13.2). Darkening began as soon as 90 seconds after sighting an antagonist and dark color state was attained 2-4 minutes later, although in some individuals maximum contrast was occasionally attained within 80 seconds.

The dark color state invariably developed during aggressive or courtship behavior. Typically, darkening was less extreme during courtship than in male-male aggressive encounters, but often the first response to a female was an intense darkening that abated slightly within several minutes. Females did not darken during encounters with males. The extent of background darkening appeared to vary with the degree of behavioral arousal and from male to male. It is possible that this reflects the tonic level of color-influencing hormones which, in turn, may vary with the animal's experience. Both individual differences and degrees of color change are indicated in Table 13.1. We must emphasize, however, that it is the quantitative contrast that is relevant rather than any subtlety of absolute color.

Color change phenomena in many reptiles are known to be under varying proportions of both neural (sympathetic) and endocrine control (melanocyte stimulating hormone and adrenal chromaffin hormones) (Bagnara and Hadley, 1973). Neural effects on chromatophores are almost immediate while endocrine effects must rely upon the vascular transport of the effecting hormone and are thus relatively slow (depending on the size of the animal). The long latency to color change in *B. fasciatus* (Table 13.2) suggests an endocrine mechanism. In our observations, the occurrence of color change was limited to periods of social interaction. It is reasonable that preparation for imminent aggressive encounters will activate the adrenal "emergency" stress response. Since it is known that the adrenal chromaffin hormones associated with acute stress, epinephrine and norepinephrine, can act at the dermal chromatophores to effect color changes in *Anolis carolinensis* (Hadley and Goldman, 1969), it is likely that a similar mechanism is acting in *Brachylophus*. This possibility is also suggested by observations of darkening in female *B. fasciatus* during parts of the reproductive phase (Cahill, 1970), a period identified in another iguanine species *(Dipsosaurus dorsalis)* as being relatively stressful (Callard, *et al.*, 1973).

Cloacal Discharge: On several occasions during the course of social interactions we observed a subject discharging the contents of its cloaca. These observations, combined with those of frequent tongue-flicking during the interactions of free-ranging animals, indicates a potential social role for chemical clues of cloacal origin. It should be noted that while *B. fasciatus* has conspicuous femoral pores, no specific behavioral pattern associated with these were observed.

Recent research on the iguanid lizard *Sceloporus* has indicated that cloacal discharge may contain chemical information of signal value. Bricks on which male *S. occidentalis* perch or bask accumulate deposits of fecal material and femoral and proctodeal gland secretions. When males are presented with soiled bricks, the lizards respond with an increased frequency of tongue-flicking and the performance of push-up displays (Duvall, 1979). Other iguanids that vary their tongue-flicking behavior in a manner suggestive of chemosensory investigation are *S. jarrovi* (DeFazio *et al.*, 1977), *Sauromalus obesus* (Berry, 1974), and *Iguana iguana* (Burghardt, 1977). The occurrence of cloacal discharge during social encounters may indicate the presence of conflicting tendencies (approach-withdrawal) that are often manifested in varying ratios of sympathetic and parasympathetic autonomic activation: A common consequence of this conflict in mammals is defecation and urination, responses believed to have given rise to a ritualized signal system in many mammals (Morris, 1956a).

Table 13.2: Tonic (Light, Resting) Color, Typical Time to Change, and Dark (Aroused) Color in *Brachylophus fasciatus*

Subject Number	. . . .Tonic Color* . . .		Time to Change (sec)	. . . Dark Color*	
	Body	Band		Body	Band
Male 1	10.0 GY 5/8	5.0 BG 6/2	210	5.0 BG 3/2	5.0 BG 7/4
Male 2	2.5 G 5/8	5.0 BG 7/4	330	5.0 BG 3/2	5.0 BG 6/6
Male 3	2.5 G 4/8	5.0 BG 7/4	170	5.0 BG 3/2	No change
Male Alpha	5.0 G 6/6	5.0 BG 7/4	80	5.0 BG 3/2	5.0 BG 5/6
Male B1	7.5 GY 5/6	5.0 BG 8/4		No change	No change
Male B2	2.5 G 5/8	5.0 BG 6/6		5.0 G 4/6	No change
Female 1	2.5 G 4/6	5.0 BG 7/4		No change	No change

*Tonic and dark body color states are indicated in standard Munsell notation. The first portion indicates hue (color) on a scale of ten for green (G), green yellow (GY) or blue green (BG). The next numbers represent value/chroma (light-dark/intensity-saturation); low value numbers are darkest, low chroma numbers are pale.

Tongue-Flicking: Tongue-flicking in *Brachylophus* is a brief (0.5-1.0 sec.) extension of the tongue from the mouth, a behavioral pattern shown in some reptiles to be important in providing chemical information in the vomeronasal sensory pits (see Burghardt, 1980). In *Brachylophus,* as in other lizards studied (e.g., DeFazio *et al.*, 1977), the tongue is typically touched to the substrate ("tongue-touch") but occasionally extruded into the air with no attempt to touch a surface ("air-lick"). Tongue-touching was consistently observed to occur more frequently in individuals displaced from their enclosure for behavioral testing than in those moving in their home cages. Characteristically, banded iguanas would move in a new habitat, pause, tongue-touch the substrate, and move again in what appeared to be chemosensory as well as visual exploration of an unfamiliar environment.

Tongue-touching was also performed by individuals in their home cages when shown another lizard, and in 3 out of 4 cases in which males were shown mirrors. It is not clear what chemosensory information might be obtained in the instances of home-cage tongue-touching but such observations support the idea that non-specific arousal may be a relevant consideration. A similar phenomenon is manifested by *Anolis carolinensis* which will increase the tongue-flick rates in home cages when an artificial "wind" moves foliage in the cage (Greenberg and Rodriguez, 1979), an effect not eliminated by severing the connections of the vomeronasal and olfactory apparatus with the central nervous system (Greenberg, unpublished data).

Snorting: In several social encounters, male *B. fasciatus* were observed forcefully expelling fluid from their nostrils. This fluid contains salts, as indicated by the gradual accumulation of a salt residue on the sides of glass cages and is presumably secreted by nasal salt glands. A similar accumulation on the sides of isolation cages indicates that snorting is not limited to times of social interaction.

Snorting has also been observed in *Amblyrhynchus* and *Conolophus* (Carpenter and Ferguson, 1977) and in three species of the iguanine genus *Sauromalus* (Norris and Dawson, 1964). Salt glands have been investigated in at least nine terrestrial iguanid species and found to aid in maintaining electrolyte balance (Peaker and Linzell, 1975).

There is no evidence that this behavioral pattern is of any significance as a signal; however, it is interesting to note that scales surrounding the nostrils provide a distinct surrounding band of yellow that visually emphasize them. It is possible that snorting at times of elevated behavioral arousal may reflect underlying autonomic activation and thus is at least a potential signal.

SUMMARY AND CONCLUSIONS

The social displays of captive *Brachylophus fasciatus* observed during staged encounters consisted of relatively brief bursts of rapid head bobs and highly individual sequences of slow bobs. Some individuals displayed a unique progressive transformation in the bobbing pattern as encounters progressed.

Display modifiers observed in males were profile changes and color changes. Profile changes enlarged the apparent size of an animal and involved movements of the ventral throat ("extended throat") and/or torso ("sagittal expansion"), and to a slight extent, the dorsal crest. Color changes involved a darkening of the green background color, increasing contrast with the vertically traversing blue-green bands.

Other behaviors occurring during social encounters were cloacal discharge, tongue-flicking, and snorting. Although these behaviors serve primarily nonsocial functions, their occurrence in a high-arousal social context suggests that they may also be to some extent under social stimulus control and could thus contribute to inter-individual communications.

Separately caged males viewing each other across the length of the reptile house (<15 m) would change to their dark color state and continually display at each other indicating that *B. fasciatus* males may be territorial and probably control relatively large, exclusive territories in nature. This is also suggested by the fact that the color difference between the tonic color state and the agonistic-related dark color state is extensive and easily perceived at a distance. Also, the displays are of long duration, a feature likely to evolve when displays are performed beyond the range of physical attack; in most lizard species studied, short distance displays tend to be of brief duration.

The most unusual characteristic of *B. fasciatus* displays is the individual-unique pattern generated by each of our subjects. The differences between each lizard's stereotyped pattern was so great that there was no way of presenting a single display action pattern graph to represent the species. The explanation for this extraordinary trait may be that the specimens came from separate demes, and their display pattern differences reflect differing gene pools. However, even though the species resides on a series of islands, it is unlikely that this is true for all individuals studied. It is reasonable to argue that display variability might arise in the absence of the need to discriminate close relatives; however, there is no direct evidence to indicate that this is true and, in fact, Rand has asserted (pers. comm.) that display variability is not increased in anoles that have been long isolated from congeners on Antillian Islands.

An alternative hypothesis is ecologically based. Given the premise that the species is moderately long-lived and strongly territorial, it is possible that the territories are stable through time (many months). The individual-unique display patterns could then facilitate recognition of nearest neighbors and be a part of a selected mechanism for decreasing aggressive behavior between recognized neighbors. Only strange males in the vicinity of a resident's territory would elicit a high level of arousal and escalated agonistic behavior.

Our observations, in concert with those of Gibbons and Watkins (this volume), have raised some intriguing questions about *B. fasciatus* ethoecology and have shown the possibilities of using the species as an experimental animal. We hope that through the collaboration with the Fiji government and with the zoological parks having breeding programs, there will be increasing opportunities to study *B. fasciatus* in greater depth.

Acknowledgments

We wish to acknowledge our deep gratitude to the Knoxville Zoological Park for allowing use of their facilities. In particular, Reptile Keeper, Chris Norris; Howard Lawler, Curator of Herpetology; the staff of the KZP Reptile Complex; and Keeper, Carol Ricketts. This study would not have been possible without their interest and cooperation.

Appendix 1

Descriptive Statistics for the Units of the A_1, A_2, and B Elements from 26 Displays by Male 1

Element	Statistic	Element Units (Intro Bob)	1	2	3	4	5	6
A_1	$\overline{X}$	1.07 s	0.85 s	0.69 s	0.50 s	0.28 s	0.69 s	0.58 s
	SD	0.08	0.10	0.06	0.10	0.19	0.09	0.09
	CV	8%	12%	9%	20%	68%	13%	16%
	N	6	13	13	13	13	13	13
A_2	$\overline{X}$	1.08 s	0.81 s	0.66 s	0.0	0.42 s	0.72 s	0.54 s
	SD	0.15	0.15	0.10	–	0.12	0.06	0.11
	CV	14%	18%	13%	–	28%	9%	20%
	N	10	13	13	0	13	13	13
B	$\overline{X}$		0.87 s	0.66 s	0.80	0.34 s		
	SD		0.12	0.14	0.11	0.07		
	CV		21%	21%	13%	22%		
	N		63	63	63	63		

Note: $\overline{X}$, mean; SD, standard deviation; CV, coefficient of variation; N, number of elements analyzed.

Appendix 2

A. Descriptive Statistics of the Introductory Bob Durations from Nine Displays by Male 2 Which Contained Six Introductory Bobs

Statistics	Introductory Bobs in a Display: 1st Bob	2nd Bob	3rd Bob	4th Bob	5th Bob	6th Bob
$\overline{X}$	1.07 s	1.07 s	1.15 s	1.19 s	1.15 s	1.12 s
SD	0.30	0.40	0.31	0.27	0.27	0.20
CV	28%	29%	27%	23%	24%	18%
N	9	9	9	9	9	9

B. Descriptive Statistics for the Units of the A, B_1, and B_2 Elements from 23 Displays by Male 2

Element	Statistics	Element Units: 1	2	3	4	5
A	$\overline{X}$	1.15 s	1.10 s	0.63 s	0.65 s	0.53 s
	SD	0.21	0.38	0.17	0.10	0.07
	CV	18%	36%	26%	16%	13%
	N	23	23	23	23	23
B_1	$\overline{X}$	1.25 s	1.46 s	0.69 s	0.86 s	0.53 s
	SD	0.33	0.33	0.14	0.15	0.09
	CV	26%	23%	20%	17%	17%
	N	33	33	33	33	33
B_2	$\overline{X}$	0.0	1.81 s	0.63 s	0.87 s	0.51 s
	SD	–	0.23	0.12	0.14	0.08
	CV	–	12%	19%	16%	16%
	N	0	17	17	17	17

Appendix 3

Descriptive Statistics for the Units of the A_1, A_2, and B Series Elements from 31 Displays by Male 3

Element	Statistic	Element Units 1	2	3	4	5	6
A_1	$\overline{X}$	0.87 s	0.75 s	0.47 s	0.0	0.66 s	0.59 s
	SD	0.16	0.13	0.12	–	0.09	0.09
	CV	18%	17%	24%	–	13%	15%
	N	16	16	16	0	17	16
A_2	$\overline{X}$	0.85 s	0.75 s	0.0	0.29 s	0.65 s	0.59 s
	SD	0.16	0.12	–	0.11	0.09	0.08
	CV	19%	16%	–	38%	14%	13%
	N	15	15	0	15	15	15
1st *B* of the Display	$\overline{X}$	0.93 s	0.77 s	0.07 s	0.75 s	0.55	
	SD	0.1	0.14	0.10	0.08	0.09	
	CV	10%	18%	154%	11%	17%	
	N	31	31	31	31	31	
2nd *B* of the Display	$\overline{X}$	1.03 s	0.4 s	0.26 s	0.82 s	0.49 s	
	SD	0.12	0.35	0.28	0.08	0.09	
	CV	12%	88%	106%	10%	17%	
	N	31	31	31	31	31	
3rd *B* of the Display	$\overline{X}$	1.1 s	0.12 s	0.56 s	0.84 s	0.48 s	
	SD	0.13	0.26	0.13	0.08	0.09	
	CV	12%	209%	24%	10%	20%	
	N	23	23	23	23	23	
4th *B* of the Display	$\overline{X}$	1.17 s	0.0	0.56 s	0.87 s	0.50 s	
	SD	0.09	–	0.1	0.09	0.06	
	CV	8%	–	17%	11%	13%	
	N	16	0	16	16	16	

Appendix 4

Descriptive Statistics for the Units of the A, B_1-B_2, and B_3, and B_3-B_4 Elements from Eight Displays by Male Alpha

Element	Statistic	Intro. Bob	Element Units 1	2	3	4	5	6	7	8
A	$\overline{X}$	1.01 s	0.69 s	0.67 s	0.66 s	0.65 s	0.50 s	0.58 s	0.44 s	0.51 s
	SD	0.24	0.18	0.06	0.05	0.04	0.05	0.05	0.04	0.06
	CV	23%	26%	9%	7%	6%	10%	10%	9%	11%
	N	4	8	8	8	8	8	8	8	8
B_1-B_2	$\overline{X}$		0.98 s	0.99 s	0.67 s	0.88 s	0.57 s	0.45 s	0.5 s	
	SD		0.15	0.09	0.11	0.14	0.06	0.09		
	CV		15%	9%	17%	16%	11%	19%		
	N		10	10	10	10	10	10		
			....*.... (1–2)							
B_3-B_4	$\overline{X}$		1.83 s		0.58 s	0.81 s	0.57 s	0.11 s		
	SD		0.22		0.07	0.05	0.07	0.19		
	CV		12%		12%	6%	13%	171%		
	N		7		7	7	7	7		

*The combination of Units 1 and 2.

Appendix 5

Descriptive Statistics for the Units of the *A* Elements from Six Displays by Female 1 and 15 Displays by Female 2

		. . . . . Element Units.								
Subject	Statistic	Intro. Bob	1	2	3	4	5	6	7	8
Female 1	$\overline{X}$		0.50 s	0.48 s	0.43 s	0.44 s	0.38 s	0.28 s	0.53 s	0.51 s
	SD		0.18	0.12	0.05	0.00	0.06	0.06	0.03	0.04
	CV		36%	34%	11%	0%	14%	22%	6%	8%
	N		6	6	6	6	6	6	6	6
Female 2	$\overline{X}$	0.55 s	0.58 s	0.56 s	0.71 s	0.74 s	0.61 s			
	SD	0.08	0.14	0.06	0.13	0.12	0.12			
	CV	15%	10%	11%	18%	16%	20%			
	N	7	15	15	15	15	15			

Appendix 6

Descriptive Statistics of the Durations of the First Four Interelement Pauses Appearing Within the Displays of Four Males

		. . . Inter-Element Pauses in a Display. . .			
Subject	Statistic	1st	2nd	3rd	4th
Male 1	$\overline{X}$	0.60 s	0.97 s	1.21 s	1.29 s
	SD	0.24	0.25	0.24	0.16
	CV	39%	26%	20%	13%
	N	20	20	11	5
Male 2	$\overline{X}$	1.44 s	1.49 s	1.52 s	1.6 s
	SD	0.26	0.33	0.31	0.14
	CV	18%	22%	20%	9%
	N	20	20	20	20
Male 3	$\overline{X}$	1.01 s	0.97 s	1.29 s	1.60 s
	SD	0.41	0.24	0.26	0.32
	CV	41%	24%	20%	20%
	N	31	31	23	16
Male Alpha	$\overline{X}$	1.08 s	1.44 s	2.00 s	
	SD	0.15	0.36	–	
	CV	14%	25%	–	
	N	8	3	2	

14

The Behavioral Inventory of the Green Iguana, *Iguana iguana*

Hansjürgen Distel*
J. Veazey
Max-Planck-Institut für Psychiatrie
Munich
Federal Republic of Germany

INTRODUCTION

Green iguanas live in the light-shade mosaic of trees along rivers, lakes and in mangrove swamps (Swanson, 1950; Moberly, 1968). They are also found in relatively open, arid areas if food resources are sufficient (Müller, 1972). Adult male iguanas have been described holding territories in which several smaller males, females and juveniles are tolerated (Müller, 1972). Females may travel over extended distances (Montgomery et al., 1973) and gather together in large groups at favorable nesting sites where they excavate burrows of 1-2 m length and 25-50 cm depth for laying eggs (Rand, 1968; Rand and Rand, 1976). In Colombia, from where our iguanas were obtained, mating takes place in November and December, egg laying in March, and hatching at the end of May, just before beginning of the rainy season (Müller, 1972; see also Harris, this volume). The behavioral correlates of sleep and wakefulness (Flanigan, 1974) and tonic immobility (Prestude and Crawford, 1970) have been described in captive iguanas.

This chapter offers a systematic description of the behavioral inventory of captive green iguanas. The classification and interpretation of behavioral events presents special problems since, for example, often the same component or elementary movement takes part in several functionally different patterns. Observers are often more inclined to emphasize the function of behavioral patterns rather than their individual components; however, restricted behavioral components may be more amenable to an experimental approach. Therefore, the behavior of green iguanas will be described at two levels: Firstly as elementary postures and movements, and secondly, as behavioral sequences with a functional significance.

MATERIALS AND METHODS

Eight large green iguanas (*Iguana iguana iguana* L.) from Colombia, 5 males and 3 females, were used in this study. Their snout-vent length ranged from 36 to 46 cm and their weight from 2.3 to 5.2 kg. They were housed together or in smaller groups in three similar indoor enclosures with a floor area of approximately 4 m^2. Eccentrically placed heat or light lamps (1,000 W) provided a thermal gradient for behavioral thermoregulation during the 12-hour day. Large branches were placed on the floor and a water tank 70 x 40 x 14 cm or water bowls were provided. The animals were usually fed lettuce, bananas and seasonal fruits.

Observations on spontaneous behavior were carried out from behind a one-way window over a period of 18 months. More than 310 hours were spent in *ad hoc* observation sessions ranging from some minutes to several hours; systematic observations were recorded during a total of 80 hours. In order to observe fighting behavior the animals were separated into subgroups of two or more. Males that had separately established a dominance relationship were then brought together. Under these conditions fighting between dominant males occurred immediately. A total of 11 fights were observed. During the most active behavioral periods, November through March, motion pictures were taken for documentation in the normal housing environment. An Arriflex 16 mm camera (24 frames/sec.) equipped with an Angenieux 10x12B zoom lens was used. For motion picture analyses of some behaviors, the housing enclosure was later elevated to camera level, to avoid parallactic distortion, and supplied with a clear front. In addition, the rear wall was made semicircular and painted with horizontal lines which served as references at heights of 10, 20 and 30 cm (Figure 14.1). A total of 38 hours of film was exposed. From this motion picture material line drawings were obtained of selected behavioral patterns and of 92 displays which has been analysed frame by frame.

RESULTS

External Features

External morphological features which appear conspicuous often have a functional correlate in the animal's behavior, although this has still to be systematically demonstrated in the green iguana.

The *basic color* of adult green iguanas is a grayish to brownish green. Three patterns of stripes can be distinguished: Bright *shoulder stripes*, usually bluish-green, on the forelegs; black *abdominal stripes* on the yellow skin of the belly, most prominent during displays (Figure 14.4); and a regular ring-pattern of yellow-black *tail stripes* (Figure 14.1). Highly aroused animals develop a *lightened* body *color* (color change 3-5 min.). Males which lose a fight (color change

0.5-2 min.), low ranking, lethargic, or cold animals have a *darkened* body *color*. A *white head* is prevalent amongst dominant males, extremely aroused or overheated animals, and sometimes in courting males. The most conspicuous color is the *red mucosa* of the opened mouth.

There are three prominent skin appendages supported by muscles or vascularized tissue: The *dewlap,* a large permanent but extendable skin-flap hanging from the throat; the *jowls,* larger in males than in females, muscular protuberances below the angle of the jaws, extended by the hyoid apparatus–as is the dewlap (Oelrich, 1956); and the dorsal *crest fold* which bears the crest spines. The apex of each jowl is covered by two *giant scales.* The larger one is circular and exceeds the eyes in diameter. Indications as to possible functions of the giant scales have not been found yet. Scattered over the neck are medium-sized *conical scales* which probably protect it from combat and copulation biting. The dorsal crest is distinguished by 1-4 cm long *crest spines* which are flanked by two rows of smaller spines.

Figure 14.1: Male green iguana in filming enclosure. The large dorsal crest spines are almost completely lost.

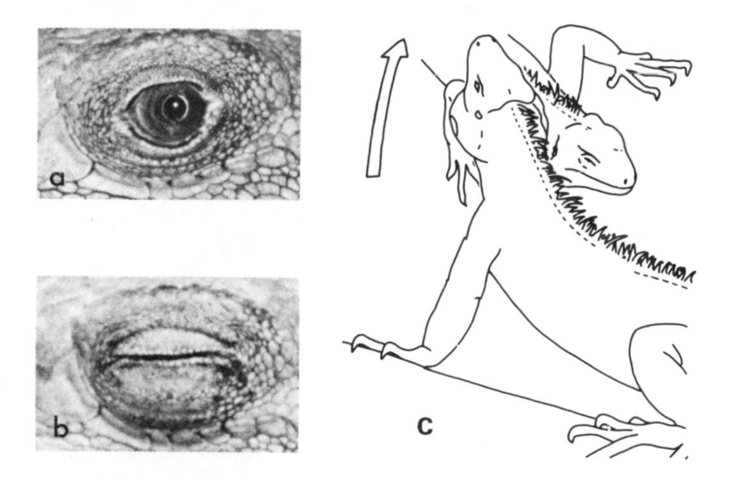

Figure 14.2: (a) Left eye dilated pupil. (b) Eye closed. (c) Eye-closing in response to head-nodding display.

Elementary Postures and Movements

Green iguanas are known to be extremely sedentary creatures, both in nature and in the lab (Moberly, 1968). In the slow flux of their behavior the different movements and postural components are easily detected.

The Head: The *head* may be held *horizontally* (1); slightly *raised* (2) as in Figure 14.1; *rested* (3) on a substrate (Figure 14.2c, upper animal); *stretched out* (4) as in Figure 14.4; *stretched up* (5) before upward escape or after drinking; and *lowered* (6), e.g., during retreat. Further, an iguana may keep the head *turned* (7) to one side for a prolonged time; *rotated* (8) around the head axis; *cocked* (9), i.e., rotated and turned; or *tipped down* (10) from the elevated shoulder girdle.

Movements which effect such head positions are termed accordingly. However, directional relations may be identified as, e.g., *head-turning toward* (11), *away* (12) or *perpendicular* (13) to a given stimulus. Most of the following repetitive head movements may also be oriented towards a stimulus. The *smooth scanning* (14) from one side to the other is usually elicited by, but not oriented toward peripheral visual or acoustic stimuli. The quicker, *jerky scanning* (15) may enhance vision of stationary objects during the intermittent pauses. An alternating *head-scraping* (16) against a substrate, e.g., a branch, may sometimes be elicited by food particles sticking of the mouth. Feeding of large pieces is supported by brief *sideward jerking* (17) toward the object. Vigorous *shaking* (18), e.g., of the adversary's limb, is observed during combat fighting. *Head-flinging* (19) is similar to shaking, but is freely swinging, and occurs as an evasive movement: with vomiting, and when attempting escape after capture.

The Eyes: In the *basic* (20) situation the eyes are open and centered in the middle of the orbit. Occasionally they may be *retracted* (21) or *protruded* (22) through the action or release of the retractor bulbi muscles. In the motionless iguana, eye movements are often the only signs of alertness: Usually, they are *conjugate eye-movements* (23) but *convergent eye-movements* (24) may also be performed when an object is in front of the mouth. *Rotating eye-movements* (25) are observed in compensation to lowered or raised head positions (Heath et al., 1969). When one eye becomes *fixed* (26) upon a stimulus, head turning may follow until the eyes are back in normal position.

Closing of *one eye* (27), i.e., the lower eyelid, is not only a reflex to tactile and visual stimuli that may become potentially noxious, but is also elicited by the display of a dominant male (Figure 14.2). *Closing* of *both eyes* (28) is an element of sleeping, resting, and tonic immobility, and is occasionally observed with pawing and digging movements. Protection of the eyes is further provided by *blinking* (29) the eyelids briefly, and by moving the *nictitating membrane* (30) across the eye. Finally, a distinct *dilatation* of the triangular *pupils* (31) can be noted during defensive display.

The Jaws: Although more exact physiological definitions would be sometimes possible, the customary functional terms are used in the following: When an animal is resting the *mouth* may be *closed* (32) or *slightly opened* (33). It is held open widely during *gaping* (34) which is part of the defensive display, and during *panting* (35). It is opened maximally during *yawning* (36) and then closed rapidly. Opening of the mouth is slower than closing also during *biting*

(37) of, e.g., food. *Grabbing* (38) of a piece of food corresponds to a single bite. *Snapping* (39), which is like rapid grabbing with immediate release, may be performed against other lizards or predators. *Catching* (40) occurs in the same situations as snapping but the adversary is not released. *Combat neck-biting* (41) is similar to snapping and directed to the back of the neck of the other animal during male combat (Figure 14.17b), while *copulatory neck-gripping* (42) is similar to grabbing and directed to the side of the neck. Both are performed from a mounted position and share the same initial orientation movements.

The Tongue: During panting and gaping the *tongue* may be partially *engorged* (43), i.e., retracted and lifted. Four kinds of movements may be distinguished: *Tongue-flicking* (44), i.e., repeated rapid protrusions directed preferentially towards prominent objects (Figure 14.3), is typically performed as: exploratory behavior, element of head-nodding display, and result of transient excitation, e.g., other passing animals. *Licking* (45) is similar but less intense or object oriented than tongue-flicking, and is in addition, an element of drinking behavior. *Mouth-licking* (46) is a simple movement towards the corner of the mouth along the upper rim. With *tongue-extruding* (47) the tongue is pressed through the slightly parted teeth and retracted quickly. This is very effective for removing particles, e.g., of food.

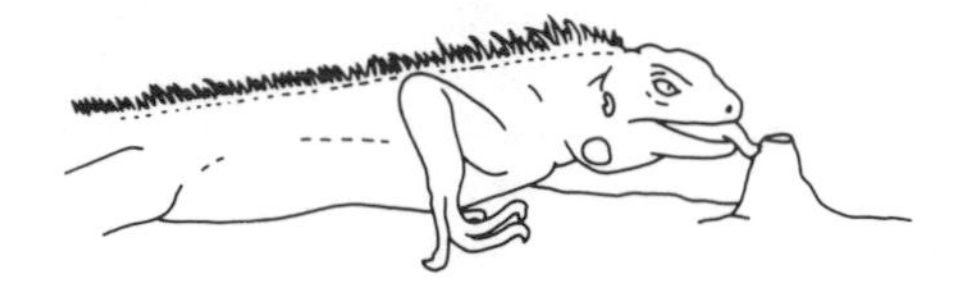

Figure 14.3: Tongue-flicking.

Figure 14.4: Body display; abdominal stripes become visible.

The Gular Region and Appendages: The *dewlap* is *extended* (48) during most displays (Figures 14.4 and 14.7), during basking (Rand, 1968; Müller, 1972), and after drinking. The *jowls* and the *crest fold* may be *extended* (49, 50) during defensive and combat displays. Further, less discernible *gular movements* (51) may be observed with swallowing, strong breathing, and sometimes with displays.

The Trunk: When iguanas are basking under a heat source the trunk is relaxed and even *extended laterally* (52). During displays, however, the trunk is *compressed laterally* (53) thereby increasing the animal's sagittal profile (Figure 14.4). Occasionally, when an animal was caught, the trunk became *expanded maximally* (54) through strong inflation and then maintained by an expiratory rest.

During defensive and combat displays, expiration after prolonged inspiratory rests may result in loud *hissing* (55) sounds. On other occasions, bursts of expiration are observed which sound like *coughing* (56) when the mouth is open, or like *sneezing* (57) when it is closed. With the latter, salt droplets may be blown from the nostrils.

The Limbs: During resting and sleeping, one or more *limbs* are *extended backward* (58); during swimming all four. With increasing states of arousal, the *limbs* are either *adducted* (59) but not supporting the trunk, or the *trunk* is *supported* (60, Figure 14.1), or *fully raised* (61) by the forelimbs (Figure 14.7). Elementary movements of single limbs are: *pawing* (62), which is a repeated scratching of the ground and occurs spontaneously, in the lab, sometimes after locomotion; or *scratching* the *pelvis* (63) region with the forelimb, after bending the trunk, and the *head* (64) and neck region with the hindlimb (Figure 14.5); and *backward-striking* (65) which is similar to pawing but directed into the air (Figure 14.6) and is characteristically performed towards a courting male which is approaching from the rear.

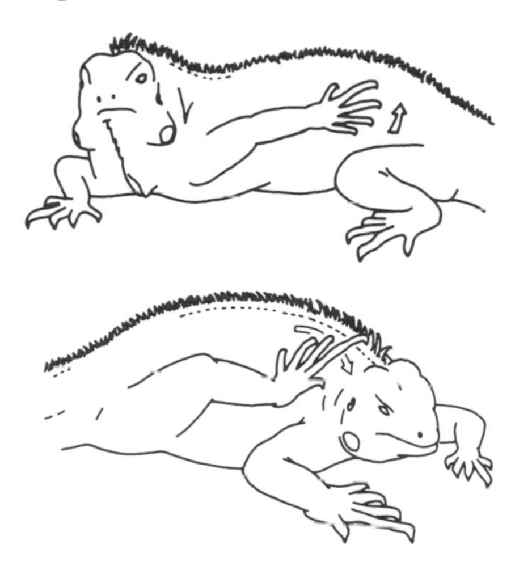

Figure 14.5: Scratching of the pelvis (above) and of the head region (below).

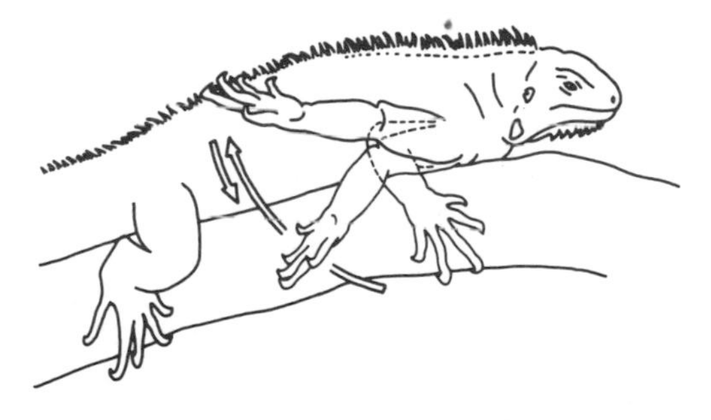

Figure 14.6: Backward-striking.

The Tail: Usually the tail is dragged on the ground; however, it may serve as a counterbalance in the arboreal environment and as a propelling force during swimming. During defecation, copulation, and fast running the *tail* base is actively *lifted* (66) from the ground. A *tail-lashing* (67) with wide sweeping movements from one side to the other and a *tail-twitching* (68) of the tip is observed during male combat. During defensive display, *tail-striking* (69) to one side is typical (Figure 14.9). It is aimed with precision against the adversary.

Complex Movements of the Head: The most conspicuous and complicated head movements may function as displays: The *stereotyped head-nodding* (70) consists of three invariable phases, (a) an initial bob with which the head is maximally thrown up, followed by (b) a pause in which the head is kept elevated, and then (c) a number of additional bobs (4-10) of decreasing amplitude. The duration of this sequence varies between 2.7 and 4.5 seconds depending mainly on the number of additional bobs. There are some differences in stereotyped head-nodding amongst individuals but usually not reliably enough to be specific for one animal (Figures 14.7 and 14.14). The *rotary head-nodding* (71) consists of rotatory movements around the long axis which rapidly alternate (5-6 cps; Figure 14.8b) and are superimposed on irregular vertical head movements (Figure 14.8a). This virtually only occurs before stereotyped head-nodding, so extending the head-nodding display considerably–up to 16 seconds; and is probably best regarded as an amplification element. The *vibratory head-nodding* (72) consists of very rapid but small movements (9-10 cps) in the vertical plane which are modulated on slower irregular vertical and sometimes horizontal movements. It is performed by males during the mating season usually as an introduction to stereotyped head-nodding but often by itself (Figure 14.8c). *Head-swinging* (73) consists of lateral head movements which alternate (1-2 cps), and on which smaller, more rapid vertical movements are modulated (Figure 14.8e). Characteristically, it is performed in response to vibratory head-nodding plus frontal approach. (See also Dugan, this volume.)

Integrated Postures: Body color and some postural elements are often associated, and may very well indicate different states of alertness and excitation: In moderately alert animals with a basic color, the head is most frequently horizontal and the limbs are adducted. This may be called the *normal posture* (74). During sleeping and prolonged immobility states, and in the case of lethargic animals, a darkened color is associated with a *relaxed posture* (75), i.e., head rested, eye(s) closed, trunk extended laterally, limb(s) extended backward. The signs of an *alerted* (76) animal are head raised and trunk supported (Figure 14.1). An *excited* (77) animal usually has a lightened color, the head is raised, and the trunk is compressed laterally and raised by the forelimbs. Finally, the *defensive posture* (78), which is part of the defensive display (Figure 14.9), consists of the head being turned toward and the body presented laterally to the adversary, mouth gaping, tongue engorged, dewlap extended, trunk compressed laterally and raised. The color may become lightened.

Integrated Movements: The simplest integrated movements are *body-turning* (79) for postural adjustment, and *body-presenting laterally* (80) to a stimulus, i.e., perpendicular to the direction. *Leaning-sideways* (81) by bending neck or tail, or tilting the trunk towards the stimulus and eventually by extending both legs on the other side. It is part of the male combat but also per-

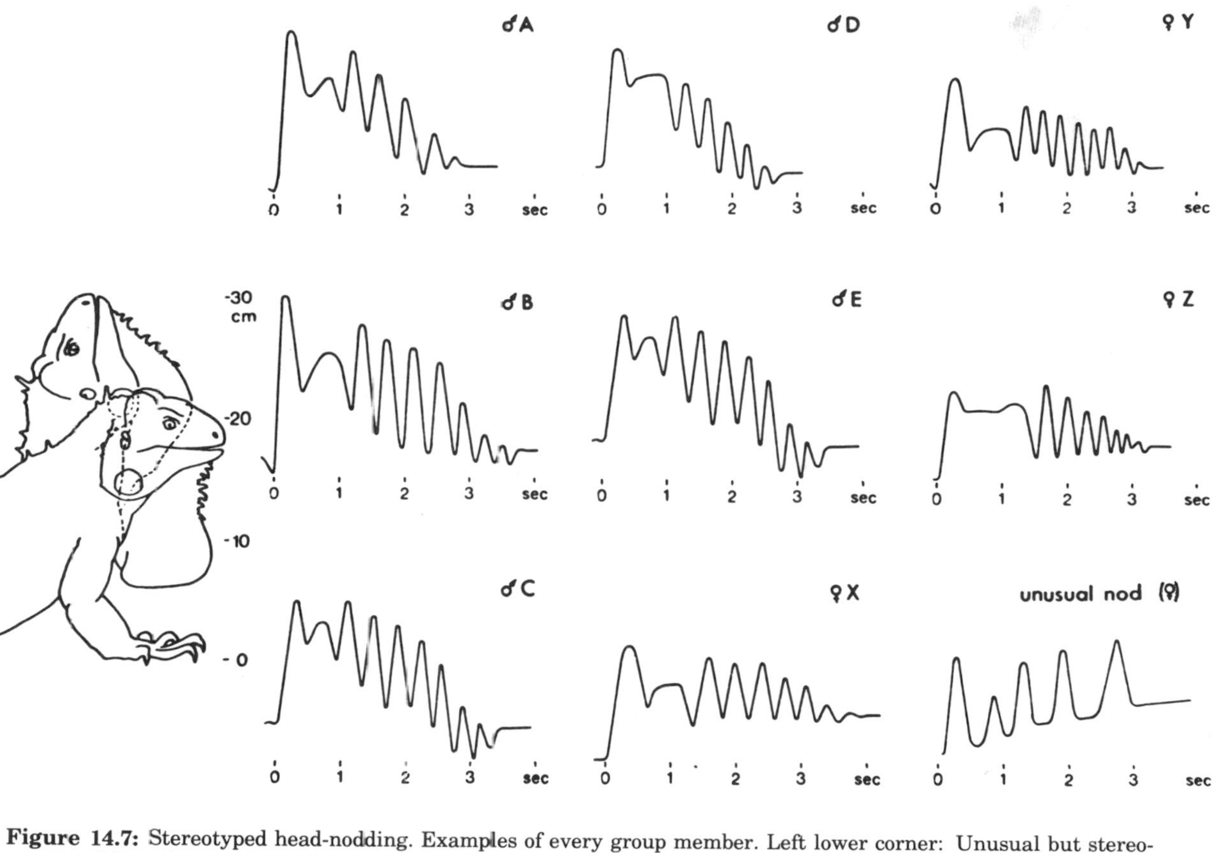

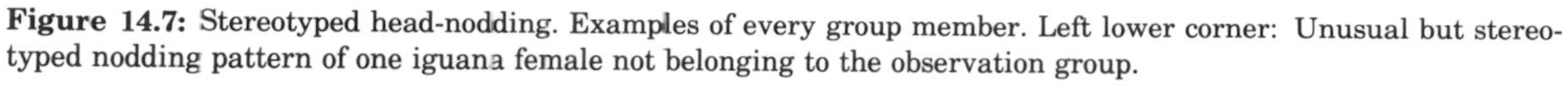

Figure 14.7: Stereotyped head-nodding. Examples of every group member. Left lower corner: Unusual but stereotyped nodding pattern of one iguana female not belonging to the observation group.

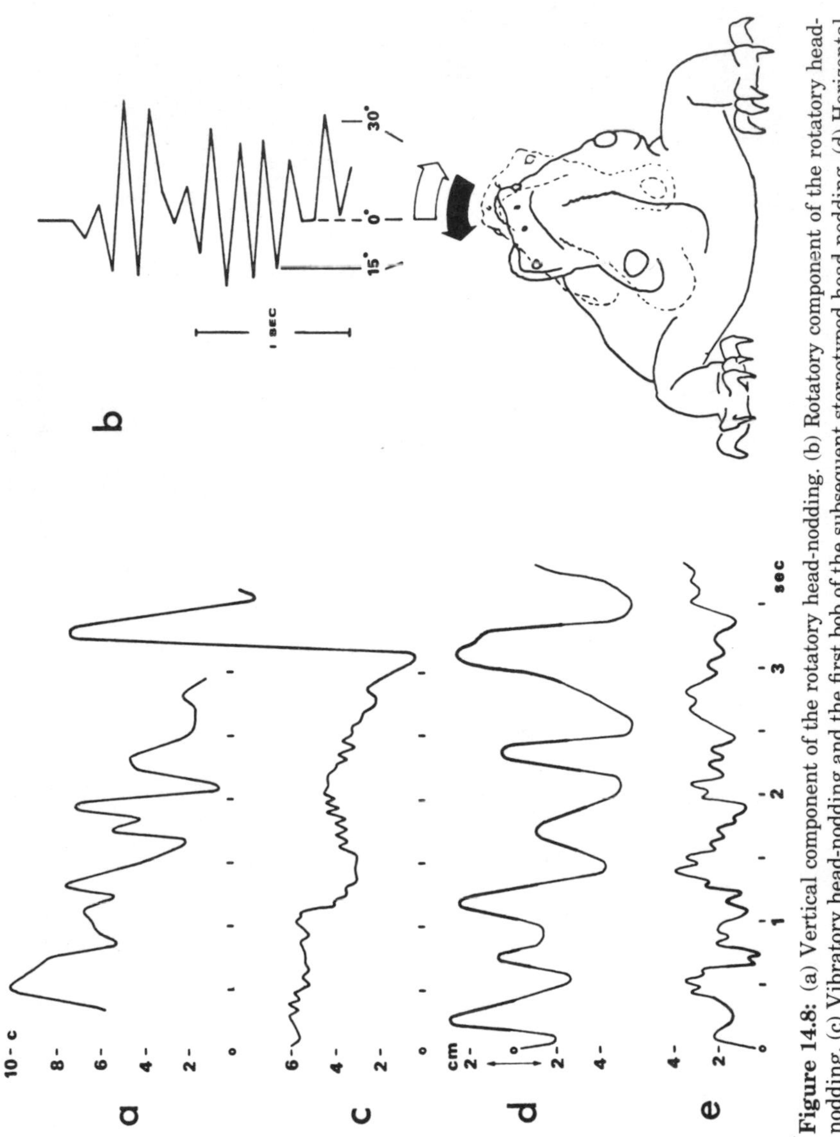

Figure 14.8: (a) Vertical component of the rotatory head-nodding. (b) Rotatory component of the rotatory head-nodding. (c) Vibratory head-nodding and the first bob of the subsequent stereotyped head-nodding. (d) Horizontal component of head-swinging. (e) Vertical component of head-swinging.

formed towards any passing animals. *Rotating axially* (82), i.e., a full turn of the body around the long axis with the tail moving in the opposite direction, is performed as defense, e.g., when caught. In connection with copulation and defecation two movements are observed: *Straddling* (83), i.e., placing the hindlimbs apart and moving the pelvis; and *pelvis-scraping* (84), i.e., scraping and shivering movements against a substrate. Under the same conditions as pawing but in more excited animals, a complete pattern of *digging* (85) movements are observed in the lab. This repeated pattern starts with 4-8 pawing movements of one, and then of the other forelimb, followed by 1-4 backward pushing movements of firstly the contralateral hindlimb, and then the other (cf Rand, 1968).

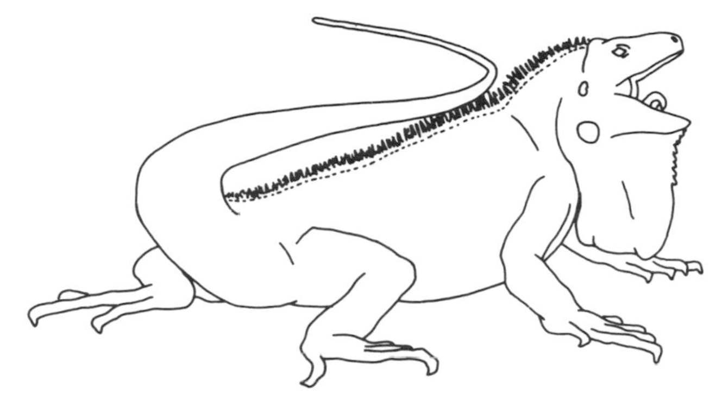

Figure 14.9: Defensive posture and tail-striking.

For defining locomotive movements, speed, the strength of the undulatory component and the contact with the ground serve as useful criteria: Pure lateral undulation with all limbs extended backwards characterize *swimming* (86) movements. *Crawling* (87), however, is a slow locomotion with body contact to the ground and with a very weak undulatory component. The more rapid *walking* (88) has some more undulation but no ground contact. The very rapid *running* (89) is defined by pronounced undulatory components and no ground contact. *Strutting* (90), an element of the combat display, is comparable to slow walking but with all four limbs extended fully. However, *jumping* (91) is completely different as both fore- and both hindlimbs are thrust out together, and as no undulatory component is present.

In addition, several locomotive elements may be distinguished by the specific stimulus situation: *Approaching* (92), i.e., crawling or walking toward a stimulus; *retreating* (93), i.e., lowering the head and moving slowly backwards or sideways away from the stimulus; *escaping* (94), i.e., turning the head and running a short distance away from the stimulus but avoiding any obstacles; and *violent flight* (95), i.e., suddenly running blindly at highest speed in any direction. Further, *circling* (96), i.e., walking or strutting (distance 1-2 m) around another animal or running when closer (<1 m); and *mounting* (97), i.e., climbing onto another animal, which occurs not only as an element of combat and mating behavior but is also observed to occur when animals crowd beneath the heat lamp.

In fighting males, the *face-off position* (98) is a parallel head-to-tail position which two animals maintain through circling; in the *T-position* (99) the head is closely oriented towards the other animal's head, neck or trunk region (Figure 14.10). In aggressive or defensive encounters a *lunge* (100) involving a sudden approach and retreat is performed as an intended or completed snapping from not further than jumping distance.

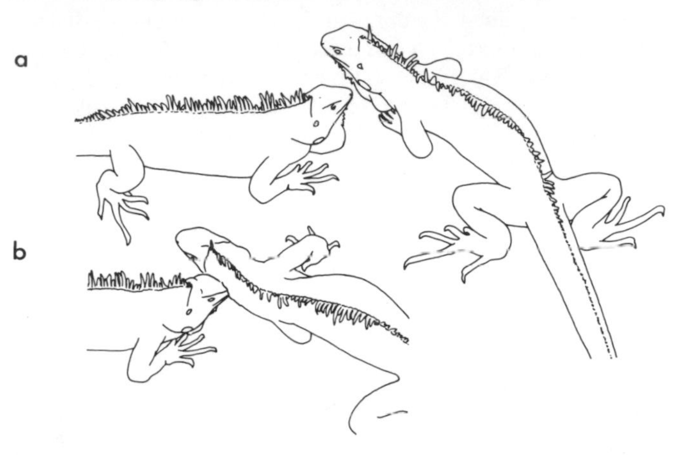

Figure 14.10: Pushing-leaning contest. (a) T-position and leaning-sideways by the left animal. (b) In addition, pushing of the right animal.

Behavioral Sequences

Basic Activity: In the morning, after warming up under the heat lamp, and in the late afternoon, behavioral activity lasts up to 45 minutes. The remaining time, the iguanas spend apparently *resting* (101), usually in a normal or relaxed posture, interrupted, however, by very short bouts of *"basic activity"* (102; Figure 14.11) which follow each other in intervals of 2-8 minutes. Dominant males perform in addition head-nodding displays (cf. Figure 14.15). Basic activity is triggered often by external events but occurs also spontaneously.

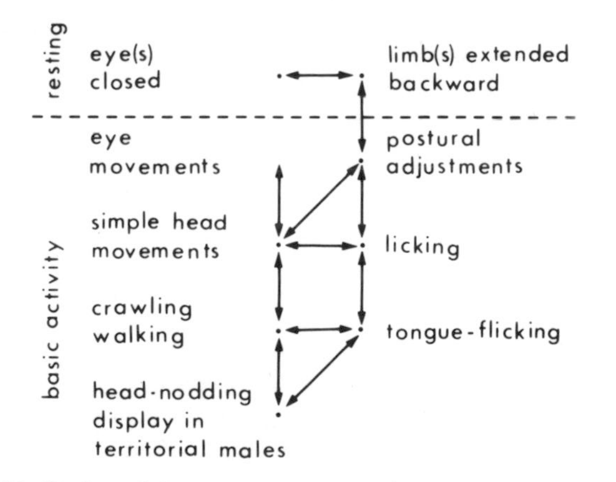

Figure 14.11: Basic activity sequence. (The arrows indicate possible sequences of behavioral elements in the given context.)

Animals are considered *lethargic* (103) when they are barely excitable by external stimuli and perform little basic activity. This is typical for freshly shipped animals. During the night the animals are *sleeping* (104) in a relaxed posture with both eyes closed.

Exploratory Behavior: Exploratory behavior is directed towards new objects in the cage and clearly observed in a new enclosure. *Exploring* (105) consists mainly of walking with the head stretched out, and with concomitant persistent tongue-flicking. Further, all scanning movements and cocking of the head may be regarded as visual exploratory behavior.

Grooming Behavior: Those behavior patterns which are self-directed and potentially remove particles sticking to the skin are categorized as grooming (Figure 14.13). Generally, grooming behavior is more common in mammals and birds–even so, scratching may be performed spontaneously.

Thermoregulatory Behavior: Thermoregulatory behavior (Figure 14.12) appears to be little elaborated, at least under laboratory conditions, since in the field additional elements are observed (Rand, 1968; Muller, 1972). Under a heat lamp a lateral extension of trunk and limbs is observed which can be regarded as *basking* (106) behavior. When overheated, the animals raise the trunk, change to a lighter color, and eventually start panting.

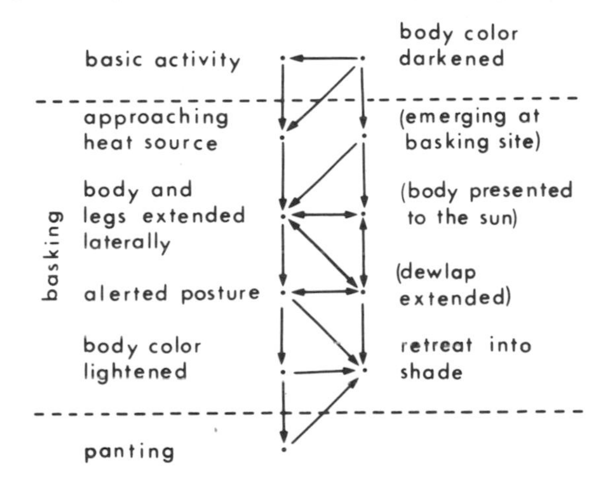

Figure 14.12: Thermoregulatory behavior sequence. In parenthesis field observations by Rand (1968) and Müller (1972). (The arrows indicate possible sequences of behavioral elements in the given context.)

Drinking Behavior: Iguanas drink only occasionally. When fed with hydrous plant material, they take in water only every other week. Two means of drinking are observed: *lapping* (107) of water on a surface and more often, *sucking* (108) with the snout submerged into water. With sucking, gular movements and, occasionally, short licking movements are visible. After drinking, the trunk is raised, the head stretched up and the dewlap extended, then tongue movements follow.

Feeding Behavior: Iguanas may also feed irregularly, although if they are not lethargic they may feed every day. *Feeding* (109) sequences are fairly simple (Figure 14.13); only when the grabbed pieces are too large the more elaborated *inertial feeding* (110) is performed. This consists of brief head-jerking toward the food hanging from one side of the mouth, and of rapid concomitant bites. Thereby the inertia of the food is used to engulf it in the jaws.

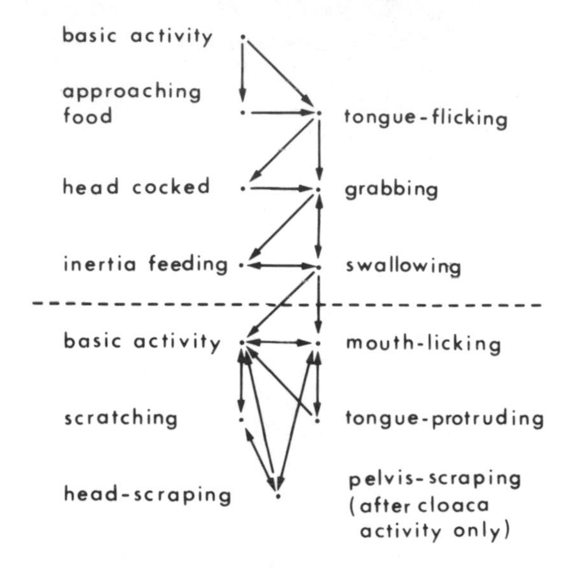

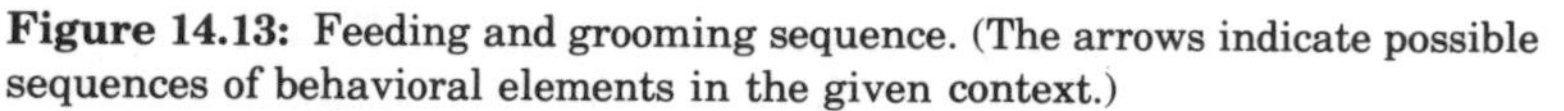

Figure 14.13: Feeding and grooming sequence. (The arrows indicate possible sequences of behavioral elements in the given context.)

Defecation Behavior: *Defecation* (111) is typically associated with water, e.g., it was performed over a large water tank or after sweeping the floor with water. The characteristic posture is with hindlimbs straddled and tail lifted. After a few steps, pelvis-scraping may follow.

Display Behavior: The most common type of display is the *dewlap display* (112). This consists of dewlap extension only and is supplemented occasionally by head-turning perpendicularly to the stimulus direction. In alerted animals it is easily triggered by movements of all kinds. With increasing excitation, *body display* (113) may follow. The trunk is raised and compressed laterally, and the crest fold becomes extended (Figure 14.4).

The *head-nodding display* (114) is usually performed at a low level of excitation. It may vary from a simple stereotyped nodding without dewlap display, to a full-blown sequence with accessory movements (Figure 14.14). For example, a dominant male may rise and walk a short distance towards an elevated point with associated tongue-flicking. Shortly before reaching the point, rotatory or vibratory head-nodding starts. If another animal has elicited the display, the body will be presented laterally and the head turned slowly towards the animal. Then, just at the end of the introductory nodding, the head is turned rapidly away and a single tongue-flick is made. However, if a display is not directed, the tongue-flick is directed straight ahead. Stereotyped head-nodding follows. The dewlap is extended either before or after the introductory movements.

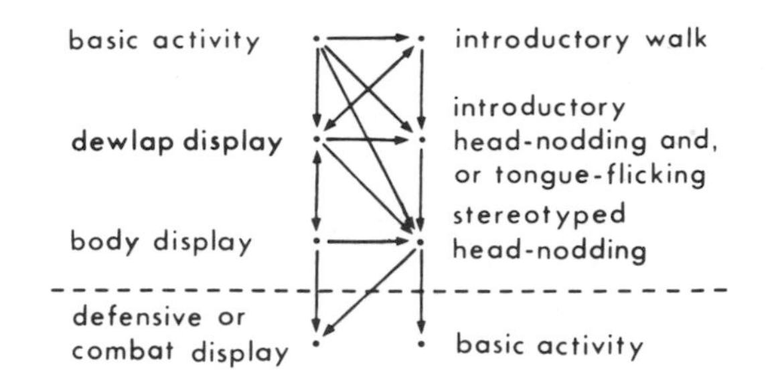

Figure 14.14: Head-nodding display sequence. (The arrows indicate possible sequence of behavioral elements in the given context.)

Head-nodding displays are performed by both sexes but more often by males. They may be elicited by other animals. However, in the case of one isolated territorial male, 30 spontaneous displays were counted within 200 minutes (Figure 14.15). Also, head-nodding display were often performed at the end of any behavioral sequence, and after a potential danger, such as a departed predator.

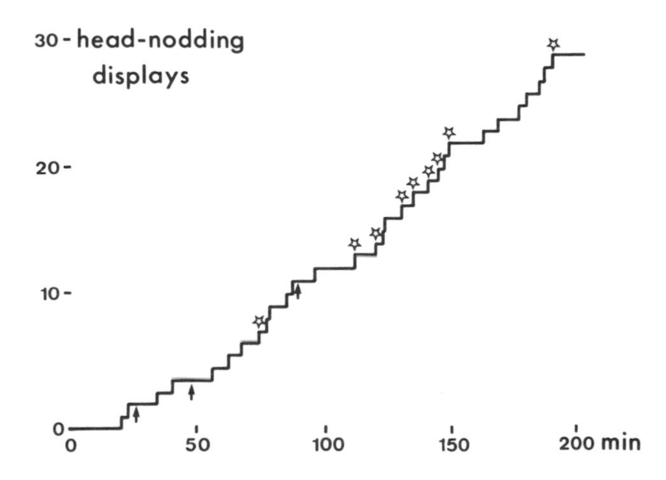

Figure 14.15: Cumulative record of head-nodding displays in an isolated, territorial male. Star symbols: Introductory head-nodding and accessory movements in addition. Arrows: Intermediate bouts of basic activity without head-nodding displays.

The *defensive display* (115) is observed in response to a potential predator represented by a dog or a man. In addition to the defensive posture described above, the animal hisses, lashes with the tail, and may snap when near enough for a lunge (Figure 14.9). Defensive displays can also be elicited toward a fairly large object, i.e., a piece of cloth (40 x 40 cm) moved rapidly towards and over the animal.

Fighting Behavior: The first phase of the extremely stereotyped male fighting behavior (Figure 14.16) begins with the *combat display* (116). It consists of strutting, circling the challenged male, dewlap and body maximally displayed, hissing, and tail-lashing. Head-nodding displays are interspersed. With the closing of a circling spiral, head and body are lowered and tail-lashing ceases.

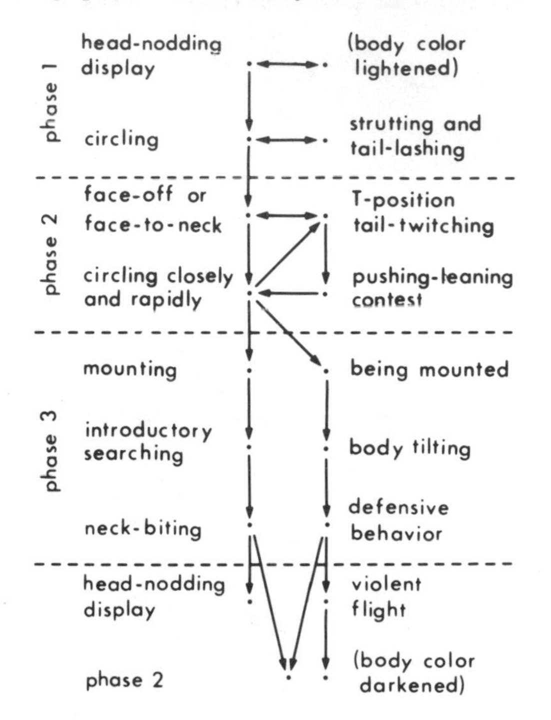

Figure 14.16: Male fighting sequence. (The arrows indicate possible sequences of behavioral elements in the given context.)

By circling more rapidly and more closely, both males arrive at a face-off or a face-to-neck position (second phase). One may succeed in mounting the other while they are circling. Usually, however, one may arrive in a more perpendicular position to the other. A full T-position is then completed through leaning-sideways and body-turning, respectively. During the subsequent *pushing-leaning contest* (117; Figure 14.10), the pushing male may back up from time to time and perform a shallow stereotyped head-nodding and then push again. At some point the close circling may be taken up again leading to another contest or to mounting. Pauses often interrupt the contest, and before action will be resumed tail-twitching may occur.

Mounting indicates the beginning of the third phase. After reaching the neck, the mounting animal starts searching movements to both sides with the mouth closed or opened slightly. The animal underneath responds by leaning-sideways toward the same side as the searching movements are directed. After several shifts of the head from one side to the other, the mounting animal opens the mouth slowly and bites the neck or back, pulling back immediately. Often this, or a second neck-biting, causes the mounted animal to flee violently (Figure 14.17). The body color of the defeated will then darken rapidly. However, if the mounted male succeeds with its defensive behavior and dismounts the other, the sequence starts again with a circling and the pushing-leaning contest. The only observed aggressive encounter between females consisted of dewlap and body display, circling, hissing with snapping attacks.

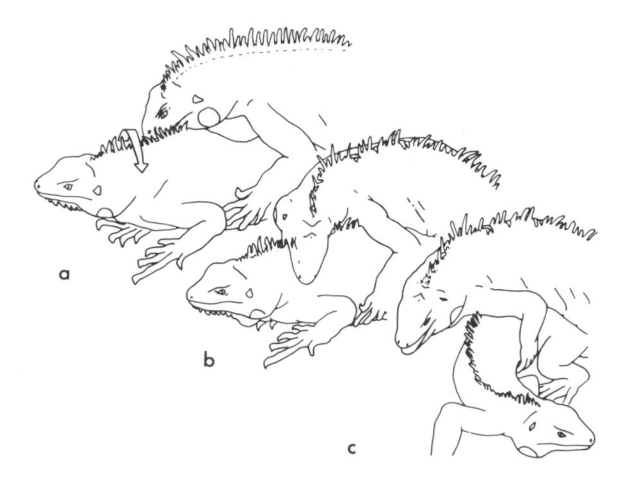

Figure 14.17: Mounting and combat neck-biting. (a) Orienting movements before neck biting. (b) Neck-biting. (c) Violent flight of the bitten animal.

Escape and Defense Behavior: In response to novel movements in the environment or noises, iguanas readily perform *freezing* (118) of their ongoing action. They may later continue or they may perform tongue-flicking or scanning movements. However, if the stimulus persists, the head is lowered and the extensor tonus decreases. Eventually, the animals may retreat, e.g., by sliding behind a branch they were sitting on. Some animals may close one or both eyes, or in extreme situations, arrive at a state of *tonic immobility* (119). Long periods of freezing are often followed by a sudden violent flight. When animals are about to escape but not to freeze they may raise the trunk and keep the head stretched up.

Besides the described defensive display, the defense against predators and other iguanas may involve lunges, snapping, and catching combined with head-shaking, or when caught themselves, head- and body-flinging, quick axial rotation, and brief attempts of violent flight.

Mating Behavior: Sexually active males approach equally females and males in the cage. Characteristically, they then perform a vibratory head-nodding (Figure 14.8b). A male which has mounted successfully proceeds to the neck and performs searching movements to both sides before grabbing firmly the neck skin on one side. Moving backwards, he starts straddling in order to achieve an optimal position for *tucking* the *tail* (120) underneath the female's tail. However, the female has to lift actively the tail base for this to be achieved (which was the only 'solicitory' behavior observed). *Protruding* the *hemipenis*

(121), ejaculation, dismounting and pelvis-scraping complete the sequence. Although many mating attempts have been observed, few were completed to the last stages. Both sexes reject a courting male before mounting with head-swinging (Figure 14.8 d,e) or backward-striking (Figure 14.6), and when mounted with scratching or other defensive behaviors.

DISCUSSION

As lizards are observed to perform a variety of typical movements and postures that can be singled out from activity of a more continuous nature, it is certainly not by chance that two recent descriptions of lizard behavior utilized a similar method of enumerating in detail the behavioral elements. When comparing the behavioral inventories of the bearded dragon (Brattstrom, 1971) and of the blue spiny lizard (Greenberg, 1977b) with that of the green iguana, it becomes apparent that a complementary movement or posture can be found for most behavioral elements. However, in both species, additional elementary postures of body, tail, feet and toes are described. These elements may all be employed in the highly adaptive thermoregulatory behavior of these relatively small lizards (Greenberg, 1976). The lack of such postural elements in the green iguana probably reflects the marginal importance of a constant thermoregulation, possibly due to size.

The elements of display behavior are widely shared among lizards and have been amply documented for other iguanid and agamid lizards (Gorman, 1968; Carpenter et al., 1970; and reviews by Carpenter, 1967; and this volume). In the green iguana, vibratory head-nodding appears to be homologous to the rapid courtship bobs of iguanid lizards (Carpenter, 1967). The stereotyped head-nodding pattern is species-specific in the green iguana as it is in other species (Carpenter, 1967; Hunsaker, 1962; Jenssen, 1970). However, no differences were found with regard to an assertive or aggressive type of the head-nodding display, except for the ritualized tongue-flick away from the opponent in an aggressive situation. The initial phase of male fighting can be reasonably regarded as a challenge display (Carpenter, 1967). Individual differences in the performance of stereotyped displays exist in the green iguana, but less distinctly than as in some anolis lizards (Jenssen, 1971; Stamps and Barlow, 1973); further, geographical variations are likely (Ferguson, 1971; Jenssen, 1971).

Observations indicated head-swinging and backward-striking signified unwillingness to be mounted. They are continued while the courting male mounts, and subsequent escape is attempted. Hence, they are regarded as rejection rather than submission signals as are similar forelimb movements in other lizards, e.g., *Lacerta* (Kramer, 1937; Kitzler, 1941; Webster, 1957) or as subordination of dominance as in *Amphibolurus* (Carpenter et al., 1979; Brattstrom, 1971).

Combat fighting of the green iguana has been rarely observed in the field; and of three phases only the first phase (the combat display), and a neck-biting intention movement which corresponds to the last phase, have been reported (Müller, 1972). The second phase, that is, the pushing-leaning contest, has never been observed. This may well reflect population differences between Müller's and our own Colombian iguanas. A sub-species difference is suggested in the observation that *Iguana i. rhinolopha (= I. tuberculata)* performs gripping of the forelimbs instead of neck-biting (Peracca, 1891, cit. Müller, 1972).

The challenging encounter of the first phase is fairly similar in all iguanid species hitherto investigated (Carpenter, 1967). However, among the most closely related species, significant elements of the second and the third phase are shared. Marine iguanas *(Amblyrhynchus cristatus)* of the Galápagos Islands contest each other with front-to-front head butting and may also mount and bite the neck of an opponent after a chase (Eibl-Eibesfeldt, 1955; Carpenter, 1966), whereas the Galápagos land iguanas *(Conolophus subcristatus)* immediately attempt to bite the opponent's neck. On occasion they also butt the opponent with the top of the head (Carpenter, 1969). The combat sequence of the green iguana appears to be intermediary to the combat sequences of these large Galápagos iguanas. Considering that these closely related species supposedly share a common South American ancestry, and also that the elements of neck-biting and head-butting are shared, suggests that such an ancestor had already developed the major components of combat behavior.

Eye-closing in the green iguana is not only an indicator of a general decrease in wakefulness (Flanigan, 1974) or a protective reflex to noxious stimuli but also a specific response to social stimuli such as a head-nodding display. This may prevent further aggression since with the closing of the lid a potential sign stimulus (the eye) is eliminated from the sight of the displaying animal (cf. Figure 14.2). A second hypothesis is that eye-closing is merely the shutting out of an unpleasant stimulus thereby allowing the animal to maintain an undisturbed state (Chance, 1962). This is consistent with the protective functions of eye-closing, and is supported by its occurrence as a pre-escape behavior during electrical brain stimulation (Distel, 1978). The only comparable cut-off behavior described in reptiles is the active closing of the pupils to social stimuli in the gecko *Ptyodactylus* (Werner, 1972).

Functional aspects of external features and behavioral elements have been reviewed in Mertens' (1946) extensive work on the defensive behavior in reptiles. We would like to emphasize the fact that many behavioral elements may take part in functionally different sequences. For example, the primary function of tongue-flicking of snakes and carnivorous lizards is the uptake of odor particles, and constitutes a decisive role in prey or mate trailing behavior (reviewed by Burghardt, 1970). In the herbivorous green iguana tongue-flicking has become a ritualized component of the head-nodding display, and is in addition, performed after any behavioral activity or after transient excitation.

Digging behavior consists of a very specific sequence of limb movements and has a function in the egg-laying of females (Rand, 1968); however, in the lab, digging movements are performed sporadically and throughout the year by both sexes. They occur, like pawing, spontaneously or after a frustrated climbing attempt of the smooth cage wall. Stereotyped head-nodding is certainly a social signal but not necessarily dependent on the presence of other animals, as Rand's field observations confirm: Nesting females display this when about to move, when emerging from a burrow, or when settling down to bask, but not when meeting other females. It seems that they often display behaviors such as tongue-flicking, digging or head-nodding spontaneously, or in toning themselves down from an elevated level of arousal (Distel, 1978).

SUMMARY AND CONCLUSIONS

The behavioral patterns performed by adult green iguanas in captivity are systematically described, and also briefly some features of the external appearance, such as coloration, appendages and scales, which may have behavioral relevance. Sixty-nine elementary postures and movements of head, eyes, jaws, tongue, appendages, trunk, limbs and tail are distinguished, as well as fifty-two more complex or integrated behavioral elements. The context of their occurrence is characterized in terms of behavioral sequences which comprise exploring, grooming, basking, drinking, feeding, defecation, displays, fighting, escape, defense , and mating behavior.

Motion picture analyses of head-nodding displays revealed two variable introductory and one stereotyped nodding patterns. Head-nodding displays were spontaneously performed, triggered by external stimuli, and oriented toward conspecific animals. Females and males which rejected the mounting approach of courting males signalled with a specific head-swinging pattern or with repeated backward-striking of one limb. The male fighting behavior was found to be highly ritualized: It consisted of (1) an initial display sequence, (2) a pushing-leaning contest between the opponents in T-position, and (3) mounting and combat neck-biting. Unilateral eye-closing was not only performed as a protective reflex, but also in response to social stimuli such as head-nodding. Functionally, this was interpreted as cutting off stimuli which would otherwise increase the probability of retreat from a maintained position.

The behavioral inventory of the green iguana as described in the present investigation is by no means complete. Nesting behavior (Rand, 1968), juvenile behavior (Henderson, 1974; Burghardt, 1977), and specific reflective responses (e.g., Heath et al., 1969) have not been covered. The restricted environment provided in the laboratory suppresses or distorts necessarily some behavioral aspects; it allows on the other hand a more intimate examination than the field. It is hoped that this inventory will facilitate field observations and be a referential framework to the behavioral physiologist who is interested in a specific pattern.

Note

*Present address: Institut für Medizinische Psychologie, Universität München

Acknowledgments

The authors are indebted to Professor D. Ploog for his support in the course of this study, and to Tony White for help with the manuscript.

15

Orientation in Dispersing Hatchling Green Iguanas, *Iguana iguana*

Hugh Drummond
Institute of Biology
Department of Zoology
Universidad Nacional Autónoma de México
México D.F., México

and

Gordon M. Burghardt
Department of Psychology
University of Tennessee
Knoxville, Tennessee

INTRODUCTION

Every year in February some 100-200 gravid female iguanas swim to the 0.3 hectare islet of Slothia to deposit their eggs in a small clearing (Rand, 1968). Slothia, which is covered with fairly dense secondary growth, lies adjacent to Barro Colorado Island (BCI), a 1600 hectare reserve of moist tropical forest in Gatun Lake, Panama (Figure 15.1). The hatchlings emerge over a three-week period in May and all depart rapidly from the islet, most of them taking the shortest route to BCI across a shallow channel (Burghardt et al., 1977). They coordinate their activities, synchronizing departures from nest-holes and movements overland and across open water.

While in the nest clearing the newly emerged iguanas generally walk slowly with frequent pauses unless disturbed by an aerial predator, often flicking their tongues onto the substrate and scanning their surroundings. Some iguanas depart alone, but most move off in groups of from two to twelve and exhibit various behaviors that could serve a communicatory function (Burghardt et al., 1977; Burghardt, 1977). Most of the hatchlings head for the SE tip of

Slothia where they explore the shore in groups and some climb reeds, giving all the while many indications of vigilance and social interaction. Eventually they drop or walk into the channel and swim across to BCI.

This immediate, rapid, coordinated and oriented dispersion of the hatchlings raises many interesting questions. The work reported here is concerned with where the iguanas go, what causes them to go, and how they find their way.

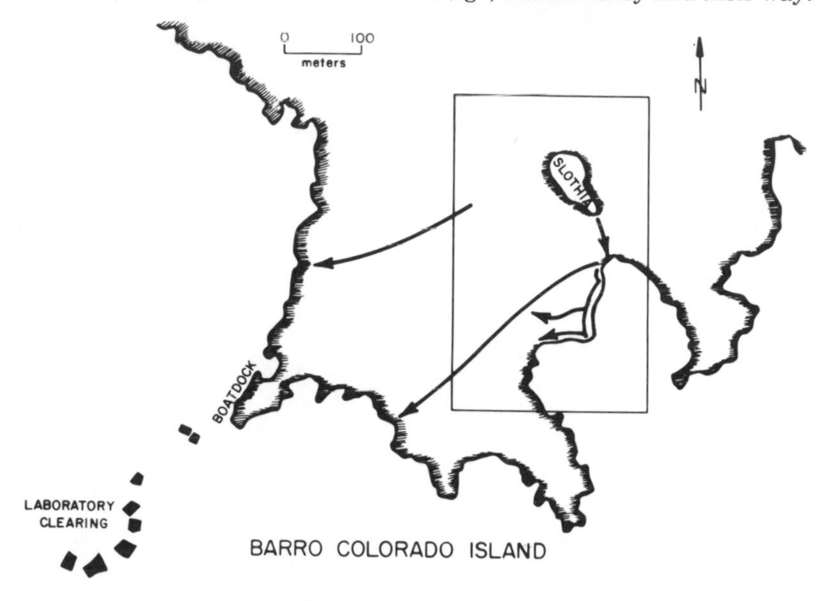

Figure 15.1: Location of Slothia nest site in relation to BCI, with some observed hatchling travel paths.

ORIENTATION AT THE NEST-SITE

Burghardt et al. (1977) reported that during one period of departure from the Slothia nest-site in 1975, 52 of 61 iguanas headed SE, the direction that would take them to the SE tip of the islet and the shortest route to BCI. However, an iguana looking SE from the nest-site sees only a mass of vegetation, as it does to the NW. Vegetation (and a blind) surround and overhang the clearing such that the only view out of the clearing lies to the SW, across the bay to more distant shores of BCI (Figure 15.2). To the NW, N, and NE the ground rises gently and the vegetation is slightly sparser, although it overhangs the clearing more. How do the hatchlings orient? To answer this question we first collected precise descriptive data on the direction faced by the iguanas at the nest hole and their subsequent movements.

Observations on Departure Direction

Methods: In 1977 the Slothia nest-site was observed from the blind for over 60 hours during the three-week emergence period, largely between 0830 and 1530 hours. 81 iguanas were seen to exit from 17 nest-holes and depart from the clearing. One or two greater anis *(Crotophaga major)* visited the clearing daily to prey on the iguanas but were often driven away to prevent disruption of the hatchlings' orientation behavior. Recorded for the 70 undisturbed departures were (1) 'head-up' direction, i.e., direction the iguana faced when its head became visible, (2) direction of movement when exiting from hole (when hind limbs became visible), (3) directions in which the iguana walked or ran in the clearing, and (4) direction in which the iguana departed from the clearing. For the first three the lizard's orientation was assigned to one of eight compass bearings by comparing it with two lengths of cord staked out on the nest-site on N-S and E-W axes. Numbered pegs were set in the ground around the irregular perimeter of the nest-site at 60 cm intervals. Departure directions were obtained by plotting on a scale map of the clearing lines connecting each iguana's nest-hole with the mid-point of the two pegs between which it departed.[1] While this method 'irons out' the departure routes of those iguanas that wandered in the clearing, it does yield a good approximation of departure routes, and enables calculation of mean angles and application of circular statistics.

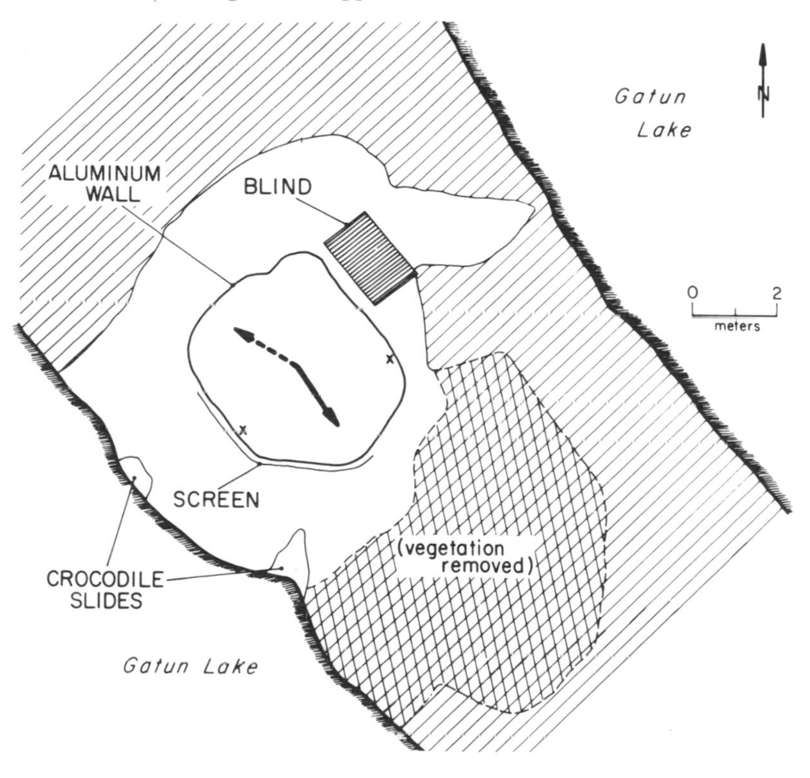

Figure 15.2: Slothia nest clearing, showing aluminum wall and cloth screen erected in 1978 and area from which vegetation was removed in February 1978. Arrows represent mean departure directions of undisturbed iguanas taking the SE and NW 'options' in 1977. The crosshatching demarcates the "SE" sector (see Table 15.1).

It would be misleading to treat the departure of each lizard as an independent datum since 42 of the hatchlings moved off in groups, and group members presumably influenced each other's orientation. Therefore, a mean angle was computed for each group, according to the method of Batschelet (1965). Lizards were regarded as members of a group if their actual paths were approximately parallel or converging and their proximity in space and time was sufficient to permit maintenance of visual contact. Note that this definition varies from the strictly temporal one used by Burghardt et al. (1977) for earlier years (1974-76).

To ascertain the visual perspectives available to iguanas at the nest-site we inspected the images in a small mirror held close to the substrate.

Results: The distribution of departure directions of singletons and groups are presented separately (Figure 15.3). Neither of the angular distributions is uniform (Singletons: U=175, $p < 0.01$; Groups: U=180.7, $p < 0.01$; Rao's Test). The singletons cluster around the SE with about 20% heading NW, while all the groups departed to the SE, except for one group heading N. With larger sample sizes it is likely that both categories would have a bimodal pattern, with a small but significant number heading NW. The groups and singletons are not significantly different (Watson's $U^2=0.0427$, $p > 0.1$) and the mean vectors of the singletons and groups taking the 'SE option' have almost identical locations (Figure 15.3).

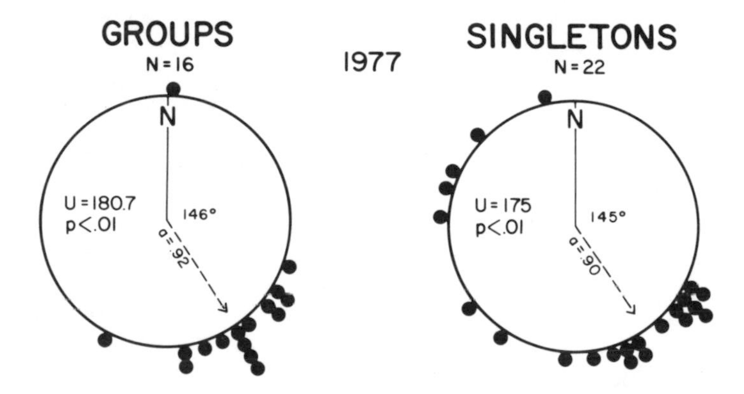

Figure 15.3: Directions taken by iguanas departing from Slothia nest clearing. Group directions are means of group members. Vectors are derived from data points on the SE arc (180°) only (i.e., iguanas heading toward SE tip of Slothia). a = length of mean vector.

The two general directions taken by the hatchlings lie toward the two ends of Slothia, so it is possible to make a meaningful comparison of groups and singletons on the basis of how many within each category selected a direction which would carry them toward the SE tip of the islet (Table 15.1). Only 1 group out of 16 failed to select that direction ($p<0.001$, binomial test), whereas one-third of the singletons did so. The difference between singletons and groups is significant ($G_{adj} = 4.27$, df = 1, $p<0.05$; Sokal and Rohlf, 1981).

Table 15.1: Departure Directions of Single Iguanas and Groups Emerging from the Slothia Nest-Site in 1977

	Entire . . .Nest-Site. . .		N half of . . .Nest-Site. . .	
	SE	NW	SE	NW
Singletons	19	9	2	6
Groups	15	1	6	1
	G_{adj} = 4.27 $p < 0.05$		$p < 0.05$ (Fisher's test)	

Inspection of the map of departure directions suggested that departures to the NW were more frequent from nest-holes located at the N end of the site, away from the central cluster of nest-holes. It proved possible to draw a line running due E-W across the site just N of the center such that four of the five nest-holes from which iguanas departed to the NW lay to its N. Taking all departures observed from *all* holes N of the line, six out of eight singletons went NW but only one out of seven groups (totalling two out of 18 individuals) did so, a significant difference ($p < 0.05$, Fisher's Exact Probability Test, Table 15.1).

Thus it is clear that the iguanas selected a direction of departure with the majority heading SE toward the nearest point of BCI and a minority moving off toward the NW end of Slothia. Further, the data support the conclusion that groups make the 'correct' decision more frequently than singletons. Very few iguanas headed SW toward the bay. These results fit the pattern of previous years where the majority of lizards also headed SE. In these years lizards that departed to the N were sometimes seen reentering the clearing and crossing to the S. Animals heading SE originally were never observed recrossing N. For example, in 1976 observations and time-lapse super 8 films showed that 27 out of 32 animals went SE. This included 7 of 8 groups and 4 out of 6 singletons.

The initial 'head-up' directions were random ($X^2 = 5.6$, df = 7, $0.5 < p < 0.7$) confirming the impressions gathered from observations in prior years. The 'head-up' directions of some holes were consistent for several individuals, possibly because of hole topography or inclination, but most holes are vertical on flat terrain and they do not tend to point in any common direction. Exit directions, however, differed significantly from random ($X^2 = 23.23$, df = 7, $p < 0.005$) (Figure 15.4), indicating that the iguanas selected a direction before leaving the nest-hole. Comparison of the exit directions of singletons and group-departing individuals did not reveal a significant difference ($X^2 = 13.0$, df = 7, $0.05 < p < 0.1$), yet only one singleton exited toward the W, NW, or N, whereas twelve group members chose that way. The evident difference between the exit and departure directions suggests that further determination of orientation took place after exiting. There is additional support for this in the frequency of changes of direction made within the circle of pegs. Those iguanas that did not dash from the exit hole to the edge of the clearing walked and ran in short spurts with intervening pauses of varying lengths. The mean number of changes in direction of 45° or more (estimated) for all lizards not disturbed by a predator was 0.67 (N=70), with some individuals making as many as four. There were, of course, many changes of lesser degree. Singletons and group members did not differ substantially with regard to either the number of individuals making changes of direction or the number of changes made. However, singletons did appear to depart more quickly from the clearing and showed a stronger tendency to make lengthy headlong dashes.

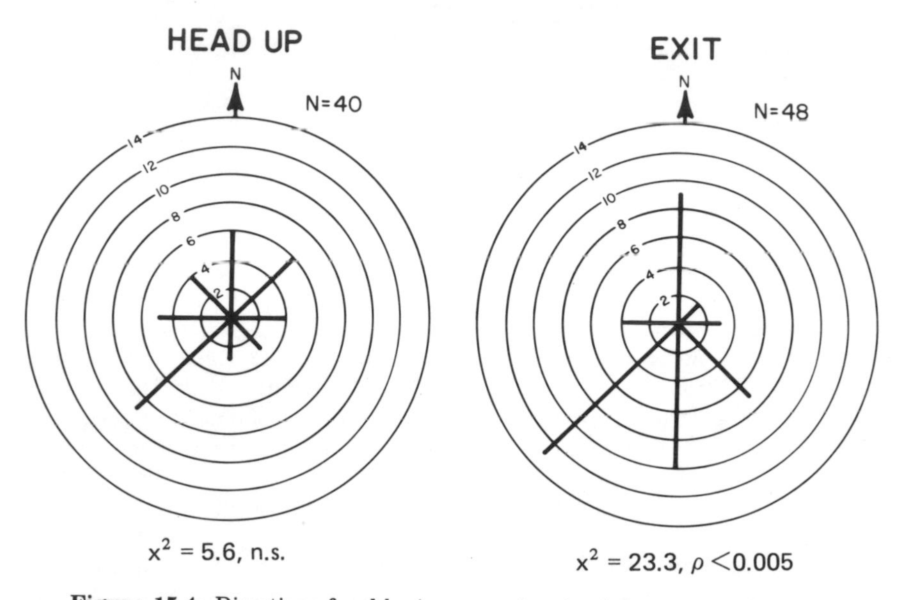

Figure 15.4: Directions faced by iguanas when head first appeared at nest-hole (head up) and at moment of exit from hole. Shows iguanas observed departing from Slothia nest clearing in 1977. Line length represents number of lizards facing that way.

Discussion: Since changes of direction in the clearing were common and could be as radical as 90° or 180°, and since comparison of 'head-up', exit, and departure directions indicates a gradual convergence on the SE and NW 'options,' we conclude that iguanas in the clearing engage in an active process of orienting. Environmental stimuli presumably elicit walking and running in specific directions as well as pausing and scanning, and subsequent stimuli may elicit locomotion in a different direction.

Iguanas departing in company with others are more likely to take the SE option than are singletons (Table 15.1), and our present interpretation is that the SE route conforms more closely to the general tendencies of the iguanas' dispersal orientation. This suggests that several hatchlings orient better than one. They might do so simply because groups linger in the clearing longer than singletons and consequently have greater exposure to the relevant stimuli, but two other possible mechanisms should be considered. The departure direction of a group may represent either a compromise of the orientation preferences of its members (cf. Hamilton 1967) or conformity of group members to the preference of an individual–a leader. Group departures we have observed implicate both of these. On occasions group orientation is achieved through the simultaneous and successive edging forward of individuals, with no permanent leader. At other times an individual walks or runs consistently in some direction and others follow. On one occasion in 1977 several hatchlings which were edging out of a nest-hole (in the northern, difficult half of the clearing) paused facing W, NW, N and NE, then turned one by one to follow an individual that walked more resolutely to the SE, all finally departing from the clearing in that direction. In 1975, a super 8 film record showed several animals near their exit holes simultaneously following to the SE an iguana crossing the clearing from the NW (see also Burghardt, et al., 1977).

The Role of the Visual Field

Green iguanas nest on several islands and islets in the waters around BCI and their offspring need to follow diverse compass bearings to reach the main island. Therefore, it is unlikely that celestial or magnetic cues initially determine the hatchlings' orientation; the use of local cues is more probable. The frequent tongue-flicking and presumed visual scanning seen in the nest clearing implicate chemical and visual stimuli, although either behavior could subserve quite different functions; e.g., nest-site imprinting, formation of groups, or predator detection.

A human observer unfamiliar with the topography of Slothia and BCI could probably determine, by looking out from the nest-site at ground level, that the most direct terrestrial route to the distant landmass (BCI) lies to the SE. According to this hypothesis, the iguanas' departure direction is determined by visual assessment of the topography of the bay area. We tested this in 1978 by erecting a screen (see below) on the SW side of the clearing to deprive the neonates of the sector of the panorama which reveals the direction of convergence of Slothia's and BCI's shores.

Another plausible hypothesis is that orientation at the nest-site is toward the masses of vegetation bounding the SE and NW sides of the clearing and that SE is the preferred direction because the vegetation is denser, greener, or closer. A significant change was made in the nest clearing in 1978. In February a large swath of vegetation to the SE of the nest-site was removed by other workers to enlarge the clearing, and when emergence started it had not grown back. Removal of this vegetation provided a direct test of this hypothesis. Furthermore, it improved the visibility, from the clearing, of the bay of BCI to the S, namely that part of the panorama which reveals the convergence of the shores.

Methods: A length of brown burlap cloth, 4.3 m long by 1.1 m high, was pinned to several wooden stakes implanted around the SW edge of the nest-site (Figure 15.2). This screen was alternately taken down and replaced throughout the emergence period until a large sample of departing iguanas was obtained. It was first erected on May 10, at least 5 days after the start of emergence.

Throughout the period of observations the nest-site was also enclosed by a 25-30 cm high wall of aluminum flashing. Installed to enable the capture and marking of emerging hatchlings (for observations of their social behavior at the channel), this fencing also manipulated potential orientation cues. For most parts of the nest-site the wall did not obstruct the view of the shores and vegetation of BCI, yet the lake surface itself could not be seen from any point in the nest-site below 5.5 cm. Numbers painted on the wall at 60 cm intervals enabled specification of departure directions. Most of the hatchlings ran straight into the wall, but whenever one veered off shortly before reaching it the initial trajectory was noted.

The observer in the blind noted for each lizard its departure direction and whether it was in a group, then removed it from the enclosure. When two or more iguanas departed as a group, a single departure was recorded in the direction of the first animal to reach the wall.[2]

Initially the departees were left to roam the enclosure. To avoid the possibility that they could then influence the orientation of subsequent iguanas, departures toward or within 1.80 m of lizards restrained by the wall were discounted. Since this caused loss of data and could still have introduced a bias, we changed policy and removed departees promptly, although this meant leaving the blind and disturbing the nest site. However, the orientation behavior of the trapped lizards is not without interest and systematic observations were made, all when the screen was not present. Since emerged iguanas concentrated in the SE sector, the observer recorded, every 60 seconds, the number of emerged iguanas in the enclosure and the number within a 5.5 m sector to the SE (Figure 15.2). Lizards more than ca. 30 cm from the wall were not counted, nor were those which did not move between censuses, since prolonged freezing (as in response to a predator) could lead to inflated scores not truly representing choice of direction.

Results and Discussion

Departure Directions: The iguanas which emerged when there was no screen almost invariably took the SE 'option' (Figure 15.5). The departure angles were not different from those of the singletons in 1977 (Watson's $U^2 = 0.060$, $p >> 0.1$), and although the scatter to the SE was greater in 1978, the mean vectors are only 5° apart. This demonstrates that proximity of vegetation is not a critical determinant of departure orientation on Slothia. Furthermore, it made little or no difference that the iguanas could not see the lake from the nest-clearing, demonstrating that the lake itself does not provide necessary visual orientation cues.

When the screen was in place the iguanas no longer oriented their departure to the SE (Figure 15.5), and their mean angle was almost diametrically opposite that of the screen-absent lizards (Watson's $U^2 = 0.270$, $p < 0.01$). Only two lizards out of eleven departed to the SE, significantly fewer than when the screen was absent ($p < 0.005$ one tailed, Fisher's Exact Probability Test). This demonstrates unequivocally that visual cues are important determinants of the orientation of emerging iguanas. Unfortunately it does not show which visual cues the iguanas respond to, and as the low emergence frequency in 1978 prevented further experiments, we are left with alternative explanations of the 'screen effect.'

The data are consistent with the hypothesis that visual assessment of major local topographic features enables the emerging iguanas to 'select' a direction that will carry them, by the most direct overland route, to a distant vegetation mass. Since the iguanas still oriented SE when they could not see the lake, it may be that dispersal travel paths do not hug the shoreline per se but rather the edge of the adjacent vegetation mass. With the screen in place there were no visual indications that the shore of BCI converged on that of the nest-site in a SE direction, and unknown cues caused the iguanas to orient to the N.

It is quite possible that departure to the SE (in response to unknown cues) was inhibited by active avoidance of the screen. However, when the aluminum wall was down we once inserted into nest-holes four iguanas which had been trapped by the aluminum wall three days previously.[3] Three walked or ran straight toward the screen, and two actually climbed over it. Their behavior suggests that the screen was not a threatening object nor even a substantial obstacle, but it is possible that they had habituated to such stimuli since their first emergence.

Effects of the blind must also be considered. This structure presents to the nest-site a rigid 2 m square dark brown facade, subtending an angle of about 40° from the center of the nest-site (Figure 15.2). The observer in the dark interior looks through a tiny hole in a panel of brown burlap cloth. Possibly the emerging hatchlings avoid the blind, but it seems unlikely that such avoidance could contribute very much to the observed clustering of departure directions. It is conceivable that the clustering is largely due to avoidance of the space to the SW and the blind to the NE, but for that to be so the iguanas would have to be giving the blind an extremely wide berth. While we cannot entirely rule out this possibility, there is little to recommend it. Olfactory avoidance of the observer in the blind is an extremely unlikely explanation in view of the fact that when an observer sat quietly in the bushes to the SE, hatchlings crawled right past and even over him.

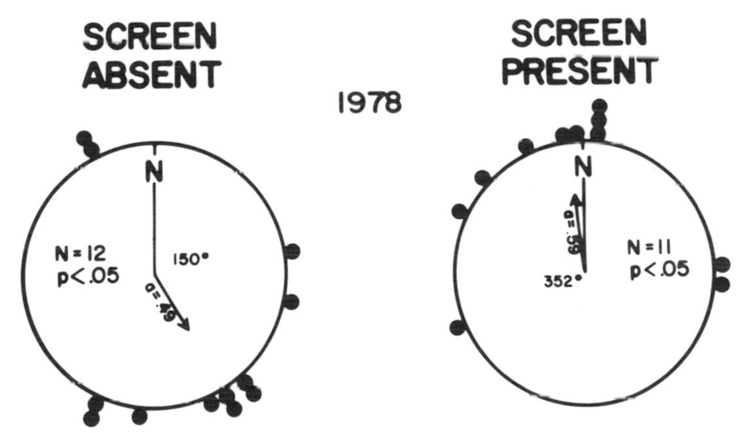

Figure 15.5: Directions taken by iguanas departing from Slothia nest clearing in 1978 in presence and absence of cloth screen. Each data point represents a singleton or the leading member of a group. a = length of mean vector.

Orientation of Trapped Iguanas: Iguanas prevented by the aluminum wall from leaving the nest-site spent most of their time running or walking along the bottom of the wall, frequently leaping at it in attempts to climb over.

Only rarely did one walk away from the wall. They concentrated their activity in restricted sectors of the wall but occasionally wandered widely in their efforts to bypass it.

Since the iguanas were often clumped and seemed to influence each other's spatial behavior we have treated each set of iguanas simultaneously present in the enclosure as a single group. This also circumvents the problem of identifying individuals. For each trial with N iguanas observed over t minutes there were Nt observations of individual locations in the enclosure, minus all observations of individuals that did not move between samples or were more than 30 cm from the wall. The proportion of the Nt observations falling in the 5.5 m SE sector (Figure 15.2) was calculated for each trial. In all eight trials well over half were in that sector although it comprised scarcely more than one-third of the perimeter (Table 15.2). Under the very conservative null hypothesis of $p = 0.5$ that the greater proportion in any trial will fall in the SE sector, the probability of all trials favoring that sector is 0.008 (binomial test, two-tailed).

Thus the hatchlings persisted in heading SE even when deprived of the view in that direction, the inevitable consequence of approaching the aluminum wall. Until we know what they orient by in the first place we cannot be sure what this implies. If our hypothesis is correct and the view over the bay provides the principal orientation cues, then the experimental subjects' persistence in orienting SE could reflect inertial guidance. However, it is also quite possible that an iguana which has selected a direction by one set of environmental cues (e.g., the bay panorama) can maintain it by reference to another set.

Table 15.2: One-Minute Scan Samples of Iguana Locations Within Nest-Site Enclosure

			Observations		
Date	Observation Period (min)	Number Iguanas	Total	Number in SE Sector	Proportion in SE Sector
5/5	12	15	180	152	0.84
5/6	77	2	128	75	0.59
5/6	110	3	166	135	0.81
5/7	21	2	42	32	0.76
5/8	142	4	162	130	0.80
5/8	32	2	23	15	0.65
5/9	45	5	109	89	0.82
5/9	4	3	12	10	0.83
Totals	443	36	822	638	0.78

SOCIAL ATTRACTION

Since the iguanas often leave the nest clearing in groups it is possible that attraction to other iguanas is involved in determining the direction taken. Two pilot experiments were performed. Harry W. Greene attempted to influence the behavior of hatchlings in the nest-clearing with stationary and moving rubber models but was unsuccessful, probably because the iguanas (and anis) were not deceived by them.

Another method was used by the second author and Greene to test social attraction in hatchlings in 1975. Iguanas were observed as they emerged from an artificial nest hole into a symmetrical rectangular box that had two identical glass-fronted compartments at either end, one with and one without a live hatchling iguana. Kiester (1979) has shown that adult *Anolis* lizards are attracted to conspecifics regardless of sex in a similar test environment.

Methods: The test apparatus was a rectangular wooden box with inner dimensions of 117 cm x 39 cm x 48 cm high. At each end a glass partition formed a compartment 13 cm wide. In each compartment a stick 50 cm x 0.6 cm in diameter was placed rising from the lower front corner to the opposite rear corner in mirror symmetrical fashion. The interior walls were painted a light blue/gray. The box rested 19 cm above the floor on blocks, and through the middle of its floor protruded a 4.1 cm diameter pipe, almost flush with the floor, through which neonate iguanas could be introduced. The floor of the box, except for the emergence hole, was covered with brown paper, as were the glass walls at each end behind the small compartments. The box was illuminated by a 150 watt bulb centered 183 cm above it. The top of the box was covered with hardware cloth and cheesecloth to minimize responses to any external visual cues.

The testing took place in the screened animal house in the BCI laboratory clearing over a 2½ hour session beginning at 2210 hours on May 15, 1975. The 10 iguanas tested were trapped earlier that day on Slothia in tubs inverted over active nest holes.

Before each trial an iguana (the 'model') was placed on the branch in one of the glass-fronted compartments. It usually remained there throughout the test. The compartment used (left or right) was balanced across the ten trials. The paper towelling on the floor was changed after every trial to remove any chemical traces.

A trial was run in the following manner. The entrance hole was covered with an inverted dish to which was attached a string that protruded through the cheesecloth directly above the hole. An iguana from the collecting bag was gently introduced into the length of curved pipe extending under the box, and a metal cap was screwed on the pipe. After 3 minutes the dish was slowly raised and pivoted to one side, and the behavior of the iguana was recorded. If the animal did not exit within 2 minutes (3 minutes for first two subjects) the cover was lowered and raised again. When one subject did not emerge after 8 minutes, the cap on the entrance hole was removed and he or she was gently prodded with the fingers in an attempt to mimic the pushing from behind often observed in naturally active nest holes. An observer sitting on a table beside the box dictated behavioral observations into a tape recorder, and another recorded times and directions taken.

Results and Discussion: Eight out of 10 iguanas approached the compartment with the model. One ran to the non-model side at 3′07″ and stayed there motionless until the trial ended at 5′30″. Another emerged from the hole at 3″ but stopped and remained motionless facing the non-model compartment until the trial ended after 5 minutes. Thus, 8 of the 9 that made a choice went to the conspecific side ($p < 0.05$, two-tailed binomial test), where they often scratched at the glass, apparently struggling to approach the model. These locomotor attempts frequently followed movements (e.g., head turning) of the model, indicating that movement increased its stimulus value.

This result provides preliminary experimental confirmation of the impression gained during field observations (see above), that hatchling iguanas tend to approach other hatchlings. Such an attraction could have significant effects on orientation at the nest-site and should be considered in any analysis of that behavior.

THE DISPERSAL ROUTE

Observations

It is difficult to obtain information on the hatchling iguanas' dispersal route after leaving the nest clearing since they are small, wary, and cryptic in coloration and behavior. Over 400 hatchlings have been paintmarked, toeclipped and released at the Slothia nest-site over five seasons (1974-1978) but few have been relocated after their departure from the islet, and predation by aerial, terrestrial, and aquatic predators appears to take a toll (Greene et al., 1978). Nonetheless, we now have information on their points of departure from Slothia, their travel paths beyond the islet, and their subsequent movements in a clearing on BCI.

Departure from Slothia: In the years 1974-1976 the swimming paths of more than 50 iguanas embarking from the SE end of Slothia were recorded. Most of these departed in groups, generally between 1000 and 1200 hours. Of these about 75% took the shortest direct way to BCI. The remaining animals usually veered somewhat westerly and hit the spit on BCI from the side or swam into the bay west of the point. Those that took the short routes crossed in a matter of seconds while the others spent many minutes in the water. Even in the long swims the directions taken by the iguanas converged remarkably.

Some hatchlings do not leave from the SE end of Slothia, but set out from other points along Slothia's shore. Burghardt el al. (1977) reported seeing a few swimming from the NE end of Slothia toward the BCI dock area, and Rand (pers. comm.) found several sleeping on stumps in the bay to the E of Slothia in the late 1960s, midway between the islet and BCI. Another was found there in 1977. It is likely that these animals were travelling from Slothia, but the significance of their locations poses problems since ships passing across (Gatun Lake frequently generate strong currents in the channel that could sweep the iguanas way off course.

Travel Path Beyond Slothia: In 1977, when the lake level was abnormally low, conditions were especially favorable for observing dispersal paths. Slothia was united with BCI by a bridge of mudflat and rock, and a 1-4 m mudbank was exposed around the bays of BCI (Figure 15.6). Observations of the 'channel' and adjacent bay of BCI were made through binoculars from a canoe lodged in the aquatic vegetation *(Hydrilla)* off the point of BCI for 19 hours, and with telescope and binoculars from the point at the SW end of the bay for 35 hours. The movement paths of all groups and individuals seen were recorded by referring to features of the nearby shore and vegetation. On ten nights the shores of Slothia and the point and bay of BCI were searched with a headlamp. Sleeping iguanas' locations were noted, and the iguanas were paintmarked and toeclipped, then released immediately at the same spot.

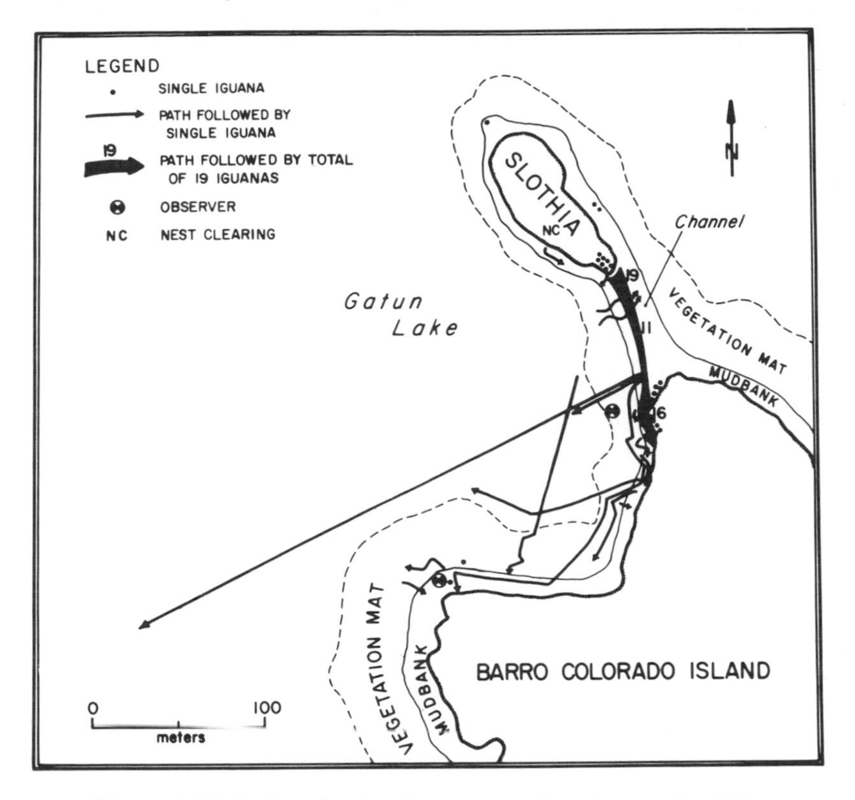

Figure 15.6: Paths taken by dispersing newborn iguanas in 1977.

Figure 15.6 shows the locations and movement paths. Four groups (N=3,4,4,8) were seen crossing the dry 'channel' and continuing S along the bay. Alternating running and walking, group members stayed together, following very similar paths and progressing fairly rapidly over the bank. However, all foundered 10-30 m from the point, where predators (a basilisk lizard, *Basiliscus basiliscus,* a Swainson's toucan, *Ramphastos swainsonii,* and greater anis) caused them to freeze and scatter. Several were seen running back toward Slothia or ascending the vegetation near the shoreline to heights of ca. 0.5-5.0 m. Even when no predators were seen (by the observer) the iguanas ascended stumps, reeds and sticks in this area, and perched with their heads oriented over the bay, much as they do at the SE tip of Slothia. No lizard that ascended higher than ca. 1.5 m into vegetation was ever seen to descend (although one dropped from its perch in a tree onto the immobile observer seated on a mudbank), but those that perched lower typically would descend and approach the water, then head S along the bank or walk out over the hydrilla. Companions would follow or choose alternative directions.

The movement paths over the hydrilla and open water reveal the same 'indecision' as the wanderings on the bank. An iguana would hold a course for several meters then, often after a pause, adopt another. Such vacillations occasionally resulted in a zig-zag course as the iguana appeared to respond alternately to competing stimuli. Those which followed short paths over the hydrilla *toward* the channel or BCI may have been arriving from another nest-site located near the boat dock.

The data of Figure 15.6 suggest that on reaching BCI all the Slothia iguanas head S around the same bay. However, lizards that went SE into the other bay would not have been seen. That some at least choose the bay to the E of the point is indicated by the two iguanas found high in the reeds on Slothia's E shore and the individual found on a stump in the E bay.

Only five of the eighteen iguanas captured and marked at their sleeping perches on the dispersal route were seen again, but the second sightings furnish interesting information and are reported individually:

(1) No. 2 crossed the channel to BCI in a group of eight on May 11 at 1325 hr and climbed 50 cm into the vegetation after a predation attempt. At night it slept in the same spot. The next morning from 0600-1015 hr no iguanas were seen there, but at 1015 No. 2 dropped about 3 m from a bush overhanging the mudbank and ran onto the hydrilla with an ani in (unsuccessful) pursuit.

(2) No. 7 was marked at its sleeping perch in the reeds on the point of BCI on May 13. Next morning at 1045 hr it descended from those reeds and joined three iguanas on the shore. After extensive peering over the bay, they moved S along its shore (there were now 5). No. 7 stopped and walked N. A bird passed overhead and No. 7 ran into the vegetation at 1130 hr. The other four also turned back. That night No. 7 was found in the reeds on the point of Slothia, having crossed back to the islet! The next day at 1157 hr it was seen back on the point of BCI. After interacting with other hatchlings it set out over the water and traced a path around the bay, vacillating between hydrilla and mudbank and finally reaching the western extremity of the bay at 1300 hr (Figure 15.6). After walking out along a fallen trunk projecting over the hydrilla and peering in the direction of the boatdock, it returned to the mudbank and ascended about 5 m into the hanging branches of a tree until lost from view.

(3) No. 9 was found in the reeds on the SE point of Slothia at night on May 13. The next day at 1045 hr it was seen moving S along the shore of BCI point with four others, including No. 7. Alarmed when the bird flew overhead at 1130 hr, it ran into the vegetation. That night it was found sleeping on a nearby perch.

(4) No. 11 was marked at its sleeping perch on BCI point on May 14. The next morning it crossed the channel in company with three others *from Slothia to BCI,* having previously crossed the other way. At 1245 hr it was seen halfway around the bay of BCI, making its way across the hydrilla.

(5) No. 13 was marked in the Slothia point reeds on May 14 and found 11 days later in the upper part of the laboratory clearing, about 250 m from shore.

Movements in the BCI Laboratory Clearing: Apart from the hatch sites and shoreline, the only intensive observations of juvenile iguanas on BCI have been in the laboratory clearing (Figure 15.1). Here many individuals have been observed and marked over several years; details of these observations will be published separately. Briefly, however, it is known that some of these animals were originally marked on Slothia prior to first observation in the clearing. For example, two hatchlings marked and released on Slothia on May 15 and May 16, 1974 were found in the laboratory clearing on July 19. There is also evidence that the groups of iguanas show some stability throughout the dispersal phase. On the evening of May 8, 1976 four iguanas were found together in the reeds near the BCI boat dock, the only iguanas found anywhere that evening. The following evening two of these animals were discovered close together in low vegetation near the workshops, part way up the hill and on the edge of the lab clearing. In the clearing the young iguanas often are also found in groups and move about together whilst foraging; for several months many are found at night sleeping in shrubs at various locations in the clearing. The numbers of animals decrease markedly by November, but it is unclear whether emigration, predation, or both are responsible.

Marked animals have been recaptured up to a year later, the following May, in the laboratory clearing. However, by the time the new flux of hatchlings arrives the yearlings are very rarely seen. Thus we feel that once the hatchling iguanas find a suitable habitat, they more or less settle down and movements that do occur are not related to the initial migratory phase.

Discussion

The Slothia hatchlings move away from the islet very quickly, most of them taking the shortest route to BCI across the channel and some crossing wider stretches of water in other directions. They embark on courses toward land or vegetation masses and show a tendency to follow a terrestrial route, hugging the lakeshore. However, they do cross areas of open water, although they usually hesitate and wander a little before doing so. Some individuals set out over wide stretches of open water. Changes and even reversals of direction are common, but the latter pose problems of interpretation because of the possibility of social influence from iguanas arriving from elsewhere. However, it is clear that fresh 'decisions' about direction are intermittently made by travelling iguanas, presumably on the basis of stimuli impinging at the decision points. The vacillation could be a consequence of the iguana responding not to one or a few stimuli but to a stimulus configuration that is complex and changes with the iguana's location. The moving iguana, in this view, is exposed to a continually changing perspective and alters its orientation in accordance with the changes. Iguanas that err and head back toward Slothia (seen only in 1977 when the channel was dry) are no doubt corrected by social tendencies when fresh departees are encountered.

Since 54 hr of observation yielded relatively few instances of travel around and across the bay and since several iguanas were seen to climb high into the vegetation but not seen to descend, we must question the representativeness of our observations. It is conceivable that the iguanas climb to the forest canopy or its lakeshore margin to take up residence there or to continue their dispersal over the top. This will only be resolved when we develop more effective tracking techniques (cf. Troyer, 1982).

In addition such tracking will enable us to measure movement over longer periods of time and determine the duration and speed of dispersal. When this has been accomplished for several nest-sites, including those on other islets and the main island, we will be in a much better position to relate the iguanas' long-distance movements to habitat topography, and to analyze their motivation. Experimental evidence implicating an endogenous factor in the rapid dispersal phase is presented below.

MOTIVATION AND DISPERSAL

Slothia clearly could not support the large numbers of iguanas that hatch there, so it is not surprising that they leave. But the rapidity and completeness of their departure are impressive. Burghardt et al. (1977) reported that only four percent of 130 iguanas marked at the nest-site in 1974 and 1975 were sighted on Slothia more than 3 days after hatching. Programmed dispersal is indicated, but is it endogenous or elicited by environmental stimuli? In the former case the hatchlings would disperse over a certain distance or for a certain time regardless of the particular environment into which they emerge; in the latter, dispersal would occur because the environment into which they emerge possesses stimuli eliciting departure or because it lacks stimuli that would inhibit departure. Both mechanisms could be operative.

Pilot Displacement Experiment

To test the hypothesis that dispersion is governed by a behavioral tendency of limited duration, 15 hatchlings that had settled in the BCI laboratory clearing were marked and released at the Slothia nest-site at 1015 hr, May 19, 1978, by inserting them into nest holes. For the remainder of that day and the following three days the channel area was observed continuously with a spotting scope from the observation point on the bank of BCI bay (Figure 15.6). Each night the shores and vegetation of Slothia were searched.

Most of these animals were never seen again, but two were seen by the reeds at the SE tip of Slothia on days two and four, and one of them was still there on day eight. This sedentariness does not, however, contrast greatly with the behavior of non-displaced animals. In 1978, 198 iguanas were captured at the nest-site, marked, and released in the same place up to 5 days later (79% within 24 hr). Of the 42 subsequently sighted at the SE tip of Slothia, 37 (88.1%) were observed there on the day of release or the subsequent day, 4 (9.6%) were observed 2 or 3 days after release, and one was seen six days after release.

However, the behavior of the two displaced iguanas during the days they remained on Slothia did indicate a motivational state contrasting with that of newly emerged, undisturbed animals. They wandered slowly in the channel area, climbing stems and feeding on small plants, and did not follow iguanas that crossed the channel. They were always as plump as those in the laboratory clearing, whereas iguanas dispersing from the nest-site always look lean and do not feed, although they occasionally will nip at vegetation. It is most unlikely that the plumpness was due to food ingested before displacement: 3 hatchlings captured in the laboratory clearing and retained unfed in indoor cages at approximately ambient temperature for 3 days became quite thin and their weight declined by 10%.

The evidence is meagre, but we believe the explanation of the behavior of the two displaced animals lies in the different tendencies of newly emerged iguanas and those that have completed the first phase of dispersal: newborn iguanas appear to be programmed to disperse very rapidly from the nest-site. Nor is this exclusively a local phenomenon, for hatchlings have been sighted swimming in regions of Lake Gatun far removed from Slothia, indicating that lizards from other nest-sites disperse with equal vigor.[4] Further, mainland nest-sites are also rapidly evacuated; thus the phenomenon is not restricted to islet habitats.

GENERAL DISCUSSION

Howard (1960) speculated that populations of many vertebrate species contain a proportion of 'innate dispersers,' animals congenitally predisposed to disperse beyond the confines of the parental home range at or before reproductive maturity. The Slothia iguanas' dispersal is away from a nest-site which lies outside the parental home range and is made by all individuals immediately after emergence from the nest-hole. Our pilot data indicate that the tendency to move away from the nest-site may be independent of the particular environment encountered (i.e., 'innate' *sensu* Howard), but we particularly need more data from Slothia and from other nest-sites, including non-communal ones. The uniqueness of this behavior among iguanine lizards is hard to assess until information on other species becomes available. Certainly dispersal away from the nest-site occurs in other iguanines (e.g., *Cyclura cornuta stejnegeri,* Wiewandt, 1978) but no published accounts reveal dispersal as rapid and directed as we have seen in *Iguana iguana.*

Dispersion has been documented in smaller iguanids but on a less dramatic scale. Blair (1960) found that *Sceloporus olivaceous* invariably emerge from the nest 'with a rush,' although they are highly variable as to when and how far they disperse; approximately 30% of the females and 20% of the males failed to disperse at all. The remainder were divided into those that dispersed immediately after hatching to the vicinity of their eventual home range, those that dispersed to establish a juvenile home range and later underwent a second dispersal to the area of the adult home range, and those that were intially sedentary and dispersed during the subsequent fall or spring. There was variability in dispersal distance too, with 71.5% of females and 63.9% of males dispersing no further than 60 m. *Uta stansburiana* hatchlings studied by Tinkle (1965) were even more sedentary, travelling a mean overall distance of less than 6 m between the times of hatching and attainment of mature size.

Indeed, to find dispersal of a newborn reptile comparable to that of *I. iguana* we have to look to the 'swim-frenzy' of the sea turtles. The rapid seaward movement of green turtle *(Chelonia mydas)* hatchlings is well known (e.g., Carr, 1963, 1972), and we now know that vigorous non-random movement away from land is sustained for at least four hours (Frick, 1976). Carr (1965) found that

20-day-old hatchlings oriented seaward when confined in a water tank but that the response was absent two months later, and Mrosovsky (1968) found that the frenzy of green turtles and hawksbills *(Eretmochelys imbricata)* declined within a single day of hatching (see also Ehrenfeld, 1974). Available evidence indicates that these turtles, like the iguanas, generally do not feed during their rapid travel, although they occasionally stop to rest (Frick, 1976). Parallels between the two species include rather narrow nest-site requirements and parental migration to the site for oviposition.

The iguanas' hurried locomotion away from Slothia is all the more striking when compared with their modest movements in the laboratory clearing (cf. Henderson, 1974).

The hatchlings' vigorous dispersal undoubtedly serves to distribute them more widely over suitable habitats. The advantages of this could lie in reduced predation and/or resource availability. We think a major reason for rapid dispersal is removal of the neonates from a zone where predators could learn to exploit them on a regular basis. Predation pressure in the vicinity of the nest-site certainly appears great enough to be an important selective force (see Greene et al., 1978, Drummond, in prep.). A similar reason has been advanced for the persistent locomotion of green turtle hatchlings, which have to traverse coastal waters where they are vulnerable to predatory fish (Frick, 1976).

Until we are able to follow larger numbers of hatchlings and collect data at other nest-sites we must be cautious in interpreting the dispersal paths reported above. Evidently the hatchlings move along shores, across bays, over expanses of water separating land masses, and into 'clearings' with low vegetation. After the first, rapid phase of dispersal they inhabit areas of secondary growth where they spend their time in low bushes or on the ground (Henderson, 1974; Burghardt, 1977), and only rarely are they found on the rain forest floor (Rand, pers. comm.). Thus their travel paths, as we interpret them, are such as to carry them to the habitat that they evidently prefer.

Carr (1955) found a green iguana nest on an ocean beach, half a mile from the nearest adult habitat. Iguanas on BCI swim to numerous islets to lay their eggs and there is reason to believe that they may journey several kilometers from home range to nest-site (Montgomery et al. 1973). Hatchlings emerging from such sites clearly need orientation skills to convey them quickly across exposed areas to appropriate habitats. The orientation of the Slothia hatchlings at the nest-site is non-random, most selecting a heading which carries them through a dense mass of vegetation to the land mass visible from the nest-site. According to our tentative interpretation, they approach the visible shore of BCI by a detour over terrain concealed from view.

There are several reasons for believing that visual cues determine the orientation. Firstly, lizards are generally believed to have good vision, and it has been shown that green iguanas can discriminate several colors, including green (Rensch & Adrian-Hinsberg, 1963). Secondly, the hatchlings' head-turning and climbing in the nest clearing and further along the dispersal route strongly suggest visual scanning. This might serve predator detection exclusively, but Kiester et al. (1975) described how several species of *Anolis* will climb a post, scan their surroundings and then approach their preferred habitat. In their choice experiments, however, they placed artificially displaced lizards on or near the top of the artificial post and no controls were employed to test for the effect of scanning from the post top; thus the case for 'post-vantage behavior' in natural habitat selection is less definitive than desirable. Our field observations strongly implicate visual scanning and 'post-vantage behavior' as components of an active orientation process. However, the head turning behavior is also consistent with the use of magnetic cues to guide movements, shown to be a probable factor in salamander (Phillips and Adler, 1978) and pigeon orientation (Walcott, et al., 1979). Thirdly, results of homing experiments on *Sceloporus orcutti* (Weintraub, 1970) and *Dipsosaurus dorsalis* (Krekorian, 1977) are consistent with homing through visual recognition of learned landmarks and therefore suggest that iguanid lizards can make adaptive directional responses to visual stimuli from landscape features. Fourthly, the experimental results reported above show that a visual obstacle can disrupt the hatchlings' orientation, and are consistent with the hypothesis that the hatchlings orient in response to visual cues which reveal the configuration of the nest clearing and the shore of the facing island. Additonal work is needed to further test this hypothesis and determine the nature of the cues (e.g., linear perspective, overlap, texture, movement parallax, etc.).

It is intriguing that hatchlings departing from the nest-site in groups orient more consistently than singletons. Hamilton (1967) speculated that groups of animals orient more accurately than individuals. He proposed that flocking in birds functions to improve navigation and suggested that "the evolution of orientation mechanisms has been related intimately to the concurrent evolution of social organization and energy acquisition systems" (pp. 57-58). Burghardt et al. (1977) and Greene et al. (1978) suggested that sociality in the hatchling iguanas serves an anti-predator function. Thus improved orientation and reduced predation could be concurrent advantages of sociality which have jointly led to its evolution. Conceivably improved orientation was the original selective pressure, and once the hatchlings started travelling together they became subject to predation pressure favoring the evolution of those behaviors which exploit group membership to reduce predation. A similar argument could be constructed for predation preceding orientation as the selective pressure accounting for the origin of sociality [see also Rand's (1968) discussion of communal nesting and the evolution of sociality in adult iguanas].

The results of the second experiment suggest that the hatchlings can maintain orientation even when deprived of the eliciting cues. This would appear adaptive in a small animal which has to move through dense vegetation. The iguanas' movement paths around BCI suggest that orienting stimuli are periodically reevaluated, and that courses are set for relatively brief periods. Frick (1976) and Ireland et al. (1978) found that hatchling green turtles can maintain a course over the open ocean by day or night even when all land is below the horizon. Mrosovsky (1978) suggested that magnetic, olfactory, or other orientation systems are calibrated against bearings taken in response to other cues and used by the turtles when the original cues become unavailable. Duelli (1975) has shown that the gecko *Hemidactylus frenatus* is capable of endogenous course control, possibly using receptors in the labyrinth. Thus one possibility is that the iguanas orient initially by visual stimuli and then maintain their bearing by inertial orientation [as Drury and Nisbet (1964) suggest for migrant songbirds]. Fischer (1961) has shown that *Lacerta viridis* is capable of orienting by an artificial sun, and the experimental results of Kiester et al. (1975) suggest that structural features of nearby vegetation can serve anoles as visual orientation cues. Thus either the sun or nearby vegetation could be used by the iguanas to maintain their orientation, although inertial cues would seem more appropriate for movement through dense vegetation.

It will be some time before we have adequate descriptive and explanatory accounts of either the hatchlings' dispersal from the nest-site or the oviposition migrations of the gravid females. Both are intriguing and possess features apparently unique to this species. A recent and rather startling discovery is that while the well-known diurnal emergence of the Slothia hatchlings peaks around midday, large numbers can emerge and depart from the clearing at night, even under an overcast and moonless sky (Drummond & Burghardt, in prep.). There is a fascinating possibility that these versatile animals emerge from the nest-hole equipped with alternative orienting systems which rely on different sensory modalities.

SUMMARY AND CONCLUSIONS

The diurnal dispersal of neonate green iguanas from the islet of Slothia in Gatun Lake, Panama was studied over several years by field observations and experiments. The majority of hatchlings departed from the nest-clearing toward the SE, a direction which carried them to the (hidden) nearest point of adjacent Barro Colorado Island (BCI); a minority headed NW. Orientation is an active behavioral process which starts at the nest-hole and generally results in convergence on one of the two principal directions. A greater proportion of groups than singletons selected the "correct" southeasterly direction, indicating that groups may orient better than individuals. Group departure directions appeared to result from both a compromise of individual orientation preferences and conformity of some individuals to the preferences of others. Experimental results demonstrated that hatchlings tend to approach other hatchlings.

Orientation cues were examined experimentally: a screen was erected to the SW of the nest-site to cut out visual cues disclosing the direction of convergence of the shores of Slothia and BCI; and removal of vegetation to the SE tested the hypothesis that the hatchlings simply approach the nearest or densest mass of vegetation. Results demonstrated that the iguanas do not approach the nearest or densest vegetation and were consistent with the hypothesis that orientation depends on visual assessment of the topography of the local bay area. However, alternative explanations could not be ruled out. Chemical cues from conspecifics may play a part, particularly during nocturnal emergence, and a bearing selected in response to one type of cue may be maintained by reference to cues of another type.

Most hatchlings dispersing from Slothia crossed the narrow channel to BCI. Beyond the channel, many were observed following a terrestrial route hugging the lake shore, but they also made some excursions over water (usually after some hesitation), sometimes for several hundreds of meters. It is not known how representative these observations are, and it may be significant that some hatchlings ascended vegetation at the lakeshore and were not seen again.

A rough comparison of the hatchlings' speed of movement away from Slothia and in a distant clearing, and the results of a displacement experiment indicated that they may have an endogenous tendency to disperse rapidly from the nest-site during the first few days after emergence, and that subsequent movements are on a smaller scale.

The principal function of rapid oriented dispersal may be to remove the vulnerable hatchlings quickly from a zone of high predation risk to a safer habitat. Movement in groups (i.e., social behavior) could have evolved to improve orientation as well as to reduce the risk of predation.

Notes

[1]The pegs were not in place at the start of the study period, so precise departure directions were not obtained for 6 lizards.

[2]Mean directions were not used for groups because the first-departing lizard was retained by the wall and might exert an unnatural influence on other group members.

[3]The first emergence of these animals was not observed, so their original orientation behavior is not known.

[4]Reports of swimming individual iguanas have come from (1) midway between BCI and Frijoles, i.e., midway between BCI and the far shore of Lake Gatun, (2) the vicinity of Isla De Lessups and Isla Orchid, north of BCI, (3) between Isla Pantera and Isla Tigre, NW of BCI in Gatun Lake, and (4) between Isla Pepper and BCI.

Acknowledgments

This work was supported by a Noble Fellowship from the Smithsonian Tropical Research Institute awarded to Hugh Drummond, and NSF Grants BNS-75-02333 and BNS-78-14196 to G.M. Burghardt. During the writing, Hugh Drummond was supported by Hilton Smith Fellowships from the University of Tennessee. Facilities and other aid were generously provided by STRI. We thank A.S. Rand and H.W. Greene for substantial assistance in the field, especially the social attraction experiment, and S. Rojas-Drummond for spending long hours observing from the blind. We are also grateful to R. Andrews for help in the laboratory, to the workers on BCI for their enthusiastic help, and to K. Adler, H. Ambrose, and A.S. Rand for their useful comments on the manuscript.

16

The Benefits of Sleeping Aggregations in Marine Iguanas, *Amblyrhynchus cristatus*

P. Dee Boersma
Institute for Environmental Studies
University of Washington
Seattle, Washington

INTRODUCTION

The Galápagos marine iguana is unique among lizards because it exploits a marine environment, feeds on algae throughout its life, and forms large sleeping aggregations at night. Adult male marine iguanas of different races range in size from 0.5 kg to over 10 kg, but regardless of size, marine iguanas frequently cuddle together at night.

The ability of marine iguanas to thermoregulate over the wide range of ambient temperatures (15° to over 50°C) has been intensively studied (Mackay 1964; Bartholomew and Lasiewski 1965; Bartholomew 1966; Morgareidge and White 1969; White 1973; Dawson et al. 1977). Body temperatures of individuals of all sizes are regulated during the day between 35° and 37°C (Bartholomew et al. 1976; Bartholomew and Vleck 1978) but fall to ambient or within a few degrees of it at night (Bartholomew 1966; Mackay 1964).

The ease with which sleeping aggregations form is clearly enhanced by the foraging ecology of these lizards. Females and young forage primarily on exposed reefs at low tides, while the larger adult males feed subtidally on the submerged algae (Boersma 1982). Their only food, marine algae, occurs in the intertidal or subtidal zones in dense patches that are probably indefensible because of both their temporal and spatial pattern of availability. Among terrestrial lizards, in contrast, food resources are widely spaced, continuously available during the day, and in many cases, defensible. Suspension of territorial behavior and long trips by individuals to the communal sleeping sites would be necessary for most lizards to form sleeping piles. Marine iguanas, by comparison, can readily aggregate because they are restricted to the coast. Nonetheless, ease of formation does not alone provide a reason for the adaptive significance of iguana sleeping piles. However, since cuddling reduces the exposed surface area of individuals, sleeping in groups may be favored to reduce nocturnal heat loss.

Methods

I measured cloacal temperatures of both solitary and piled iguanas during the night to determine relative rates of cooling. Individuals were grasped behind the head and at the base of the tail and removed from the pile. Occasionally other iguanas were awakened and ran from the pile, but generally iguanas were not awakened until they were captured. Characteristics such as large spine size, bulges at the base of the tail, and bright coloration distinguished males from females. Occasionally the hemipenes of males were visible. Body temperatures were taken from 10 January to 4 March 1972 at Pta. Espinosa, Fernandina, with a multi-channel Yellow Springs telethermistor. In addition, I made general observations on iguanas on Santa Cruz, Genovesa, Isabela, Fernandina, Española, Santiago and Santa Fé Islands during five trips to the islands between 1970 and 1978.

RESULTS

Pattern of Formation of Sleeping Pile

Marine iguanas bask in the sunlight, frequently touching each other; but at night they sleep in crevices in the lava flow, on rocky shores or sandy beaches, alone or in dense aggregations. Pile formation begins around 1600 hrs when the air starts to cool rapidly and is complete by dark. Marine iguanas are not normally active at night (Bartholomew 1966; Carpenter 1966; personal observations), and remain in the piles until morning. Sleeping piles are conical, and individuals usually have their heads toward the center with their tails pointed outward, like the spokes of a wheel (White 1973; personal observation).

Individuals either cuddled in sleeping piles or slept alone in crevices or places where rocks shielded them. Solitary individuals and those on the outside of a sleeping pile were usually large males, while iguanas near the center were immature or female iguanas. Young iguanas avoided the outsides and tops of the sleeping piles.

Body Temperature as a Function of Position in the Sleeping Pile

If sleeping piles act to retain heat, then early in the morning iguanas near the center of the pile should have the highest body temperature. I ranked iguanas by their relative position in the sleeping pile from the center to the outside and then by body temperature from the highest to the lowest. In piles of more than 10 iguanas, those nearest to the center were significantly warmer than iguanas on the outside or periphery of the pile (Table 16.1). Regardless of sex or size, iguanas that were covered (center) were warmer in all but one instance than iguanas sleeping alone (Table 16.2). Iguanas sleeping on the outside of a pile were significantly warmer than solitary sleeping individuals (X^2: $p< 0.001$).

Three to four hours after sunset all individuals were still above ambient air temperature, in contrast to Bartholomew's (1966) finding that iguana temperatures are virtually the same as air temperature two to three hours after sunset. (Bartholomew, pers. comm. took the temperatures of solitary individuals sleeping in exposed sites and in vertical cracks.)

Table 16.1: Spearman Rank Values

These values show iguanas closer to the center of the sleeping pile have higher body temperatures than those further away. As pile size increases, the relationship between central position and higher Tb is strengthened.

Sleeping Pile Size	Number of Iguanas Measured	Level of Significance, r_s
6	5	0.68
6	5	0.80
10	10	0.58*
12	11	0.67**
20	6	0.98**
50	15	0.54**

*p = 0.05
**p = 0.02

In the early morning, iguanas in the middle of the pile were warmer than those on the outside (Table 16.2). Thus, the same pattern of warmer iguanas in the middle of the sleeping pile holds from a few hours after sunset to just before sunrise.

Body Temperature as a Function of Pile Size

Sleeping piles may keep iguanas warm, particularly the smaller iguanas in the center of the pile. To determine if piling slowed heat loss, I measured the body temperatures of iguanas in different sized sleeping piles just before sunrise when their body temperature should be the lowest. The size of an iguana will affect its cooling rate, so to examine the importance of pile size independently of lizard size, I recorded the temperature of the warmest iguana in the pile. The temperature of the warmest iguana in the pile increased with pile size (Figure 16.1). Sleeping with one or two individuals slowed the cooling rate but cooling was even slower in the larger sleeping piles. Although there may not be much difference in heat retention between large piles, joining the larger pile may still be advantageous. The surface-to-volume ratio decreases with pile size, so large piles have a proportionally smaller outside edge. An individual that joins a larger pile will therefore, on the average, be more likely to be covered.

Body Temperature as a Function of Lizard Size

The physiological mechanism that retains heat should be similar between the sexes, but adult males should cool slower than females because of their greater mass. Between 1930 and 2300 hours the mean male temperature (27.6 ± 0.2°C, n=60) was significantly warmer than the mean female tempera-

Table 16.2: Internal Body Temperature of Marine Iguanas Sleeping at Night in Piles at Pta. Espinosa, Fernandina

Iguanas were divided into three categories by sleeping position within the group (center, inside, and outside). Center iguanas had ¾ or more of their body covered by other iguanas. Inside iguanas were covered by at least one iguana, and outside iguanas were not covered by other iguanas but were in body contact. (All temperatures were in °C.) Tb = Cloacal temperature.

Date and Time	Pile Size	Substrate Temperature	Air Temperature	Iguana Position in Sleeping Pile: Center $\bar{X}_{TbC}$	±SE	N	Inside $\bar{X}_{TbI}$	±SE	N	Outside $\bar{X}_{TbO}$	±SE	N	Solitary $\bar{X}_{TbS}$	±SE	N	$\bar{X}_{TbC} - \bar{X}_{TbO}$	$\bar{X}_{TbO} - \bar{X}_{TbS}$
1/28/72	12	25.7	24.6	29.9	0.2	6	28.8	0.0	2	28.6	0.4	3				1.3	1.8
1930	10	26.2	24.3	28.1		1	28.1		1	27.6	0.1	3				0.5	0.8
to	8	25.2	24.3							26.9	0.1	3					
2100	6	25.2	24.2	28.7	0.2	3	28.7		1	27.9		1				0.8	1.1
	5	25.0	24.1	28.6		1				26.3		1				2.3	- 0.5
	20	25.0	24.1							29.1	0.1	3					
	S												26.8	0.3	12		
	Sample means			29.3	0.3	11	28.6	0.2	4	27.9	0.3	14	26.8	0.3	12	1.4	1.1
2/20/72	15			28.0	0.4	3	27.1	0.5	2	27.8		1				0.2	1.0
2200	8	27.0	26.8	28.9		1	28.1	0.2	4	27.4	0.3	4				1.5	0.6
	3			28.2		1				28.4		1				-0.2	1.6
	2									27.3	0.2	2					0.5
	S	27.6	26.6										26.8		1		
	Sample means			28.2	0.4	5	27.8	0.3	6	27.5	0.2	8	26.8		1	0.7	0.3
1/31/72	50	26.4	26.0	27.4	0.1	4	27.0	0.1	8	26.4	1.0	3				1.0	-1.5
2230	20	26.5		31.0	0.5	3				29.3	0.1	4				1.7	1.4
	10	26.8	27.0	29.8		1				28.2		1				1.6	0.3
	7	27.0	24.3							28.4	0.1	2					0.5
	2	24.8	27.8							27.9	0.3	6					0.0
	S												27.9	0.2	7		
	Sample means			29.0	0.7	8	27.0	0.1	8	28.0	0.3	16	27.9	0.2	7	1.0	0.1
2/24/72	20	24.6	23.4	25.2	0.4	2	24.5	0.1	2	24.2	0.2	6				1.0	1.2
0435	18	24.8	23.8	25.8		1											
	6	24.3	23.4	24.4		1	24.7	0.1	9	23.9		1				0.5	0.9
	3	24.3	23.4				25.3		1	24.4	0.1	3					1.4
	2		23.4							24.0	0.2	3					1.0
	S												23.0	0.1	11		
	Sample means			25.2	0.4	4	24.8	0.1	12	24.2	0.2	13	23.0	0.1	11	1.0	1.2
Entire sample				28.4	0.2	28	27.0	0.2	39*	26.8	0.2	48	25.7	0.3	31	1.6	1.1

Note: Position of iguanas in sleeping pile: I = inside; C = center; O = outside; S = solitary.

*Iguanas included where pile size was unknown.

ture (24.8 ± 0.2°C, n=43, t test = $p<0.01$). However, large size is not enough to keep an iguana warm throughout the night. In the early morning, regardless of size, solitary individuals had the lowest temperature of any sleeping iguana. Five of eleven solitary individuals sleeping on the sand and on exposed lava near the sea were as cold as the air.

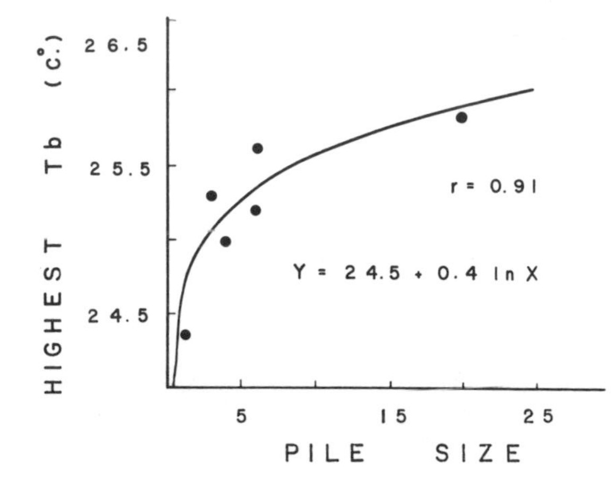

Figure 16.1: The Tb at dawn of the warmest marine iguana in a sleeping pile increased with sleeping pile size.

DISCUSSION

Sleeping in aggregations reduces the rate at which the body temperatures of marine iguanas fall during the night so that individuals within piles have body temperatures 2°-3°C above solitary individuals at dawn. Likewise, Myers and Eells (1968) found that snakes in groups cooled more slowly than solitary individuals. Such higher nocturnal body temperatures mean that metabolic rates at night are substantially higher for iguanas sleeping in piles than for those sleeping alone. What advantages might override this energetic cost of sleeping in aggregations?

Predation

Lizards are more active at higher temperatures, so that any increase in temperature may be advantageous in avoiding predators. The Galapagos Hawk *(Buteo galapagoensis)*, Short-eared Owl *(Asio flammeus)*, and a variety of other avian predators, as well as snakes, prey on marine iguanas (Beebe 1924; Carpenter 1966; Boersma 1982). As marine iguanas grow larger they become immune to terrestrial predators (Carpenter 1966). Thus, young and females are attacked more often than males (Boersma 1982). Sleeping piles may be an adaptation to avoid predation. An indivdual that cuddles reduces its likelihood of being attacked because of either the physical or numerical protection of being in a group. Furthermore, an iguana may be more mobile if it is in a group because it will cool more slowly. At low body temperatures, land iguanas *(Conolophus pallidus)* are less mobile and more vulnerable to predation (Christian 1978). If the major advantage of sleeping in a pile is safety, then large iguanas, which are relatively immune to predators, should be more likely to sleep in the open. I often saw large males exposed on the lava or sand sleeping alone but never young.

Rate of Food Processing

Another advantage of sleeping together and retaining heat is faster and more complete digestion. Avery (1973) found that the lizard *Lacerta vivipara* processed food more rapidly at higher temperatures. The frequency of digestive contractions decreases with decreasing temperature for at least two other lizards (Mackay 1968). The time food stays in the stomach appears to be determined by the body temperature of the lizard (Diefenbach 1975).

Having the gut empty may be beneficial in the following ways. Marine iguanas forage mainly at low tide. The intensity of foraging depends on the tide cycle, and more iguanas go foraging during times of lowest monthly tides because of the increased reef exposure (Boersma 1982). Each day an individual will forage once or twice or not at all, depending on the tide cycle. Algae processing time may be important, at least during the lowest tides, when foraging bouts are frequent. If an iguana filled its stomach just before sunset, slept in a pile where it maintained a body temperature of 25°C and went foraging at a 7 a.m. low tide, its stomach would be empty, assuming they digest food similarly to *Lacerta vivipara.* If this same iguana, with a body temperature of 20°C, slept alone its stomach would not be empty when it went foraging. Furthermore, storage capacity may be particularly important to smaller iguanas that have smaller stomachs and less storage room. If storage room is limiting, then smaller iguanas should always sleep in sleeping piles during the lowest tides of the month.

Digestive Efficiency

The digestive efficiency of a herbivorous lizard *Dipsosaurus dorsalis* increased from 54% at 33°C to 63% at 37°C, to 70% at 41°C (Harlow et al. 1976). Harlow et al. (1976) found that lizards kept at 28°C did not pass food beyond the stomach and other lizards follow a similar pattern (Harwood 1979). Janzen (1973) assumes that basking by vegetarian lizards is necessary for food processing. The digestion in herbivorous marine iguanas may likewise be more rapid at higher temperatures. Efficient digestion that results in more energy per foraging effort will be particularly important to young iguanas because they must reach a critical size before they are unsuitable as prey.

The only disadvantage of maintaining a higher body temperature at night is the increased metabolic cost. The rate of oxygen use for a 1000 g lizard with a body temperature from 20° to 37°C is nearly linear (Bennett and Dawson 1976). A higher body temperature at night will increase metabolism, but this cost may be minimal or completely offset by faster and more efficient food digestion.

Testing Between Hypotheses

If maintaining a higher body temperature is advantageous only in food processing, then iguanas with empty stomachs will incur metabolic cost by maintaining a higher body temperature. If enhanced mobility is unimportant, then iguanas with empty stomachs should always sleep alone.

Four predictions should be true if the main value of cuddling or piling at night is related to food processing. First, iguanas with empty stomachs should not cuddle. Second, large adult males should gain less from sleeping in aggregations than females because males feed subtidally and can forage whenever they wish. Females and young that forage on the exposed reef should benefit the most. Third, benefits from higher body temperatures should be greatest when low tides occur at dawn rather than later in the morning. Fourth, iguanas that are building storage reserves rapidly to defend a territory or lay eggs should sleep in piles. Therefore, I expect a lunar cycle pattern in the sizes of sleeping aggregations and the proportions of individuals using them. What limited data I have support these predictions.

First, during the breeding season (December-March) males defend territories and sleep alone more frequently than females. While defending a territory, the males fasted for a number of days so that their stomachs were presumably empty.

One male that for three months always slept alone alternated between sleeping deep in a crevice and on the surface of the lava. The variation in his sleeping position was probably related to his foraging pattern.

Sleeping pile formation appears related to the tidal cycle. During the middle of the tide cycle, iguanas foraged less and presumably had empty stomachs. Aggregations composed mainly of females near the tidally exposed reefs moved inland to crevices and lava tubes for a number of days and fasted (Boersma 1982). Females in these protected sites frequently slept alone and stayed in the crevices during the day. Moving inland to protected sleeping sites is probably mainly an adaptation to reduce the metabolic rate when they are not foraging. In contrast, the predominately male aggregation on the tip of Pta. Espinosa, Fernandina, retained its normal position and did not move inland, although some males, presumably with empty stomachs, frequented crevices or cavities during the heat of the day to lower their metabolic rate. On two occasions, males that had been defending territories and sleeping alone slept in piles with other males after leaving their territory and foraging. When the lowest tides of the month occurred and iguanas grouped along the shore where they had emerged after foraging, the largest sleeping piles formed and both males and females joined them.

SUMMARY AND CONCLUSIONS

Young and female marine iguanas do not always sleep in piles, so predation cannot be their most important function. Furthermore, the pattern of pile formation is related to the foraging rhythm. The main advantage of cuddling or sleeping in piles is probably faster and more efficient food processing. Other lizard species may accrue the advantages of a sleeping pile by digging a burrow. I would expect lizards that form less dense aggregations may elevate their body temperature to process food at night by not only using a burrow but by sharing it. Huey (in press) has suggested that because of the lag in cooling of deep soil, burrows are warmer at night than other sites. Consequently, lizards can stay warmer at night in a burrow. Behavioral adaptations that increase body temperature at night should be seen most commonly in herbivorous species (1) when the lizard has a full stomach, and (2) when building storage reserves rapidly is necessary.

Marine iguanas cannot dig burrows because of the terrain, but sleeping piles and to a lesser extent, deep crevices, act much like burrows to conserve warmth so that the iguana's internal temperature is higher than the ambient temperature. The main advantages marine iguanas gain by cuddling are (1) faster processing; and, (2) more efficient digestion of food.

Acknowledgments

I thank Sally J. Cloninger and Chris Kjolhede for help in the field. Gordon Orians, Ray Huey, Paul Colinvaux and Jerry Downhower have given generously of their ideas; and Gordon Orians, Barbara Peterson and Ray Huey critically read drafts of the manuscript and improved it. I gratefully acknowledge the assistance and cooperation of the Charles Darwin Research Station during the course of the study. The study was supported by N.S.F. Grant GB-29065X.

Section V

Social Organization

Iguanines are unusual among lizards in several aspects of their social behavior. First, although they may defend territories strongly the territories tend to be for mating rather than for feeding, a point emphasized by both Dugan and Wiewandt (Chapter 17) and Ryan (Chapter 21). Territories may be held year round (Caicos ground iguana, Iverson 1980; green iguana, Dugan, Chapter 18) or only for mating season (marine iguana, Carpenter 1966b). The mating territory may contain sunning perches, food, or burrows that are necessary to the females, but these are not uniquely available there, and sometimes are completely absent, so that the female must spend part of her time elsewhere. When a territory does contain resources such as food necessary to the male, the defense of these is secondary to the primary function of mate defense.

Males of several iguanines exhibit more than one strategy. The Mona Island ground iguana may choose between being a long term dominant in an inferior area or a temporary dominant in an area with many females (Wiewandt, Chapter 7). Adult green iguana males may be dominant territory holders, peripheral territory holders, or sneaking "rapists" (Dugan, Chapter 18). Land iguana males on Fernandina may be dominant or peripheral (Werner, Chapter 19). Ryan has modeled the relationship between male strategy and environmental conditions and population parameters. He suggests that long range cost-benefit ratios of these strategies must be about equal; those with high risk and high energy cost also have high returns in terms of mates. In some cases the low cost and low gain alternative is taken by smaller and presumably younger individuals but this is not necessarily true (e.g., land iguanas). Multiple strategies may be more common in iguanines than in other lizards because of their low annual mortality, which makes it likely that they will have a chance in more than one breeding season. Multiple strategies are open also to females and we find that there are considerable differences between individuals in the same population in, for instance, how much they defend their nest (Christian and Tracy, Chapter 20).

Iguanines appear unusual among lizards in that they associate with one another in ways that have no apparent sexual or dominance component. This is recorded for at least green iguanas where young animals move together in small groups called chuletas (Burghardt et al., 1977), and in which adults out of breeding season may move, feed, or sun together (Dugan). Marine iguanas may sleep in aggregations that seem to help regulate their temperature (Boersma, Chapter 16). Giant *Sauromalus* share burrows, frequently in heterosexual pairs, even during years when no breeding occurs (Case, Chapter 11). Sometimes these associations serve some immediate function as in marine iguanas. Sometimes they probably serve to help watch for predators. Sometimes individuals may be sharing information about food or other aspects of the environment in ways that might involve social facilitation (Greenberg, 1976) or conspecific cueing (Kiester, 1979). Sometimes we just do not know.

Iguanas spend long periods apparently doing nothing. One of their commonest responses to threat is to freeze immobile, or to inch very slowly around their perch and out of sight. Casual observers tend to conclude that because they do not see them do much that iguanas are simple. As the results of long term studies begin to accumulate we are finding that these long lived reptiles may sometimes be slow but they are not simple. After 18 months of watching a group of iguanas on a hillside in Panama Bev Dugan said, "Iguanas are not dull; they are just very subtle."

17

Socio-Ecological Determinants of Mating Strategies in Iguanine Lizards

Beverly Dugan[1]
Department of Psychology
University of Tennessee
Knoxville, Tennessee

and

Thomas V. Wiewandt[2]
Department of Ecology and Evolutionary Biology
University of Arizona
Tuscon, Arizona

INTRODUCTION

Background

During the past two decades, a pronounced shift in emphasis has developed among field biologists: instead of simply describing and classifying elements of social behavior, attempts are being made to identify environmental conditions responsible for their evolution. Such studies of the adaptiveness of behavior gained tremendous momentum in ornithological circles during the sixties, and mammalogists followed suit in the seventies. The growing body of comparative data on the adaptiveness of vertebrate social organization, unified with modern theory in population biology (largely derived from research on social insects), gave birth to the "new" fields of sociobiology and behavioral ecology (see Wilson, 1975 and Krebs and Davies, 1981). This line of thinking has been slow to catch on among herpetologists, perhaps because reptiles and amphibians are seldom thought of as social creatures (see Burghardt, 1977) and because details about their life histories are known for so few species. This is unfortunate, as reptiles occupy a critical position in vertebrate evolution, and cause-effect relationships should be easier to identify in animals with less complex behavioral repertories. We predict that much of this unrealized potential in herpetology will unfold in the eighties.

Before proceeding further, readers should become acquainted with three terms commonly used by behavioral ecologists and which appear throughout this chapter: mating system, mating strategy, and territory. A *Mating System* is the manner in which the reproductive members of a population apportion their energy and/or activity for mate acquisition. Mating systems are characterized on the basis of the ecological and behavioral potential to monopolize mates (Bradbury and Vehrencamp, 1977; Emlen and Oring, 1977). Mating systems may consist of one or more *Mating Strategies,* behavior employed by the *individual* in obtaining mates. The word strategy does not imply that a conscious decision is made by the individual who employes it. Since behavior is under selection to maximize the reproductive success of individuals, inter- and intra-individual differences in mating strategies can be expected under differing ecological and/or social conditions. It is particularly important when studying long-lived species to examine this inter- and intra-individual variability in mating strategies in order to understand the structure of the population's mating system. The term *Territory* denotes a fixed area that is defended from members of the same species by advertisement, threat or attack (Brown, 1975).

Rationale

Studies of lizard behavioral ecology are still in the descriptive stage. Detailed information on mating systems among lizards is scarce, and the mating strategies of individuals are rarely addressed. Among field investigations that have the most to offer are those of Rand (1967) on *Anolis lineatopus,* Stamps and Crews (1976) on *Anolis aeneus,* and Trivers (1976) on *Anolis garmani.* Data on the iguanines, among the more unusual lizards, are even more rare. Berry's (1974) field study of the chuckawalla, *Sauromalus obesus,* Dugan's (this volume) on the green iguana, *Iguana iguana,* Werner's (this volume) on the Galapagos land iguana, *Conolophus subcristatus,* and Wiewandt's (1977, 1979) on the Mona Island ground iguana, *Cyclura stejnegeri,*[3] are the few among many published papers on iguanines that treat reproductive strategies in detail.

Mating behavior cannot be considered distinct from other aspects of a population's ecology. Since differential reproductive success defines selection, mating systems should be particularly responsive to environmental variations, and should therefore provide insights into the diversity of selective pressures operating on populations. In this review, we hope to integrate mating systems theory with what is known about life-history and reproductive behavior of iguanines in a manner that will suggest avenues for future research.

Iguanid Mating Systems

If one sex invests more in the production and care of offspring than the other, the sex that makes the greater investment becomes a *limiting* resource for members of the opposite sex. Disparate parental investment leads to competition for mates among members of the *limited* sex (Bateman, 1948; Fisher, 1958; Trivers, 1972). This situation often results in polygamous mating systems. Among mammals, the female assumes most of the parental care, and polygyny is common. In birds, on the other hand, feeding and defense of the young often require both parents, and monogamy is more common. (For further discussion of factors favoring polygamy, see Orians, 1969; Williams, 1975; Wilson, 1975.)

Other than nest construction, egg-laying, and in a few cases nest-guarding, parental care is lacking among iguanid lizards. Since females invest more in gamete production than do males, polygyny is expected to be widespread in this group. Competition among males for females leads to greater activity on the part of males than females in securing mates, and greater variance in male reproductive success.

In the typical iguanid mating system, male territories overlap the home ranges of several females. Female activities center around resources essential for survival (food, shelter, basking sites); male activities center around access to females and other resources (Stamps, 1977). Although polygyny is widespread among iguanids, the manner in which mates are acquired varies with ecological demands specific to a population. Iguanines differ from all other iguanids in their life history characteristics and ecological role, and deviations from the typical iguanid pattern of mate acquisition reflect these differences.

LIFE HISTORY CHARACTERISTICS

Ecologically and evolutionarily meaningful interpretations and predictions about mating systems/strategies are closely tied to an understanding of other life history attributes. We therefore begin by discussing several key factors which influence reproductive behavior in iguanine lizards: (1) diet, (2) body size, (3) longevity, and (4) annual breeding schedule.

Diet

With only one possible exception known (*Brachylophus fasciatus,* see Cahill, 1970; Gibbons and Watkins, this volume), iguanines are primarily herbivorous (Iverson, this volume). They are, however, opportunistic, and most species take animal matter when and where it can be easily obtained.

The relationship between social structure and distribution and abundance of food is poorly understood for lizards of this subfamily. Among insectivorous iguanids, aggression and territory size appear related to food availability and intraspecific competition for food (Simon, 1975; Stamps, 1977). Turner et al. (1969) found a strong correlation between home range size and body size in lizards. The same relationship exists in birds and mammals, and it has been suggested that larger animals need larger home ranges to meet their energy requirements (Schoener, 1968, 1977; McNab, 1963).

Data from iguanines do not conform to this prediction however. During the non-breeding season, home ranges of large male green iguanas were found to be approximately one-third the size of those of other males and females (Dugan, this volume). In a year of food shortage, home range sizes did not differ significantly between male and female chuckawallas (Nagy, 1973). Nagy suggested that since home range size did not increase as vegetation became less available, home range size in this population was not related to food availability and energy requirements.

Patterns of Food Utilization: Because plant foods often differ appreciably in quality, in quantity, and in local, seasonal, and annual availability, one would expect iguana movements and social behavior to be patterned accordingly. Where preferred food items are concentrated and superabundant, or patchy and shifting with time, their defense is not theoretically feasible (see Brown, 1964). Energetic feasibility of defending a food supply also diminishes in densely vegetated habitats where surveillance is difficult.

Absence of food-based territorial defense does in fact appear to be a common pattern within the Iguaninae. Lizards may congregate near a food source and make short-term forays away from centers of activity (basking sites and retreats) to feed. Evans (1951) described a high density colony of *Ctenosaura pectinata* in Mexico that lived in a rock wall adjacent to a cultivated field where they fed on newly planted beans. Boersma (in press) found marine iguanas patchily distributed along the coast of Isla Fernandina, the concentration of iguanas varying with water depth, bottom gradient, and algae production. Feeding aggregations of dozens of desert iguanas *(Dipsosaurus)* were seen in flowering creosote bushes *(Larrea divaricata)* by Cowles (1946).

An iguana may shift its center of activity for several days or weeks to follow seasonal fruiting and flowering patterns. Mona *Cyclura* usually forage in areas adjacent to retreats; however, longer-term shifts in location sometimes occur when preferred fruits are available elsewhere (Wiewandt, 1977). This pattern probably reaches its extreme in green iguanas. An iguana may remain in the same tree from one day to several weeks. Trees contain basking sites, shade, sleeping perches, display posts, and food. An individual may travel through a large area, moving from tree to tree, over the course of several months. During the fruiting season of the wild plum *(Spondias)*, long-range movements are clearly determined by food availability (Dugan, 1980).

In relatively open habitats where food is more evenly dispersed and easier to guard, iguanas may profitably defend a feeding area. While territorial male *Conolophus subcristatus* suspend all feeding during the mating season, food availability and shelter for females both appear related to the quality of a male's breeding territory (Werner, this volume). Similarly, multi-purpose territories offering food, shelter, and access to females are sometimes maintained by male *Sauromalus obesus* (Berry, 1974). In this species, food defense appears to be a consequence of, rather than a function of, territorial defense (Ryan, this volume).

Constraints on Activity: Problems associated with processing plant foods in reptiles have been discussed by a number of authors, most recently by Rand (1978), Auffenberg (this volume), Iverson (this volume), and Nagy (this volume). While more data on time and energy budgets, digestive efficiency and costs, and foraging habits are needed, what is now known indicates that restricting energy expenditures is especially important to iguanine lizards (for example, see Rand and Rand, 1976). Time budget figures for *I. iguana* (Dugan, 1980) and for *Cyclura stejnegeri* (Wiewandt, 1977) show that these species are sporadically active and normally spend over 90% of their waking hours lying or sitting motionless. Even during the mating season, *C. stejnegeri* males were inactive between 82 and 88% of the time, *I. iguana* males approximately 80% of the time. Thus, viewing the evolution of iguanine lizard behavior from the perspective of energy optimization appears particularly useful.

Body Size

Physiological Considerations: Physiological relationships between diet and lizard body size have been examined by Pough (1973) and Wilson and Lee (1974). Presumably herbivory is energetically more efficient for larger lizards, since energy requirement per gram of body weight decreases with increasing size. Activity requires, however, more total energy for a large than a small lizard.

The relationship between body size and energy expenditure should affect the energetic feasibility of competition and the form competition will take. Large body size offers a competitive edge during agonistic encounters and increased resistance to periods of drought and food shortage. Population-specific growth characteristics are, however, largely influenced by such factors as the relative abundance and distribution of high-energy foods, the duration of the activity season, and predation intensity (see Wiewandt, this volume). Behavioral correlates will remain obscure until more meaningful analyses of food environments and utilization efficiencies are possible.

Susceptibility to Predation: Accompanying large body size, an animal may gain partial (Van Devender, 1978) or total immunity to predation. Adults of large insular species in particular often have no natural enemies (Case, this volume; Wiewandt, this volume). Because susceptibility to predation affects life expectancy, different patterns of social behavior are likely in populations under varying degrees of predation intensity.

Longevity

Iguanines are among the longest-lived lizards. Estimates of longevity range from less than 10 years for *I. iguana* (Müller, 1968) to more than 40 years for *Cyclura pinguis* and *C. stejnegeri* (Carey, 1975; Wiewandt, 1977). Two important correlates of a long life span are an extended growth period and iteroparity (Tinkle, 1969).

Extended Growth: Iguanines are reproductively mature before normal adult size is reached, and growth continues, although slowly, throughout adult life. As in many other vertebrates (Trivers, 1972) large male size and social status are related and are important correlates of mating success among iguanines. Males may be excluded from the breeding population for a large proportion of their adult lives. Where breeding animals are grouped, smaller males in the population might be expected to adopt alternate strategies for obtaining mates (Emlen and Oring, 1977; Gadgil, 1972; see "Consequences of Male-Male Competition" below). Multiple mating strategies among males have been documented in *I. iguana* (Dugan, this volume), *Cyclura stejnegeri* (Wiewandt, 1977, 1979), and *Conolophus subscristatus* (Werner, this volume). Insofar as mating success is related to size, status, age, and experience, ontogenetic changes in mating strategy are to be expected throughout adulthood (see Ryan, this volume).

Iteroparity: Those individuals that survive to adulthood often live to reproduce many years. In order to maximize reproductive success, an optimal mating strategy should balance immediate prospects of reproductive success against long-term prospects (Pianka, 1976). For example, nest sites were scarce on Mona Island, and competition among nesting female *C. stejnegeri* was intense. In such an ecological and social setting, one might expect reproductive effort to increase steadily with age (see Wiewandt, 1977, 1979, this volume). There consequently should be an optimal allocation of energy to growth *vs* reproduction.

Since the amount and manner of reproductive investment differ between the sexes, one might expect the allocation of energy to differ between males and females. As in most large reptiles, an increase in clutch size accompanies increasing female body size (and presumably age). The cost of egg production presumably retards female growth relative to that of males. Adult males grow faster than females in *S. obesus* and *C. carinata* (Berry, 1974; Iverson, 1979). Thus, sex differences in energy investment may affect sexual dimorphism. For example, Van Devender (1978) monitored growth and reproduction in two populations of *Basiliscus basiliscus*. Although females grew and reproduced at roughly equal rates in both areas, male growth rate and reproductive pattern were site specific. In an area of high population density, young males reproduced later and grew at a faster rate than in the low density population. Delayed reproduction and faster growth rates in the high density population presumably reflected the intensity of male-male competition for mates. By delaying reproductive maturity with associated behavior, males may shunt another season's energy into growth. Thus, growth and reproductive strategies may be responsive to social as well as environmental cues.

Breeding Schedules

All iguanines are seasonal breeders. Although the timing of breeding differs among species and populations, the major environmental pressures on breeding schedules seem to be (1) optimal conditions for incubation and hatchling emergence, and (2) insurance of an adequate food supply for either offspring, adults, or both. These factors, plus the modifying influence of social behavior, are discussed in detail in Wiewandt (this volume).

The timing of reproduction is significant in that it ultimately determines the temporal availability of females and affects the duration and form of male-male competition for females. The average ratio of fertilizable females to sexually active males present at any given time during the breeding season–termed the "Operational Sex Ratio" by Emlen and Oring (1977)–is one important determinant of mating systems. The greater the imbalance in this ratio, the greater the degree of polygamy expected (also see Bradbury and Vehrencamp, 1977). If, for example, females of an iguana population mate asynchronously, say during several months of the year, one might expect males to maintain breeding territories throughout that period, resulting in a relatively stable social order. Under such circumstances, great disparity in the mating success of individual males is apt to occur. Where females are more synchronized in their receptivity, males must make a more concentrated breeding effort, resulting in seasonally intensified male-male competition and less variance in mating success. Data from other groups of vertebrates support these predictions (e.g., see Wilson, 1975; Emlen and Oring, 1977; Wells, 1977b), but as yet little pertinent information is available for the Iguaninae. In testing such predictions, data on the duration and synchrony of female receptivity are critical, and largely lacking.

Iguanines appear to be quite variable in degree of breeding synchrony, and details of mating activity vary accordingly. The breeding activity of *C. stejnegeri* is shorter than that reported for any iguanine to date. Over one year's time, female receptivity was limited to a 13-day period (Wiewandt, 1977). Male-male competition for breeding rights was keen, particularly in an area of high iguana density. While five of the seven males resident in the study area mated at least once, variance in male mating success was high. Year-round territoriality and frequent contact between the sexes were positively correlated with mating success, and may have facilitated rapid mate acquisition. In contrast to the Mona iguana, the receptive periods of female green iguanas were staggered over a six week period. Large males established temporary breeding territories and polygyny was extreme; with a 1:1 sex ratio, approximately 60% of mature males did not breed in each of two seasons (Dugan, this volume). Instead of the higher frequency of aggressive encounters observed among *C. stejnegeri* males, competition between *Iguana* males took the form of increasing display rates.

Part of the variability in breeding synchrony among the iguanines is attributable to constraints imposed by climate. Irregular reproduction among desert iguanines is probably a response to harsh environmental regimes (see Wiewandt, this volume). In good years, a small proportion of *Dipsosaurus* females may produce two clutches (Mayhew, 1971), and some chuckawallas reproduce only once every two years (Berry, 1974). There is evidence from captive *Brachylophus*, on the other hand, that this species is more like other tropical iguanids in its reproductive phenology. Captive banded iguanas will copulate throughout the year, and females lay about five eggs every four months (Cahill, 1970; Arnett, 1980). Recent observations by Gibbons and Watkins (this volume) indicate that females mate shortly after oviposition, may lay more than one clutch per season, and apparently spend most of their lives in a gravid condition. Males in the wild probably remain in reproductive condition much of the year and, other factors permitting, maintain breeding territories accordingly.

MATE ACQUISITION BY MALES

Male Competition

Among polygynous species, an individual male's reproductive success ultimately depends on how many females he can mate with. How this is achieved depends on the ecological and social environment of the population.

The form of intermale competition reflects the nature and distribution of resources attracting females relative to the numbers of competing males. Competition may center on any number of resources, such as food and/or shelter, with access to females being so determined. In order to understand the mating system, the major factor(s) governing intermale competition must be identified.

"Competition" for Food: While food distribution often influences patterns of movement, only infrequently is it an object of territorial defense in iguanine lizards. For reasons already put forth under "Diet," it is apparently seldom profitable for male iguanas to seek access to females via control over food resources. Nevertheless, proximity to the best food and water throughout the year may be an important determinant of male reproductive success. Even if the resource is not defended, frequent contact with females passing nearby may significantly enhance a male's breeding potential. This appears to be true of *I. iguana* (Dugan, this volume), *C. stejnegeri* (Wiewandt, 1977), and *C. subcristatus* (Werner, this volume).

Size dimorphism is claimed to be a mechanism for reducing intraspecific competition for food in many iguanids (Schoener, 1977; Stamps, 1977). Although most iguanines are sexually dimorphic in body size and head size, there are few data indicating that males and females exploit different food resources or different parts of the same food type. The marine iguana is an apparent exception: Males feed on submerged reefs, females on exposed reefs. Large size presumably reduces the cost of submerged foraging. Adult size and sexual size dimorphism are less on islands where there are no offshore reefs (Boersma, in press).

Indirect food competition may be reflected in growth rates. Dominant male chuckawallas and Caicos ground iguanas grow faster than subordinates (Berry, 1974; Iverson, 1979). Whether the difference results from the dominant's priority of access to quality resources, or whether an individual is dominant because of a superior ability to utilize resources, is unknown. A longitudinal study of species with highly variable hatchling growth rates would be enlightening.

Competition for Retreats: The activities of most iguanines are retreat-centered. Retreats may be centers of social activity, may provide shelter from climatic extremes and escape from the sun, predators and conspecifics. Retreats also provide nesting sites for *Cyclura carinata* (Iverson, 1979). Locations near food and water seem to be important. In areas where rock cavities, tree holes, or animal burrows are not available, the lizards may dig their own, substrate permitting. Retreats are of minimal or no importance to most populations of adult *Amblyrhynchus, Iguana,* and *Brachylophus.*

Although few detailed data on the distribution and use patterns of retreats are available, it appears that females select retreats for their comfort, safety and proximity to food and water (Wiewandt, 1977, 1979). In many populations of iguanine lizards, choice retreats appear to be the proximate object of male-male competition. In *Conolophus,* males must provide retreats for themselves and prospective mates during the breeding season. Territorial males dig one or more holes that they share with females. Availability of existing holes and loose soil for excavation of new burrows, plus proximity to vegetation that provides food and shade, appear to be important components of a male's territory (Werner, this volume). Similarly, territorial male *C. carinata* defend a burrow which is also used by females. Natural crevices were scarce on Caicos, and the iguanas were forced to dig their own. Juveniles and subordinate males frequently shifted burrow locations, but territorial males defended a home burrow year-round (Iverson, 1979).

Ctenosaura clarki in Mexico exhibit a unique form of burrow usage. This is a small iguanine (125-250 mm SV) that lives in natural cavities in logs, trees or cacti. Individuals are sometimes found in the same tree, but never in the same hole. Their solitary habits apparently derive from a relatively specialized escape behavior. When disturbed, an individual retreats into a hollow trunk and blocks the entrance with its spiny tail. All individuals fit their particular hollow so that the tail effectively blocks the opening. Individuals change their hiding places as they grow larger (Duellman and Duellman, 1959). This situation suggests that competition among males for trees containing suitable retreats for one or more females may occur.

Competition for Mating Territories: Competition for mating territories differs from types of competition previously discussed in that a mating territory *per se* is established with no apparent connection with defense of food or retreats. The mating site is the proximate object of male-male competition. Resources critical to females are not economically defensible, and a mating territory provides a place within which to court and mate undisturbed. Examples of this pattern are known for two species: *I. iguana* and *A. cristatus.* Male green iguanas defend tall, conspicuous trees that have limited access (little canopy interlock), and are therefore relatively easy to defend. Food may or may not be present in the territory; tall, dead trees are frequently used. Males of *A. cristatus* defend a small area on the lava rocks (Carpenter, 1966; Eibl-Eibesfelt, 1961; Trillmich, 1979; Boersma, in press). Territories are separate from aggregations of females and bachelor males, and do not contain food.

The Form and Intensity of Competition: Population density and habitat features play an important role in shaping the form and intensity of competition among males. Although we discuss these factors separately, it should be borne in mind that they often occur in combination and their separate effects cannot be parceled out. Relationships between habitat and competition in insectivorous lizards have most recently been reviewed by Schoener (1977) and Stamps (1977).

Under conditions of high intensity competition, it may not be energetically feasible to sustain territorial defense for more than a few weeks of the year. Wiewandt (1977, 1979) monitored use of rock crevices in Mona iguanas and found patterns of territorial behavior related to density and habitat complexity. In an area where many favorable retreats were naturally clustered and density of females was high, males maintained territories only during the breeding season and defense centered around the network of burrows used by females. Less than 20 meters away, year-round defense of appreciably larger courtship/breeding territories occurred where retreats were more dispersed and easier to guard. Aggressive interactions and male-male interference during courtship and copulation were more frequent in the high density than the low density area. The one male that attempted long-term defense in the complex habitat was quickly defeated when other males began vying for territories early in the breeding season. From the standpoint of energy optimization, short-term defense of a small territory was the only male reproductive strategy effective for monopolizing a structurally complex area that attracted many females.

Typically, territorial species shift to dominance hierarchies or social tyrannies when population densities increase (see Brattstrom, 1974). At relatively low densities (7-8 per ha) chuckawalla males have been observed to maintain non-overlapping home ranges (Johnson, 1965) and all-purpose territories (Prieto and Ryan, 1978) throughout the activity season. Berry (1974) observed a social tyranny in a high density population (about 14 per ha). Variations in social structure and the frequency of aggressive interactions reported in different populations of *Sauromalus* may be related to costs of territoriality (see Ryan, this volume), some of which are density dependent.

Habitat severity may drastically alter competition. Chuckawallas may curtail reproduction in bad years (Berry, 1974; Nagy, 1973). Territoriality is likewise absent. Of even greater interest is Berry's observation that a certain proportion of adult male chuckawallas do not undergo testicular recrudescence during the spring breeding season. It is conceivable that in a lizard whose food supply is dependent on erratic precipitation, males with a low probability of immediate reproductive success increase over-all fitness by shunting a season's energy into growth. It would be of interest to know what segment of the male population curtails reproduction, and whether or not this is a socially and/or climatically induced phenomenon.

Consequences of Male-Male Competition

In his theory of sexual selection, Darwin (1871) attempted to explain characters such as brilliant coloration, morphological adornments, and display behavior that are of benefit in mate acquisition. Such characters may result from intersexual selection, in which the mate preferences of one sex select for certain attributes in the other sex, and intrasexual selection, in which individuals with attributes advantageous in competition for mates incur the greatest reproductive success. In this section, we are primarily concerned with intrasexual selection.

The degree of sexual dimorphism generally increases as the variance in reproductive success of the limited sex increases (Trivers, 1972; Emlen and Oring, 1977). Extrapolating to the iguanines which tend to be polygynous, one would expect sexual dimorphism to be common. However, factors other than sexual selection may influence such differences between the sexes (Wiewandt, this volume), and iguanines vary in the degree of sexual dimorphism and in the way it is expressed.

Displays: In general, data on the form, variability and use of displays among iguanines are lacking compared to that available for other iguanids (but see Carpenter, this volume). All iguanines seem to employ typical iguanid challenge displays (lateral presentation, side-flattening), but the form and use of headbob displays appears to be highly variable.

The green iguana's signature bob is highly stereotyped. Although all displays analyzed conformed to a species characteristic pattern, variability in display performance was greater among individuals than within individuals (Dugan, 1982). Similarly, Berry (1974) found individual differences in display pattern among chuckawallas. In contrast Wiewandt (1977) found such extreme variability in *C. stejnegeri* bobs that no core display could be identified. Three display units, the toss, the roll, and the toss-roll occurred in all possible combinations reflecting both context and level of arousal. Similarly, Werner (this volume) noted that ritualized behavior is not common in *Conolophus subcristatus*.

Headbob displays appear to be important in regulating interactions in mainland iguanas; males of both *Sauromalus* and *Iguana* display during advertisement and in aggressive interactions (Berry, 1974; Dugan, 1982). In *Cyclura carinata* and *C. stejnegeri,* on the other hand, headbob display rarely occurred in aggressive interactions (Iverson, 1979; Wiewandt, 1977). Male Mona iguanas and Caicos iguanas skipped low-intensity threat preliminaries (head displays), and moved quickly into the chase or lateral challenge and attack phases.

The difference in display variation and use between *Conolophus* and the cycluran iguanas, and the chuckawallas and green iguanas, may be related to different selective pressures on mainland and insular species (Wiewandt, 1977). Animals may be most vulnerable to predation while preoccupied with other activities, such as fighting. Where greater benefits are derived from immediate escalation, low intensity displays are more likely to diminish in importance in environments with light predation pressure, such as insular environments. Also, one would expect a lizard's displays to include discrete signal elements whenever a number of related species live sympatrically, thereby promoting species recognition. Coloration, ornamentation, and body form, and display sites in some cases, may also be cues for species identity (Rand, 1961; Ferguson, 1966; Rand and Williams, 1970; Webster and Burns, 1973; Williams and Rand, 1977). Where there is geographic isolation and no possibility for ambiguity in species recognition, one would expect greater display variability in the population.

Size Dimorphism: Size dimorphism varies from little (*C. stejnegeri,* Wiewandt, 1977; *Brachylophus,* Cogger, 1974) to extreme in *C. carinata,* and *Ctenosaura similis* (Iverson, 1979; Fitch and Henderson, 1977). Because of the complexity of factors affecting this character, the degree of dimorphism seen in a population may be very difficult to interpret (see Wiewandt, this volume).

The manner in which mates are acquired may affect differences in size between the sexes. For example, if larger males in a population have greater reproductive success, either because large size is of benefit in intrasexual competition for mates, or because females mate preferentially with larger males, then one would expect intense selection for large male body size. The degree of polygyny should influence the intensity of sexual selection, and therefore the degree of size dimorphism (e.g., Van Devender, 1978). In other words, the greater the variance in reproductive success among mature males, the stronger the selection for large male body size (see section on Iteroparity).

Accessory Structures: Males of many species have enlarged heads, long dorsal spines, and various facial adornments. Many of these characteristics appear to come into play during challenge and combat. Use of the head in contests of strength, either in combat or threat, appears to be common among iguanines. Horns, spines, or cushioning bulges on an iguana's head, neck and front legs may enhance display and serve as defensive weapons to disengage an opponent's hold and afford protection during combat. The enlarged jowl musculature in males of some species appears to provide an advantage in this context (e.g., *C. stejnegeri,*Wiewandt, 1977, 1979). Long nuchal spines and dorsal crests certainly increase the animal's apparent size and ferocity during a lateral challenge display. Green iguanas are dimorphic in dewlap size (Müller, 1968). Display plays an important role in this species, and the long, pendulous dewlap increases the conspicuousness of these long distance signals (Dugan, 1982).

Sexual Mimicry: In most, if not all, iguanines young males are for the most part identical to females in behavior and external appearance (e.g., *C. carinata,* Iverson, 1979; *C. stejnegeri,* Wiewandt, 1977; *I. iguana,* Dugan, this volume). Delaying the onset of dimorphism may be a means for young males to conceal their sexual identity. Young *C. carinata* males were observed to wander through territories of large males unmolested (Iverson, 1977). Similarity to females might allow a young male to remain in or near breeding adults without being challenged, monitor the activities of the group, and breed opportunistically. Opportunism of this sort occurs in at least one iguanine, *I. iguana* (Dugan, this volume).

MATE ACQUISITION BY FEMALES

Female iguanas invest large amounts of energy in egg-production, and do not increase reproductive output by acquiring multiple mates. Therefore, mate quality rather than quantity must be the dominant consideration in female mate acquisition. The female's criteria in selecting a mate will largely depend on the male's ability to monopolize resources used by females.

Resource Control

A male may demonstrate his desirability by controlling an area that is habitually used by females. Thus, when females show strong site attachment, they mate with the male that controls the area. Some large male *C. stejnegeri* shared burrows year-round with females. When one of these territorial males was removed, the resident female remained at the site and mated with his successor (Wiewandt, 1977).

When control of female activity areas is not possible, the male's ability to provide a place to court and mate may be of primary importance. This seems to be the case in *A. cristatus* (Carpenter, 1966; Boersma, in press; Trillmich, 1979) and *I. iguana* (Dugan, this volume). Some female green iguanas visited several male territories before remaining in one, usually for the duration of the breeding season.

Data on the behavior of individual females are largely lacking; thus, the female's role in pair formation is not well understood. Certain males of all iguanines control territories during the breeding season, and therefore control resources used by females, be they shelter and food, shelter only, and/or simply a place to mate. However, rarely does a male's defense give him exclusive rights to a specific female(s), as would be the case if female activity areas were entirely nested within the territories. In most species studied to date, females encounter several males, either because (1) female home ranges overlap those of several males, (2) territory ownership changes, or (3) a territory may be invaded opportunistically during the temporary absence of the resident male. Territorial male chuckawallas controlled retreats that were used by females, but Berry (1974) noted that females moved through the territories of several males before copulating with one.

Moderately high male-female encounter rates prior to mating suggest that factors in addition to the male's ability to monopolize resources used by females influence female mate choice. Among the factors females may use in selecting partners are the quality and/or quantity of a male's courtship (see "The Role of Courtship") and the quality of the defended resource.

Territory quality may be determined by the availability of suitable retreats, proximity to preferred foods and water, the amount of interference from competing males, and the number of other female residents. Subtle differences in these factors may be difficult to detect and measure, and evaluating their influence on female mate choice would require closely monitoring the behavior of larger numbers of females throughout the breeding period. This approach has been successfully employed in some studies of birds (e.g., Shepard, 1975), mammals (e.g., Downhower and Armitage, 1971), and amphibians (e.g., Wells, 1977a; Howard, 1978).

The Role of Courtship

Prior to copulation, individuals must locate members of the opposite sex, identify those individuals as suitable sexual partners, and synchronize the motor patterns involved in copulation. Schein and Hale (1965) classified the three types of stimuli involved in sexual behavior as (1) broadcast stimuli, (2) identification stimuli and (3) synchronizing stimuli.

One might expect the relative emphasis placed upon the first two categories to reflect different ecological demands among populations of iguanines. Thus, large arboreal green iguanas spent long periods of time 'broadcasting' (Dugan, this volume). Broadcast stimuli played a minor role in ground-dwelling populations of *Sauromalus* and *C. stejnegeri* (Berry, 1974; Wiewandt, 1977).

Identification stimuli reveal not only the individual's species, sex, and reproductive condition, but also may provide additional information on which a female can base her selection of a mate. We predict that as the environmental potential for accruing females through resource control diminishes, the role played by female choice based on male courtship performance should increase. Although meaningful interspecific comparisons are not possible at this time, it is known that both the quality and quantity of male courtship are important to females. Differences in courtship "finesse" were seen in *Cyclura stejnegeri* and *Sauromalus obesus* (Wiewandt, 1977; Berry, 1974). Observations of Mona iguanas indicate that such differences in style may be critical to a male's mating success. Moreover, extended courtship accompanied the year-round territoriality of some males, the more successful ones, and may have functioned to establish and/or maintain a bond with females walking through or residing in the territory (Wiewandt, 1977). In both *Iguana iguana* and *Conolophus subcristatus* courtship behavior was seasonal but began well ahead of copulation (Dugan, this volume; Werner, this volume). In *Conolophus*, such activity generally increased as mating time neared, and male breeding success was positively correlated with courtship frequency. In the only iguanid lizard whose social behavior and reproductive endocrinology have been investigated simultaneously, *Anolis carolinensis*, female receptivity was facilitated by the amount of male courtship received (Crews, 1974).

Multiple Copulation

Multiple copulations have been observed in the field among *C. stejnegeri,* (Wiewandt, 1977, 1979), *Conolophus subcristatus* (Werner, this volume), and *Iguana iguana* (Dugan, this volume), and in captivity among *Brachylophus* and *I. iguana* (Arnette, 1978; Peracca, 1891). This is in striking contrast to *Anolis carolinensis,* in which mating induces prompt termination of female receptivity (Crews, 1973). Receptivity extending beyond the first copulation or two per reproductive cycle is rare among vertebrates (see Trivers, 1972) and may be interpreted in three ways. (1) The cost to the female of rejecting males exceeds its advantage (Parker, 1974). This situation may apply to female green iguanas that are "overcome" by small males (Dugan, this volume), but does not account for other observations. *I. iguana* females readily mate several times with territorial males. Female Mona iguanas were observed to move from territory to territory and mate with as many as three different males (Wiewandt, 1977). (2) It may be to a female's advantage to increase the genetic variability of her offspring. It is often assumed that a female should limit herself to the services of a single male (Trivers, 1972; Williams, 1975). However, if a female can select from among several of a group of high status males, it is not clear why she should not diversify her reproductive investment in this manner. In *A. cristatus*, a species that has seemingly little to gain from a strategy of multiple copulation by virtue of its tiny clutch size (2-3 eggs), females mate only once per reproductive season (Trillmich, 1979). (3) The female may be attempting to insure adequate fertilization. Insuring fertility may be particularly important if males mate many times during a short breeding season. One male Mona iguana mated 11 times in a 13 day period, and each copulation decreased in duration as the season progressed (Wiewandt, 1977). In some birds and insects the concentration of spermatazoa drops during successive ejaculations, and copulation time may be shortened as well (see citations in Trivers, 1972). If frequent copulation does lower a male's fertility, one might expect females to compete for males. In *C. subcristatus,* female-female aggression is most intense during the breeding season (Werner, this volume). Receptive females were observed to drive other females from a male's territory. Males of this species are restricted to 2 copulations per day. Intolerance of other females may reflect an attempt by the aggressor to protect herself from an overspent mate.

Implications of Multiple Copulation

Sperm Competition: Multiple copulation before fertilization and the ability to store sperm are preadaptations for high levels of sperm competition (Parker, 1974). Definitive data on the occurrence of sperm competition among lizards are lacking, but evidence does suggest that it is a distinct possibility.

Anatomical evidence on the presence of seminal receptacles in a number of lizard families indicate that sperm storage may be widespread in this group (Cuellar, 1966; Fox, 1963). Delayed fertilization has been reported in *Uta stansburiana* (Cuellar, 1966). Good genetic evidence that sperm competition does occur in at least some reptiles was found in the garter snake *Thamnophis sirtalis* by Gibson and Falls (1975).

Postcopulatory Guarding: Parker (1974) noted two theoretical considerations that are necessary for the evolution of postcopulatory guarding: (1) an overlap period during which sperm from ejaculates of different males can compete for a female's ova and (2) a high male/female encounter rate relative to the duration of the overlap period. Research on insects suggests two possible mechanisms by which sperm competition may operate (Parker, 1970). (1) Success at fertilization is a function of the relative number of sperm from each male present within the female. The evolutionary result of such a mechanism involves selection for frequent copulations and high levels of sperm production in males. (2) The last male to mate fertilizes most of the eggs. Here, the timing of copulation, rather than the number of sperm produced, is more important to male reproductive success. In either case, males should prevent other males from having access to females with whom they have mated. Behavior interpreted as postcopulatory guarding has been observed in *C. stejnegeri* and *I. iguana* in the field, and *C. carinata* in captivity (Wiewandt, 1977; Dugan, this volume; Iverson, 1979).

Whether a female "chooses" to or is forced to mate with several males, multiple insemination may provide the ultimate proving ground for males, via sperm competition and/or postcopulatory guarding.

SUMMARY AND CONCLUSIONS

A mating strategy is the behavior employed by the individual in mate acquisition. The available data indicate that within the framework of a mating system (usually polygyny), the manner in which individuals acquire mates is highly variable, both between species and within populations of iguanine lizards. This variability can be attributed to: (1) life history characteristics–diet, size, longevity, and breeding schedules; and (2) the ecological and social environment of the individual.

The form of competition among males for females reflects the distribution and abundance of resources used by females relative to the size and status of competing males, and ultimately determines male reproductive success. The female's criteria in choosing a mate primarily depend on the male's ability to monopolize resources used by females. Multiple copulations in at least 3 species *(Cyclura stejnegeri, Iguana iguana,* and *Conolophus subcristatus)* may function to insure fertility, but may also be viewed as an extension of female choice.

The variety of ecological situations in which iguanines occur accounts for much of the variability observed in social structures, mating systems, and mating strategies. However, most of the data reported to date are fragmentary and do not permit adequate assessment of factors underlying the reproductive behavior of a population.

Measurement of variables that determine the operational sex ratio are particularly important, and largely lacking. Components of the OSR include: (1) the proportion of males in the population that are reproductively active, (2) the onset and duration of receptive periods of individual females, and (3) the amount of time required to service each female, which includes not only copulation duration, but also the duration of all courtship preceding each copulation. Determination of breeding synchrony requires investigation concentrating on restricted geographic areas. Finally, the proportion of males that actually have access to and mate with females is critical to an understanding of mating systems, but has rarely been reported.

Additional problems that require special attention include the following: patterns of resource utilization by all sex and size classes during the non-breeding season; differential growth rates; competitive strategies of individual males during the breeding season; the relative importance of display in advertisement, territorial defense, and courtship; the relative importance of female choice and the criteria used by females to select mates; the form and function of pre- and postcopulatory interactions between males and females; and the relationship between changes in reproductive physiology, external morphology, and behavior. Care should be exercised when interpreting behavioral observations made in captivity; Berry (1974) noted "little if any" similarity in courtship between pen-mated chuckawallas and that of field animals.

Selecting terminology to describe iguanine mating systems is also problematical and requires special caution. The use of traditional classification, i.e., monogamy, polygamy (including polygyny and polyandry), and promiscuity, masks the complexities in such systems and obscures rather than suggests explanations for diverse breeding habits. This shortcoming prompted Emlen and Oring (1977) and Bradbury and Vehrencamp (1977) to develop new ecological and evolutionary schemes for classifying avian and mammalian mating systems, emphasizing the manner of and environmental potential for mate monopolization. Emlen and Oring suggest that this classification might prove useful in treating other vertebrate groups.

Most iguanine mating systems appear to fall broadly under the category of "resource defense polygyny," defined as a system where males acquire multiple mates indirectly, by defending resources essential to females (see "Competition for Retreats"). However, applying such a term excludes consideration of female choice independent of resources defended by males. It is apparent from patterns of space usage by males and females, and from female movements prior to copulation (see "Habitat Resources") that females are not simply passive occupants of a male's territory. The importance of competition among members of the limited sex (i.e., normally males in polygynous systems) in understanding mating systems has long been recognized. The greater caution in mate selection attributed to the limiting sex is not only a correlate of, but may also in part determine (via feedback through sexual selection) various facets of the mating system (Trivers, 1972). At present, no single term adequately reflects the diversity of mating strategies being uncovered within iguana mating systems.

We feel that because of their uniqueness, because they provide a number of opportunities for comparative studies, because many are insular reptiles established in discrete, low-diversity ecosystems amenable to comprehensive study, and because the status of many populations is in jeopardy, the iguanines deserve more detailed investigation. Particularly in the area of behavioral ecology, long-term field studies of marked individuals and plenty of patience are indispensable.

Notes

[1]Current address: Rt. 15, Box 367, Gray, TN 37615.
[2]Current address: 3436 N.E. 2nd Ave., Ft. Lauderdale, FL 33334.
[3]*C. stejnegeri* = *C. cornuta stejnegeri* of Schwartz and Carey, 1977. See Note 2 in Wiewandt, this volume.

Acknowledgments

The authors wish to thank Gordon Burghardt, Hugh Drummond, A. Stanley Rand, Mike Ryan, and Kentwood Wells for reading an earlier draft of the manuscript and offering many helpful comments.

18

The Mating Behavior of the Green Iguana, *Iguana iguana*

Beverly Dugan[1]
Department of Psychology
University of Tennessee
Knoxville, Tennessee

INTRODUCTION

The main purpose of the investigation reported here is to provide a broad based description of the mechanisms and adaptive strategies of mate acquisition in the common green iguana, *Iguana iguana*. A complete understanding of the various ways in which reproduction is achieved by the individuals of a population requires information on the organism's ecology and social structure during the non-breeding season. Thus, these topics are also addressed.

Various aspects of the reproductive biology of the green iguana, *I. iguana*, have been rather thoroughly described. In Panama, the site of my study, egg laying occurs in February and March. Female iguanas lay a single clutch of between 2 and 6 dozen eggs annually, depending on the female's size (see Rand, 1968; Rand and Rand, 1976, for detailed accounts of communal nesting). Females may migrate as much as 3 km to find suitable nest sites (Montgomery et al., 1973). Egg laying is timed so that eggs incubate under optimal conditions and hatchlings emerge at a time when food is abundant (Rand, 1972). The young hatch in early May at the start of the rainy season, and rapidly disperse from the nest site (Burghardt et al., 1977; Drummond and Burghardt, this volume). In contrast to the information available on egg laying, incubation, and hatching, detailed observations of green iguana mating behavior are lacking. Peracca (1891) first described mating in captive iguanas. Müller (1972) presented a general account of green iguana social and mating behavior in NE Colombia.

METHODS

The Study Area

Flamenco island is a 13.5 ha island (maximum elevation 75 m) at the Pacific entrance to the Panama Canal (Figure 18.1). Flamenco is 3 km offshore; the Fort Amador Causeway connects the island to the mainland. The surface area consists of either rocky cliffs or a thin layer of rocky topsoil, depending upon the slope of the particular location. Scrubby second growth vegetation covers the island. The dominant trees include *Cecropia peltata, Spondias mombin, Sapium* sp., and *Bursera* sp. Vines belonging mainly to the families Convolvulaceae and Sapindaceae are abundant. Trees occur in clumps along the grassy slopes.

Flamenco is uninhabited; the Panama Canal signal station is the only active installation. Since access to the island had been restricted for several decades, hunting pressure had been much lighter than in other parts of Panama, and the iguanas readily habituated to an observer. Observations were confined to a 5.5 ha area on the northwest side of the island. Flamenco's slope on this side allowed an observer to watch the iguanas at or above the level at which they occur in trees.

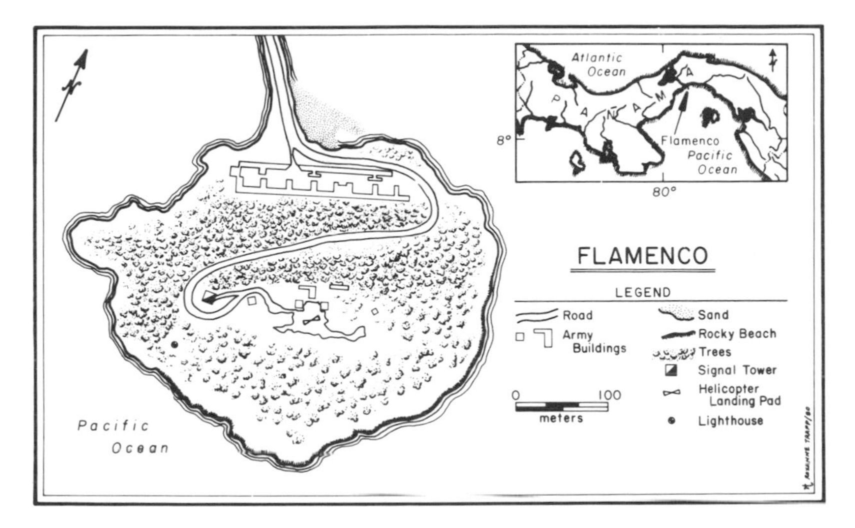

Figure 18.1: Map of Flamenco Island. Inset shows Flamenco's location with respect to the mainland.

Data Collection

Iguanas were captured with the aid of Panamanian field assistants and their hunting dogs. One hundred and fifty-one animals were captured between July and December, 1978. Each iguana was weighed to the nearest gram, measured (snout to vent = SV, and total length), marked, and released at or near the point of capture. All lengths are reported as SV length in cm. Animals were marked by toe-clipping, freeze-branding or painting, and with colored plastic tags attached with monofilament to both sides of the neck and tail base. Capturing and marking seemed to affect neither the iguanas' movements nor their behavior.

Over 1,000 hours of field observations were made from October, 1977 through January, 1979. Focal animal samples (Altmann, 1974) and Super-8 mm film provided detailed behavioral data. Information on individual movements and home range size was obtained from censuses of the study area, capture-recapture records, and radio-telemetry.

The locations of marked iguanas were plotted on an aerial photograph of the island. Parts of trees or groups of trees that the animals were known to have used were circled, and the outer perimeters of the circles were connected. No attempt was made to estimate home range volume. Home range size was computed by two methods: (1) by weighing cut-outs of each home range, and (2) by counting the number of squares on a piece of graph paper each home range enclosed. Measurements were then compared to the appropriate 1 ha scale. The two methods gave results witin 3 m^2 in all but two cases. The two measures were averaged for the final estimate.

Between August, 1978 and January, 1979, the gonads from 52 animals (25 males, 27 females) were examined to assess reproductive condition. Outside of the mating season, animals were obtained from a variety of sources within 50 km of Flamenco. During the mating season (Oct.-Jan.) only iguanas (19 females, 15 males) from Perico Island, approximately 0.5 km from Flamenco and similar in topography and vegetation, were used.

RESULTS

Size and Sexual Dimorphism

Males grow to a larger size than females (Figure 18.2). The largest male captured (45 cm, 3.78 kg) was 11 cm longer and 1.91 kg heavier than the largest female (34 cm, 1.87 kg). The mean female weight was 74% of the mean male weight. Males averaged 1.53 kg (s.d. = 1.04); females averaged 1.14 kg (s.d. = 0.34).

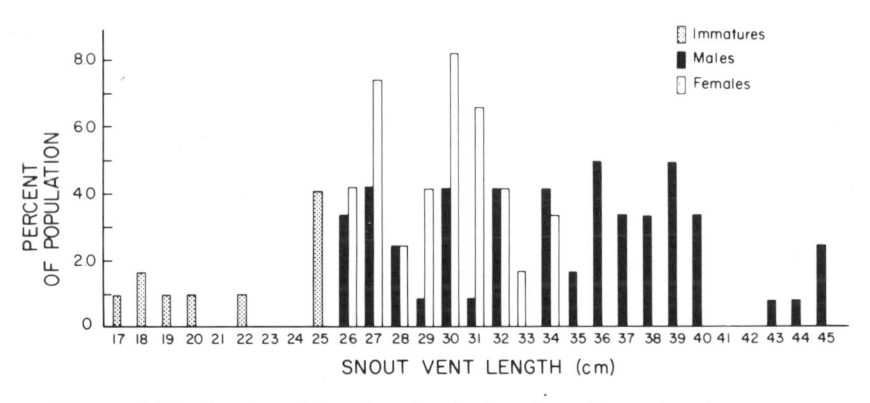

Figure 18.2: Snout-vent lengths of male, female and immature iguanas, expressed as a proportion of total iguanas captured (N = 151).

In addition, males have larger heads, longer spines on the dorsal crest, and larger femoral pores. Males and females also differ in color. Females, and medium and small males, vary from bright to dull, dark green. The largest males are dull gray, gold or tan. The male's color intensifies during the breeding season, at which time most large males vary from bright gold to red-orange. Breeding coloration is at least partly related to social factors. One territorial male changed from bright orange to dull brown within a few hours after having been deposed by a larger male. Two days later he regained both his territory and his bright coloration.

Secondary sexual characteristics develop slowly in mature males. Enlargement of the femoral pores coincides with the onset of reproductive maturity (see below). The distribution of immatures in Figure 18.2 indicates the stage at which femoral pore enlargement occurs. At 31 or 32 cm (in the 4th or 5th year), other secondary sex characteristics, such as size of the head and dorsal crest, begin to be evident. Rates of color change are highly variable, but all males greater than 36 cm are predominantly gold or tan.

Characteristics of the Population

Since large males were more likely to be sighted than other members of the population, census data were not used to determine population composition. Records from captured iguanas were considered to be a reliable estimate (Figure 18.2). I assume that individuals in a particular sex/size class were no more likely to be captured than members of any other (see Dugan, 1980). Few juvenile iguanas were seen on Flamenco, and no attempt was made to locate them. Thus, the data that follow do not include juveniles.

Flamenco supports between 36 and 50 adult iguanas per ha (45-70 kg/ha) (Dugan, 1980). Of 143 mature iguanas that were captured on Flamenco, 65 were female, 78 were male. The sex ratio did not differ significantly from 1:1 ($X^2 = 1.2$, df = 1, $p > 0.20$).

Three classes of mature males, each differing in the development of secondary sex characteristics, were recognized: small, medium, and large. Femoral pore enlargement is the only character that distinguishes small males (26-31 cm, 35% of total males) from females. Medium males (32-36 cm, 27% of total males) exhibit all secondary sex characteristics except body color, which is in various stages of change among males in this group. All secondary sex characteristics are well-developed in large males (37-45 cm, 38% of total males). Since the range in female body size was restricted (26-34 cm, Figure 18.2), and since size-related physical and behavioral characteristics were not apparent, this sex class was not subdivided into size classes.

Reproductive Maturity

Animals smaller than 26 cm were considered to be immature. The smallest males with enlarged testes and the smallest females with yolking follicles measured 27 cm. Although few animals below this size were collected, three small females (20, 21 and 24 cm) collected during the breeding season had undeveloped ovaries. These data, plus juvenile growth records from Barro Colorado Island, Panama, indicate that iguanas in Panama reach reproductive maturity in the third year (at approximately 30 months), between 25 and 27 cm. Although probably rare, recent evidence indicates that a few female iguanas in Panama may be reproductively mature in their second year (Rand, personal communication). Of 60 gravid females captured on Slothia (a small island near Barro Colorado Island in Gatun Lake, Panama) in February 1980, one of these had been marked after hatching in May 1978.

Social Structure During the Non-Breeding Season

Movements and Home Range Size: Except for occasional descents to the ground for basking, for tree-to-tree movement, for escape from disturbance, and for females for traveling to nest sites, iguanas remained in the tree tops. An iguana might stay in the same general location from one day to several weeks. A single tree could meet all of an iguana's short-term requirements–food, thermoregulatory sites, and sleeping perches.

Long-range movements were prompted, at least in part, by local changes in food availability. *Spondias mombin* fruits during August and September. Some trees bearing this fruit contained 10-15 animals for the duration of the fruiting period, after which the iguanas moved to other locations. Large congregations of iguanas also were observed late in the dry season (March and April) when many of the trees were leafless. The most frequently eaten food item at this time was an abundant vine *(Cardiospermum grandifolium)*. Why the iguanas were clumped in a small proportion of vine-bearing trees is unclear. Perhaps exploitation of a known resource is more efficient than searching for new food items (see Auffenberg, this volume), particularly at a time of potential food shortage. Social facilitation of feeding as described by Greenberg (1976) may occur; groups of iguanas frequently fed simultaneously. These large groups dispersed early in the rainy season when a variety of new leaves became available.

The movements of all sex and size classes overlapped in both space and time (Figure 18.3). The home range sizes of 8 females, 9 small and medium males, and 10 large males (Table 18.1), were compared using a one-way analysis of variance. A significant difference in home range size was found among the 3 groups ($F = 4.69$, $df = 2,24$, $p<0.02$). The Student-Newman-Keuls test revealed that the home ranges of large males were significantly ($p< 0.05$) smaller than those of females and other males. This result is opposite the relationship usually observed between home range size and body size in lizards (Turner et al., 1969). This finding does not reflect sample size bias, since more sightings over longer periods of time were obtained for large males than for other members of the population. Although the home ranges of large males were smaller than those of other iguanas, each large male's home range contained at least two seasonally preferred food resources, plus a year-round supply of food.

Social Interactions: Iguanas were found singly and in small groups throughout the year. However, I concentrated on larger groups in order to maximize the amount of information obtained from observations.

Large congregations usually consisted of 1 large male, 1-3 medium males, 1-3 small males (when recognized) and 4-6 females. It was not unusual to observe two large males in the same tree at the same time, although they never remained in the same location for more than a day. Only two agonistic encounters between large males were observed outside of mating season. In one of these, the males alternated signature bobs (Dugan, 1982; also see display-action-pattern graphs presented in Carpenter, this volume and "stereotyped nodding" in Distel, this volume) accompanied by lateral compression; in the other a face-to-face pushing battle also occurred.

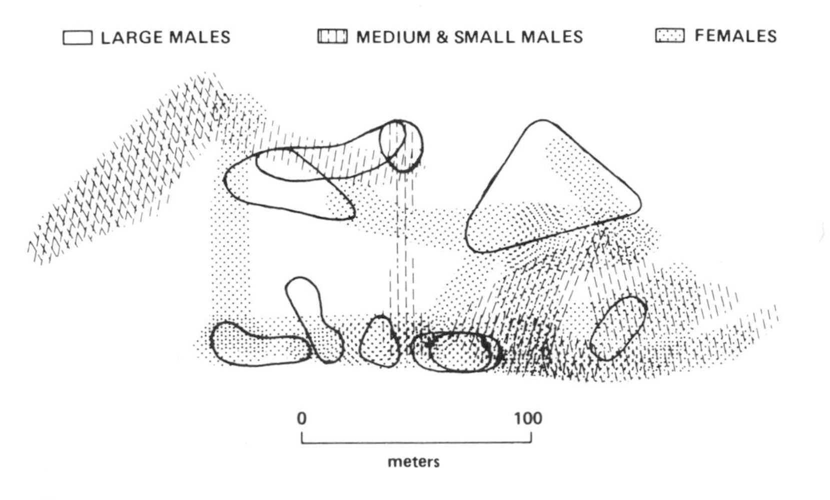

Figure 18.3: Home ranges of 10 large males, 9 medium and small males, and 8 females. This sample is only a small portion of the iguanas in the study area, and thus under-represents home range overlap.

Table 18.1: Mean Home Range Size of Iguanas for Which at Least Three Recordings per Animal were Made Over a Minimum of Two Months

Group	Home Range Size (ha) Mean	Home Range Size (ha) S.D.	Number of Sightings Mean	Number of Sightings Range
Large males (N = 10)	0.080	0.073	41	21–71
Medium and small males (N = 9)	0.220	0.113	9	3–28
Females (N = 8)	0.245	0.181	9	3–22

Interactions between large and medium males were rarely observed, but the differences in their behavior were striking. Medium males avoided large males at all times. Large males displayed more than medium males throughout the year (Figure 18.4). Only large males had access to conspicuous perches that were used as display posts. Conspicuous perches were also preferred as basking sites by females, and large males and females basked in close proximity. At all times of the day during the dry season, large males were more likely to be found close to one or more females than were medium males. Distance to the nearest female was recorded in 23 scan samples (March 24-April 6) during the dry season when visibility was best (Figure 18.5). Distance estimates were classified into one of 4 categories: 0-1.5 m, 1.5-3 m, > 3 m, solo. Large males were less likely to be solitary and more likely to be near one or more females than were medium males ($X^2 = 30$, df = 3, $p < 0.001$). Although heavy foliage sometimes interfered with observations during the wet season, my impression is that this finding would generalize to the rainy season.

Encounters among smaller iguanas were brief (consisting of lunges and chases), and usually occurred during changes in location (moving from sleeping perches to basking sites, basking to feeding locations, to sleeping perches, etc.). These encounters may have functioned to establish dominance hierarchies in transient groups. Some mechanism for regulating interactions in such high density situations would be advantageous. However, the dense foliage and large number of unmarked animals prevented the determination of relationships between individuals.

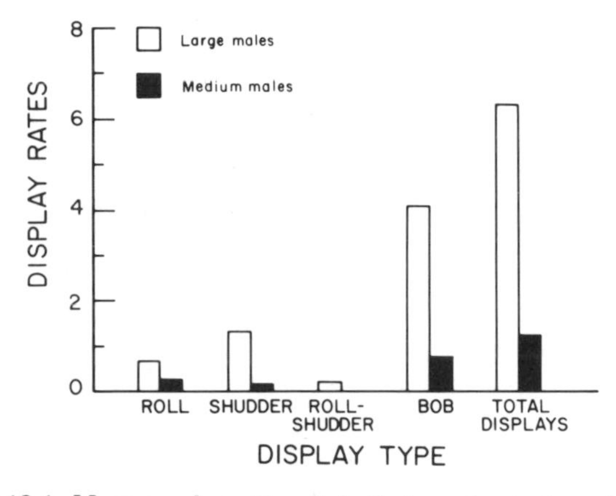

Figure 18.4: Mean number of headbob displays observed per 15 minute observation interval for large and medium males (large males, 543 intervals; medium males, 149 intervals). Bob = Signature bob. Observations were made during both the breeding and non-breeding seasons.

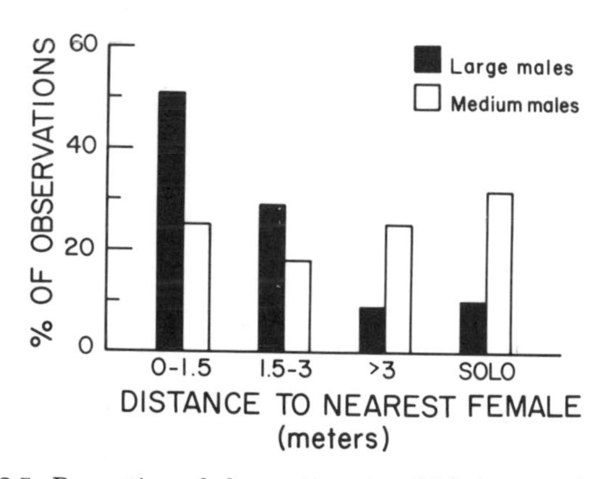

Figure 18.5: Proportion of observations in which large and medium males were recorded at varying distances from a female (> 3 = female in same tree but further than 3 m from male). Based on 23 scan samples.

Mating Season Activity

Reproductive Phenology: The major physiological and behavioral changes between mid-October and late January are presented in Figure 18.6.

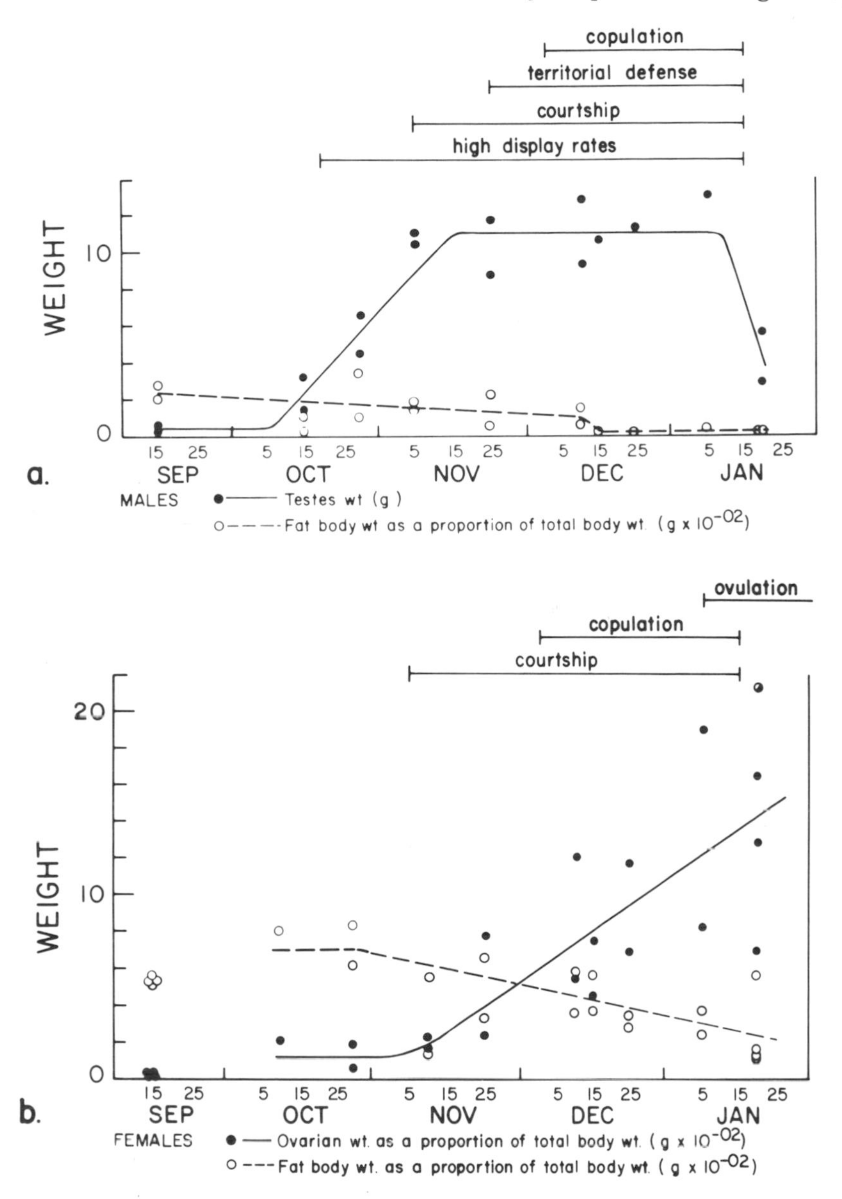

Figure 18.6: Changes in gonad and fat body weights from September through January, 1978-1979. Curves were fitted by eye. Solid bars above each graph denote the period during which the specified behavior occurred. ◑ = female with oviducal eggs.

Testes growth begins in the last two weeks of October and increases at a steady rate until early November, at which time a plateau is reached. Follicular growth and yolk deposition also begin in mid-October, but growth is slow relative to that of the testes. The result is that males reach reproductive readiness before females.

In polygynous mating systems, one would expect males to be more active than females in securing mates (Trivers, 1972). The depletion of fat stores by males in mid-December reflects the cost of this activity (Figure 18.6). Females' fat reserves decrease as ovarian weight increases, but fat reserves are not depleted until shortly before egg laying in February or March, reflecting the female's greater investment in gamete production (see Derickson, 1976, for a discussion of lipid utilization and reproductive activity in reptiles).

After the first week in November, variance in female ovarian weight is greater than variance in male testes weight. Part of the variation among females is related to size (Fitch and Henderson, 1977). Part of the scatter however, is attributable to a 2-3 week variation in reproductive development among females. Stigmata appear in the ovaries immediately prior to ovulation, and were first noticed in one of two females examined on 4 January. Stigmata were present in 3 of 4 females examined on 20 January; the fourth female contained oviducal eggs. Ovulation begins in early January after ovarian follicles exceed 20 mm in diameter, and continues through January. In 1978-1979 ovulation began 5 weeks after the first observed copulation and 3-4 weeks before the onset of nesting.

Ovulation and a sharp drop off in testes size in late January coincide with the breakdown of territories. Termination of female receptivity and reduced testosterone levels in males presumably underlie the change in social structure. However, the cues that elicit this rather rapid and drastic change are not obvious.

The general pattern of physiological and behavioral changes does not reflect the diversity of mating strategies seen in *I. iguana*. Although all males examined had enlarged testes regardless of size, not all males established territories. Male iguanas in each size class adopted different strategies to obtain mates. Due to the complexities of such a system, I will first discuss territory establishment by large males, and then interactions between females and territorial males. The behavior of medium and small males will be discussed separately.

Territory Establishment: An increase in display rates by large males, rather than an increase in agonistic encounters, marked the onset of breeding activity. Only two agonistic encounters between large males were observed at this time. Large males tended to be well-spaced and conspicuous. Display rates increased in mid-October, and were maintained at a high level till late January. Significant differences in display rates between months can be determined by inspection of Figure 18.7. The vertical lines represent $\pm$ 1 s.d. x $\sqrt{2/N}$, where N equals the number of observation intervals. Non-overlapping lines indicate a significant difference ($p = 0.05$) between any two pairs (see Burghardt, 1969). Significant increases in display rates were observed between October and November and other months during the non-breeding season, and between October and

November, and December and January, when mating activity was most intense.

Most large males had established territories by late November. Medium males were no longer observed in the same tree with large males. Territorial males alternated periods of rest with display bouts, courtship bouts, and territorial patrols. Patrolling consisted of a series of short perch changes through the tree, each perch change being followed by a signature bob. After late November, large males were rarely observed to feed. By late December, many of them appeared to be emaciated.

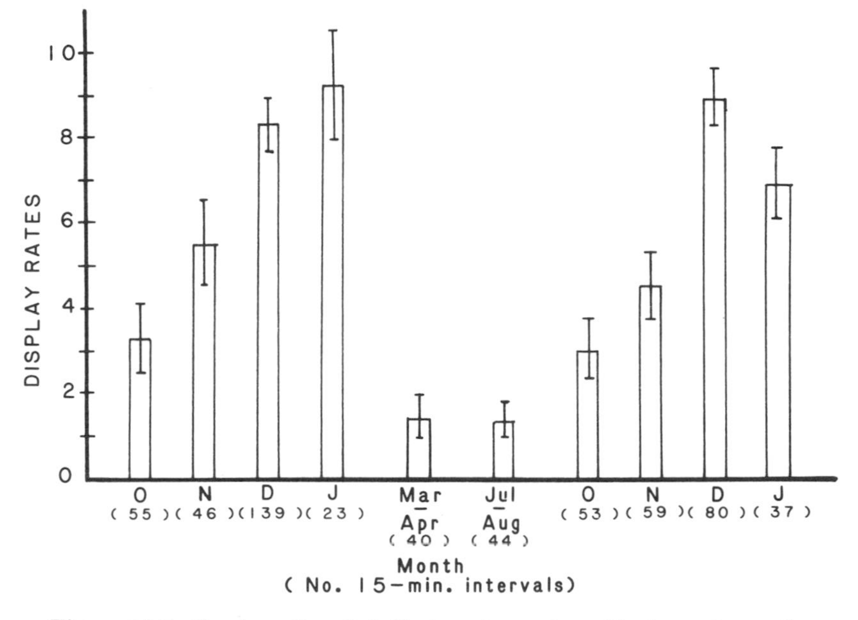

Figure 18.7: Number of headbob displays observed per 15 minute observation interval: Large males only, 1977-1979. Vertical lines represent ± 1 s.d. x $\sqrt{2/N}$, where N = the number of intervals. See text for explanation.

Observations in 1977 suggested that large conspicuous trees were preferred mating territories. In order to assess territory characteristics, the following measurements were made on all trees (N = 76) of adequate size to be used by adult iguanas in a 0.7 ha area: diameter at breast height (DBH), estimated height, estimated canopy volume, estimated percent canopy interlock with other trees, and estimated percent vine cover. Significant differences between trees that were used as territories in at least one of two seasons (N = 9) and those that were not were found for only two of the variables, DBH ($t = 6.55$, df $= 74$, $p < 0.001$) and canopy volume ($T = 4.04$, df $= 74$, $p < 0.001$). Due to different growth patterns, tall trees may have small canopies, and trees with large canopies may be relatively short. Territories tended to be taller ($\overline{X} = 17.0$ m, s.d. $= 3.28$) than other trees ($\overline{X} = 10.95$ m, s.d. $= 5.26$) but the difference was not significant. Tree species was also recorded. There was no apparent tendency for certain tree species to be preferred over others.

In other parts of the study area that contained fewer trees, the canopies of mating territories interlocked little or not at all with neighboring trees. Mating territories, then, provided conspicuous perches, a good vantage point from which to survey the surrounding area, and those trees with few access routes (little canopy interlock) were easily defended from intruding males. Food was not an essential component of a mating territory. Of 19 territories observed closely in both years, 6 did not contain food. The average of daily recordings of the number of females per male were compared for territories with and without food. The mean number of females per male in trees with food was 2.5, and 2.7 in trees without food. Males in the latter were no less successful in attracting females. Females in territories lacking food moved singly or in groups to nearby areas to feed in late morning. Males followed receptive females to feeding areas and stayed within a few meters of the female(s), displaying frequently. Territorial males were rarely observed to feed during these excursions.

Mate Acquisition: Beginning in early November, large males were regularly observed with one or more females. At this time, males moved to trees that contained females, and females visited displaying males. Pair formation occurred in one of (at least) three ways.

(1) Males moved into trees already containing females, and established territories there. A solitary female was joined by a male in late October, 1978. The two remained together for the duration of the breeding period and were observed copulating on 20 December. A second female resided in the territory from early December until the end of the breeding period.

(2) Males left their territories to court females in nearby trees. One female was courted by two different males for the last two weeks in November, 1978. At the end of November, she moved to the territory of one of these males, remained there throughout the breeding period, and copulated with him at least 3 times between 29 December and 11 January.

(3) Individual females visited several males before taking up residence in a territory. Two females that were observed closely, one in 1977, the other in 1978, visited and were courted by three different males. The female observed in 1977 was courted by each male at least twice. She moved into the territory of one of these on 13 December, and copulated at least twice with him between 3 January and 18 January. The second female copulated with the territorial male at least twice between 26 December and 11 January, 3 weeks after having settled into the territory.

These and other observations indicate that individual females were courted for at least 4 weeks by one or several males before becoming receptive (i.e., copulating), and females were receptive for at least 15 days, possibly longer.

Courtship and Copulation: In the most common courtship sequence, the male approached the female from behind, performing a shudder bob (rapid, low amplitude vertical head movements) before, during or after the approach. The female then moved the posterior two-thirds of her tail to one side (tail arch). After the tail arch, the male stopped and performed a signature bob. This sequence (recorded as one courtship bout) was repeated from one to more than a dozen times, with pauses between each approach. Females retreated before tail arching in long sequences of courtship bouts. Females did not tail arch when the courting pair were face to face. Rather, the female remained prone while the male repeatedly shuddered and bobbed before her. Preliminary analysis suggests that receptive females were more likely than non-receptive females to perform shudder bobs and signature bobs in response to the male's shudder approach (Dugan, 1980). Receptive females neither retreated nor tail arched as the male approached. The male moved up on the female from behind and secured a neck grip. At this point, the pair sometimes walked slowly together, possibly a maneuver to find a secure position on what often appeared to be precarious perches. The male then twisted his tail under that of the female, and inserted a hemipenes. Copulation lasted from 2.5 to 12 minutes ($\overline{X} = 7.4$, s.d. = 2.6, N = 26). Males were never observed to mate more than once per day. Individual females were observed copulating from one to five times each during the receptive period, usually with the same male.

As noted previously, a minimum of four weeks of courtship preceded copulation. At least the last two weeks of the courtship period were spent with the male with whom the female mated. After copulating 2 or 3 times, the amount of courtship immediately prior to copulation decreased dramatically. Interactions between one pair observed in 1977 illustrate this point (Figure 18.8). The proportion of 15 minute observation intervals prior to copulation in which courtship occurred decreased from 7 December, when the first mating was observed, until 12 December, the last day this pair was observed. The mean number of courtship bouts in those intervals that courtship did occur follows a similar trend.

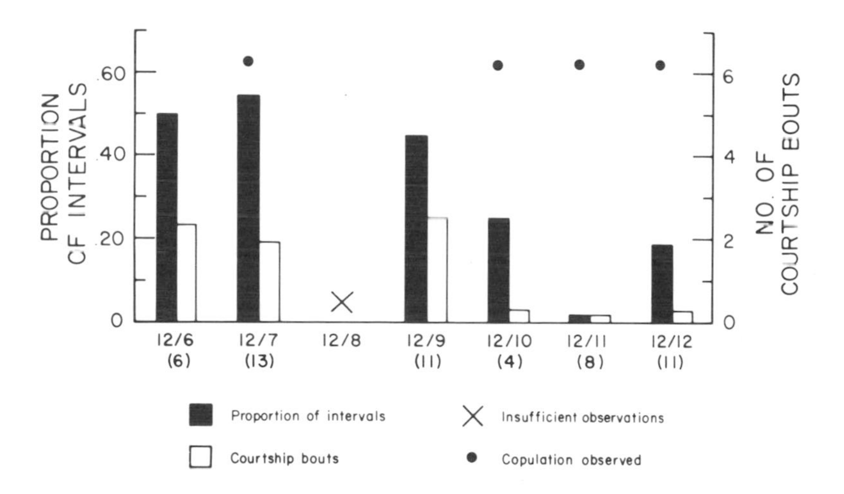

Figure 18.8: The mean number of courtship bouts and the proportion of 15 minute observation intervals in which courtship occurred. Dot above bars indicates that a copulation was observed on that date. Data from a male-female pair observed in 1977. Number of intervals are in parentheses below dates.

Each mating territory contained from one to four females. In addition to displays, territorial patrol and defense, courtship made a significant demand on the male's time. It is therefore of interest to know how a male apportioned his time among the resident females. Data from two territories in which more or less continuous observations were available, and that contained at least two females during observation periods, are presented in Figure 18.9. It can be seen that the males engaged in more courtship bouts with non-receptive than receptive females. Thus, it appears that the amount of time that the territorial male spent with each female depended in part on her reproductive state. Periods of receptivity among females in a single territory did overlap. In these cases, males courted each female daily, but mated with each on alternate days.

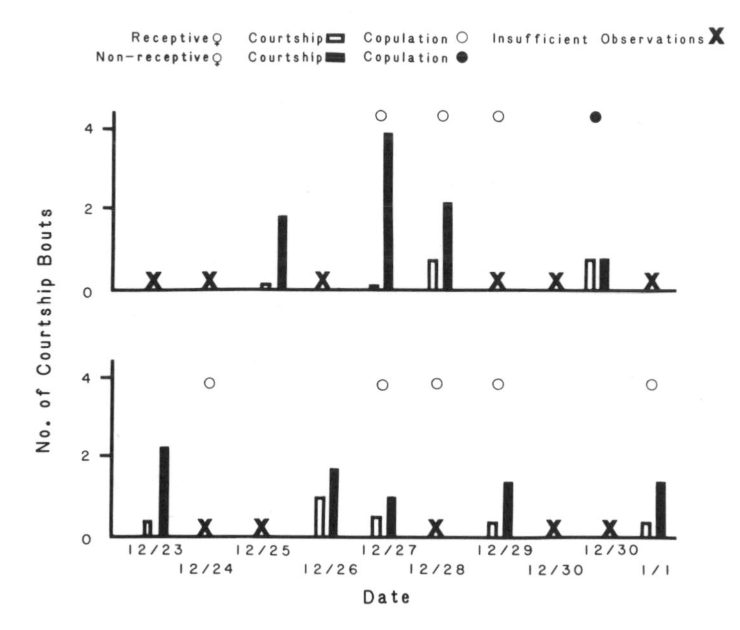

Figure 18.9: Mean number of courtship bouts per 15 minute observation interval observed between a territorial male and each of two females. The two graphs represent data from 2 territories observed in 1977. The top male is "Y-shoulder," bottom male is "Pig." Circle above bars indicates copulation on that date.

Although the pattern described above was typical, males varied in their courtship behavior. One male in particular (Macho) never persisted in courting any one female for very long. Rather than forming a "bond" with each female in succession, he alternated vigorous courtship bouts among his three females. Table 18.2 compares the number of matings per observation hour for the two males whose courtship patterns are presented in Figure 18.9 (Pig and Y-Shoulder) and Macho. Presumably, Macho inseminated each resident female before the mating season ended, but he probably did not copulate as many times with each resident female as did Pig and Y-Shoulder with females in their territories.

Table 18.2: Records of Number of Females in Territory and Number of Copulations from Three Territories Observed in 1977

Summary of Observations	Territorial Males Pig	Territorial Males Y-Shoulder	Territorial Males Macho
Hours observed*	20	9	11
Period	12/6-1/3	12/23-12/31	12/12-1/9
$\bar{X}$ No. females observed in tree	3	2	3
No. of females mated	4	2	1
No. of copulations per observation hour	0.70	0.44	0.09

*Since 24 of 25 observed copulations occurred between 1000 and 1500 hours, only observations made during those hours are included.

The Behavior of Medium Males: One 36 cm male in 1978 did show intensified color, displayed, and patrolled a tree that had been a territory in 1977. However, other iguanas were never observed in the tree with him. Another 36 cm male also exhibited territorial behavior, and was observed to copulate with the two females in his territory. All other medium males observed in both years remained on the periphery of a mating territory for periods of time varying from one week to the duration of the breeding period. They were rarely observed displaying, feeding, and were never observed patrolling. Medium males sometimes ventured close to a territory. The large male displayed, then chased the medium male if the latter did not retreat. Peripheral males did court females moving into and out of a territory. However, copulations never resulted from the observed courtship bouts.

In order to determine whether a medium male would take over a vacated territory, a large male (38 cm) was removed from his tree on 15 December, 1978. Within two days, the medium male (34 cm) that had been observed in the area for 2 weeks prior to the manipulation had moved into the territory, displayed from the post used by the previous owner, and patrolled the tree daily. However, the medium male was never observed with either of the two females that had accompanied the previous resident. After 2 weeks the large male was released in the area. He regained his territory within 24 hours, and the medium male was not seen again.

The Behavior of Small Males: Because of their "discreet" habits and similarity to females, small males could move into and out of territories without being challenged. Some small males remained in territories for brief periods of time (i.e., a few hours), others for several days at a time.

Small males entered territories at a distance from, but within sight of, a large male without eliciting a response from the large male. They then typically moved to a clump of leaves at the end of a branch and remained motionless. Small males did not, to my knowledge, enter a leafless tree while the territory owner was present. If a patrolling male moved to within 2 m of a small male, the small male immediately dropped from the tree. However, if a small male was caught unaware by the patrolling male he would approach to within one meter and immediately lunge and head-butt the interloper from the tree.

Small males attempted to copulate with females in the territory when the territorial male was occupied elsewhere. In such attempts, the small male ran toward the female, leapt on her back, and tried to secure a bite, presumably on the neck, usually wherever he landed. In all 15 attempts observed the female was uncooperative and typically thrashed violently, threatened with mouth open, turned toward and attempted to bite her suitor. Large males were nearby and rushed to the female's defense in 5 of 15 observed attacks. Of the remaining 10, small males copulated successfully three times.

Size, Status and Mating Success: A total of 69 attempted copulations (i.e., male secured bite on neck or dorsum) were observed in both years. I assume that these observations are an adequate sample of copulatory activity in the study area for the following reasons: (1) most observed copulation attempts occurred on conspicuous perches, (2) copulations that occurred in less noticeable locations were easily sighted because of the conspicuous nature of the copulatory posture, (3) the struggle that accompanied copulation attempts by small males was quite conspicuous, (4) most copulations exceeded 5 minutes in duration (see "Courtship and Copulation"), and (5) activity in at least 2 territories could be monitored from any observation post in the study area.

Small males comprise 35% of the male population (Figure 18.10) and performed 6% of observed copulations. Medium males, 27% of total males, performed 4% of the observed copulations, and large males, 38% of total males, performed 90% of the observed copulations. The distribution of observed unsuccessful and successful copulation attempts among males in the three size classes differed significantly from the expected distribution (Table 18.3). This variance in male mating success was due in part to interference by territorial males, the reaction of females to small males, and differential access to females. Small males were significantly more likely to attempt copulation than were medium males (Kolmogorov-Smirnov test, $p < 0.02$). It appears that this difference in copulatory encounters is due to a higher frequency of unsuccessful encounters by small males, rather than a difference in frequency of successful encounters (Table 18.3). It should be noted, however, that the medium male that held a territory in 1978 performed both copulations observed for medium males.

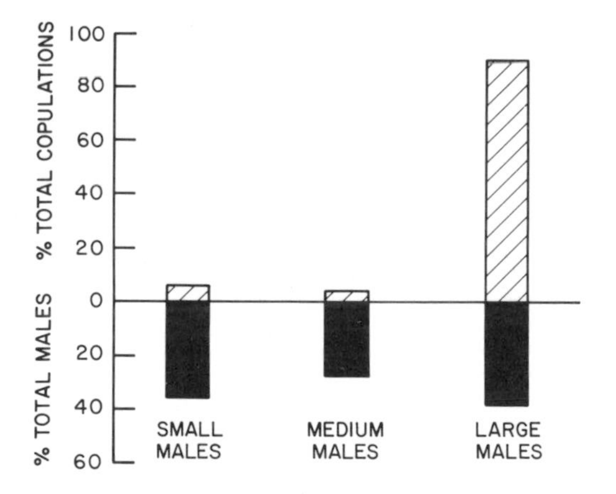

Figure 18.10: Top: percentage of observed successful copulations that were performed by small, medium, and large males. Bottom: percentage of total males on Flamenco that were small, medium and large.

Table 18.3: Number of Attempted and Successful Copulations Observed in Both Years.

Observations	Males: Small	Medium	Large	p*
Unsuccessful attempts	12	0	8	0.01
Successful attempts	3	2	44	0.01
Total attempts	15	2	52	0.01

*Probability values (p) are results of Kolmogorov-Smirnov tests of difference between observed and expected frequencies. Expected values were based on percent of total males in each size class.

DISCUSSION

Male Breeding Strategies

Large Males: Large males are distinguished from medium and small males, and all females, by their relatively confined movements during the non-breeding season. By virtue of their age, size and social rank, large males are able to establish themselves in an area where their own foraging efficiency can be optimized, and where a supply of females passing through the area is assured. In addition, a large male's home range contains one or more suitable mating trees. Since virtually all courtship and copulation occur in a territory, territory ownership is critical to a male's reproductive success.

Confined movements, plus maintenance and communication of dominance status (selecting conspicuous perches, high display rates) outside the breeding season define the large male's long-term strategy. This strategy (1) allows him close contact with females passing through his area (possibly important in mate acquisition at a later date), (2) insures his access to a mating territory in November, and (3) reduces the frequency of encounters and dominance struggles with other large males (by advertising one's location, mutual avoidance is facilitated). It may be of particular importance to a large, herbivorous lizard to avoid aggressive encounters and thereby minimize energy expenditures (see Dugan and Wiewandt, this volume).

During the breeding season, large males defend not only the territory, but also females resident within the territory. Territorial males were observed to defend females from attempted copulations by small males, to follow receptive females from the territory to feed, and to attempt to herd females from territories that were invaded by larger males. The time lag between copulation and ovulation, and the fact that female iguanas mate several times, suggest that sperm competition may occur in this species (Dugan and Wiewandt, this volume). If so, then direct defense of females may be interpreted as post-copulatory guarding (see Parker, 1974).

Medium Males: Because of their interactions with large males, medium males cannot move into territories and mate opportunistically as do small males. Nor have medium males achieved the size and status that is related to large male success. By remaining on the boundaries of mating territories, medium males could monitor the activity of the breeding group. Observations indicated that medium males were occasionally successful in territorial establishment, and did take over vacated territories. However, this strategy brings minimal returns because (1) the rate of territory turn-over is quite low, and (2) females may not remain in territories that medium males do take over. In addition, copulatory attempts by medium males appear to be less frequent than such attempts by small and large males. Thus, mating success is relatively low for males in this size class.

Females did move close to peripheral males, and medium males did court females outside territories. Why did medium males not try to improve their mating success by copulating "by force" as did small males? At least two explanations are possible. (1) The risk of attack from a large male did not outweigh the benefits to be gained from such attempts. This explanation is only a partial one, since large males were never observed to interrupt courtship bouts that occurred between medium males and females outside territories; attacks from a large male would only be likely if a large male followed a female from the territory. (2) A medium male may gain a territory within the lifetime of those females with whom he interacts, and it may not be in the medium male's interest to "alienate" females. Learning influences long-term foraging movements in *Cyclura carinata,* a related iguanine (see Auffenberg, this volume), and it seems likely that experience also plays a role in social relationship among lizards (Burghardt, 1977, reviewed learning in reptiles).

Small Males: The interactions observed between large and small males indicate that a small male was tolerated in the territory only until his sexual identity was discovered by the territory owner. Large males did not appear to recognize small males from a distance. Close range recognition may have been based on subtle visual cues, or a difference in responses between small males and females. Femoral pore secretions may have also provided chemical cues regarding sexual identity. Chemical cues communicate sexual identity in *Sceloporus* (Duvall, 1979), and although the function of the femoral pores is unknown, it has been suggested that they do function in chemical communication (see Cole, 1966; Wiewandt, 1977). The small male's strategy, then, involves not only morphological similarity to females, but also behavioral patterns that permit him to avoid large males, and thus remain in the territory and breed opportunistically.

The contribution of such a strategy to a male's fitness may be of particular importance in a long-lived species such as *I. iguana.* Observations of two males dating from July, 1976, indicate that a male does not reach 37 cm SV before its 7th year, and that large males may maintain a "sphere of influence" for at least 2½ years. The low mating success in medium males, the number of years between reaching reproductive maturity and achieving large male status, and the possibility of death during the non-reproductive years, provide an explanation for why the small male strategy occurs in green iguanas. However, the mechanisms underlying small male mating behavior are not known. The difference in the mating behavior of small, medium, and large males may be due to ontogenetic changes. It would certainly be to a male's advantage to mate as frequently as possible in early adulthood. Alternatively, certain males may stop growing at an early age, in which case small male behavior would be related to size, but would be independent of age. Such genetically determined mating strategies have been reported in other vertebrates (e.g., red deer, *Cervus elaphus,* Darling, 1937; ruffs, *Philomachus pugnax,* Hogan-Warburg, 1966; Van Rhijn, 1973). Unfortunately, sufficient growth data to assess the latter hypothesis were not available.

Female Breeding Strategies

The pattern of female movements and the nature of interactions observed between small males and females indicate that female iguanas are selective in mate acquisition. Females mate preferentially with large males. A complex of attributes probably convey the large male's suitability as a mate. His large size reflects his age, and therefore his ability to survive to an old age. His size also gives him a competitive advantage over other males. Territory ownership not only reflects this superiority, but also offers the female a place to court and mate with minimal disturbance. In short, a male's age, size and territory may all communicate his genetic quality (see Howard, 1978). However, since a female's home range overlaps that of a number of large males, she encounters a number of suitable mates, but usually mates with only one. Courtship behavior may provide additional evidence of a male's fitness as a mate (see "The Role of Courtship").

Females violently protested all copulation attempts by small males. Unlike large males, small males have neither proved their ability to survive to an old age nor their ability to hold a mating territory. Nor do they provide the morphological and behavioral (courtship) stimuli that large males provide. Cox and LeBoeuf (1977) suggested that female protest during mounts is a mechanism of female mate choice. The female elephant seal's violent reaction to being mounted attracts other males in the vicinity, male competition for the female ensues, and the female thereby maximizes the probability of mating with a high-ranking male. Similarly, female iguanas protest mounts by small males and the territorial male interrupts the copulation attempt if he is nearby. If the large male is not in the vicinity (as is usually the case), the small male's ability to hold on to the female during the struggle may be an additional "test" to which small males are subjected. If the ability to capture and hold on to females can be passed on to the offspring, then it may be to the female's advantage to copulate with a persistent small male.

The Role of Courtship

Courtship may function in species and sex identification (e.g., Morris, 1956b). *Ctenosaura similis* is sympatric with *I. iguana* over much of its range, and may share the iguana's arboreal habitat (pers. obs.). However, the two species differ morphologically and have very different headbob displays (pers. obs.; Carpenter, this volume). Both characteristics may be cues for species recognition (e.g., Jenssen, 1970, 1977; Williams and Rand, 1977; Rand and Williams, 1970). It is doubtful, therefore, that proper species identification requires such a lengthy courtship period. The large male's rapid response to intruding males argues against the need for a long courtship period to insure proper sex identification.

Courtship prior to pair formation may provide a female with a portion of the information on which she bases her choice of mate. In November, before stable relationships are established, male-female encounter rates are high. Females are exposed to the display and courtship of several males, either because they visit males, males visit them, or because displaying males are conspicuous. The components of male display affect female mate selection in other iguanid lizards. Females given a choice between males performing either normal bob patterns or altered bob patterns showed significant preferences for "normal" males (*Anolis nebulosus,* Jenssen, 1970; *A. carolinensis,* Crews, 1975). Male display rates have been shown to be an important correlate of male mating success in ruffs (Shepard, 1975) and of male sexual competence (number of spermatophores produced) in newts (Halliday, 1976).

After a female has established residence in a male's territory, the pair engages in courtship for at least two weeks before copulating. Extended precopulatory interactions after mate selection has presumably occurred may cement bonds of familiarity between mates. The establishment of such a bond should benefit the female if prolonged courtship restricts the number of females a preferred male can mate with, thereby minimizing the fitness of other females. In addition, continuous attention from a male may confirm her earlier choice of a fit mate. This would decrease the likelihood that a female will mate with a male who pretends to be highly fit but is not (see Fisher, 1958; Williams, 1968). If bonds of familiarity are a prerequisite to copulation, this should benefit the male if the probability of desertion by a receptive female is reduced. If a bond of familiarity must be established before copulation occurs, and if the receptive period is the optimum time for insemination, then it would be disadvantageous for a female that is receptive or near receptivity to seek another partner. Desertion appears to be uncommon among receptive females; only one female iguana was observed to switch mates after becoming receptive. Guarding against female desertion should be particularly important in species where females mate several times during a single breeding period, and where sperm competition may occur.

The long courtship period may be important in facilitating receptivity in female iguanas. This has been shown to be the case in *Anolis carolinensis* (Crews, 1974). That the amount of courtsip was reduced after females became receptive suggests that this same phenomenon may occur in *I. iguana.*

The latter observation also suggests that a successful male should optimize the amount of time spent courting each female, depending on her reproductive state. This prediction is borne out by the comparison of 3 males in Table 18.2. Pig and Y-Shoulder courted non-receptive females more than receptive females. They were observed to copulate more often than Macho, whose courtship bouts were briefer and more evenly distributed among his females on a daily basis. The female with whom Macho was observed mating had copulated 5 times with Pig, and once with a small male. Since sperm competion is a possibility in *I. iguana,* the number of times a male copulates with each female, and when he copulates with her, may be important components of his reproductive success.

One might expect several resident females to compete for the male's attention for the following reasons: (1) females mate preferentially with large males who comprise approximately 38% of the male population, (2) a male mates with only one female per day, and (3) a long period of courtship and frequent copulation appear to be important to females. Females were observed to interrupt courting pairs, and dominance relationships among females were observed in some territories. Unfortunately, the data are too few to determine whether or not female aggression is related to competition for males.

SUMMARY AND CONCLUSIONS

The social behavior of *I. iguana* was studied for 16 months on Flamenco Island, a small island on the Pacific Coast of Panama. During the non-breeding season, iguanas have large overlapping home ranges. Large males confine their movement to an area that is roughly one-third the size of the home ranges of other members of the population.

Iguanas are seasonal breeders. Mating activity begins in the last half of October, with increasing rates of display by large males. Male-female encounter rates are high in November. By early December, large males have established mating territories containing 1-4 females. Copulation occurs during the first 6 weeks of the dry season, from December through mid-January.

Male breeding strategies are related to size and the development of secondary sexual characteristics. Small males move into mating territories and mate opportunistically with females. Medium males remain on the edge of territories, and may assume ownership of vacated territories. Large males maintain their status in a circumscribed area year-round, and thus insure priority of access to suitable mating territories in late November.

Females mate preferentially with large males. Extensive courtship with several males before females have taken up residence in a male's territory, and immediately prior to copulation, suggest that courtship is an important factor in female mate choice, in facilitating female receptivity, and in establishing bonds of familiarity between mates. Females mate several times during a single breeding period. Multiple copulation has been reported in only two other iguanines, *Cyclura cornuta stejnegeri* (Wiewandt, 1977) and *Conolophus subcristatus* (Werner, this volume).

Behavioral-ecological studies of a large number of animal species have indicated that methods of mate acquisition are related to the social and ecological environment (see reviews in Brown, 1975; Emlen and Oring, 1977; Wilson, 1975). The spatial and temporal distribution of resources critical to females in large part determines the form and intensity of male-male competition for females and the degree and form of female mate choice (Trivers, 1972; Bradbury and Vehrencamp, 1977; Emlen and Oring, 1977). Males may monopolize mates indirectly, in which case resources essential to females are the proximate object of competition, or directly, with females being the proximate object of competition (e.g., harem defense). When neither resources nor females are economically defensible the importance of female choice in mate acquisition is expected to increase (Emlen and Oring, 1977).

Male iguanas do not control females directly; at any time of the year, a male gains access to females only when their movements overlap in space and time.

Year-round defense of resources is not energetically feasible. Basking sites, shade and sleeping perches are abundant and widely distributed. The patchy distribution and seasonal availability of food resources, plus the structural complexity of and poor visibility in a tree or group of trees, make food defense impossible (see Brown, 1964).

During the breeding season, large males do defend locations from which to display, court and mate. However, indirect control of females is minimal even for territorial males. Females are free to, and many do, travel from territory to territory. Competition among large males for females takes the form of increased display rates rather than increased levels of aggression. The overlapping movements of all sex and size classes outside the mating season, plus the complexity of the habitat, require that a male advertise his presence to both males (who should avoid him), and females (who should visit him) in a highly redundant manner, including selection of conspicuous trees and conspicuous perches in those trees, frequent performance of conspicuous displays, and color intensification.

The energy expenditure required of the large male strategy would not pay off for small and medium males because (1) large males, by virtue of their size, have priority of access to choice mating territories, and (2) females mate preferentially with large males. However, all males are more likely to encounter receptive females near breeding groups than elsewhere. In such a situation, males excluded from the breeding population can be expected to develop alternate strategies (Gadgil, 1972). Peripheral males have been observed in a wide variety of species (e.g., fish, Warner et al., 1975; amphibians, Wells, 1977; Howard, 1978; birds, Wiley, 1974; mammals, LeBoeuf, 1974). It is not always clear whether subordinate males remain near breeding groups to intercept and copulate with females or to take over vacated territories (Perrill et al., 1978; Wells, 1977). In *I. iguana,* which of the two subordinate strategies is more feasible depends on the size and appearance of the individual.

Since males have minimal control (either direct or indirect) over females, then the importance of courtship in assessing mate quality can be expected to increase (see Williams, 1966). The long courtship period prior to copulation may benefit females by providing the opportunity to review several males before being inseminated, to insure the fitness of a chosen mate, and also by limiting the number of other females that a preferred male can mate with, thereby minimizing the fitness of other females. The cost of courtship persistence to the male, in terms of time that could be spent obtaining other mates, depends on the number of other receptive females available to him.

The breeding synchrony among females in part determines the degree of polygyny (Emlen and Oring, 1977; Trivers, 1972). If courtship and copulation time is large relative to the total amount of time that mates are available, then the potential for monopolizing several females is low. The moderate asynchrony among female iguanas permits males successful in territorial establishment to mate with several females. The temporary consort relationship between a male and several successive females reflects this asynchrony. Whether males contribute to asynchrony by their selective attention to certain females, or whether females vary in their responsiveness to stimuli that induce receptivity and males consort more with females soon to be receptive, is unclear. However, in view of the length of the courtship period prior to copulation, staggered periods of receptivity are of benefit to males in that several mates can be obtained in the six week breeding period. The degree of asynchrony should primarily be constrained by the duration of the optimal nesting and hatching period, a source of potential cost to both males and females.

Note

[1]Current addesss: Rt. 15 Box 367, Gray, TN 37615.

Acknowledgments

Based in part upon a dissertation submitted in partial fulfillment of the requirements for the Ph.D. degree at the University of Tennessee, Knoxville. This research was supported by NSF grant BNS-77-11344, and by grants from the Smithsonian Tropical Research Institute, Sigma Xi, and the University of Tennessee Department of Psychology to B.A. Dugan, and by NSF Research grants (BNS 75-02333 and BNS 78-14196) to Gordon M. Burghardt. Nick Brokaw kindly identified the plants on Flamenco. Gordon M. Burghardt aided with data collection during the 1978 mating season. Special thanks go to A. Stanley Rand for his support, advice, and stimulating discussions, and help with all aspects of the field work, and to César Abrego, Epifanio Aizprúa, and Jona Aizprúa, for catching iguanas. Gordon M. Burghardt, Hugh Drummond, George Middendorf, A. Stanley Rand, Wayne Van Devender, and Paul Weldon provided helpful comments on earlier drafts of the manuscript.

19

Social Organization and Ecology of Land Iguanas, *Conolophus subcristatus*, on Isla Fernandina, Galápagos

Dagmar I. Werner
Charles Darwin Research Station
Isla Santa Cruz
Galápagos, Ecuador

INTRODUCTION

My field studies on the land iguana, *Conolophus subcristatus* were designed to investigate the complex interrelation between ecological factors, social structure and behavior of this large, herbivorous iguanid. Under the specific conditions offered on Isla Fernandina, land iguanas seemed ideal for studying the effects of an almost complete lack of interspecific resource competition and minimal predator pressure on behavior and social structure. The field studies were started in 1976 and have been carried out for three years. One year and eight months were spent on Isla Fernandina. During this time two complete seasons of premating, mating, and post mating development were observed: from January 1978 through June 1978 and from February 1979 through the beginning of August. Five shorter visits of four to six weeks were spread over the entire yearly cycle.

This preliminary report was written largely in the field prior to detailed data analysis. Therefore the results presented here may be modified by the quantitative data once it is available, but the basic ideas should not change.

GENERAL AND BACKGROUND INFORMATION ABOUT *Conolophus*

Distribution and Classification

The genus *Conolophus* is endemic to the Galápagos Islands. Two species have been distinguished in the last species revision (van Denburgh, 1913). *C. pallidus* occurs only on Santa Fe, *C. subcristatus* on Fernandina, Isabela, James, North Seymour, Plazas and Santa Cruz Islands. Various populations became extinct through man-introduced feral mammals. Most of the remaining populations are in danger of extinction. The land iguanas on North Seymour were introduced from Baltra in 1932 and 1934. The remaining population on Baltra Island was exterminated by soldiers based there during World War II. The only land iguana population that is relatively undisturbed is the one on Isla Fernandina where I carried out the main part of my studies.

I have the impression that most isolated populations have reached the subspecies level at least, and that a new classification is urgently needed.

Description

All *Conolophus* populations show distinct sexual dimorphism in size. Males are between two to three times as heavy as females. Size varies from island to island. The largest iguanas are found on North Seymour (originally Baltra) where a male may weigh as much as 12.2 kg and the smallest ones on Plazas, where a male usually does not weigh more than 5 kg. On Isla Fernandina males average around 7 kg and females around 3.5 kg. It is notable that the extremes in size are found on the smallest islands on which *Conolophus* occurs. These islands (Plaza, North Seymour, Baltra) show the least diversity in food and the greatest food limitations. The populations are exposed to the same predators on these islands and it should prove interesting to test the food availability and predation hypotheses as determinants for body size (see Case, this volume) on these land iguana populations.

Lost tails cannot be regenerated in *Conolophus*. Tail loss frequency and regeneration ability in *Sauromalus* populations are discussed by Case (this volume). The reduced tail break facilitation in the insular gigantic forms, *S. hispidus* and *S. varius,* is attributed to the low predation pressure these populations are exposed to. The observations on *Conolophus* fully support this argument. Land iguanas have even fewer predators than *Sauromalus* and seem to have lost tail break facilitations all together. It is impossible to pull off a Land Iguana's tail. A firm tail certainly is advantageous for intraspecific interactions. Many males and females show bite scars on their tails inflicted by conspecifics, but they still have complete tails. It is interesting that in both cases, *Sauromalus* and *Conolophus,* the loss of tail break facilitation goes hand in hand with the loss of the regeneration ability that *Conolophus* lacks almost completely. There were only two individuals out of about 200 with incomplete tails which showed some regeneration growth.

Weather Conditions on Isla Fernandina

The weather conditions are unpredictable on the whole Galápagos Archipelago. Descriptions of general weather conditions can be found elsewhere (Wiggins and Porter, 1971, Werner, 1978).

The cold season (about May through December) on Fernandina did not, in the three years of my investigations on this island, show the typical garua weather (drizzling rain) as described for other islands. The warm season may or may not be accompanied by regular rainfall, but heavy rainfall did occur in four successive years (1976-1979).

Depending on the altitude, the amount and time of rainfall varies between the coast and summit. The weather side, the southeast flank of the volcano, apparently receives much more rain, since we often saw heavy downpours there from our position on the west flank, where it did not rain. The least amount of rain falls at the coast and the highest occurs above an altitude of 600 m. The amount of yearly rainfall is not nearly as important as the frequency with which it occurs. As much as 30 mm of rain may fall within a three hour period, after which rivers form in the usually dry canyons. They dry out again about two hours after the rain stops. Around three weeks after the last rainfall the vegetation starts to dry out. The coloring of the island's lower regions turns into brownish gray if it does not rain for several months as was the case in 1978.

Temperatures vary according to altitude. In the warm season the daily maximum on a sunny day is about 33°C at the coast, 30°C at an altitude of 300 m and 27°C at 1200 m altitude. The daily minimum is about 10° to 12°C below the daily maximum. In the cold season the average temperatures are 4°-7°C below those of the warm season. The lowest temperature I ever recorded was 8°C at an altitude of 1200 m.

Cloud cover is irregular, but clouds usually form at an altitude of 500-700 m and also shade lower parts of the island. The rim may be cloud-free or foggy. The top of the island was cloudless during and after egg deposition (June, July) in 1978 and 1979, while an almost permanent cloud belt formed at 600-900 m altitude.

Ecological Situation of Isla Fernandina

Fernandina is a shield volcano which reaches an altitude of 1495 m above sea level. The island's diameter is 32 km. The crater is 5 km in diameter and 900 m deep. Fernandina is among the most active volcanoes of the world; four eruptions were recorded in the past six years.

The island shows a wide range of ecological conditions. This is due to the following main factors:

- Different altitudes and subsequent differences in weather, flora and fauna.
- The weather side of the island (southeast) is exposed to more wind and a higher amount of rainfall.
- Frequent volcanic eruptions keep changing the island's surface since its origin about one million years ago.

About two-thirds of the island's surface consists of bare lava or lava fields that are sparsely vegetated by first settlers (e.g., *Bracchycereus* which grows in the cracks of lava rock). Vegetation oases surrounded by recent lava flows are numerous at the foot of the volcano, where otherwise vegetation is nearly totally absent. Overgrown areas become predominant at altitudes of 500 to 700 m. The top of the island is covered by vegetation to about 60%. Plant communities differ not only according to different altitudes and weather conditions, but also according to the substrate that consists of lava of different ages and consequent degrees of erosion. About the third part of the island's surface, the northwestern region, was covered with ash deposits during a violent eruption in 1968.

This part is now in the process of being repopulated by plants and animals.

The types of plant communities range from scarcely overgrown lava fields and the ash fields which show equally little but totally different plant growth to scrubby areas as well as junglelike forests which may be found in altitudes above 450 m above sea level. Some grassland is present at the top of the island.

There are some 200 species of plants on Fernandina. Endemism restricted to Fernandina is relatively low due to the island's proximity to Isabela.

Besides shore birds and various colonies of seabirds the following land birds occur on Fernandina: the Galapagos Hawk *(Buteo galapagoensis)*, two owls *(Tyto alba* and *Asio flammeus)*, the Mockingbird *(Nesomimus parvulus)*, the Large Billed and Vermillion Flycatcher *(Myiarchus magnirostris* and *Pyrocephalus rubinus)*, the Galápagos Dove *(Zenaida galapagoensis)*, 8 species of Darwin finches of the genera *Geospiza, Camarhynchus, Cactospiza, Platyspiza* and *Certhidea*, the Galápagos Martin *(Progne modesta)*, the cuckoo *(Coccyzus melacoryphus)* and the Galápagos Rail *(Laterallus spilonotus)*. The following land mammals are present: An endemic rice rat *(Nesoryzomis narboroughi)* and a bat of the genus *Lasiurus*. The following reptiles are found: The marine iguana *(Amblyrhynchus cristatus)*, a lava lizard of the genus *Tropidurus*, a snake of the genus *Dromicus*, and a gecko of the genus *Phyllodactylus*. Besides these forms no other land vertebrates were found on Fernandina.

Nutrition and Food Competition

According to the rule in iguanines, the land iguana is mainly herbivorous, but feeds on animal matter if easily accessible, e.g., caterpillars or dead grass hoppers. Juveniles have been observed to leap after grasshoppers.

The number of plant species eaten exceeds 50, but land iguanas are highly selective in choosing certain plants and plant parts. The favorite plant consumed depends largely on the availability in the environment of the iguanas' location. In my main observation area, i.e., at an altitude of 300 m above sea level, the highest preference was for buds and flowers of the vine *Ipomea alba* (morning glory), and the highest percentage of plant matter consumed by the iguanas (at least 90%) was the leaves of the same plant. This coincides with the fact that *Ipomea alba* is the most abundant plant in that area (after rainfall, reproductive season). *Ipomea alba* grows only at an altitude of 450 m and below and depends on a good layer of soil without rocks. At higher altitudes where *Scalesia* is predominant the iguanas mainly feed on the soft leaves and flowers of this plant. In the various ecological settings of the island, different plant species are predominant and iguanas have different favorite food items. During the feeding process the iguanas are highly selective. A plant part may be eaten at one but not at another time. On their way to the mating territories and after leaving these for egg deposition females have been observed to feed exclusively on *Sonchus oleraceus* for as long as one week. This plant does not occur in high densities; it covers less than 1% of the feeding areas used during the reproductive season. No observation indicates a correlation between sexual dimorphism in size and feeding habits in *Conolophus;* both sexes feed on the same plant matter. Sexual dimorphism in size is certainly not correlated to a reduction of intraspecific food competition in *Conolophus,* as suggested by Stamps (1977) and Schoener (1977) for other iguanids. Dry fecal matter is occasionally eaten, mainly by females. Dirt grains may be licked up. One iguana which had recently died was dissected. It showed one-third of the colon's volume filled with dirt. This presumably helps digestion.

Food competition by other vertebrates is virtually absent. None of the other vertebrates occurring on Fernandina are herbivorous with the exception of two species of Tree Finches *(Camarhynchus),* which feed on the fleshy fruit of some plants land iguanas also include in their diet. These finches feed mostly in the higher parts of the plant or bush which land iguanas would be unable to reach. The distribution of the Tree Finches also is restricted to high elevations and overlaps with the land iguana only in part of its distribution range. I think that the Tree Finches are a negligible factor in terms of food competition.

Predators

Land iguanas have two native predators: A snake of the genus *Dromicus* and the Galapagos Hawk. The snake seems to feed mainly on individuals of hatchling size. At hatching time, in October, the entire hawk population of the island seems to concentrate near the nesting areas. These are situated inside the crater and the hawks dive for the emerging hatchlings. The hawk also preys upon larger juveniles, while adults are totally free of predators. Males show no fear reaction to hawks; females may flee into burrows or other hiding places, or show a threat posture accompanied by wiggling the tail.

Age at Maturity and Life Span; Population Composition

The land iguana seems to be a very long lived reptile; I assume that an individual which has reached adulthood has a life expectancy at least of somewhere between 20 to 40 years. Extrapolation of measurements of growth rate indicate that females reach adulthood between 7 and 10 years and males have attained adult size at an age of 11 to 16 years. The number of juveniles older than six months is strikingly low. On counts in all types of habitats the number of juveniles was never higher than 1 per 10 adults and usually only 1 per 30 adults is seen. Hatchlings are abundant near the crater rim after the hatching season (October), but by about six months after hatching time they are seen on extremely rare occasions. This most likely is correlated to the concentration of nest sites inside the crater (see next paragraph) and the distribution of the hatchlings from there over the whole island. The low number of juveniles may be due to the small survival rate of hatchlings during their hazardous trip from the nesting to feeding areas.

The sex ratio is close to 1:1. On almost all counts, except during nesting time, the number of males was somewhat higher than that of females. This data is certainly due to the higher visibility and lower fear level of males, who not only by their coloration and size are more obvious but also use more exposed locations than do females.

Nest Sites and Nesting

For nesting the females leave the mating grounds. On the observed part of the island, the western flank of the volcano, at least 95% of the female population migrated to the top of the island from where the females descended into the 900 m deep caldera for nesting. We equipped 13 females with long distance tracking devices and followed them on their way to the rim. All females descended into the crater where they could not be located any further for logistic reasons. We observed several thousand females circling the crater at the rim searching for the descent. Females marked at the shore were subsequently found inside the crater. This means that some females walk a distance (aerial) of over 15 km. They ascend 1400 m to the crater rim from where they descend 900 m into the crater, at the floor of which there is a sizable nesting area, which was found by Tui de Roi (personal communication). Probably many females nest behind loose boulders in the crevices of the crater wall. A second nesting area (N2, Fig. 19.1) was discovered on a platform inside the crater.

For nesting the females dig burrows very much like those described for the green iguana, *Iguana iguana* (Rand, 1968, 1972). In N2 we measured nest temperatures and it became obvious that nests were only built in places heated by fumarole activity where hot vapor emerges from a crevice a few meters in length. Nest temperatures of two nests measured here during a day-night cycle in June show very little fluctuation and are much higher than in comparable areas lacking fumarole activity. At a depth of 38 and 31 cm temperatures of 33.5° to 35.0°C and 32.2° to 34.2°C respectively were recorded. A nest away from fumaroles showed a temperature range of 26.2° to 30.4°C at a depth of 41 cm. The difference between the temperatures in the two types of nests becomes more extreme during the cool season when a nest away from fumaroles shows a minimum of 23.1° and a maximum of 26.8°C during a day-night cycle at a depth of 32 cm (measured in September). The volcanic heat may be the reason for the great effort made by females to reach the nesting areas. But it also could be that weather conditions are a factor for nesting inside the crater and that not all the nests are constructed near fumaroles. The sky above the crater usually is cloudless during nesting and incubation time while the flanks of the volcano are surrounded by a cloud belt, during approximately 70% of the day time. Evidence (egg shells) for nests not inside the crater was found at the shore in old earth and at an altitude of 700 m. Both locations would not have provided enough ground for substantial nesting. Extensive searches revealed evidence of three nests only. It is noteworthy that the soil consistency in my main observation area would allow nest construction and areas suitable for nest construction cover a large part of the west flank of the volcano yet are not being used for nesting.

There is a distinct laying peak in the first two weeks of July, and nesting takes place during six weeks at the very most. Females defend their nests for several days and possibly weeks. In N2 the whole surface of an area of about 150 m by 40 m was covered by nests separated by a distance of 1 to 2 m. On a single day we counted 350 females in that area. The low number of dug-out eggs from previous females was striking.

All land iguana populations observed lay one clutch per year (Fernandina, Santa Cruz, Cartago-Bay on Isabela, Plaza, Barrington). The clutch size on Fernandina varies between 7 and 23 eggs. Hatching was observed in October, about three and a half months after laying.

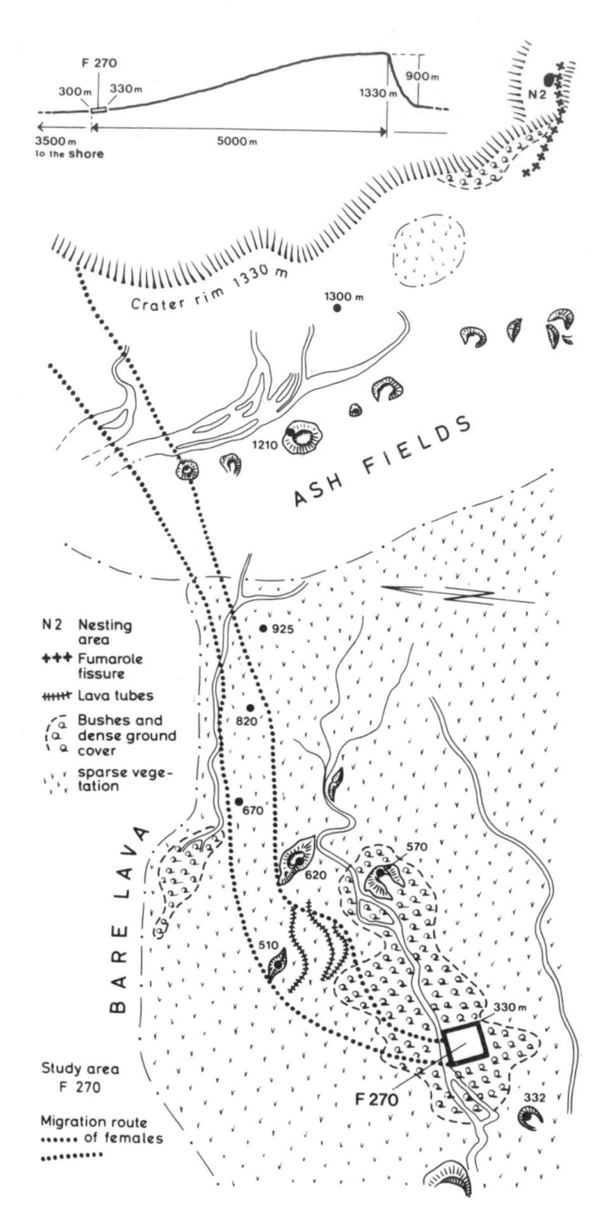

Figure 19.1: Survey of upper part of the western flank of Isla Fernandina, with the main study area F270, and the walking route of the females from F270 to the crater rim for egg laying. Twelve out of thirteen females used the outlined route on their way to the rim. Depending on weather conditions they needed between three and ten days to reach the rim. A cross section of the island is provided indicating the altitude of the rim at the point of arrival of the females, depth of the crater and the location of the F270 study area. A nesting area (N2) is indicated where the nests are heated by fumarole activity. Study area F270 is 200 x 160 m.

THE SOCIAL STRUCTURE OF *Conolophus*

The social structure varies according to ecological conditions because the behavior of the land iguanas changes under different circumstances.

Most observations about the social structure were made in one main study area called F270 in the following.

The Main Study Area, F270

The main study area is located at the west flank of the volcano 330 m above sea level. The area was devastated by a mud flood during the eruption in 1968 which destroyed most of the original vegetation and deposited very fertile ash in part of the area. All land iguanas found here have come from at least three kilometers away. The size of the study area is 200 m by 160 m (32,000 m^2) and it represents a wide range of ecological settings which are common on the whole island. They can be described by five main parameters:

A. Food plants.
B. Bush cover (shade).
C. Ground consistency.
D. Open area (sunny).
E. Slope.

The ecological settings found in F270 are described in Description of Mating Areas and Male Territories.

Number of Territories in the Study Area F270

Between 16 and 20 males have established territories at the mating peak in F270. 30 ± 5 females mate with the territory holders.

Changes in the Social Structure Within the Yearly Cycle

There are two main phases in the social structure of the land iguana. The reproductive phase occurs in the warm and the non-reproductive phase in the cold season. The two phases differ primarily in the following ways:

- Both sexes are more active in the reproductive phase.
- Certain behavior patterns occur only in one phase.
- Spacing patterns are distinctly different in the two phases. In the reproductive phase individuals search out each other and remain close together though separated by a critical distance which depends on circumstances. In the non-reproductive phase close distances are avoided and individuals chase one another away.
- Different habitats are used in the two phases.

Reproductive Season

The social structure in the reproductive season is extremely complex. A series of mating strategies (term used in the sense of Dugan and Wiewandt, this volume) is employed by the males for obtaining females. Individual differences in behavior are pronounced in both sexes. Individual flexibility in switching from one strategy to another one, if the first one fails to attract females, can clearly be demonstrated for males. Females are much more subtle in their behavioral expression so that it is hard to define distinct strategies for them. The social system of the land iguana cannot be classified by the common terminology because the application of a single term obscures the complexities of the system, as Dugan and Wiewandt (this volume) emphasize. Basically females choose males and males use a series of tactics to obtain females. Part of the male population is territorial (T-males), another part is non-territorial (N-males) and there are also males who switch between territoriality and non-territoriality, potential territorial males (P-males). Females occupy home ranges that are several times as large as one male-territory. They show no territorial defense.

Arrival in the Mating Areas: Three months before mating the adult population starts to concentrate in the mating areas. These areas show a combination of distinct ecological features which is provided on only 3% of the island's surface. Thus during the mating season the entire adult population concentrates in this small area.

A few territorial males remain in the mating area throughout the non-reproductive season and use the same territory in subsequent years. Of the other males that have left the area after mating in the previous season, the territorial ones are first to arrive. Arrival time varies; the first territorial male may arrive as early as mid-January, the last one as late as at the end of March. The beginning of territorial defense also is variable between January and March. A male who has arrived in January may start defending his territory as late as March. Non-territorial males (N-males) arrive much later than the territorial ones (T-males). Their highest concentration is directly before and during mating time. In contrast to territorial males N- and P-males come and go. It is therefore difficult to determine an exact ratio of territorial versus peripheral males (N- and P-males). At mating the ratio between the territorial and peripheral males fluctuates between 3:2 and 2:1 (territorial:peripheral). Females start gathering in the mating areas about three weeks after the males, and their number increases until three weeks before copulation time. Then the number of females is slightly larger than that of T-males.

Description of Mating Areas and Male Territories: The mating areas must provide satisfactory conditions in the following respects.

–Thermoregulation.

–Food for females (males do not feed or feed little during premating and mating time).

–Burrow construction.

Male territories constitute the central part of the mating areas. Since males do not feed during the mating season the presence of food is not necessary in a territory. Food however, must be available close by for the females.

In the main study area about one-third of the available area is occupied by male territories. Territories provide the following combinations of the ecological features mentioned earlier in this chapter.

A. Food plants: An extensive feeding area is present inside the territory or no farther than 30 m.
B. Bush cover: Bushes may be lacking or cover up to 60% of the area.
C. Ground consistency: An area large enough to construct at least one burrow is present in a territory. This requires a minimum area of 3 m x 3 m of soft soil.
D. Open area: Open area may cover 50% of the area of a male's territory. In areas with dense ground cover, enough unvegetated area for thermoregulation is created by the dirt heaps which form in front of the burrows constructed by the males.
E. Slope: Areas with a steep east or west slope (more than 20°) are not being used for territories. Southern slopes in the F270 study area were not steeper than 20°, an inclination still allowing territorial defense.

Microclimatic factors, such as exposure to wind certainly influences the quality of a territory, but no exact measurements were made.

Excluded for territories are: areas with rocky ground, areas too distant from feeding places, regions with dense bush cover and areas with steep slopes. According to these criteria the distribution of the territories in the F270 study area was patchy. There is a series of adjacent territories, but there are also isolated territories.

Territory types vary according to ecological circumstances. In F270 there are extensive feeding grounds in which females concentrate for food consumption. Male territories with neighboring territories have very well defined border lines. Territory size varies greatly. Of the 16 (to 21) territories in F270 the smallest one covered an area of 250 m^2 and the largest one 1600 m^2. In other parts of the island territories are established in areas where food is thinly spread. Here no distinct territorial limit between neighbors can be defined and the size is several times larger than that observed in F270. These territories could also be called home ranges, because only a central part (burrows) is defended. Outside the defended part the males may defend an individual distance, or they avoid contact with other males.

The quality of territories can be ranked according to the combination of ecological features mentioned earlier in this chapter. The territory of highest quality is one that provides a feeding ground which extends beyond its limit. All or a large part of the area consists of soft soil for the construction of a number of burrows. A few bushes provide shade during the hot midday hours. Females from other territories cross it for reaching a feeding area or they feed inside. However, mating success is not only correlated to territorial quality, because the male's behavior is at least as important as the ecological setting of his territory.

The central part of a territory is the area where one to eight burrows are constructed and maintained by the male. The number of burrows used in a territory reflects the activity level of the male and is correlated to mating success.

The Male System: Two basic types of males can be distinguished, the territorial (T-males) and the non-territorial males (N-males). T-males establish and defend their territory, which they never leave, except when chasing another individual, or on other exceptional occasions (see Male Competition for Territories). N-males concentrate near the territories shortly before and during mating time and take chances to rape females. Their behavior is very much like that described for the S-males in the green iguana (Dugan, this volume). N-males do not attempt to acquire a territory and show no territorial behavior. They are not necessarily smaller than T-males and some of them appear to be older than T-males. They may remain in the mating area for one day or stay there throughout the mating period. After mating they leave the area. Several N-males returned to the same mating area in subsequent years. Territorial males do, or intend to defend the same territory in subsequent years. If a territorial male loses his territory he becomes a P-male. He may behave like a N-male, or he may try to acquire another territory.

Early in the mating season the territorial limit is defined according to ecological criteria and tradition (e.g., the location of intact burrows constructed in the previous season is important). During the first month of the reproductive season, territorial boundaries are relatively unstable. They are redefined depending on the choice of area by newcoming T-males. Later in the season, there are fights and shifts of territory limits triggered by the location of females. Thus a territory is not an unchanging unit but in contrast is a very dynamic entity.

The Female System: Females are much more unstable in their choice of a defined area than males. They return to the same general area, but most females do not choose the same home range as the one occupied in the previous season. Upon their arrival in the mating grounds they go on inspection tours checking a number of territories. After this tour which lasts several days a few females stay with one male. Most females change territories (i.e., spend the night in a burrow of a different male) after a few days or weeks. The amount of time spent inside or outside the territory depends on the availability of food. Females are not concerned about territorial limits and pass them unmolested (exception see Courtship and the Importance of Burrows). If several females share one male's territory they may gather in small groups in front of his burrow(s) where they spend the night with or without the male.

Mate Acquisition and Behavior: In the land iguana's social system, as in any polygynous system, males concentrate their efforts on fertilizing as many mates as possible, whereas females invest their energy in the production of a high number of good quality of gametes. Accordingly different is the intrasexual competition for mates. Territory defense by males starts three months before mating, while females show competitive behavior for males only shortly before and during mating and even then aggressive (threat and chase) encounters are rare and very short.

The two types of males (T-males and N-males) in the land iguana invest drastically different amounts of behavioral energy in acquiring mates. T-males court females and defend territories, N-males perform neither of these activities. T-males defend areas that provide for the females and their own needs (burrows, thermoregulation) offering access to nearby food for the females. A T-male may assemble as many as 7 females in his territory, but no male was observed to mate with more than 4 females in a single season.

Territorial Defense and Female Influence on Territory Limits–Early in the mating season males establish territories irrespective of the presence of females. Once the territories are established almost all females have arrived in the mating area (end of March). Now males shift the territory borders according to where the females are. Females are unpredictable in their permanence in a specific territory and a male whose territory was populated by several females at one time may be left alone a week later. With approaching mating time, competition for females increases, so do the efforts of males to retain the females inside their territories by increased courtship behavior. Males whose territories are not visited by females show an increased amount of display behavior. They approach the limits of other males' territories in the direction of a female. The neighboring male is being provoked to a display or a fight. These fights may result in a shift of the territorial boundary. In some cases a female left her original male and joined the fighting neighbor, whereupon fighting between the two concerned males stopped.

Some females seem to be undecided about which of two males they want to join. It was repeatedly observed that a female chose a territory limit as a feeding area. Both males intended to approach the female and ended fighting, while the female stayed nearby continuing to feed. This was repeated during several consecutive days until the female joined one of the two males. If the territory limit was shifted during such a fight the female joined the male who enlarged his territory.

The ownership of a territory is demonstrated by the male's obvious presence there. He excavates burrows and patrols the central part of the territory where his burrows are located. While walking the male gives the assertion display (Carpenter, 1967). This behavior corresponds to the undirected territorial behavior described for *Tropidurus delanonis* (Werner, 1978). Males approaching the territorial limit are chased away (N-males) or are challenged by displays at the border. Displays may turn into fights. These are short early in the mating season, but may become extended later on. The most persistent fights were observed in males who had owned a territory in a previous mating season, that was now occupied by another male, and by males whose territory failed to attract females. In both cases the male attempted to acquire part of or the entire territory of another male. In one case two males fought for up to eight hours per day on more than 40 days in a two month period.

Fights are largely unritualized. A fight normally lasts for half an hour to several hours. Usually fights are settled without serious injury. Extended fights end when one male is seriously injured or when he has lost all teeth which are the final weapon to control the opponent.

Individual Differences in Behavior and Mating Strategies–Individual differences are pronounced in both sexes. This is partly due to ecological differences in the occupied areas, but in comparable situations individuals may react quite differently.

These differences can best be demonstrated for the T-males. Besides territory defense, burrow construction and courtship behavior, males can employ a series of additional strategies for obtaining mates. The following list indicates strategies observed in 15 T-males. The number of males who used the respective strategy is indicated in brackets.

A. Acquiring a female by conquering a neighbor's burrow [2]: If no female has joined a male by four to six weeks before copulation peak he may intend to approach a neighbor's female. At the limit of the territory he is stopped by the neighboring male; subsequently there is a fight. Fights are continued until a decision is reached. This may take several weeks. If the intruding male wins the fight, the female(s) that used the concerned burrow remain there and become his mate(s).

B. Enticement behavior [1]: The only female in the territory of male 688 (M-688) leaves his territory about 6 weeks before copulation time. M-688's neighbor (M 403) has one stationary female and another one which comes and goes. M-688 repeatedly approaches his neighbor's stationary female with display bob, is stopped at the territory limit and subsequently both males fight. After a few days of repeated fighting at the border line M-688 approaches the stationary female showing the shudder bob. At the territorial limit he turns around and continuing the shudder bobs he walks back to his nearest burrow. This sequence is repeated several times a day and after 5 days the addressed female leaves M 403 and joins M-688 remaining in his territory until mating time.

The shudder bob has been described for many iguanids (Carpenter, 1977; Dugan, this volume) as introductory behavior before copulation. The land iguana also performs it in this context. But only one male performed this bob for "seducing" a neighbor's female.

C. Stealing a neighbor's female [3]: This behavior was performed shortly before and during copulation time by males who did not include females in their territory (2 cases) but also by a male who already had females. The male enters another male's territory, grabs a female by a neck fold, and carries her to the center of his territory where he rapes her in front of one of his burrows. The females remained, in all instances observed, with the new male at least for one week.

D. Following a female [2]: This is a common behavior in the green iguana (Dugan, this volume) but it is shown only rarely in the land iguana. One male whose territory was not chosen by a female four weeks before copulation time, left his territory and followed a female who had recently arrived in the mating area. The female chose a burrow outside any male's territory and the male established a new territory around this burrow.

The only female of another male left his territory for feeding in areas not covered by territories. This male left his territory and closely followed the female, spending some nights outside his territory. After a few days both iguanas had returned to the original territory.

E. Herding [1]: Six weeks before mating time a female who had visited several territories entered the territory of a male who had no female at that time. He courted the female intensely. The female tried to leave the territory and was prevented from doing so because the male cut off her way by walking between her and the territory limit, courting the female constantly. After two hours of this "game" the female tried to run out of the territory. The male ran after her, grabbed her by the neck fold and raped her on the spot (without carrying her to his burrow). The female stayed in the territory for one day.

F. Giving up the territory and raping females: Raping females is typical for N-males. If however, a T-male loses his territory and cannot acquire another one he may stay in the mating area and take chances to rape females. One male whose territory was almost constantly void of females gave up the territory three weeks before mating time and showed the typical behavior of N-males. At mating time he raped females.

G. Drop in activity level [1]: A male who became the owner of a territory, that was very successful in the previous season, was not visited by a female during the first two weeks after the acquisition. During this time his activity level dropped so drastically that he did not even court approaching females (three weeks before copulation). This male did not leave the territory and did not mate in that season.

Male Competition for Territories–An interplay of several factors determines the amount of competition for a certain territory.

- –Ecological qualities of the territory, including distance from feeding grounds.
- –Location of the territory within the territorial system.
- –Number of females inside the territory.
- –Number of neighboring territories with and without females.
- –Number of territorial males (P-males) who have no territory and intend to acquire one.

In the F270 study area there is a total of 16 (to 21) territories (Figure 19.2). The main block consists of 8 (to 10) adjacent territories. An isolated group consists of 3 adjacent territories (temporarily an additional group of 2 adjacent territories formed), and there are 5 solitary territories. The main block is located on an area of ash deposits of a mud flood during the 1968 eruption, providing a thick layer of adequate soil for burrow construction. Uphill, toward the crater, vegetation becomes abundant and territories in this direction provide feeding areas. Downhill, there is scarce vegetation and territories lack food. The area beyond these territories (outside the study area) still provides soft soil, but it is not covered by territories, presumably because the distance to feeding areas is too far (more than 40 m). There was one central territory in this block in 1978. (It was subdivided into three territories at mating time. These were maintained in 1979. In the following I will consider this area as one territory.) All other territories are peripheral.

Food is abundant in two of the territories forming part of the small group. The third one offers food in its vicinity. Solitary territories either offer abundant food or some food is scattered over the surrounding area.

The following gives an overview of the number (in brackets) and type of territories represented in F270:

Tc [1]: Central territory of the main block. Territory provides 30% shade and abundant food.

Tpf [3]: Peripheral territory of the main block providing 20 to 30% shade and abundant food inside or at its limit.

Tpl [6]: Peripheral territory providing 0 to 10% shade and little or no food.

Tsg [3]: Territory of a small group. 10 to 40% shade, food abundant inside or nearby.

Ts [5]: Solitary territory. 10 to 50% shade, food abundant or scattered.

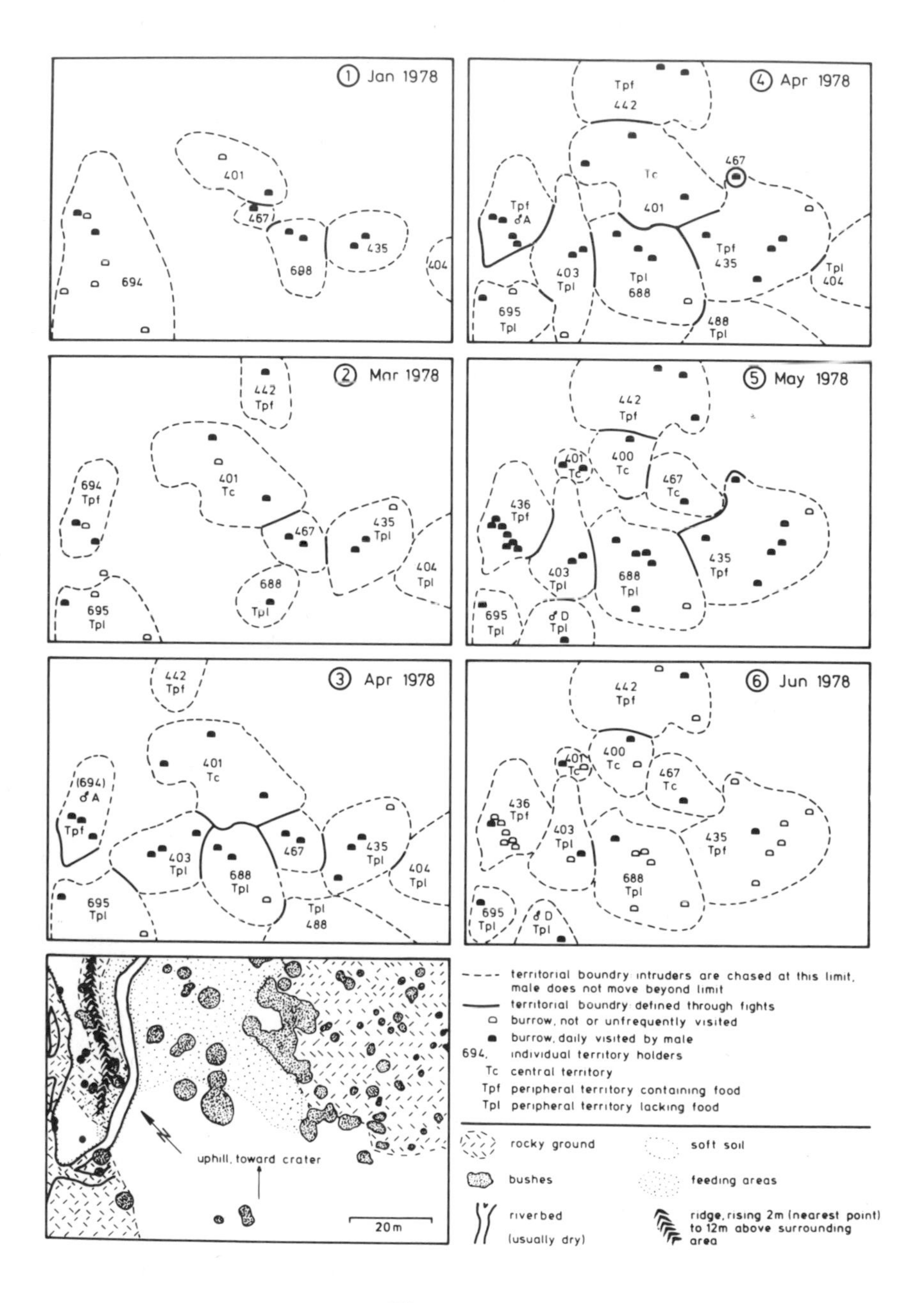
① Jan 1978
② Mar 1978
③ Apr 1978
④ Apr 1978
⑤ May 1978
⑥ Jun 1978
territorial boundry intruders are chased at this limit, male does not move beyond limit
territorial boundry defined through fights
burrow, not or unfrequently visited
burrow, daily visited by male
694, individual territory holders
Tc central territory
Tpf peripheral territory containing food
Tpl peripheral territory lacking food
rocky ground
soft soil
bushes
feeding areas
riverbed (usually dry)
ridge, rising 2m (nearest point) to 12m above surrounding area
uphill, toward crater
20m

Figure 19.2

Figure 19.2: Ecological setting and distribution of territories in the main territorial block in the F270 study area, in the 1978 reproductive season. The ecological setting (left bottom) shows the areas providing soft soil and rocky ground, bushes, and feeding areas. The surface area has an inclination of 5 to 10%. There is a ridge rising up to 12 m above the surrounding area with steep slopes on both sides. This ridge was frequently visited by adult males without territories (P- and N-males) presumably to observe the activities in the mating area.

The changes in the territorial constellation are indicated in 6 stages:

(1) January 1978: Transition from non-reproductive to reproductive phase. There is a loose agglomeration of six territories at the limit of which some males start to display and to fight. No females are present in the area yet.

(2) End of March 1978: Two more males have arrived in the area, the location of two territories (467 and 688) is shifted. General increase in size of territories. Most females have arrived in the area.

(3) Mid April 1978: Four weeks before mating peak. The territories are tightly packed and fights are frequent. The location of some territories is shifted (688 and 442). Male 694 loses his territory to a newcomer. All females have arrived.

(4) End April 1978: Situation directly before the beginning of matings. Changes in territorial limits are due to female location. Male 688 has conquered a burrow with a female from male 403. After 6 weeks of almost continuous fighting male 435 has conquered male 467's territory. 467 retreats into a burrow outside the territorial complex where fights continue. Male 442 has increased the size of his territory.

(5) Mid May 1978: Mating peak. Male A loses his territory to the former P-male 436. Male 467 loses his burrow to male 435 and subsequently attacks male 401 who has up to five females in his territory. 467 conquers one burrow within four days. Meanwhile P-male 400 invades and defends the other half of 401's territory. 401 retreats into that area of his territory that was least visited by females where he is not attacked any longer. All females of this territory leave the battle field and join neighboring males.

(6) Beginning of June 1978: Shortly after mating phase, most females have left the area for egg laying. Very few fights and displays occur. Males occupy one burrow. Note that the number of used burrows increases from January to May.

In two subsequent years 1978 and 1979 the Tc was visited by more than one and up to 5 females at any one time, from one month before, until half way through mating time. All 3 Tpf had at least one female (up to 7) with the exception of one territory in the 1979 season. Of the Tpl one territory was unsuccessful in both years (the same owner), all others had none or two females most of the time; two of the 6 Tpl temporarily were visited by more than 2 females. Thus there seems to be a general tendency for more females in the Tc and Tpf than in the Tpl.

Females were definitely attracted to males who showed a high level of display and courting behavior, so that variations in the number of females in the different territory categories can be related to the male behavior. The observations show that the activity level of those males who failed to attract females was by a multiple factor lower than that of those males who showed a high amount of display and courting activity. This is best demonstrated by the observation that one owner of a Tpf mated with 4 females in one year and a different owner in the following year failed to attract females altogether.

Male competition for a certain territory is directly related to the number of females it contains. Since the activity level of a male strongly influences the number of females he gathers in his territory, he will increase the competition potential for his females by a high amount of display and courtship behavior. Competition for a certain territory type is summarized in the following list by indicating the amount of fighting.

Tc: Fights and displays are frequent and at mating time the owner of the Tc lost his territory to a male who had attempted to obtain a territory for two months. The territory was subdivided into three smaller units occupied by three males in 1979. Neither the original nor the new owner mated in the 1978 season.

Tpf: One of the three Tpf was a battle place throughout the premating and mating season in 1978. Three different males took turns in owning the territory. This territory was frequented by up to 5 females. In the early part of the 1979 mating season the territory changed its owner three times. The last owner failed to attract females for his lack of activity. Neither females nor males paid any more attention to it. The second Tpf was a battleground between two males during 6 weeks in 1978, when the 1977 owner attempted to regain his territory. The 1978 owner maintained the territory in that year and in the following one.

The only neighbor of the third Tpf was the owner of the Tc, whose territory always contained females. Even though displays and fights at the limit were frequent the Tc holder never attempted to acquire part of the Tpf's territory. Fights with peripheral males were rare and short.

Tpl: Owners of Tpl were almost never challenged by peripheral males trying to acquire a territory. Fights here occurred between neighbors of which one owned a female and the other one did not. In three cases such periods of heavy fighting resulted in the loss of one or more burrows (females) and in one case the whole territory was lost to the neighbor.

Tsg: Even though ecological conditions in these territories are superior to those of the Tpl, the number of females here is not higher. Fights occurred between neighbors, like the Tpl situation. One of the fights resulted in the shift of the territorial limit and the acquisition of a neighbor's female.

Ts: The number of females in these territories is one or two. Fights are extremely rare.

N-males and P-males are centered in the main territorial block. In a single season no more than about 8 P-males were observed to fight with a territory holder. This resulted in the loss of one territory in 1978. No fights between peripheral males was observed in either Tsg or Ts. Competition for territories thus is greatest in areas with a large number of adjacent territories and concomitantly a large number of females. There is almost no competition for solitary territories. The data suggest that there is no significant difference in the mating success of the holders of the listed territory categories, if the number of males who held a specific territory in a season is divided by the number of females who were fertilized in it.

Courtship and the Importance of Burrows, Mating and the Behavior After Mating: Courtship behavior starts 8 to 10 weeks before mating. The amount of courtship generally increases as mating time approaches. But courtship may be much more intense during the first days after a female's arrival in a male's territory than it is directly before mating time. The amount of courtship behavior differs not only between individuals but also varies within the same individual in different years. A high rate of courtship behavior attracts females and the presence of females influences courtship behavior.

Males construct the living burrows and except for fights all social activity centers around the burrows. Increased burrow construction and maintenance behavior is coupled with an increase in display and courtship. As a general rule active males construct and maintain a large number of burrows and at the same time attract more females. Burrow inspection by the females is the prime activity upon their arrival in the mating area. Both sexes spend the nights in the burrows and remain about one hour at their entrance after emerging in the morning and before retreating in the evening. Most of the males' courtship behavior is performed near burrows. Males perform a specific behavior pattern by which they seem to indicate the ownership of their burrows to the females. During the mating phase females almost always stay at the entrance of a burrow. Directly before, during and after the mating phase females may spend several hours of the day inside a burrow or they even do not emerge for as long as five days.

The number of females a male mates with is more or less proportional to the number of females he had in his territory before mating. The highest observed number of females a male mated with was 4 in 1978 season in a Tpf territory. Like *Cyclura cornuta stejnegeri* (Wiewandt, 1977) and *Iguana iguana* (Dugan, this volume), *Conolophus subcristatus* females may seek multiple copulations with different males. No observed female mated with more than two males. The males never mated more often than two times per day, using a different hemipenis for each mating.

Throughout the premating season females show an appeasing posture when approached or courted by a male. This posture is distinctly different from the rejection posture shown after the mating phase. Both postures allow the females free passage through any territory without being raped by the territory holders. During the mating phase which lasts for about one week the females show neither of the two postures. Territory holders approach the females with a distinct premating ritual before copulation. Peripheral males, i.e., N-males and P-males, show no introductory behavior to copulation. They also do not respect the appeasement and rejection postures of the females shortly before the mating phase until after it is over. They rape females at any given chance. The females' only way to avoid unwelcome partners is flight into a nearby burrow. Peripheral males sneak into the territories, hiding from the females and territory holders behind vegetation. As soon as the territory owner walks away they surprise the females, grabbing them by the neck fold before they can run away. A female is raped if she is unable to drag herself into a burrow, forcing the male off her back. I estimate that about 5 to 10% of the successful matings are accomplished by peripheral males.

All 15 closely observed females refused to mate in 1979. In spite of this, they reproduced and laid fertile eggs. This suggests that the land iguana like other iguanids (Cuellar, 1966) is capable of sperm storage. The females probably used sperm from the previous mating season. No differences in behavior and social structure could be observed between the two mating seasons in 1978, when most females mated at least once, and in 1979 when all observed mating attempts were rejected by the females. The only difference observed was the lower amount of fighting, display and courtship behavior of males in 1979. In this year, at mating time the females switched from the premating appeasement directly to the postmating rejection posture. Males repeatedly performed the premating ritual. After being refused each time, some of the territory holders attempted to rape their own females, a behavior not observed in the previous season. About 10 rapes by territory holders were observed in 1979. Probably none of these rapes were successful, because the copulations never lasted longer than 12 seconds, whereas true copulations last for 30 to 120 seconds. The rapes were directed only at a few females. In both years, 1978 and 1979, mating (attempts) started in the last days of April and were observed during 25 consecutive days in 1978. After that time the females gather in groups as large as 7 individuals in front of burrows, except when they feed or hide inside the borrows. The amount of courtship behavior of males now drops and by the time females leave the area, at the end of May, the males show almost no more display, fighting and courtship behavior. They now start to feed and mainly thermoregulate in front of their burrows.

The females leave the mating area for egg deposition one by one and not in groups. According to weather conditions 13 radio tracked females needed between 3 to 10 days to traverse the aerial distance of 6 km and an altitude of 1000 m in order to reach the crater rim. From there we were unable to follow the tracks of their final venture, when they descended the steep crater walls in order to reach the nesting areas inside. Only one of the closely observed females from the F270 study area was observed laying on July 10th, about 7 weeks after the theoretical (1979 season) mating peak.

Non-Reproductive Season

For males the non-reproductive season begins after copulation time. Male activity now drops drastically and their interest in the females still remaining in their territories declines steadily until no more courting behavior is seen.

This is about the point at which the females leave the area to nest. Males who have held territories in which food remains abundant may remain there throughout the non-reproductive season, using them as home ranges and using one of the burrows for spending the night. Males in whose territories food is inadequate, will leave and make excursions of up to several kilometers to visit the feeding areas at higher elevations. They return sporadically. Thermoregulation does not seem very important, because iguanas may spend the nights in burrows which cool down to almost air temperature or they sleep in the open. Except for eating and sporadic wanderings the males show an extremely low level of activity. Males who were observed in the main study area did nothing all day but move back and forth from the sun into the shade, thermoregulating, in some cases walking no more than 5 m.

The males avoid each other now and those staying in their former territories defend the sleeping burrow and an individual distance of about 8 m. Males chase the females who no longer perform the appeasement gesture.

The movements of the females after egg deposition have not yet been observed. Since they cross half the island to reach the nesting sites, I assume that they exploit feeding areas on their way back to the mating area. They avoid close contact with each other and with the males. As males do, they may spend the nights in suboptimal burrows and usually do not share a burrow with a male.

In lower regions rainfall may be negligible from March until December. Therefore food is much scarcer in the cold season than in the warm period of the year. In the reproductive season when only 3% of the island's surface is exploited for food males feed little if at all for as long as 3 months. In the non-reproductive season they have first access to food by chasing the females. During the non-reproductive season when food is scarcer the population is distributed over a much wider range.

There are few observations concerning juveniles since their number is very low. They may maintain a home range the year round. Juvenile and non-territorial males seem to wander considerable distances and it seems that they are the pioneers in populating suboptimal areas. I have found three non-territorial males at distances of 6 to 10 km away from the original point of capture.

DISCUSSION

The social structure of the land iguana, *Conolophus subcristatus,* on Isla Fernandina conforms with the basic patterns found in the iguanines so far studied, the green iguana, *Iguana iguana* (Müller, 1972; Dugan, this volume), the ground iguanas, *Cyclura cornuta stejnegeri* (Wiewandt, 1977), and *C. carinata* (Iverson, 1979) and several species of the chuckawalla group, *Sauromalus* (Berry, 1974; Prieto and Ryan, 1978; Case, this volume). Polygyny, the mating system in land iguanas, is a typical pattern for species with disparate parental investment (see Dugan and Wiewandt, this volume; Trivers, 1972). Plasticity of social organizations is documented for several lizard species (e.g., Noble and Bradley, 1933; Brattstrom, 1974; Stamps, 1973; see also Wilson, 1975). Individual plasticity in the application of different mating strategies depending on ecological conditions as well as on the momentary situation is pronounced in the observed land iguana population. The different mating strategies observed in *C. cornuta stejnegeri* and the green iguana, and the variable social structure depending on ecological conditions in the chuckawalla, *Sauromalus obesus,* give evidence for a high adaptive potential to varying factors in the ecosociological context in iguanine lizards.

The ecological niche of the observed land iguana population differs in one main aspect from that of all other observed iguanine populations. Fernandina offers a wide range of ecological settings, from jungle to desert, whereas the environment of the other iguanines is relatively uniform, and food may not be available in dry years. On Fernandina rainfall is likely to occur because of the island's altitude, and if lower regions become dry there may still be food at higher elevations. This may be the reason why the males can invest all or most energy stored in the fat bodies into competition for females during the reproductive season, and why the land iguana males show the highest activity level of all observed iguanines during this season.

Ritualized behavior is not prominent in the fighting contests of the land iguanas. Rarely do males introduce fights by displays, and like *C. cornuta stejnegeri* and *C. carinata,* males proceed quickly to physical challenge. This suggests that there is a general difference in the type of aggressive encounters between island and mainland populations; mainland *Sauromalus* and *Iguana* use a high amount of display behavior in their fighting contests. Predator pressure, as Dugan and Wiewandt (this volume) suggest, is certainly related to this difference: in populations with low predator pressure (island), intensely fighting males are not selected out by predators. Land iguana males totally lack predators. This, together with food not being a limiting factor, allows the males extended and energy consuming fights.

The environment of the Mona and the Fernandina iguana is similar in one main respect. On both islands only a small portion is suitable for territories. These are the areas that provide rock caves (Mona) or soft soil for burrow construction (Fernandina) in addition to food nearby. Competition for mates thus centers in small areas and mating success is a direct correlate to a male's ability to monopolize resources (territory). The variability of the behavior of Mona iguanas has been documented by Wiewandt (1977). In the Fernandina iguana behavioral qualities are of such great importance that the defense of an ecologically good territory is no guarantee for a male's mating success. The adaptive potential of iguanines seems to be carried to an extreme in the Mona and the Fernandina iguanas which both use a great variety of mating strategies depending on ecological circumstances and differences between individuals.

Males establish territories every year. In spite of this, in one of the three years of observation the females refused to copulate. It seems that food had been very sparse in the non-reproductive season preceding this particular mating season. No rainfall occurred (at the crater rim) from mid-April until the end of October 1978, and possibly longer. Due to lack of food the males most likely were unable to restore the fat bodies on which they rely in the following mating season. Depleted fat bodies did not allow competitive activities. Evidently the females refused to mate for the lack of behavioral criteria on which they rely for male choice.

An indirect measure of a male's activity level is the number of burrows he constructs or maintains. The burrows may be of importance to both sexes in respect to thermoregulatory needs. They are important to the females as hiding places from males with whom they do not want to mate. Especially during mating time the females stay in the close proximity of a burrow into which they flee or try to, when surprised by a non-territorial male attempting a rape.

Food is not a prerequisite for a territory; there must however, be a feeding ground nearby for the females. A male feeds little or not at all during the reproductive phase even if his territory contains food. By not feeding the males allow more females to exploit the feeding areas inside or near their territories and in this way increase their chances of mating with more females. On the other hand, males have first feeding rights during the non-reproductive season, when they chase the females away.

For nesting the females leave the mating grounds. Probably more than 95% of the female population of the western flank of the volcano walked to the rim and descended into the crater. Incubation inside the crater must offer selective advantages that outweigh the effort and risks for the females to reach it. No conclusive explanation can be given at this point.

Land iguanas lay one clutch per year, consisting of 7 to 23 eggs. Competition for nesting places is pronounced and the females guard their nests. I suspect that only a very small percentage of the hatchlings survives because of the hazards while leaving the crater. Moreover, once they have reached the rim they may have to search for days or weeks before they find food. The number of hatchlings and juveniles we encountered during extensive searches was extremely low.

The land iguana is the only reptile I know of, that combines sperm storage with multiple inseminations. Moreover, sperm storage in this case is not necessary because of a potential unavailability of mates, as the observations show. It very much looks like a female's fitness is improved by a diversified gene combination and is governed by her ability to mate with males whose genome optimally combines with hers.

SUMMARY AND CONCLUSIONS

The social structure and ecology of the land iguana, *Conolophus,* was studied for three years. *Conolophus* is herbivorous and the sexual dimorphism in size is not correlated to intersexual food competition. Both sexes feed on the same plant matter.

The social structure of the land iguana is highly complex and differences in the individual flexibility of mating strategies and in behavior are pronounced. The evolutionary background for this is seen in the wide niche the land iguana can exploit. Fernandina is a volcano that offers a great variety of ecological settings and there are virtually no vertebrate food competitors on the island.

In the reproductive phase males establish territories while females choose a male to mate with. A male's mating success not only depends on the quality of his territory, but also on his behavior. Competitive pressure is a function of the number of females a male gathers in his territory. Territories with a high number of females are more likely to be lost than those with a low number of females. Early in the mating season territorial boundries are defined according to ecological criteria and to the location of burrows that were constructed in the previous mating season. Later on the presence of females is the main factor for shifts in the territorial limits. Males who have no females inside their territory may cause fights by intruding into another male's territory, trying to conquer a burrow (burrows) and at the same time the female(s) that inhabit(s) it (them). There are various other strategies males may use to obtain females. About one-third of the male population consists of peripheral males who rape females at any given chance.

Females stay for a varying length of time in the territories of males. Some females stay with the same male throughout the mating season, others stay only for days or weeks with the same male. Their choice of a male is based on: the quality of his territory, the male's behavior and probably also on the number of females competing for the same male. Females seek multiple inseminations with different males. They reproduce every year. In one year none were observed to copulate but they laid fertile eggs. The largest number of females a male was observed to mate with was four.

The high investment of behavioral energy of males competing for females must be seen in relation with the predictability of food supply on Isla Fernandina. All other studied iguanines show a much lower activity level during the reproductive phase than the land iguana. In their environment (low islands) food may become scarce in dry years. The total lack of predators for male land iguanas is another factor allowing extended and intensive fights. Ritualized behavior is not prominent in fights and it seems that the lack of predator pressure is the responsible factor.

For nesting all females from the western part of the island leave the mating areas and migrate to the crater rim. Nesting areas were found inside of the 900 m deep crater. The nests in the studied nesting areas are heated by fumarole activity.

Males do not feed during the reproductive phase. They have feeding precedence over females during the non-reproductive phase and may chase them. In the reproductive phase the adult population concentrates on 3% of the island's surface, in areas that are suitable for territories. Soft soil and food for the females are the main requirements for mating areas. In the non-reproductive phase the population is distributed over a much wider range.

Acknowledgments

The entire field studies were generously supported during three years by the National Geographic Society.

The cooperation and assistance of the Charles Darwin Research Station and the Galapagos National Park were vital for the success of the field work.

Space and facilities for data evaluation were generously offered by the Florida State Museum in Gainesville during four months.

Among the students and assistants who participated in the field work I owe special thanks to Monique Altenbach, Fanny Rodriquez, Hanspeter Stutz and Udo Hirsch. The work would not have come about without the dedicated help of my Ecuadorian field assistants Oswaldo Chappy, Felipe Gomez, Washington Jaya, Francisco Jaya, Simon Villamar, Wilson Suarez and Bolivar Reinoso.

I am grateful to Adolf Pormann for his constant interest and support of my research. Valuable suggestions on the manuscript were received from Stanley Rand. Zdena Planeta and Marianne New generously helped with the preparation of the manuscript. The final maps were prepared by Helga Birke and Hermann Kacher of the Max-Planck-Institut für Verhaltensphysiologie, Abt. Wickler at Seewiesen.

20

Reproductive Behavior of Galápagos Land Iguanas, *Conolophus pallidus,* on Isla Santa Fe, Galápagos

Keith A. Christian[1]
and
C. Richard Tracy
Department of Zoology and Entomology
Colorado State University
Ft. Collins, Colorado

INTRODUCTION

The two species of land iguanas are found in the Galápagos Islands. *Conolophus pallidus* is endemic to Isla Santa Fe, and *C. subcristatus* is found on the islands of Fernandina, Isabela, Santa Cruz (nearly extinct), and South Plazas. A small population lives on Seymour, but these iguanas were introduced from the now extinct Baltra population. *C. subcristatus* also were formerly found on Santiago. *C. pallidus* is lighter in color than *C. subcristatus* and it lacks the yellow and reddish colors about the head (see Werner, this volume). The adult males of the Santa Fe iguana attain lengths in excess of 50 cm SV and masses in excess of 7,000 g; whereas females of this species average 44 cm and 4,300 g. The diet of these herbivorous lizards consists of a variety of shrub leaves and flowers as well as the pads, fruit and flowers of *Opuntia* cactus.

As part of a larger project on the determinants of space utilization by the Galápagos land iguana *(Conolophus pallidus),* we studied patterns of behavior associated with reproduction and nesting in these large lizards. Throughout the non-breeding season (December through August) males and females were found to occupy large overlapping home ranges. The home ranges are, in fact, spaciotemporal (Wilson 1975) feeding territories in which the iguanas defend only those portions of the territory within which they happen to encounter an intruder. However, without respect to location, adult males supplant females and adults supplant juveniles. Hatchlings are largely ignored by adults, which are more than two orders of magnitude larger. The principal daily activities of iguanas on their home ranges include feeding, thermoregulation and social interactions when two individuals happen to meet. The amount of time spent in these daily activities varies seasonally and the details of seasonal changes in time budgets, home range sizes and thermoregulation will be presented elsewhere (Christian and Tracy, in prep.).

Nesting aggregations have been reported for the Galápagos marine iguana *Amblyrhynchus cristatus* (Carpenter 1966), the green iguana *(Iguana iguana)* (Rand 1968 and see also Rand and Robinson 1969; Sexton 1975; Rand and Rand 1976; Burghardt et al. 1977; Greene et al. 1978) and the Mona Island ground iguana *(Cyclura cornuta stejnegeri)* (Wiewandt 1979); this topic has been reviewed by Wiewandt (this volume). Nesting aggregations of Galápagos land iguanas have not been reported previously (but see Werner, this volume). However, Darwin (1845) described an area on the island of Santiago (James) in which the density of land iguana burrows was so large that the area must have been the site of a nesting aggregation of *Conolophus subcristatus*. Unfortunately, land iguanas have been extinct on Santiago since before 1906 (Thornton 1971).

The principal nesting aggregation of Isla Santa Fe was discovered on 8 October 1978, at which time females were already present and digging burrows. Nesting activities peaked in mid October, and then gradually declined until they ended during the third week of November. In mid October, over 200 nesting females were counted in the nesting area of approximately 2 hectares. Since nesting activity had already begun when the site was discovered, and because there was constant turnover of females in the area, the total number of iguanas nesting in the 2 ha plot was certainly much greater than 200. A total of eight additional nesting aggregations were discovered on Santa Fe, but all were smaller than the principal.

Different colors of paint were applied to the tail, back, or neck of the iguanas to allow individual recognition. In order to minimize trauma, the marking was done at night by simply dribbling the paint on the desired body region while the iguanas slept. Most of the iguanas slept through the procedure.

In excess of 5,200 person-hours of observations of land iguanas were made between 26 June 1978 and 23 April 1979. Of these, over 750 person-hours of observations were made of nesting females between 8 October and 3 December 1978. Over 40 nesting females were marked at the nesting site, and detailed data were collected for 21 of these.

STUDY SITES

Santa Fe is an arid island that covers 2413 ha. The island has many cliffs and steep ridges that generally run East-West. Site I (Figure 20.1). was established along a North-facing cliff face, and the dominant plant species were the tree-form cactus *Opuntia echios*, Palo Santo trees *Bursera graveolens*, and two species of shrubs, *Cordia lutea* and *Lantana peduncularis*. This site was used principally for studies of land iguanas during non-nesting activities, but it is relevant to discussions of nesting because it was the home of females that were later observed nesting elsewhere on the island, and because four of the five pairs of iguanas that were observed during courtship were at this site.

Observations were made at Site II during the nesting season and then again during February 1979 when the hatchling iguanas emerged. This site is a flat area of loose cinder soils, bordered by a more typical rocky area resembling Site I. The border is created by a distinct rocky ridge which is 1-3 m higher than the flat, relatively open area of soil where the nests were made. The species composition of plants for this rocky area is essentially the same as described for Site I above, but the nesting area itself is dominated by the shrub *Encelia hispida. Opuntia, Bursera, Cordia* and *Lantana* are also found in the area, but at a much lower density than in the surrounding areas. *Encelia,* which typically reaches a height of 1 m, is restricted to soils of loose texture (deVries 1973), which is also where the iguanas nest. As with green iguanas (Rand 1968), the land iguanas help maintain the clearing in which they nest by digging up and trampling the small plants each year.

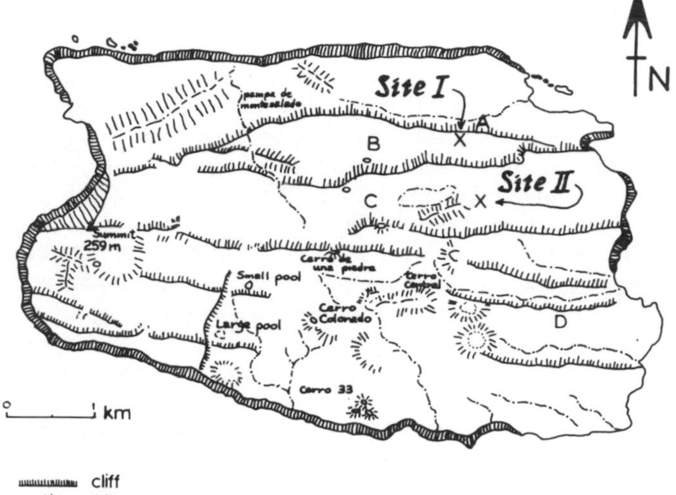

Isla Santa Fé (Barrington)

Figure 20.1: Map of Isla Santa Fe, Galápagos. Site I was used for year-round studies of home range, and Site II, which is the largest communal nesting area on the island, was used for studies of female nesting behavior. Positions A through C show the locations of smaller nesting areas. One female, after completing her nesting activities, walked directly from Site II to point D in a single day (see text for details).

COURTSHIP

Courtship took place from mid August to mid October. During this time, females left the home ranges to which they are attached most of the year (Christian et al. in prep.) to choose a male with which to mate (n=7). Some males also moved into areas adjacent to their normal home ranges, apparently to recruit females (possibly as many as 6 of 10 males observed). Approximately 44 hours of observations were made on 5 courting pairs of iguanas. Once a pair was formed, the female stayed in the immediate vicinity of the male's burrow. During this period (which averaged about 2 weeks, but one pair remained together for 7 weeks) both sexes slept in the same burrow, and spent a minimum amount of time feeding. Normal thermoregulatory behavior also was apparently curtailed, particularly by the female who spent much of the courtship period inside the burrow or just outside the entrance. Throughout courtship, females remain very dark colored. It is not known whether this coloration is a part of the courtship display, or due to lowered body temperature. During the courtship period the females were unusually skittish, and the males were very defensive of the area immediately around the burrow. Other portions of the male's home range were rarely, if ever, patrolled while the female remained near the burrow.

Displays between pairs occurred intermittantly throughout the day. The display pattern begins when the male paces around the female with a characteristic strut and head bobs that are given at a rate greater than the very common threat display. The strut involves a step in which the back of the wrist of one of the forelegs is dragged across the ground, and this is followed by a quick, vigorous head bob. This sequence, step, drag, bob, is repeated rhythmically as the male paces around the female. This characteristic strut was not observed in any context other than courtship.

The female's response to the male display is a characteristic head bob which is performed with her mouth open and her head raised nearly perpendicular to the ground surface. She typically walks very little while the male paces around her. The head bob with the mouth open was observed in females at other times of the year; it was always displayed by a female to a male and never between females.

In two cases, two females were courted simultaneously by a single mate. In these instances all three iguanas slept in the same burrow for several nights. It is certainly possible that males mate sequentially with different females since the females leave the males for the nesting area after copulation, but such sequential mating was not observed. One copulation was observed, and it followed the characteristic pattern observed in other iguanids (Carpenter 1978). The "bite hold" of the male produced some bleeding from the female's neck.

Territorial disputes between males occurred throughout the year. Disputes usually involved head bobbing displays followed by the retreat of one of the animals. These disputes occasionally developed into fights involving wrestling and biting of legs, tails, and in particular backs and napes. During the year of field study more than 15 fights between males were observed on Santa Fe. These typically lasted more than an hour and in two cases over four hours. No serious wounds resulted in the fights observed, but bleeding flesh wounds were common.

MOVEMENT TO NESTING AREA

Only one observation was made of a female leaving her mate's burrow following copulation. The female walked directly from the burrow to the area in which she eventually laid her eggs. The male followed her for a short distance to the boundary of his territory, then he resumed normal precourtship (i.e., solitary) activities. On route to her nest site (which in this case was only 200 m away, at Point A, Figure 20.1) the female encountered another male who initiated courtship displays (the strut/bob display) with her. She rejected his advances and attempted to avoid him. He persisted, however, and followed her for a distance of about 40 m, displaying the entire time before returning to the area of his burrow. The female reached her nest site in less than an hour. She started the exploratory phase of nesting (below) the next day, and she laid her eggs 15 days later.

Other females travelled greater distances to nest. Four females with home ranges at Site I (Figure 20.1) nested at Site II, approximately 1 km distant. Observations of females returning to their home ranges after nesting indicate that some travel even greater distances (see below). One female from Site I was seen at her mate's burrow at Site I only two days before she was discovered digging a nest burrow at Site II; this again suggests that the females move directly from the area of mating to the nest site. Green iguanas, on the other hand, apparently nest several months after mating in Costa Rica (Hirth 1963) and 4-6 weeks after mating in Panama (Dugan this volume).

Females do not necessarily choose the closest nesting area. Four females from Site I travelled to Site II to nest even though two other nest sites (A and B; Figure 20.1) were closer. One of these closer sites (A) was small (approximately 60 m^2), but the other (site B) was more than 1 ha. The determinants of nest site selection are not known. Females might (1) find their nest site by random search each year, (2) discover a nest site by random search the first time that they nest and then return to the same site in subsequent years, or (3) return to the site from which they hatched. The evidence does not support the first of these possibilities since the females apparently walk directly to and from (see below) the nesting areas. We have no evidence that allows us to distinguish between (2) and (3).

NESTING

Rand (1968) identified four phases of nesting in the green iguana: (1) exploration, (2) digging, (3) egg laying, and (4) nest closing. We will describe these four stages and two additional phases that are important in *Conolophus pallidus:* (5) defense of the nest, and (6) recuperation.

Exploration

Many aspects of this stage, which includes selection of a nest site, are similar to those described for the green iguana (Rand 1968) and for the marine iguana (Carpenter 1966). However, female green iguanas apparently are not aggressive to other females during this stage. This is not the case for marine iguanas (Carpenter 1966) or for land iguanas. We observed in excess of 200 aggressive displays, chases and fights between nesting females. Agonistic encounters were commonplace for iguanas in all phases except recuperation. Fights were similar to those between males, but of shorter duration (lasting only 10-15 minutes). All aggressive encounters were accompanied by the species-specific head bob display (Carpenter 1969, this volume) which was invariably performed by both participants in an encounter.

As with green iguanas (Rand 1968), female land iguanas in their exploratory phase often dig false nests (i.e., burrows in which eggs are not laid). False nest digging by land iguanas is more extensive than that of green iguanas (Rand 1968), with the false nests of land iguanas often reaching the size of the final nest burrow in which eggs are laid.

Many seemingly aberrant activities occur during this phase, some of which are associated with false nests. Many false nests were abandoned after several hours had been spent in construction. However, the false nest often would be filled in by yet another lizard. In one instance, a female left the burrow she had been digging and challenged another female that was digging her own burrow. A vicious fight followed, and after several minutes the first iguana succeeded in supplanting the second. She then entered her newly won burrow, stayed for less than five minutes, and then returned to her original burrow and resumed digging. In another instance, a female dug a burrow, filled it in, dug it out again and then walked 20 m and started digging another. These activities took most of the day, and by sunset she had dug only a shallow burrow. At this time she returned to her original burrow, but during her absence the burrow had been filled in by still another female. She then slept in a nearby burrow that had been dug and abandoned several days earlier by still another iguana. The next morning she dug a burrow in precisely the same place where she had dug the two successive burrows the day before, and on this occasion laid eggs in this burrow.

These behaviors seem counter to those expected if exploration behavior attempts to minimize energetic expenditures during nesting (see Rand and Rand 1976, and Rand and Rand 1978). Why, then, do females dig false nests? Since the same site is often ultimately used for nesting by another (or even the same) iguana, it does not seem that false nests are rejected because of edaphic features (see Rand 1968). Even if the false burrows were defective, this would not explain why a female would expend the energy to cover up a defective hole that she (or even more difficult to explain, some other female) had dug. One plausible explanation for these "exploration" activities is that females are simply not physiologically ready to lay upon arrival at the nesting area, and that digging may act as a stimulus to the physiological changes prerequisite to laying (Rand 1968).

Rand (1968) noted that green iguanas in the exploratory phase are shy and skittish, i.e., easily frightened by sudden noises or movements. Female land iguanas are not nearly as shy during this phase as they are during courtship, and even though there are a lot of movements and social displacements, these appear to be driven by overt aggressiveness rather than by skittishness.

Digging

Land iguanas' method of digging is similar to that described for marine iguanas (Carpenter 1966) and green iguanas (Rand 1968). It involves digging with the forelegs in the early stages of excavation, pushing accumulated dirt away with the hind legs as the hole gets deeper, and in the late stages, using the front legs and pectoral region to push dirt out and away from the burrow.

It takes approximately 4 hours for a female to dig a nest burrow, but this varies widely and is dependent on the number of aggressive encounters she has during the process. Females spend the night in completed or partially completed burrows during this phase of nesting.

Completed burrows are typically 1.5-2 m long, and gradually slope to a depth of 0.5-0.75 m deep at the terminal chamber where the eggs are deposited. Since no nests were excavated, we do not know the details of the architecture of this egg chamber.

Egg Laying

In five instances we observed females remaining inside their burrow for over 24 hours after partially sealing themselves inside by pushing dirt from within up into the burrow entrance. Upon emergence these females were visably shrunken, and they immediately filled in the burrow and began defending their nest (see below), suggesting that eggs had been laid. In three instances, however, females spent as little as two hours in the nest before they emerged, covered the nest, and began to defend it.

Although no nests were excavated, we estimate that the clutch size of *C. pallidus* to be about 10. This is based on 5 observations of hatchlings emerging from their nests, and it assumes that there is essentially no egg mortality.

Nest Closing

After egg laying each female scratches, kicks, and shoves dirt into the burrow. This takes approximately 1 hour, depending on the number of interruptions due to social interactions.

Captenter (1966) and Rand (1968) report that marine and green iguanas pack dirt into the burrow by ramming it with their head or snout. Land iguanas were not observed using their heads or snouts. Instead, they packed dirt by patting it with their forepaws. It is possible that they use their heads while filling in the depths of the burrow where we were not able to observe them. The late stages of nest closure are similar to that of the marine iguana (Carpenter 1966) and the green iguana (Rand 1968). All three species exhibit "surface filling" in which the area surrounding the nest is continually scratched and churned up until the exact location of the nest is difficult to determine.

Defense

Rand (1968) reports that green iguanas surface fill for some time, and then broaden their range of activity until they lose "all special interest in the site of the concealed nest." Apparently the spent females quickly become less aggressive and make no attempt to defend their nest sites after the surface filling (Rand 1968). Eibl-Eibesfeldt (1966) reports that female marine iguanas remain in the vicinity of their nests and fight other females that approach. He had no data on how long the females stand guard, but he speculated that they defended nests for several days. Recently Wiewandt (1979) has reported that female ground iguanas *Cyclura cornuta stejnegeri* guard their nests for 1 to 20 days after egg laying (see also Wiewandt, this volume). Defense of the nest site could prevent another female from inadvertently digging up the eggs and kicking them out on to the ground, as occurs in the green iguana (Rand 1968 and Sexton 1975).

There is a great deal of variability in the duration of defense behavior by female land iguanas after they lay their eggs. Of the 21 nesting females for which we have detailed data, 6 left the nesting area immediately after completing their nests. Four other females spent only 2 or 3 days vigorously defending their nests, and then these 4 females left the nesting area without any recuperation (below). Of the remaining 11 females for which we have data, 6 spent 1 to 3 weeks after completing their nest burrows defending their nests and recuperating; the other 5 females spent more than a month defending and recuperating at the nesting area. The 11 females that participated in both the defense and recuperation phases typically vigorously defended (i.e., spending most of the day in this activity) their nests for about a week, and then their defensive behavior gradually declined over the second week as they entered the recuperation phase (below). One female (Marty) was observed still defending her nest one month after she laid her eggs. She laid her eggs during the first week of November 1978 and spent over a week vigorously defending the site. She began to spend less time each day at the nest, but she would occasionally return, scratch the soil around the nest and chase intruding females. After about three weeks, she was spending virtually all of her time eating and resting in the rocky border of the nesting area about 15 m from her nest. However, during the first week of December a very late female came into the nesting area and began searching for a nest site. She was the only iguana digging in the area this late in the season, but she happened to start digging exactly over Marty's nest. By the time Marty had discovered the intruder, the new burrow was well under way, but Marty initiated a fight and chased the newcomer away. Marty then filled the partial burrow with dirt and smoothed the area with surface filling.

The nesting territories were approximately 2.5-3 m in diameter. The details of the use of space by both nesting and non-nesting iguanas will be presented elsewhere (Christian and Tracy, in prep.) Females defending a completed nest typically were dominant over females in earlier phases of nesting. During the nest defense stage, females gradually move from the nesting area and begin to sleep in rock crevices areas adjacent to the nesting site. This continued until the females eventually returned to their normal home ranges.

Recuperation

The transition between the defense and recuperation phases is gradual, but in recuperation a lot of time is spent feeding and little in nest defense. Until the recuperation phase the females have spent several weeks without actively foraging, and they were observed to eat only when a prime food item (i.e., *Opuntia* pad or fruit) was directly in their path. The activities of the recuperation phase include basking, a great deal of feeding, and resting in the shade during the heat of the day. The iguanas in this phase are markedly less aggressive.

The length of time spent in the recuperation phase before they journey back to the original home range is variable. One female from Site I was still at the nesting area at Site II five months after she had laid her eggs. However, this may represent a home range shift rather than an extended period of recuperation.

THERMOREGULATORY BEHAVIOR DURING NESTING

Rand (1968) suggested that the thermal environment restricted nesting activities of green iguanas on cool rainy mornings and during periods of very bright sun. A census of thermoregulatory behavior by nesting land iguanas was taken to discern any pattern of behavior relative to the thermal environment of the nesting area.

The high density of iguanas at Site II allowed the simultaneous observation of thermoregulatory behavior by a large number of iguanas. Observations were made at three check points along the rocky ridge bordering the nesting area. At each half hour the positions of all iguanas that could be seen from the three check points were recorded. Specifically, the iguanas were characterized as being in the sun, the shade, or the semi-shade (defined as the diffuse shade produced by leafless shrubs). In addition, soil surface temperatures (in sun, shade and semi-shade), air temperatures (at 10 and 200 cm), and a qualitative description of sky conditions (sunny, cloudy, partly cloudy) were recorded.

On a typically sunny day (Figure 20.2A), the iguanas remained in the open basking in the morning and late afternoon, but during the middle of the day, when the substrate remained above 50°C, they retreated to shade or semi-shade. The smaller total number of iguanas seen during midday has two explanations: first, some lizards retreated to their burrows if available, and second, some females retreated to the rocky area adjacent to the nesting site where deep shade (especially from *Opuntia* trees and rock overhangs) was more abundant. Iguanas that remained on the nesting area but lacked a burrow into which they could retreat (in particular, iguanas in the defense phase) typically sat at the base of the small, leafless *Encelia hispida* shrubs. As the day progressed they would readjust their positions to follow the shifting semi-shade. On several occasions, females were observed sitting upright with their forefeet propped up in the limbs approximately 15 cm off the ground. These observations were made during the hottest part of the afternoon, and presumably this posture reduces heat stress since much of the body is at a height where air temperatures are lower and wind speeds are greater than at ground level (see Porter et al. 1973).

On cloudy days when soil surface temperatures were much lower, basking activity lasted longer in the morning and started earlier in the afternoon, and lizards spent most of the day in the open (Figure 20.2B). Fewer iguanas were counted during the cooler early mornings and late afternoons. During midday, however, the iguanas were busy with their nesting activities; only a few retreated to the semi-shade and none were seen in the midday shade. Apparently the iguanas retreat to their burrows to avoid the heat of hot days and the cool of cloudy days.

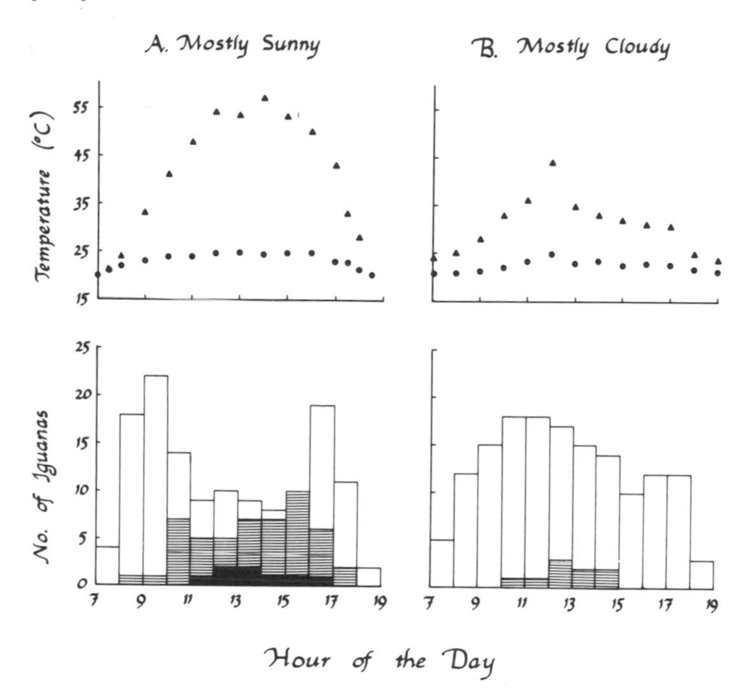

Figure 20.2: The number of iguanas in the open (open bars), in semi-shade (hatched bars) and in the shade (dark bars) for different times of the day on 28 October (A) and 30 October (B). The top graphs show that 28 October was a sunny day as can be seen by the high soil surface temperatures (triangles), and that 30 October was a cloudy day as can be seen by the low soil surface temperatures. The dots in the upper graphs represent air temperatures.

PREDATION AND CANNIBALISM ON EGGS

Sexton (1975) has described three circumstances in which the eggs of green iguanas are vulnerable to predation by black vultures: (1) during the process of digging their own nests, iguanas inadvertently dig up eggs of other females, (2) vultures take eggs out of the nest before it is covered by the iguana, and (3) eggs are laid on the surface because the female has been disturbed or for some reason is not in her burrow. Parallels to these situations exist for nesting land iguanas. On 6 occasions eggs were found on the ground and appeared to have been dug up and kicked out by a nesting female. Although we did not see the eggs being dug out of the nest, it seems inevitable given the density of iguanas in the area.

On one occasion, a Galapagos mockingbird *(Nesomimus parvulus)* was seen dragging eggs out of an open nest. The reason that the nest was open and undefended is not known.

On two occasions, we found eggs that apparently had been laid on the surface. The actual laying of the eggs was not observed, but there was no evidence of digging in the immediate vicinity. These eggs were directly in front of our observation post, and it is believed that any activity more obvious than a female simply dropping an egg behind her would have been noticed.

Most females completely ignored eggs lying on the ground. One female, however, was observed to display what seemed to be an inordinate interest in exposed eggs on four different occasions. This female was being watched closely since she contained a temperature transmitter at the time. In each case, when she saw the eggs from distances of 10-15 m, she ran directly towards them. In route to the eggs, she crossed vigorously defended nesting territories, and was chased by the residents. On one occasion she ran to an egg (which we believe had been dragged out by a mockingbird), picked it up, broke it in her mouth, and swallowed it before she was chased away by a territorial female. In another instance the egg-eating female swallowed the egg whole as she was being chased. When this egg was passed four days later it was still intact. In another instance, this female was attracted to an egg by the actions of a mockingbird that was attempting to break the egg open with its beak. The iguana ran to the egg, but the mockingbird picked it up in its beak and flew away. The iguana chased the mockingbird for about 10 m, and in the process she crossed the territories of two other females who chased her in turn.

Since all observations of egg cannibalism were of one individual, it is not known if the habit is widespread among land iguanas, or whether we observed the behavior of an aberrant female. Certainly not all iguanas ate eggs commonly since we observed many individuals that ignored exposed eggs.

MOVEMENTS BACK TO THE HOME RANGE

Montgomery, et al. (1973) monitored the post-nesting movements of green iguanas from the nesting aggregation, but the original home ranges of these animals were not known so it is difficult to interpret the data. As discussed above, one female land iguana from Site I had not returned from the nesting area even after five months. However, this seems to be exceptional, since 8 of 10 females whose pre-nesting home ranges were known returned to their original home ranges after nesting. By mid November virtually all activity had ceased on the nesting area itself, and by late November there was essentially no activity in most of the adjacent area as well.

Although our data are scant, females apparently take a very direct path back to their home ranges once leaving the nesting area. One female was seen nesting at Site II only two days before she returned to her original home range at Site I. One female was followed after she left the nesting area for a distance of more than 1.5 km. Her original home range was not known, but she was followed because she was carrying a temperature transmitter. She covered the distance between Site II and point D in Figure 20.1 between the hours of 1215 and 1645, despite the fact that she travelled up and over three steep ridges, one of which was over 100 m high. The interesting aspect of her walk was the remarkably straight path she followed, even at the cost of scaling steep rock faces rather than going around them. When an obstacle could not be climbed she detoured and then immediately returned to her straight-line course. The cues for orientation and navigation by land iguanas are not known. During her walk, this female's body temperature reached 39.9°C, very close to the panting threshhold for this species (Christian and Tracy, in prep.).

HATCHING AND EMERGENCE

At the time of hatching, *Conolophus pallidus* have a mass of about 40 g and a SV length of about 10 cm (n=5). They have a cryptic dark brown mottled pattern over a tan background.

Four instances of natural emergences were observed. These occurred between 0820 and 1015 hours. In one case, 10 hatchlings emerged during a period of 1.5 hours, but the presence of the observer probably influenced this time period. (Several hatchlings reentered the burrow when they saw the observer.) In another case, probably more typical, nine hatchlings emerged over a period of 20 minutes. One nest was unintentionally caved-in by a walking investigator, and 10-12 hatchlings immediately ran out of the artificial hole. This suggests that the baby iguanas remain inside the nest for some period between hatching and emergence.

By the end of February scores of emergence holes dotted the nesting area. The emergence holes dug by the hatchlings are typically 5-7 cm in diameter; some of the tunnels lead straight up from the nest chamber, while others are gently sloped. The hatchlings probably take the path of least resistance in digging their way out.

We observed hatchlings of three clutches basking immediately after emergence. Within one to two days of emergence the hatchlings apparently disperse from the nesting area to other parts of the island having more food and cover.

Predators of hatchling land iguanas include snakes (*Dromicus* spp.) and the Galapagos hawk *(Buteo galapagoensis)*. Only one instance of predation by a snake was observed; the snake struck and seized the hatchling by the head. Hatchlings are extensively preyed upon by hawks (we observed 40 successful acts of predation by hawks). The details of this will be published elsewhere (Christian and Tracy 1981).

Burghardt, et al. (1977) reported social groups of newly hatched green iguanas, and Greene, et al. (1978) suggested that one of the functions of these groups is to aid in the detection of predators. We observed five cases in which hatchling land iguanas were associating and moving as groups, with an average of seven hatchlings in a group. Because hatchling land iguanas are skittish, we were not able to observe the interactions within these groups, but it appears that these groups are not as socially sophisticated as a group of green iguana hatchlings.

The hatchlings on Santa Fe emerged during February, coinciding with the beginning of the heavy rains on this island. By the end of the first week in February 1979, the vegetation was lush and in bloom. The hatchlings thus, emerged at a time of abundant, succulent food, and one might postulate that the breeding season has been adjusted to synchronize hatchling with the rains. However, populations of *Conolophus subcristatus* on other islands do not follow this pattern (Howard Snell, in prep.). The environmental factors influencing the breeding cycle are apparently complex, and it may be that different factors have different effects on the various islands.

DISCUSSION

Rand (1968) suggested that the nesting aggregation of green iguanas on Barro Colorado Island, Panama occurs because the site is the most attractive nest site in the area rather than because of a mutual attraction among iguanas. In particular, he suggested that the island site that he studied provides protection from nest and egg predators (especially mammals) (see Rand and Robinson, 1969). Similarly, the aggregation of *C. pallidus* is probably not the result of mutual attraction among females either, but the presence of soil suitable for nesting is most likely the attracting feature as is the case for *Cyclura cornuta stejnegeri* on Mona (Wiewandt 1979). Most of Santa Fe is extremely rocky, and digging is very difficult (pers. obs.). There are disadvantages to communal nesting which include intraspecific interference with nesting activities (Sexton 1975) and the attraction of predators at the time of emergence (Christian and Tracy 1981).

One obvious feature of the nesting behavior in *C. pallidus* is the amount of variability among the females in many aspects of nesting behavior. Despite the variability, the formation of nesting territories which exist for an extended period and the fact that females maintain their claim to the nest site even if they leave for some time indicate that the behaviors associated with communal nesting are more developed in *C. pallidus* than in the green iguana. The situation resulting in communal nesting in green iguanas apparently has been in existence for less than 60 years (Rand 1968), but Santa Fe land iguanas have probably been nesting communally for many centuries.

The defense of the nest site, after the eggs have been laid, is a form of parental effort (Low 1978) on the part of female *C. pallidus* that is somewhat rare among lizards. Thus, *C. pallidus* exhibits reproductive behavior that is typically iguanid, but with features that are unusual among reptiles.

SUMMARY AND CONCLUSIONS

A nesting aggregation of in excess of 200 female land iguanas *(Conolophus pallidus)* was studied from early October until the end of November 1978. Detailed observations were made of 21 individually marked females that were nesting in the flat, relatively open 2 ha nesting area.

Before migrating to the nesting areas, females left the home ranges in which they are found most of the year to choose a male with which to mate.

There was some movement of males during this time, but their movements were more restricted and apparently represented attempts to solicit females. Pairs remained together for approximately 2 weeks, during which time they remained close to the burrow of the male. Courtship displays involved characteristic head bobs by both sexes, and a stereotyped strut by males.

Females apparently migrate directly to a nesting site although it may not be the nearest nesting area. Females walk at least 1.5 km in order to reach their nest site, and even greater distances are probably traveled by some females.

During the exploration phase of nesting, chases and fights were common as females started and abandoned nest burrows. Some females remained in the nest for over 24 hours while laying eggs, but others took as little as 2 hours.

Most females defended their nest site (2.5-3 m in diameter) after they had closed the nest burrow, for periods ranging from a couple of days to over a month.

During February the hatchlings emerged from the nests, and if they were not disturbed, they dispersed from the nesting area to other parts of the island containing more food and cover within one or two days.

Predation upon hatchlings by Galapagos Hawks in the relatively open and unprotected nesting area was extensive.

During the nesting period, females largely curtailed feeding (due in part to the limited area used for nesting and the relative absence of food in the area). However, the nesting females did exhibit thermoregulatory behavior. Defense of nest sites gradually decreased as females spent more time feeding in areas adjacent to the flat nesting site. The time spent feeding and recuperating from nesting activities ranged from no recuperation period to more than a month. After recuperation, females returned to their former home ranges where they spent the rest of the year.

Note

[1]University of Wisconsin, Madison, Wisconsin

Acknowledgments

David Socha spent six weeks observing nesting females, and Sylvia Harcourt and Janet Shur also made significant contributions to the field work on Santa Fe. This study could not have been accomplished without their help, and we extend great thanks to them.

The Charles Darwin Research Station provided invaluable logistical support. We acknowledge the partial financial and logistic support of Earthwatch, Inc., W.P. Porter and Dean Cilly. This is contribution No. 281 of the Charles Darwin Foundation for the Galapagos Islands.

21

Variation in Iguanine Social Organization: Mating Systems in Chuckawallas *(Sauromalus)*

Michael J. Ryan
Neurobiology & Behavior
Langmuir Laboratory
Cornell University
Ithaca, New York

INTRODUCTION

Sauromalus is one of the few genera of lizards for which there are comparative data on social organization. Previous studies indicate that social organization varies not only among populations but within a single population among years. An understanding of this variation requires not merely a description of the types of social systems but an evalutation of their dynamic nature.

Social plasticity is well known in lizards (e.g., Rand, 1967b; Stamps, 1975) and was first documented when species territorial in the field formed dominance hierarchies in the laboratory (Noble and Bradley, 1933). This phenomenon is widespread in lizards and is not merely a laboratory artifact, as was demonstrated by Evans' classical field studies of *Sceloporous gramnicus* (1946) and *Ctenosaura pectinata* (1951) and more recently by Stamps' (1973) work with *Anolis aeneus.* They showed that species may be territorial in one situation but form dominance hierarchies in another.

Hunsaker and Burrage (1969) suggest that there is a continuum from territoriality to dominance hierarchy. Many authors contend that this shift in social organization is due to the inability of males to defend territories under crowded conditions (e.g., Brattstrom, 1974). This is a plausible explanation for the phenomenon observed in the laboratory. However, it does not necessarily explain the existence of dominance hierarchies in the field and is even less applicable in explaining the more subtle variations in social organization which have been reported (e.g., Rand, 1967b).

In this chapter I review the previous reports of social behavior in chuckawallas *(Sauromalus).* I will attempt to identify the function of territoriality (*sensu strictu,* Williams, 1966; see below) in these lizards by considering the behavior in terms of its potential costs and benefits. I then consider which factors affect costs and benefits, how they might be responsible for the observed social plasticity in *Sauromalus* and propose a predictive theory of social organization in these lizards.

SOCIAL ORGANIZATION

Berry (1974) investigated the ecology and social behavior of *Sauromalus obesus obesus* near Lone Butte, California. In 1969 and 1971 only the larger males or tyrants were territorial. These territories were aggressively defended against other large males and overlapped the home ranges of females, juveniles and subordinate males. There was a high level of aggression as subordinate males constantly challenged the territorial dominants. A stable dominance hierarchy, based on size, was maintained among subordinate males. Berry classified this a tyrant-subordinate system.

Chuckawallas feed primarily on the shoots and flowers of desert annuals. This food source may show radical fluctuations from year to year due to changes in the amount of rainfall. In the winter of 1969-1970 there was below average rainfall in California. Consequently, in 1970 the food supply was low and lizards were relatively inactive. Berry reported that aggressive interactions were only 15.6% of those in 1971 and there was virtually no courtship or copulation. During this year a dominance hierarchy formed around a rock pile and several nearby perennial sweet bushes *(Bebbia juncea).* The chuckawallas fed on the flower heads of these bushes.

In summary, Berry reported that in 1969 and 1971, when food was plentiful, only the larger adult males were territorial. Subordinate males aligned in a dominance hierarchy based on size. However, during the 1970 breeding season, when food was scarce, there was no mating and no territorial defense and some males formed a dominance hierarchy centered on a food source.

During the same drought year of 1970, Nagy (1973) studied a population of chuckawallas *(S. o. obesus)* on Black Mountain in the Mojave Desert, California. After all the annuals died in May, the lizards shifted their diet to perennials. He noted there was reduced activity, infrequent aggression and no reproduction. There was also extensive overlap of home ranges, both between and within sexes, and no territorial defense. Nagy reported similar home range sizes for males and females (males = 0.20 ha, females = 0.17 ha). In previous studies where male territoriality was observed, the home range of males was much larger than that of females (e.g., males = 0.57 ha, females = 0.17 ha; Johnson, 1965).

Prieto and Ryan (1978) described social organization of the Arizona chuckawalla, *S. o. tumidus,* in Organ Pipe National Monument, Arizona. In that study, all adult males defended well-defined territories that usually conformed to the boundaries of groups of rocks. These territories overlapped the home ranges of females and juveniles but were defended only against other adult males. Males introduced into the territories were challenged immediately by residents. I refer to this system in which all adult males defended territories as "strict territoriality" to distinguish it from the tyrant-subordinate system reported by Berry.

Prieto and Ryan also observed the social behavior of an enclosed population of these chuckawallas. The adult males did not defend territories. One lizard was dominant and occasionally defended his basking site. He aggressively challenged other males and on one occasion disrupted a copulating pair and attempted to mate with the female. When the dominant individual was removed, another lizard immediately became dominant. There were several dominants during the study but never more than one at a time. They were frequently challenged and occasionally supplanted from their dominant position. The social structure in the laboratory enclosure was similar to that within a single territory of the tyrant-subordinate system.

Two species of giant chuckawallas, *S. hispidus* and *S. varius,* occur on islands in the Gulf of California. Case (1978, this volume) reported that these animals were not territorial and exhibited no obvious social hierarchy. Reproduction was infrequent, occurring only in the rainiest years and even then only 30% of the females were gravid. Case also noted that resources on the island were abnormally high in wet years and predation pressure low in comparison to mainland populations of *S. obesus.*

This brief review shows that there is a wide range in the expression of territorial behavior among chuckawallas (Table 21.1). The remainder of this chapter will attempt to define the function of territoriality in these lizards and consider the factors responsible for the observed variations in social organization.

Table 21.1: Variation of Social Organization in *Sauromalus*

	Species	Location	Situation	Social System
Berry (1974)	*S. obesus obesus*	California	Field 1970*	Dominance hierarchy
			Field 1971	Tyranny—subordinate
Nagy (1973)	*S. o. obesus*	California	Field 1970*	No apparent social organization
Prieto and Ryan (1978)	*S. o. tumidus*	Arizona	Field	Territoriality
			Lab	Despotism
Case (1978)	*S. hispidus* and *varius*	California	Islands	No apparent social organization

*Drought year in California.

THE ADAPTIVE SIGNIFICANCE OF TERRITORIALITY

The adaptive significance of any trait is best considered in terms of its potential costs and benefits. This concept was elegantly applied to the study of territoriality by Brown (1964) with his theory of economic defendability. Simply stated, this theory predicts that an animal will defend a territory only when the benefits accrued from possessing a territory exceed the costs of defense and maintenance.

Obvious costs of territorial defense in chuckawallas are time and energy expenditure and exposure to predators. Rand and Rand (1976) demonstrated that time and energy considerations are important in determining the intensity of dispute interactions in nesting *I. iguana*. In chuckawallas, energy considerations may be even more important in years of low food supply. Nagy (1973:93) suggested that "chuckawallas apparently abandon costly social behavior and reproduction in years when succulent food sources are scarce."

Territorial defense by highly conspicuous displays probably increases the male's exposure to predation. Berry observed a large number of potential predators in her study area and noted that chuckawalla remains were found in coyote scats. She also described various defensive reactions of chuckawallas to coyotes and soaring birds, some of which were known predators. Prieto and Sorenson (1975) demonstrated that chuckawallas respond to post-anal gland secretions of potential snake predators. The evolution and maintenance of these behaviors suggest that predation might be an important selection pressure.

Rand (1967a) discussed the potential benefits of territoriality in iguanid lizards. He suggested that a selective advantage is accrued to territorial individuals by increasing the possibility of (1) offspring survival, (2) securing environmental resources, (3) mating, and any combination of these benefits.

The Function

In an attempt to understand why there is variation in social organization we must first consider the selection pressures responsible for its evolution. Williams (1966) emphasized the necessity of distinguishing between *function* and *incidental consequences* of adaptation. The function of a causal mechanism implies that the mechanism was shaped by selection for the goal attributed to it. Fortuitous effects of the mechanisms are considered incidental consequences, even if these effects are advantageous to the individual. For example, if territoriality has evolved because of the selective advantage accrued to the individual by obtaining a specific, critical resource, then the defense of other resources which happen to occur in the territory is an incidental consequence and not a function of territoriality.

I will suggest the function of territoriality in chuckawallas by examining each of the benefits proposed by Rand in terms of potential costs and benefits. The hypothesis that the function of territoriality is to achieve a specific benefit will be rejected if alternate means of attaining this benefit, without incurring the costs of territorial behavior are apparent.

Offspring Survival

Parental care in lizards usually is restricted to prehatching investment, although some limited care of neonatal offspring has been reported (Evans, 1959). There are no obvious examples of parental care in chuckawallas and it is not known how nest site selection affects hatching success. Also, there are no suggestions that the male's territory influences juvenile mortality. Therefore, there is no direct evidence that territoriality in chuckawallas affords any benefit in offspring survival.

Securing Environmental Resources

A careful examination of chuckawalla ecology reveals at least three important environmental resources contained in the territory: food, basking sites and rock crevice retreats. All of these resources are of obvious advantage to the territory holder, as suggested by Berry. However, by considering the spatial and temporal distribution of these resources in concert with the behavioral ecology of the lizards, it can be determined if these resources are economically defendable: that is, if costs are minimized and benefits maximized through territoriality.

Chuckawallas are xeric herbivores. Consequently, they are especially subject to the vagaries of climate as manifested by fluctuating food supplies. As previously mentioned, social organization is affected when food is scarce (Berry, Nagy). Therefore, it might be predicted that food defense is an important function of territoriality.

In 1970 when a tyrant-subordinate system was present, Berry noted that food sources were utilized readily by subordinate males in "free zones" outside of the territories. Prieto and Ryan also showed that food sources were not contained exclusively in the territories. In fact, territorial males only descended from their rocks and left their territories during feeding. These were the only occasions when adult males were observed in close proximity with no aggressive interactions. Furthermore, in Nagy's study when food supplies started to decrease in May, there was no increase in home range size, as might be expected if the territory was food based (Simon, 1975). Females and juveniles also might be expected to partake in defense of food resources if food was critical. It seems likely that the acquisition of a food source is an incidental consequence and not a function of territoriality.

Raised areas were utilized for basking, display and the lookout stations. Lizards were often seen basking in groups, but these groups never contained more than one adult male (Prieto and Ryan). Rock crevices were utilized for sleeping and were especially important for predator avoidance. By inflating its body a chuckawalla can increase its volume by 58% (Salt, 1943) and safely lodge itself in crevices. Basking sites and rock crevices were abundant outside the territory and there was no obvious difference in quality among sites on and off the territory. It does not appear necessary to incur the costs of territorial defense to obtain these resources. This is also suggested by the fact that males did not attempt to exclude females, juveniles or subordinate males from these sites. For obvious reasons, a male might not exclude a potential mate from utilizing these resources. However, if resources are critical enough to warrant the costs of defense, we might expect the exclusion of subordinate males.

Territorial defense does not appear to minimize the costs of obtaining environmental resources. Therefore, resource defense is probably a consequence of territoriality and has had little importance in the evolution of this behavior.

Mate Acquisition

In most species the asymmetries of parental investment have resulted in mating systems which exhibit male-male competition and female choice (Williams, 1966; Trivers, 1972). Therefore, current theory predicts the evolution of behavioral strategies which maximize the male's ability to compete for females.

Is territoriality a means by which chuckawallas increase their ability to compete for females? Berry, and Prieto and Ryan, reported that males' territories overlapped the home ranges of females and juveniles and they were not aggressively challenged by territorial males. This is consistent with the hypothesis of a mate acquisition territory. Exclusion of females and pre-reproductive males would accrue no benefits if the primary function of the territory was mating, but it would be sure to increase the costs of territorial defense. In Berry's study, only large males were territorial and the smaller, nonterritorial and subordinate males were chased from the territory when detected by the tyrant. This is also consistent with the mate acquisition hypothesis since all reproductive males are potential competitors. Berry stated that possession of a territory not only increases the territory holder's possibility of mating with a female but also decreases the possibility of female-subordinate male matings. The later point is especially important in species which store sperm, as do most iguanid lizards (Cuellar, 1966).

These studies suggest that a selective advantage accrues to territorial males because of their potentially increased ability to fertilize females. Therefore, mate acquisition seems to be the primary function of territoriality while defense of various environmental resources is probably only an incidental consequence of this adaptation.

VARIATION IN SOCIAL ORGANIZATION

If mate acquisition is the function of territoriality, as suggested in the previous section, then we expect males to adjust their territorial strategies in such a way as to maximize the number of mates they acquire. There are two ways in which a territorial male can increase his ability to fertilize females: increasing the size and the intensity of defense of the territory. A larger territory will overlap the home ranges of more females and therefore should increase the male's accessibility to females. An increase in the intensity of defense also increases the male's probability of successful fertilization. A male that reduces the access of females to other males, either by excluding all males from the territory or by behaviorally dominating them, decreases the chances of female cuckoldry and hence reduces potential sperm competition.

This is not to imply that females play no role in mate selection. There are no data that indicate the extent of female choice in chuckawallas, or any other iguanines (Dugan and Wiewandt, this volume). However, if females do exercise considerable mate choice, there are reasons to believe they would preferentially select males which vigorously defend large territories (cf. Werner, this volume).

Factors Affecting Costs of Territoriality

In the absence of associated costs, males should intensively defend territories of infinite size. Obviously there are costs incurred by territoriality, including time and energy expenditure and exposure to predation. As a male attempts to enhance his access to mates by increasing the size and defense of his territory, there is an increase in associated costs. Therefore, a male should adjust the potential costs and benefits of territoriality by employing a strategy (e.g., a specific size and intensity of defense) which maximizes mate acquisition.

If costs of territoriality among individual males and populations were identical, we would expect little variation in male mating strategies. I suggest that variations in social organization can be understood by examining changes in the costs of territoriality. To do this, it is helpful to consider cost in the context of a probability function. For example, the potential cost of predation, being devoured by a predator, is identical for two individuals in different populations. However, the probability of incurring this cost is determined, in part, by the number of predators present in the area, a factor which is sure to vary among populations. In the following discussion, "cost" is considered a product of the potential cost and the probability of incurring that cost.

Size-Dependent Cost and Alternative Mating Strategies

There are important differences in the expression of territoriality among males of the same population. Berry suggested that the ability to maintain a territory is size dependent. Not only should larger males dominate in physical combat but they should be involved in fewer interactions which culminate in combat, since the probability of a contest escalating is inversely related to the degree of asymmetry in size among the contestants (Maynard Smith and Parker, 1976). Therefore, the costs of territoriality will increase more slowly with increases of size and defense of territory for larger males. Males of different sizes will employ different territorial strategies to maximize mate acquisition.

Not all sexually mature males defend territories (Berry). It should not be assumed that these males are incapable of maintaining a territory. The asymmetrical costs of territoriality might be such that for smaller males a territorial strategy is not an economically feasible means for acquiring mates (Figure 21.1). This might especially be true for long-lived species with indeterminant growth where a nonterritorial strategy should increase the possibility of survival to the next breeding season and allow a more rapid growth rate.

Nonterritorial males capable of territorial defense are not necessarily forfeiting the possibility of mating. Although it was not a common occurrence, Berry observed subordinate males attempting courtship with females. The possibility of these males fertilizing a female may be small but the costs are negligible. This nonterritorial strategy may represent a viable alternative for mate acquisition. Size-dependent mating strategies have been reported in other species with indeterminant growth (e.g., bullfrogs, Howard, 1978).

Asymmetrical Costs Among Populations

Differences in costs of territoriality among populations and within the same population among years, might similarly account for observed variation in social organization among populations. It is necessary to examine population parameters which might influence costs of territoriality.

Extending the previous discussion, the distribution of male sizes within the population ultimately will influence social organization. With an increase in the variance of male size we predict greater variation in the size and intensity of defense of territories and a larger number of males employing alternative mating strategies. As previously discussed, predation pressure is another parameter influencing costs which is sure to vary among populations. However, these two parameters are not subject to rapid fluctuations and may be of little importance in explaining shifts in social organization among years within one population.

The more interesting population parameters concern the ratio of fertilizable females to available males (operational sex ratio, OSR; see Emlen and Oring, 1977, for a detailed discussion of this concept) and how these females are distributed in space. As the OSR becomes skewed toward males there is increased competition for mates and hence greater costs of territoriality. Also, given a constant OSR (the precise ratio is unimportant) mate competition will increase with the variance of females per territory. Competition should be most intense when there are a large number of a sexually active males and relatively few receptive females and the females are clumped in space.

Distribution of male sizes, predation pressure, OSR, and the spatial distribution of females probably are all important parameters influencing costs of territoriality. The latter two parameters might be more relevant in explaining rapid shifts in social organization because of their relative instability (see below).

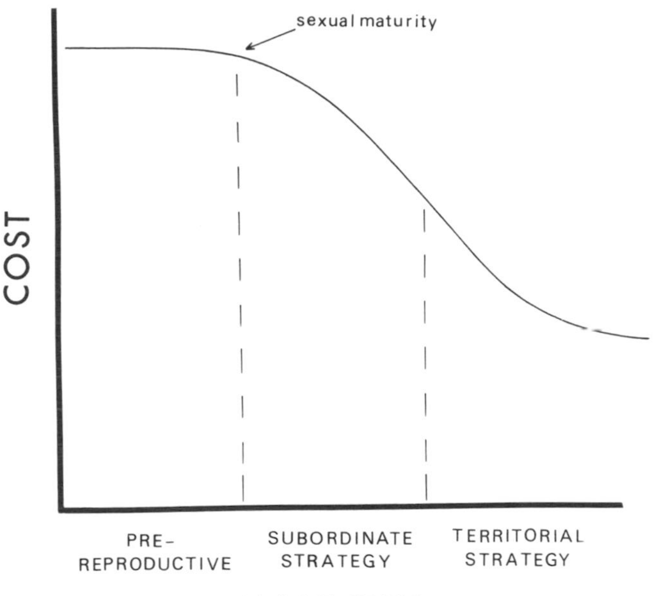

Figure 21.1: Costs of territorial maintenance should be inversely related to male size. This figure represents how asymmetrical costs of territoriality, among males of different sizes, might influence male mating strategies.

Environmental Parameters and Fluctuating Costs

Berry and Nagy showed that following winters of below average rainfall there was a drastic decrease in food supply. In these years there were fewer sexually active females and lizards of both sexes clumped around available food sources. Variation in rainfall manifested by fluctuating food supplies seem to influence both the OSR and the spatial distribution of females. Figure 21.2

summarizes the division of territorial costs by illustrating how environmental, population and individual parameters might interact to determine male mating strategies.

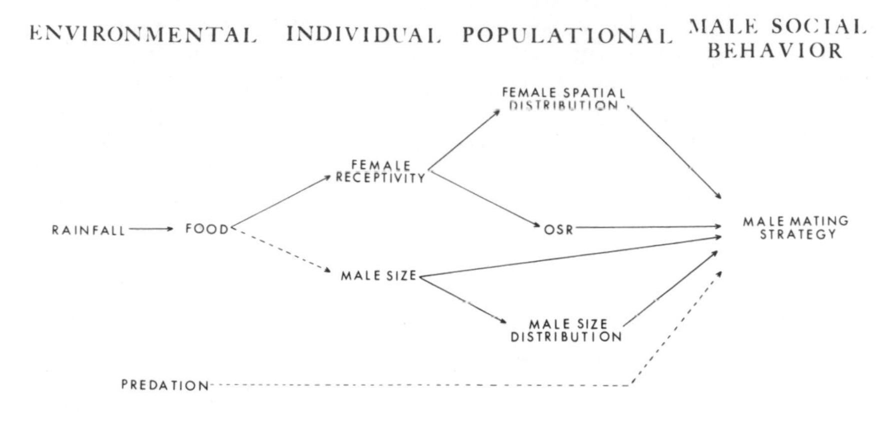

Figure 21.2: Interactions of environmental, individual and populational parameters determine male mating strategies. Solid lines represent factors which may be responsible for rapid shifts in social organization. Dashed lines represent less direct effects.

Territorial Costs as Predictors of Social Organization

I suggest that certain factors causing fluctuations in costs of territoriality are responsible for the observed variation in chuckawalla social organization. This is a *post hoc* explanation of how chuckawalla social systems are organized and is in a large part speculative. However, it can also serve as a predictive theory which can be subject to further testing if the proper data are collected.

Figure 21.3 demonstrates how an increase in costs might affect social organization. When costs are minimal territorial behavior should be widespread, as in the strict territorial system (Prieto and Ryan). As costs increase territoriality tends to become a nonadaptive stragegy for smaller males (see also Figure 21.1). These males then switch to the less costly subordinate strategy (tyrant-subordinate system, Berry).

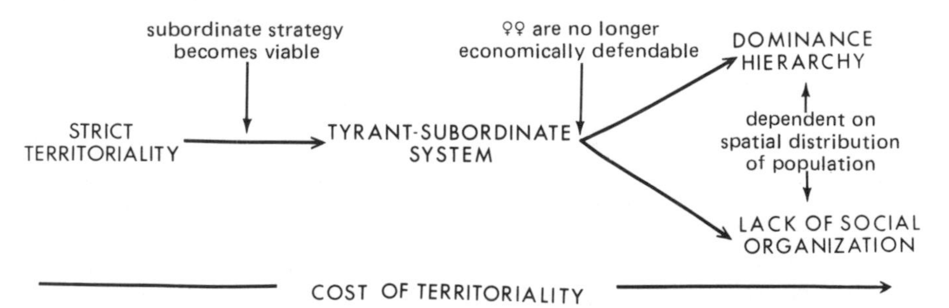

Figure 21.3: Costs of territoriality influence social structure. This figure illustrates how increased costs might be responsible for the observed variation in *Sauromalus* social organization.

A drought preceding the 1970 breeding season resulted in a drastic decrease in annual plants, the primary food of chuckawallas (Berry, Nagy). Territoriality was not observed in either of the two populations. Berry reported a dominance hierarchy around a food source and Nagy noted an almost total lack of social interactions. There are two possible causes for these types of social organization. As less food is available the relative energetic investment in territorial defense increases. Consequently, the costs increase and territorial behavior may no longer be an economically efficient strategy. The second cause relates to the effect of decreased food availability on the females' reproductive physiology. Nagy reported that only one of 18 females had enlarged follicles. Such a drastic decrease in fertilizable females would greatly skew the OSR toward males, causing an increase in male competition and an additional increase in territorial costs.

Territorial behavior was no longer a viable strategy for obtaining mates in the populations studied in 1970. Why did one population exhibit a dominance hierarchy (Berry) while one showed an almost total lack of social interactions? Dominance hierarchies are a result of continual dominant-subordinate interactions between individuals. When chuckawallas were enclosed in a laboratory a dominance hierarchy emerged (Prieto and Ryan). In the tryant-subordinate system a dominance hierarchy, similar to the laboratory situation, was present among individuals on the same territory.

Dominance hierarchies do not occur if individuals are distributed in such a way that interactions among individuals are minimized. The distribution of lizards should be influenced by resource distribution. Berry reported a dominance hierarchy centered on a rock pile which contained the only sweet bushes in the area. In Nagy's study there were few social interactions, even though lizards were in close proximity on several occasions. There is the possibility that Nagy observed few social interactions because there was already a well-formed dominance hierarchy. However, Nagy did observe aggressive interactions during apparent defense of a rock site. There are no comparable data on resource distribution for the two studies, but perhaps the highly clumped distribution of the food source noted by Berry facilitated the rate of individual encounters resulting in the formation of a dominance hierarchy.

There is little information on the social behavior of the insular giants *S. hispidus* and *S. varius*. Case (this volume) reported that these animals were not territorial and groups of adults, usually a male and one or more females, sometimes shared the same burrows and rock crevices. He also noted that recruitment is quite low in these populations. There is no reduction in clutch size but breeding takes place only in very wet years and only a small portion of the females become gravid.

Case (1978) suggests that resources are abnormally high and territoriality is not profitable (but see Case, this volume). However, if territoriality is a mate acquisition strategy perhaps the low reproductive rate is responsible for the lack of territoriality and the high resource abundance accounts for the absence of a dominance hierarchy. Precisely why there is a low reproductive rate in these populations is enigmatic. But it should be noted that when Case discusses an abnormally high abundance of resources he is referring to perennials, a food source which mainland chuckawallas eat only after the supply of annuals has been depleted. Perhaps the low rates of reproduction are caused by a less than optimal diet.

SUMMARY AND CONCLUSIONS

Previous studies indicate significant variation in *Sauromalus* social organization. An analysis of the costs and benefits of chuckawalla territoriality suggests that territories function (*sensu strictu,* Williams) in mate acquisition. Male size, predation pressure, OSR, and the spatial distribution of females are important parameters determining the cost of territorial maintenance. The latter two factors are influenced by rainfall and food and undergo rapid shifts due to the vagaries of the desert climate. Fluctuating costs of territoriality might explain the observed variation in chuckawalla social organization.

Acknowledgments

I would like to thank K. Adler, B. Dugan, D. Mammen, J. Phillips, A.S. Rand, J. Stamps, and P. Trail for comments on the manuscript. I am especially grateful to A.S. Rand for his many helpful discussions of this subject. Portions of this paper were prepared while I was supported by a predoctoral fellowship from the Smithsonian Institution.

Section VI

Conservation and Management

As with many, perhaps most, tropical animals, iguanine lizards are endangered. They are large, which means both that they are conspicuous and that their populations are relatively small. Many species occur on small islands as parts of fragile systems that are easily disrupted by habitat modification, particularly by the introduction of exotic competitors or predators.

The current status of all iguanines cannot be presented here–we do not know enough. But we do have some information from Chapter 22 by Fitch, Henderson, and Hillis, from which it appears that *Ctensaura similis* and *Iguana iguana* in Mexico and Central America are still common but rapidly becoming less so. Many ground iguanas are threatened with extinction as the table provided by Wiewandt depressingly catalogues. Banded iguanas discussed in Chapter 23 by Gibbons and Watkins, are common only on small islands and Case (Chapter 11) notes the decreasing populations of the giant chuckawallas that he attributes to capture by people. Other previous chapters also contain pertinent conservation and management information.

The threats to the continued existence of iguanines vary from species to species, but overall one of the most important is habitat modification by humans both directly and indirectly. Such modification is due to diverse influences. For example: tourist developments in the Caicos destroy ground iguana habitats; the disappearance of vacant lots in Panama City eliminates places for urban green iguanas to breed; cutting of mangroves in Mexico destroys iguana habitat; hydroelectric dams flood the sand bars on which iguanas breed. Perhaps most important is the increasing rate at which forest is being converted into pasture. Even in areas where humans are not directly changing the environment the introduction of exotic animals (such as goats that change the vegetation and compete directly with iguanines for food, pigs that dig up their eggs, cats and dogs that prey directly on them) have negative and sometimes disastrous effects on populations, particularly those that have evolved without mammalian predators and competitors.

Known Extinct and Living Forms of Cycluran Iguanas (based on Wiewandt, 1977)

Species*	Distribution	Status	Explanation for Status	Outlook for Future	Source
C. mattea	St. Thomas, V.I.	Extinct—Recent	Overhunting by pre-Columbian Indians suspected		Miller, 1918; Barbour, 1919
C. portoricensis	Puerto Rico	Extinct—Recent	Overhunting by pre-Columbian Indians suspected		Barbour, 1919
C. cornuta onchiopsis (*C. nigerrima*)	Navassa I., Haiti	Extinct—Recent	Unknown—man, cats, and goats are suspect		Thomas, 1966; Patton, 1967
Unnamed	Great Exuma, Bahamas	Extinct	Unknown		Allen, 1937
Unnamed	New Providence, Bahamas	Extinct—Pleistocene	Unknown		Etheridge, 1966
Unnamed (possibly a *Cyclura*)	Barbuda, Lesser Antilles	Extinct—Late Pleistocene	Unknown		Etheridge, 1964
C. collei	Jamaica and Goat Is.	Nearly extinct	Overhunting by man since 1700; predation from introduced mongoose; developed by U.S. Navy.	Very poor—last specimen taken in 1969; subsequent sightings in 1970, 1973, and 1978	Lewis, 1944; Woodley, 1971, 1980; Lewis, pers. comm. 1976
C. nubila lewisi (*C. macleayi lewisi*)	Grand Cayman	Endangered	Population sparse and juveniles extremely rare in 1937-1938; numbers reduced by hunting	Poor, though apparently still surviving; may be in competition with introduced iguanas	Grant, 1940; Lewis, 1944; Carey, 1975; C.B. Lewis, pers. comm. 1976
C. n. caymanensis (*C. m. caymanensis*)	Little Cayman and Cayman Brac	Declining	Dogs, cats, and hunting apparently responsible on Cayman Brac; feral cats pose a threat to juveniles, which are now rare, on Little Cayman	Unfavorable on Cayman Brac; much better on Little Cayman	Grant, 1940; Carey, 1966; M.V. Haunsome, pers. comm.**, C.B. Lewis, pers. comm. 1976

Species*	Distribution	Status	Explanation for Status	Outlook for Future	Source
C. n. nubila (*C.m. macleayi*)	Cuba, Isle of Pines, and small nearby keys	Declining	Formerly widespread—suffered initial decline from hunting; range still rapidly shrinking (cause unstudied)	Unknown	Barbour, 1919; Auffenberg, 1975
Cy. c. cornuta (*C. cornuta*)	Hispaniola and peripheral islands	Common, but threatened	Widespread in Haiti and Dominican Republic but faced with accelerating habitat alteration by man goats, and burros, plus predation from dogs, cats, pigs, and mongooses	Appears good in some areas, bad in others; depends largely upon control of domestic and feral animals plus the nature of human population growth and development	Haiti: P. Meylan, pers. comm.***; D.R. Wiewandt and Gicca, pers. observ.***
Cy. c. stejnegeri (*C. stejnegeri*)	Mona I., Puerto Rico	Common, but threatened	Limited natural habitat is being decimated by goats and considered for development; pigs are preying on eggs and cats taking juveniles	Good if human use is carefully regulated, feral animals brought under control, and nesting areas given special protection	Wiewandt (1977, 1978)
C. ricordii	S.W. Dominican Republic	Common, but threatened	Populations are highly localized; small natural range of species being disturbed by man, goats, burros, dogs, and/or cats	Good only if domestic and ferral animals are controlled and an active conservation program accompanies development	Wiewandt and Gicca, pers. observ.***
C. pinguis	Anegada, V.I.	Endangered	Habitat destruction by livestock especially goats and cattle; predation from dogs, cats, and possibly pigs	Presently small and senescent population might recover if domestic animals are controlled and development projects deferred; extinction highly possible	Carey, 1975; Wiewandt, pers. observ. 1978

Species*	Distribution	Status	Explanation for Status	Outlook for Future	Source
C. c. carinata	Turks and Caicos Is. (on 15–20 cays and islands)	Locally endangered to common, but vulnerable.	Throughout the Bahamas Archipelago, all forms have been hunted for food, "sport," or the pet trade. Most populations are confined to tiny cays, thus while some are still relatively dense, all are precariously small. Those cohabiting islands with man are also declining from habitat modification or predation by dogs and cats. Historical and paleontological records suggest many populations have already been extirpated or severely reduced in size.	Highly variable, from poor to good, depending largely upon effectiveness of government sponsored conservation programs.	Barbour and Noble, 1916; Auffenberg, 1975, 1976; Carey, 1975, 1976; Carey, pers. comm.; Iverson, 1978, pers. comm.***
C. c. bartschi	Bahama Bank (Booby Cay)				
C. r. rileyi (*C. rileyi*)	Bahamas (San Salvador and vicinity)				
C. r. nuchalis (*C. nuchalis*)	Bahamas (Acklins Bight)				
C. r. cristata (*C. cristata*)	Exuma Chain (White Cay)				
C. cychlura figginsi (*C. figginsi*)	Exumas (on several cays and islands around Great Exuma)				
C. c. inornata (*C. inornata*)	Exumas (Leaf Cay and SW Islands Cay)				
C. c. cychlura (*C. baeolopha*)	Bahamas (Andros Is.)				

*As denoted by Schwartz and Carey (1977); formerly accepted names appear in parentheses.
**1975 Royal Society/Cayman Island Government Expedition.
***Research sponsored by New York Zoological Society, through a grant to Walter Auffenberg (1975–1976).

Hunting iguanines, for food and sport is widespread for green and spiny-tailed iguanas. Capturing of ground iguanas and giant chuckawallas for the pet trade is adversely affecting many populations.

What can be done to preserve iguanines? As the species are so different in demography and ecology there is no simple panacea. Still, some things can be suggested: the most satisfactory is a system of strict reserves maintained in their undisturbed form. This has been instituted and seems to be working in the Galápagos. It seems the only solution for small populations in fragile systems particularly those that have evolved for long periods in isolation (e.g., banded iguanas, the island chuckawallas and at least some ground iguanas). Probably the greatest threat in reserves on small islands is that from introduced exotic animals. Again, the Galápagos, in their efforts to reduce or eliminate goats, provide an example of what can be done, but also of the great efforts required.

Throughout the range of all iguanines the levels and sorts of human activity that affect them must be monitored and controlled. Effective laws controlling hunting and trafficking in both dead and live animals must be passed and, more importantly, must be enforced. The latter is the more difficult. Environmental impact statements are being required for large engineering projects, at least those funded by the World Bank. Smaller scale activities do not usually require them, although when they are repeated many times they can have an even greater impact then "blockbuster" projects. Since each one is small they are harder to detect early on and harder to control. One of the ways that field biologists can help directly in the conversation effort is by bringing developing threats to the attention of the appropriate and often very overworked local authorities. Conservation departments should be monitoring human activities and population levels of at least key organisms as they affect the local flora and fauna. Another way in which biologists can help is by suggesting appropriate species to check and ways to census them.

Beyond control measures taken to keep things from getting worse, there are some ways in which the lot of iguanas might be improved. Small changes in the environment such as the providing of suitable nesting sites around the margins of new reservoirs to replace those drowned by the rising waters should be encouraged. Protecting critical sites when developments such as camp grounds are located needs consideration.

Iguanines are long lived and have been bred in captivity (e.g., Braunwalder, 1979). Green iguanas have been raised through several generations (Mendelssohn, 1980). They and spiny-tailed iguanas are very prolific and should be considered in any scheme to manage or ranch wild animals in Central and South America. However, a great many techniques will need to be worked out before this can be done successfully on a large scale. We doubt that iguanas could ever compete with chickens, pigs or tilapia as a major protein source, but as a speciality food and to supply animals for the pet and laboratory trade, semicaptive breeding is worth considering.

The conservation prospects for iguanine lizards have both negative and positive aspects. From the pessimistic side the most discouraging factor is the continuing increase in the numbers of people in the third world and their desire for the "benefits" of western material culture. As oil prices rise, underdeveloped countries must export more to purchase the imports that their increasing population needs and wants. Increasing exports means increasing domestic production, and for most countries in the tropics this means an increase in agricultural production (sugar, cattle, etc.), which puts increasing pressure for the development of remaining wild land.

On the optimistic side there is an increasing awareness on the local level, by government officials, and by the people themselves that their environment is changing for the worse and that conservation of it is important. If these new attitudes can produce action in time to save the iguanines much more than these lizards will be saved. For the plight of the iguanines lizards is a very small part of a developing picture of environmental degradation throughout the tropics.

22

Exploitation of Iguanas in Central America

Henry S. Fitch
Biological Sciences
University of Kansas
Lawrence, Kansas

Robert W. Henderson
Vertebrate Division
Milwaukee Public Museum
Milwaukee, Wisconsin

and

David M. Hillis
Department of Herpetology
Museum of Natural History
University of Kansas
Lawrence, Kansas

INTRODUCTION

Large size, open habitat and a tendency to perch in conspicuous places have rendered the iguanine lizards particularly vulnerable to persecution by man. As a result of hunting, habitat modification, and wanton killing all species have been reduced and some have been exterminated. Many that still survive are now endangered and face an uncertain future. Those with small geographic ranges and specialized habitat requirements are in the direst straits. The two species which perhaps enjoy the most secure future are the green iguana, *Iguana iguana,* and the spiny-tailed iguana *Ctenosaura similis.* Both are widespread in the Neotropical region, occurring in many kinds of habitats and many different countries. Nevertheless, even these two widespread and adaptable species have undergone drastic reduction in recent years because of heavy exploitation, and their future has become cause for concern. This chapter deals with past and present trends and future prospects in these two species and their near relatives in Mexico and Central America.

METHODS

Casual observations on the habits of iguanas and their exploitation were made by Fitch in Costa Rica, 1967-1971, and by Henderson in Belize, 1970-1971. In 1976, with a grant from the Banco Central of Managua, Fitch and Henderson undertook field work in Nicaragua from late January through March. The study included frequent visits to major markets to examine, count, measure, weigh and sex the stocks of iguanas there, and to interview vendors concerning the sources, prices and numbers of the animals they stocked. We located professional hunters and studied their methods of hunting and marketing the animals. We interviewed farmers and peasants in various parts of the country to determine the amount of hunting for home consumption and the methods employed. Also, we visited the market in San Salvador, and we studied the behavior of spiny-tailed iguanas in a natural population at Belize, and Henderson followed up with further field work there in 1977. In February and March 1979, supported by the International Fund for Animal Welfare, Fitch and Hillis worked in Mexico, Guatemala, El Salvador, and Honduras. Market places of major cities were visited to determine sources, numbers and kinds of animals sold and trends of prices. We interviewed conservationists and concerned government officials in each country to determine population trends of the animals and amount of exploitation, past and present. We visited iguana habitats to make firsthand observations and to interview country people regarding their use of the animals.

RESULTS

Reduction in Numbers and Ranges

Both *Iguana iguana* and *Ctenosaura similis* are becoming scarce where they were abundant, and are disappearing from large areas. All evidence that can be assembled from firsthand observations and from the testimony of numerous witnesses indicates that the reduction was relatively gradual, but has gained momentum. It is now accelerating steadily. The reduction is both directly and indirectly a product of human population pressure.

Both animals are well adapted to withstand exploitation, and were able to do so in the past by virtue of their high reproductive potentials. But a combination of overhunting and habitat destruction have tipped the scales against them in recent decades, with heavy use of agricultural insecticides also probably contributing to the decrease, at least in the spiny-tailed iguanas. Our data are not complete for either species but are sufficiently extensive to show widespread geographic trends.

Mexico: Exploitation is mostly in Oaxaca, Chiapas and the Yucatan Peninsula. Over much of Chiapas the climate is arid and *Iguana* is absent but formerly it was extremely abundant in the narrow strip of mangrove forest along the Pacific Coast (Alvarez del Toro, 1960). Although the mangrove stands are still fairly extensive the iguanas have been drastically reduced by overhunting (to an estimated 5% of the original population, Alvarez del Toro, pers. comm.). Those that are now taken are adolescents or small adults. Few survive to full maturity because of hunting pressure. By way of contrast, in the Gulf Coast lowlands iguanas are still abundant, and they remain abundant in parts of the

Yucatan Peninsula where the habitat is favorable. In the central valley of Chiapas the spiny-tailed iguana has been almost eliminated, but it is still moderately abundant in thinly populated areas of rocky foothill slopes and thorn scrub woodlands.

Guatemala: In Guatemala, both overhunting and destruction of habitat have been important in the drastic reduction of iguana populations. In the sparsely populated Caribbean lowlands of Alta Verapaz and Peten, the animals have been little affected, but in the heavily cultivated Pacific lowlands both types, especially green iguanas, have been reduced to remnants. On the Pacific versant the best iguana habitat was in the coastal mangrove forests, but cutting has reduced these forests to about 7% of their original extent. Most parts of the Pacific coastal plain and foothills have been subjected to intensive agriculture with the growing of such crops as cotton and sugar cane, and with heavy use of insecticides. The use of insecticides has escalated and a Government 5-year plan to reduce and regulate them has not been effective. Spiny-tailed iguanas, formerly abundant throughout the Pacific lowlands, remain common only in three relatively small coastal areas (see map) where there are large cattle ranches. There they are not exposed to insecticides, habitats have been much less altered than in cultivated areas, and hunting pressure has been lower. The reduction in both numbers and area occupied has resulted in a steady decline of the number sold. Twenty years ago there were sometimes lots of hundreds where now they are few or none. As the supply decreases and the price rises, former consumers are abandoning this food source.

Belize: From 20 March to 16 April 1971 Henderson counted approximately 300 green iguanas in the Belize City market; all or most had been brought from near Benque Viejo, Cayo District near the Guatemalan border. Green iguanas also were observed in the markets at Cayo Town, Cayo District, and Stann Creek Town, Stann Creek District. In early March 1976 no iguanas were seen in a week of daily visits to the Belize City market and we were informed that few iguanas were sold there. In April 1977 over a three week period about two dozen iguanas were seen in this market. Only gravid females were observed marketed in Belize, and the eggs were much more esteemed than the flesh. Over much of Belize the habitat is not favorable for spiny-tailed iguanas, and they are highly localized. Seemingly they are not used as food because of the widespread superstition that they feed on human corpses.

El Salvador: El Salvador, like southern Mexico and Guatemala, had extensive stands of coastal mangroves, but these have been largely destroyed and the populations of green iguanas that once thrived there also have mostly disappeared. There also were iguanas in gallery evergreen forests along streams. They were fairly common as recently as 1974, but since then both forests and iguanas have almost disappeared. The alluvial lowland forests have been reduced to perhaps 1% of their original area and within these remaining stands iguanas have been reduced to perhaps 1% of their original density. In southeastern El Salvador, Volcan San Miguel, more than 7000 feet high, with areas of bare lava rock, brushland and forest, dominates an extensive area with some iguanas, expecially spiny-tailed, still surviving. As recently as 6 years ago spiny-tailed iguanas were abundant. They cost 80 cents apiece then and were shipped out by truckloads. They are still the major game animals in the San Miguel area, but are generally absent from most of their original range in El Salvador.

Even where the lizards are not common they are still hunted intensively for local consumption. Because protein is scarce, the reward of a meal will justify the considerable effort involved in finding and killing an iguana. Hence iguanas are subjected to relentless hunting pressure even where their populations are so sparse that commercial hunting is unprofitable.

Honduras: In Honduras, the human population is relatively sparse and exploitation of animals has been less extensive than in neighboring countries. Overhunting and habitat destruction have been serious only in the Pacific lowlands–a small part of the country. In the area around Choluteca numbers have been reduced by perhaps 90%. There has been extensive cutting of gourd trees, called "Jicarales." These small, gnarled and often hollow trees provided abundant shelter for spiny-tailed iguanas over large areas of open, xeric woodland. Unrestricted use of insecticides also is suspected to have had a part in the reductions that have occurred in some areas. Spiny-tailed iguanas occur widely over the Pacific lowlands and dry interior valleys of Honduras. Green iguanas occur on both coasts but mainly in the warm, humid Caribbean lowlands. Most of northern Honduras is sparsely populated and not easily accessible. Iguanas captured cannot readily be shipped out; also hunting pressure for home consumption is slight. Over most of the country neither kind of iguana is in danger of extermination or drastic reduction.

Nicaragua: Both spiny-tailed and green iguanas are widely distributed in Nicaragua, the former occurring especially in the arid Pacific Coast versant and the latter on the Caribbean Coast and along major rivers. Ramirez (1968) noted that the country's iguana population was being reduced by mass shipments to El Salvador where the animals had already been depleted. Villa (1968) wrote that the large scale exploitation, an estimated 150,000 eaten annually in the country, threatened both species with early extinction there. In 1976 we found numbers holding up well in some parts of the country, but undergoing drastic reduction in other parts. In the Pacific Coastal strip from Diriamba, and Nandaime south to the Costa Rican border we found that the animals were so scarce that they were no longer hunted. However, most country people interviewed in that area had hunted spiny-tailed iguanas regularly in the past and had abandoned the practice only in recent years when hunting was no longer rewarding. All agreed that overhunting was responsible for the disappearance of the animals. In the northwestern part of the country we found spiny-tailed iguanas still abundant though probably less so than in the past, and they were being heavily hunted. Over much of the country's sparsely populated and relatively inaccessible areas the animals presumably remain in somewhere near their original abundance (Henderson and Fitch, 1978a, 1978b).

Costa Rica: In Costa Rica, the human population is concentrated on the Meseta Central, and the Nicoya Peninsula in the northwest. Spiny-tailed iguanas occur mainly in the dry Pacific lowlands and green iguanas in the Caribbean lowlands. There is no commercial exploitation to our knowledge, but local hunting is heavy in some areas and iguana populations have declined. Hirth (1963) noted that at Tortuguero on the northeast coast the numbers of green iguanas were diminishing rapidly. Forty gravid females were taken in one season by local residents, along a 4-mile stretch of beach, and in the same area at least seven nests were known to have been destroyed by dogs.

Panama: In Panama, Swanson (1950) noted that iguanas were extremely abundant, and he observed that vendors in the marketplace of Panama City had dozens of the animals in stock strewn over the street beside their stalls. However, Tovar (1969) observed that the former numbers had been reduced to a trickle. Sale of iguanas was made illegal in 1967 and the ban was well enforced. In the spring of 1979 the ban on sales was lifted because of a beef shortage.

Methods of Capture

Where possession of firearms is common, as in southern Mexico, Nicaragua and Costa Rica, shooting with a .22 rifle is a favorite method of taking iguanas. Some hunters merely use the animals for target practice, shooting from a vehicle at the lizards on roadside boulders or fence posts, but not attempting to retrieve or use those that are shot. Spiny-tailed iguanas that are shot often escape into nearby burrows, regardless of whether or not their wounds are mortal. In areas of southwestern Nicaragua where there had been intensive rifle hunting, the few remaining adult spiny-tailed iguanas seemed extremely wary in 1976 perhaps from the combined results of individual learning and selection against the unwary. Young hunters are often highly skilled with slingshots and use them to stun iguanas, knocking them from trees or immobilizing them on the ground.

Market hunters are skilled at capturing iguanas alive and intact. A favorite device is a pole up to 3 m or more long, equipped with a drawstring and a noose at the tip, which is slipped over the animal's head to secure it from a tree or from the ground. A meat bait, preferably calf's liver or spleen, is usually used to distract the animal's attention and make it reach through the noose. To catch spiny-tailed iguanas a string or wire noose is placed over the entrance of the burrow into which the animal has escaped.

For catching iguanas the hunter may carry a shovel and dig out the quarry after seeing it take shelter in a burrow. Excavation may also be done with a pointed stick and the hands. A typical spiny-tailed iguana burrow has only one entrance, and is between one and two meters long. Many of the green iguanas taken are gravid females and some of these are caught on the ground while digging their nest burrows. They are especially vulnerable when at work on the burrow with the head and forebody inside so that the approaching hunter is not seen, but the hindquarters or tail protrude from the hole. Female green iguanas descend from the trees and travel to favorable open sites for egg-laying, such as sandbars along streams. Handicapped by the burden of eggs, and risking travel on the ground in strange places, these females are unusually easy prey for the hunters.

A hunter is often accompanied by one or more trained dogs, which assist in the capture of iguanas. The dog sometimes runs down lizards flushed in the open, and is especially useful in outrunning those that drop from trees when disturbed. Also, the dog may help with digging out burrows, and may retrieve animals shot and wounded by the hunter. With the help of trained dogs, even small children may hunt iguanas successfully, contributing to their families' subsistence (Fitch et al., 1971).

For agricultural workers, Sunday is the traditional day for the hunt. Hunting iguanas is an important form of outdoor recreation for many persons.

Routes of Commerce

In 1979, town markets in southern Guatemala had mixed lots of spiny-tailed and green iguanas, the former from nearby areas, the latter from the Caribbean lowland of El Peten in the northern part of the country. Some Guatemalan iguanas are shipped illegally to El Salvador. Iguanas captured in western Belize were sold in the market at Belize City in 1972. In Nicaragua, there was a well organized system of distribution in 1976. A group of professional hunters in the villages of Palo Grande, Somotillo, and Villaneuva, all near the Honduran border on the northwest, supplied many of the spiny-tailed iguanas and some of the green iguanas for the markets throughout the country. Other sources were San Francisco de Carnicera on the northern shore of Lago de Managua, and San Carlos on the shore of Lago de Nicaragua near its outlet into Rio San Juan. The animals from San Francisco were assembled there by a dealer who obtained his stock from approximately 20 professional hunters in areas to the north and west of the town. San Carlos was also a collecting point and the stock there consisted entirely of green iguanas originating from the San Juan River area, some perhaps from as far east as the coast. Once weekly on a prearranged day, one or more collectors made the rounds in a pick-up vehicle to the outlying villages where the professional hunters lived, delivering their stock to the city markets. Since most of the animals originated in the northwestern part of the country near the Honduran border, they were especially abundant and cheap in the markets of towns in the northwestern part of the country, and were scarcer and higher priced in towns to the southeast.

El Salvador is a major importer of iguanas from the surrounding countries, and probably more are assembled at the market in San Salvador than at any other one place. Most animals come from the south and east, from Honduras and Nicaragua, but both countries have prohibited exportation for some years and the flow is irregular, subject to interference by the authorities.

Between El Salvador and Honduras especially, there is mutual antipathy because of fairly recent wars and border disputes, and in 1979 there was still no open commerce. Instead, trucks with various sorts of merchandise, including iguanas, parked near the border and their cargos were carried across by pedestrians. In the case of iguanas, lots of several dozen at a time may be carried in gunny sacks or in large wicker baskets. The bearers are permitted to cross through the regular check stations, perhaps through pre-arrangement with bribed officials, and the cargo is delivered to other trucks waiting on the Salvadorian side of the border. The legislation against export by Guatemala, Honduras and Nicaragua, and the political tensions between these countries and El Salvador obviously curtail the iguana trade without stopping it.

The small town of Santa Rosa de Lima in east-central El Salvador, approximately 10 miles from the Honduran border, is of key importance in the iguana trade. Shipments trucked from Honduras and Nicaragua are hand-carried across the border at Amatillo, for redistribution to markets throughout El Salvador. Every Friday several hundred iguanas pass through the markets. Some are consumed locally, and larger numbers are reshipped by truck or bus. The principal market places selling iguanas in El Salvador are shown on the map in Figure 22.1. Many of the iguanas bought in El Salvador pass through the market place at Santa Rosa de Lima after importation from Honduras, although large numbers also originate from La Union and Piedras Blancas, and some enter the country across its western border, from Guatemala.

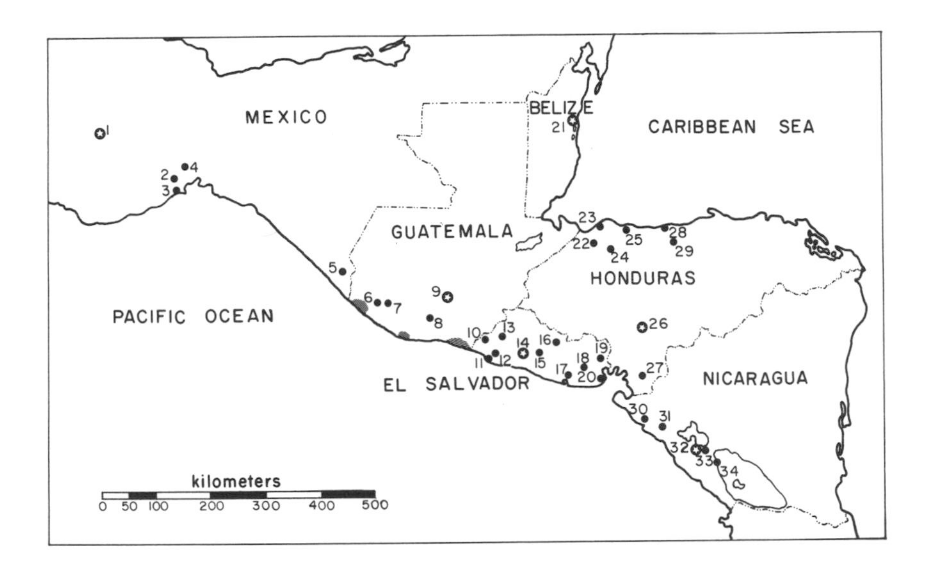

Figure 22.1: Map of Middle America showing main mercados which sell iguanines. Shaded areas in Guatemala indicate undisturbed iguanine habitat. Numbered mercado localities are as follows:

Mexico: 1. Oaxaca, 2. Tehuantepec, 3. Salina Cruz, 4. Ixtepec, 5. Tapachula. *Guatemala:* 6. Retalhuleu, 7. Mazatenango, 8. Escuintla, 9. Guatemala City. *El Salvador:* 10. Ahuachapan, 11. Acahutla, 12. Sonsonate, 13. Santa Ana, 14. San Salvador, 15. Cojutepeque, 16. Sensuntepeque, 17. Usulatan, 18. San Miguel, 19. Santa Rosa de Lima, 20. La Union. *Belize:* 21. Belize City. *Honduras:* 22. San Pedro Sula, 23. Puerto Cortes, 24. El Progreso, 25. Tela, 26. Tegucigalpa, 27. Choluteca, 28. La Ceiba, 29. Olanchito. *Nicaragua:* 30. Chinandega, 31. Leon, 32. Managua, 33. Masaya, 34. Grenada.

Another receiving point is the village of Piedras Blancas on the Salvadorian side of the Río Goascorán at its delta in the Gulf of Fonseca. Loads of iguanas carried by persons on foot or on horseback to the river bank on the Honduran side are brought across by boat to the Salvadorian side. Some are brought by boat for considerable distances upstream or downstream. The winding course of the river, with a network of channels through the delta, makes it difficult for the authorities to monitor the importation of lizards in small lots to the collecting point, Piedras Blancas. There, illegally taken Honduran iguanas cannot be distinguished from those of Salvadorian origin. Iguanas trucked from Nicaragua to El Salvador have to cross the intervening area of Honduras, with risk of delay or interception. A more direct route is from Potosi on the Cosiguina Peninsula, across the Gulf of Fonseca, to La Union in southeastern El Salvador. In early March, 1976, we determined that large shipments were being made regularly. By early 1979 the flow seemed to slow to a trickle, as we failed to find any of the animals at La Union, but were told that some were still brought by boat. Evidence that shipments across the Gulf of Fonseca had dwindled to insignificance was found in the fact that the iguanas we saw being sent from Santa Rosa de Lima included bagfulls loaded on a bus for La Union.

Prices

Until fairly recently the flesh and eggs of iguanas could be obtained with little effort by country people and were available at low cost in city markets. In recent years prices have risen steeply as the animals have become scarcer and less available so that cost is no longer competitive with that of other meats such as poultry, pork, or beef. In 1979 there was a wide price range, depending on supply and demand, distance of transport, risk of confiscation, and other factors. Table 22.1 shows retail prices at the city markets we visited in 1979. The highest price encountered was $9.11 for an adult male spiny-tailed iguana *(Ctenosaura pectinata)* at Salina Cruz, Oaxaca. Hunters usually received much less than was paid by consumers, the difference representing profits of the vendors and various middlemen.

Table 22.1: Retail Prices of Iguanas in Major Markets in February and March, 1979

Place	Price per Green Iguana	Price per Spiny-Tailed Iguana
Mexico		
Arriaga	–	30-50 pesos ($1.36-2.28) 50-65 pesos ($2.28-2.95)
Salina Cruz	–	150 pesos ($6.82, ♀) 200 pesos ($9.11, ♂)
Guatemala		
Mazatenango	3 quetzales ($3) 1.50 quetzales ($1.50, small)	–
El Salvador		
Santa Ana	3.50 colones ($1.40, ♀ without eggs)	–
San Miguel	4 colones ($1.60, ♀♀ and ♂♂)	4 colones ($1.60, ♀♀ and ♂♂)
Le Unión	6 colones ($2.40, ♀ with eggs) 4 colones ($1.60, ♂)	–
San Salvador	8, 9, 12, 13 colones ($3.20, 3.60, 4.80, 5.20) 2, 2, 2.50 colones ($0.80, 0.80, 1) 7, 1.50 colones ($2.80, large ♂; 0.60, small ♂) 10, 10, 8-10 colones ($4, 4, 3.20-4, ♀ with eggs)	2.50, 3, 3, 4 ($1, 1.20, 1.20, 1.60) 5 colones ($2)
Honduras		
Tegucigalpa	2.50-6 lempiras ($1.25-3)	2.50-6 lempiras ($1.25-3)

In Nicaragua in 1976, the hunters usually received 2 ($0.28) or 2.50 ($0.35) Cordobas apiece in northwestern Nicaragua and 3 or 4 Cordobas elsewhere. "The buyer usually paid 5 C at the Chinandega mercado, 8 (7-10) at Managua and 8-12 at Masaya and Granada. At the latter two cities, there was often an adjustment of price according to the animal's size, whereas at Chinandega the price tended to be uniform even though one animal might be as much as 5 times the bulk of another" (Fitch and Henderson, 1978:499).

In the central market of San Salvador in 1976 a large iguana cost the equivalent of 80 cents U.S., but in 1979 prices were generally from 3-6 times that much.

Protective Legislation

Iguanas have some legal protection in all, or nearly all, of the countries where they occur. In Mexico the open season is 10 November to 31 January, with a limit of three animals per day and a possession limit of six. Commercial exploitation is illegal. Enforcement of game laws and regulations is the responsibility of the *Federal Direccion General de la Fauna Silvestre,* and state officials are not involved. Actually there is little or no enforcement by Federal officials, at least for iguanas. Time and amount of hunting is determined largely by the availability of the animals and inclination of the hunters.

In Guatemala, all wildlife is legally protected by Decree 8-70, General Law of Hunting, but in actual practice there is hardly any protection. Wherever iguanas are available to local populations they are hunted for home consumption, and there is commercial exploitation in some areas. Because of stringent government regulations, few civilians own or carry firearms in Guatemala, to the benefit of the iguanas. Even slingshots are technically illegal, but the ban on them is not enforced. They are sold by the hundreds in markets and are often used for hunting iguanas.

In Honduras, exportation of iguanas is legally banned, and shipments bound for El Salvador are intercepted and confiscated from time to time. Taking of females was banned for the winter breeding season of 1978-79, and apparently was enforced, at least in the market at Comayguela. In Nicaragua, commercial hunting was banned in March 1976, during the Lenten season when iguanas are eaten more than at other times. The lizards disappeared from the markets, but we were told that many vendors continued to obtain and sell them surreptitiously, keeping them out of sight under the counters. We are not informed of recent events in Nicaragua, but the 1979 overthrow of the Somoza regime, with widespread destruction of lives and property of the Nicaraguan people, could only have increased hunting pressure on the iguanas as an emergency food source.

In Panama, iguanas have had some legal protection for at least ten years, but they have continued to become scarcer (Swanson, 1950; Tovar, 1969).

Cultural Background of Exploitation

In some parts of their range iguanas are not hunted or eaten. Attitudes toward them are based on cultural differences some of which date back to ancient times. For example, in northern Colombia the flesh of iguanas is not eaten, but egg-bearing females are hunted. When captured, they are cut open and divested of their unlaid eggs, then released to survive if they can (Harris, this volume). In most areas of sympatry the spiny-tailed iguana is preferred over the green as food. However, in Mexico the spiny-tailed iguana is widely reputed to be coprophagous and is held in relatively low esteem so the green iguana is preferred. Spiny-tailed iguanas are abundant about the outskirts of Belize City, but are not hunted there. They are especially in evidence at the city cemetery, perching on the tombstones (see Figure 22.5, bottom) and this has given rise to the superstition that they feed on corpses in the grave. They are called "wish-willies" and are avoided by the English-speaking black creoles, who consider the green iguana a delicacy.

Cultural differences exist in the methods of hunting, marketing and utilizing iguanas. In most major markets of Nicaraguan cities, live iguanas were displayed and sold in large numbers, but in the indoor market of Leon several vendors featured the dressed carcasses, and few or none of the live animals could be found for sale.

In the markets of southern Mexico and southwestern Guatemala a vendor sitting beside a heavy iron pot of several gallons capacity, suspended over a flame was a common sight. The pot contained an iguana stew consisting of chopped sections of the animals–sometimes an entire head, feet with claws, or pieces of body and tail with skin and spines. These chunks floated in a heavy brown broth. We did not determine whether the entrails had been removed. Some vendors were observed to be doing a brisk business, dispensing small servings of the stew on banana leaves or in gourds. Servings cost the equivalent of $0.45 to $1.35 in U.S. currency; more than the food value would seem to justify. No doubt customers were induced to pay the high prices because of the supposed medicinal properties of the stew.

The mystique of medicinal value seems to be a major factor behind the trade in live iguanas and the stew made from them. Various human ailments and particularly impotence, are believed to be cured or relieved by the flesh (especially that of the spiny-tailed iguana). Hence the purchasers willingly pay much more for these lizards than they would for equivalent amounts of fish, poultry, pork or beef.

Some of the educated Latin Americans with whom we discussed iguanas seemed reluctant to express enthusiasm over their edible qualities, or even to admit having partaken of the flesh. They seemed to consider iguanas uncouth and ugly animals, hardly suitable fare for civilized folk, and better left to unsophisticated aborigines.

Scattered references suggest that in various cultures in Mexico, Central America and South America, iguanas have been an important food source since prehistoric times, and that some of the customs that we observed in the 1970's are based on long standing traditons. For example Pagden (1975:145) cited an early account of the Mayas in Yucatan (between 1549 and 1563), which described in some detail their use of iguanas. The account stated of iguanas that: they are ". . . a most remarkable and wholesome food. There are so many of them that they are a great assistance to everyone during Lent. The Indians fish for them with snares which they fasten in trees and in their holes. It is remarkable how they endure hunger for it often happens that they remain alive for twenty or thirty days after capture without eating a morsel and without becoming thin."

In the Chontal text, written 1612-1614, and cited by Scholes and Roys (1968:404) it is recorded that, about 1599, Indians of Pedro Tzakum-May at Holha killed others from Campeche who had left their villages to collect wax and hunt iguanas. Scholes and Roys (1968:30) wrote of the Maya Chontal Indians in Tabasco that "Much game was hunted in the forests, including deer, peccaries, rabbits, armadillos, coatis, iguanas, wild turkeys, and curassows." However, their culture was primarily agrarian, with cacao a major crop.

On the Pacific slope of southern Mexico, iguana hunting figured in various cultures. The Huichol and Cora, two related but distinct tribes in the Sierra Madre Occidental of Jalisco and Nayarit, hunted deer, peccary and iguana (Grimes and Hinton, 1969). The Huave of the Isthmus of Tehuantepec hunted iguanas [= ctenosaurs?] in thorn forest (Diebold, 1969). The Amuzgo of southeastern Guerrero hunted iguanas with guns (Ravicz and Romney, 1969). The Zapotecs' town was divided, having the south section occupied by the poorer peasants and the hunters of iguana and wild boar [peccary?] (Nader, 1969). The Cuitlatec, primarily of Guerrero, used a whistle for hunting iguana (Drucker, et al., 1969; also see Lazell, 1973, for an account of use of whistles to hunt iguanas in the Lesser Antilles). The Tequistlatec of Oaxaca, though essentially agrarian, also hunted iguanas (Olmsted, 1969). Although ctenosaurs are not specifically mentioned in any of these accounts, the statements about iguanas may include them or may be based upon them exclusively (e.g., Diebold, 1969).

Numerous accounts indicate that in Central America iguanas were similarly important in the aboriginal culture and diet. Borhegyi (1969) wrote that in the uplands of Guatemala the pre-Columbian diet included lizard, especially iguana, and the eggs of iguana, turtle and turkey. Spores (1969:971) wrote of the southern Zapotec that they took for food "a great variety of wild animals–armadillos, wild pigs, rabbits, iguanas, and birds;" the Mixteca Baja who had a culture based on permanent agriculture, also ate flesh of deer, rabbits and iguanas. Edwards (1969) mentioned the bow-and-arrow hunting of iguanas by Indians of the Gulf of Fonseca. Lothrop (1948) mentioned the Panamanian Indians' custom of preserving the flesh by smoking, but Stout (1948) implied that iguana hunting was relatively unimportant a "secondary activity" in the economy of the San Blas Cuna.

In South America according to Gilmore (1950) iguanas were an important and highly prized food source for aborigines and were hunted with the bow and arrow. Smole (1976:182) noted that the Yanoama of Venezuela take lizards ". . . particularly the highly prized iguana . . . ." Kirchoff (1948) noted that iguanas were among the animals hunted in the Venezuelan llanos. Hernandez (1948) wrote that the Achagua and their neighbors in Venezuela and eastern Colombia used iguanas as food. In Colombia Hernandez (1948) noted that iguanas were among the animals hunted in the northern lowlands and in the Cauca Valley where such game supplemented farming for subsistence. Gillin (1948) stated that in Guiana iguana was one of the more frequently used animal foods, and he mentioned shooting and snaring as means of securing them. Also, he mentioned collecting of their eggs by the Pomeroon River Caribs. In Brazil according to Lipkind (1948) the Caraja Indians are passionate hunters who kill

Figure 22.2: Above. Mass of spiny-tailed iguanas and a few green iguanas trussed on sidewalk beside vendor's booth, Chinandega, Nicaragua, February 1976. The sack at right center contains more of the lizards. Below. Close-up view of same group of animals shown above. Photos by R.W. Henderson.

iguanas at every opportunity, but some of these Indians do not eat them. Yde (1965) described hunting of the Brazilian Waiwai. These Indians shoot iguanas with bow and arrow from small trees and shrubs along the river, and dig eggs from nests on the riverbank after locating them by probing with stick or bush knife. Latchem (1922) noted the keeping of live iguanas by Brazilian Indians who tethered them to stakes by a cord tied around the neck or through it (dewlap?). Metraux (1946) wrote that some Indians of the Chaco region, the Pilaga and Mataco, train dogs to hunt iguanas and other game. The iguana is considered a game animal of minor economic importance. "It has become fashionable among acculturated Indians to wear rings made of segments of the tail skins of iguanas" (Metraux, 1946:279).

Figure 22.3: Vendor in booth, with a small lot of spiny-tailed iguanas on display on the sidewalk, Chinandega, Nicaragua, February 1976. Photo by R.W. Henderson.

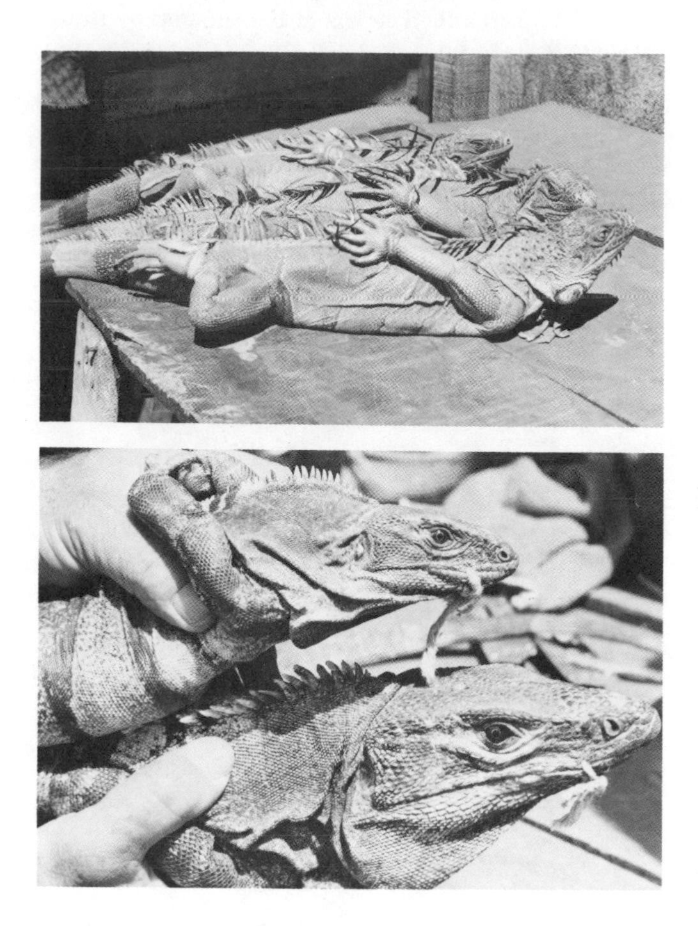

Figure 22.4: Above. Iguanas immobilized by tying feet with twine, on display at mercado at Chinandega, Nicaragua, February 1976. Below. Adult male and female spiny-tailed iguanas with mouths sewn shut, at mercado at Chinandega, Nicaragua, February 1976. Photos by R.W. Henderson.

Figure 22.5: Above. Weighing a large male spiny-tailed iguana, its mouth sewn shut, at the mercado in Chinandega, Nicaragua, February 1976. Below. Adult female spiny-tailed iguana basking on gravestone, cemetery at Belize City, February 1977. Photos by R.W. Henderson.

The Treatment of Captured Iguanas

Related to the problem of conservation is the way in which iguanas are treated after being captured. Some of the traditional practices in handling and transporting captured iguanas seem to cause unnecessary suffering and also probably cause unnecessary mortality with resulting loss to the dealers. The harshest treatment of the animals occurs in the course of the same commercial operations which are depleting populations and ought to be abolished in order to preserve iguanas as game and as a supplementary food source to country people.

Iguanas destined for city markets are usually captured intact and are kept alive because refrigeration is generally lacking and the carcasses would deteriorate rapidly at the high prevailing temperatures. Cruelty is involved in sewing the lips together (in spiny-tailed iguanas, not usually in green iguanas) to prevent their biting the handler, in tying the feet to immobilize the animal, in carrying or leaving masses of iguanas stuffed into large sacks or baskets, so that those on the bottom must support the weight of several layers above, in cutting open gravid females to remove the eggs, and in keeping the animals without food or water for long periods from capture until butchering.

Some vendors are willing to cut open the live female when the customer demands the eggs (a favorite delicacy) without wanting to purchase the entire animal. Such stripped females, obviously in dying condition, were seen from time to time in markets.

Captured iguanas are immobilized by tying the hind legs (and sometimes also the front legs) together behind the back. The traditional method was to pull off a claw with pliers and draw out the tendon of the toe which was then attached to the tendon of the opposite toe. In Panama the claw is pushed through the tendon of the opposite toe forming a toggle (A.S. Rand, pers. comm.). In Mexico and Central America in 1976-79, we rarely saw iguanas trussed by their tendons; instead their feet were tied with string but this method is more damaging to the animal. Often the string is so snugly tied that circulation is impaired, and gangrene results. Even rescued and released iguanas are liable to be permanently crippled, and die as a result. Because of such experience, the Honduran Ministry of Agriculture in 1979 sold confiscated lots of iguanas at the city market rather than attempting to rescue and release them.

Some of the suffering inflicted by the iguana trade could be easily remedied. Fewer iguanas per bag would ease the pressure on those remaining and reduce losses to the dealer. If trussed iguanas would drink, watering them would also reduce mortality and benefit the dealer. Opening live iguanas to remove eggs can best be prevented by convincing country people that the females so treated will subsequently die. Tying legs and sewing mouths can be stopped only when other techniques, which are equally effective and do not involve much extra trouble, are developed. Such alternatives need to be devised before legislation to prohibit the traditional practices is enacted. Education, and the development of new techniques must be coupled with enforceable legislation to effect change.

DISCUSSION

Exploitation of iguanas for human food began in ancient times, and both spiny-tailed and green species remained abundant over most of their ranges after hundred of years of hunting. Both these giant lizards are preadapted to be game animals by virtue of their high reproductive potentials. A large harvestable annual surplus can be produced where natural mortality factors are reduced. Both species are efficient converters of natural vegetation into high grade protein suitable for human food. As primary consumers, both are capable of attaining high population density and biomass.

In relatively recent times, a rapidly increasing human population combined with the advent of modern roads and vehicles, has led to hunting pressure greater than the animals can sustain. Live animals captured are easily transported to consumers in towns and cities. Numbers of iguanas have declined sharply. Improvident use of agricultural insecticides has occurred rather extensively in some areas, and is frequently blamed by Central American ecologists and officials as another important factor in the reduction of iguanas, although proof is lacking. The spiny-tailed iguana would of course be much more affected than the green iguana, because it is more often found on or adjacent to agricultural land. Destruction of habitat and widespread use of firearms for hunting have hastened the downward trend.

Checking and reversing the trend of decrease in iguana populations is of tremendous practical importance to the countries, governments and communities concerned. In not providing adequate protection or management to allow iguanas to increase and fulfill their potential, several nations are each losing tremendous amounts of high grade protein food annually. None of the countries concerned can well afford the waste that is involved. Most of them have taken some action (see section on Protective Legislation, p. 405) but more is acutely needed. Theoretically it would be easy to reverse the downward trend of iguana populations with a small amount of management and protection. However, in actual practice it will be extremely difficult to implement the needed management and protection. Hunting iguanas is a deeply rooted tradition that cannot be readily changed by legislation or education. Country people existing on a subsistence level will not willingly relinquish their rights to capture and eat the lizards at every opportunity. Legislation is seldom entirely effective even after it is passed, because there are too few enforcement officials, sometimes indifferent or corrupt. In addition there is a widespread disrespect for the law in human populations with no tradition of conservation, accustomed to exploiting nature.

Already there is some awareness on the part of the general public that iguanas have been depleted through overhunting. Similarly, concerned officials are aware that protection is needed. A concentrated and integrated program of legislation, enforcement, and education might shift the balance, in some key areas at least, so the dwindling numbers could again increase. Country people might be taught voluntarily to abstain from hunting for a while if they could be convinced that depleted populations of iguanas would increase from scarcity to abundance within a few years, producing an ample yearly harvest.

Compared with *Iguana iguana, Ctenosaura similis* has an even higher reproductive potential (average of 43.4 eggs per clutch vs 30.5 in *Iguana*). It has the additional important advantage that it thrives in disturbed and altered habitats. It may attain high numbers in a cemetery, a corral with rock walls, a suburban lot, an old lava flow, or a pasture with a few large hollow trees. Hence there is plenty of habitat capable of supporting spiny-tailed iguanas even in parts of Central America densely populated by humans. Green iguanas, however, are more exacting in their habitat requirements, and ordinarily need large trees near streams or lakes. It seems that the spiny-tailed iguana would be the more promising subject of the two for attempted restoration. Other chapters in this volume document the numerous differences between *Iguana* and *Ctenoaura* not only in habitat, but in size, sexual dimorphism, growth, social systems and mortality factors. These render the two species quite different problems for programs of protection, restoration, and ultimate cropping on a basis of sustained yield. Eventually somewhat different sets of laws and regulations may have to be devised for each.

Both iguanas have status as game animals. Because of their high reproductive potentials they have generally withstood hunting pressure better than associated species of game birds or mammals including quail, chachalaca, turkey, dove, rabbit, peccary and deer. The most promising future for them seems to be as game. In addition to the harvesting of edible flesh, iguanas provide recreational benefits to many hunters, which will be lost if overhunting continues or if hunting is monopolized by commercial interests.

A possible alternative to cropping them as game is that of "farming" iguanas, raising them on a large scale commercially, for food. This might be feasible if natural populations continue to decline and prices continue to escalate. Enclosures for breeding and rearing might include natural vegetation that would provide much of the food. However, territoriality and fighting (Alvarez del Toro, 1960; Peracca, 1891; Dugan this volume; Dugan and Wiewandt, this volume) would pose a major problem, making it difficult to rear the animals in crowded enclosures, and the long period of growth (probably at least two years, Müller, 1968, 1971) before the animals reached marketable size would make it difficult for the grower to compete with other kinds of meat such as poultry or fish.

A compromise system having some promise is that of maintaining breeding stock in large outdoor enclosures, to produce large numbers of hatchlings, and distributing these to restock depleted areas where habitat conditions remain favorable. The degree of success would depend upon exerting some control over mortality factors that affect hatchlings and juveniles, so that these are held to a consistently low level–less than that in natural populations. It would involve identifying the chief predators on the young and eliminating or neutralizing them: Domestic and feral dogs and cats, especially, would need to be monitored for predation on the immature iguanas.

RECOMMENDATIONS

In order to reduce harvest to the extent that depleted populations can survive and make annual gains, we suggest the following measures. (1) Sponsorship by the OAS of a committee on iguana conservation, which would undertake to supervise international cooperation in such projects as monitoring commercial shipments of the animals from one country to another, and initiating arrangements for procurement of breeding stock to be reintroduced in suitable habitat where the animals have been extirpated by overhunting. (2) Adoption of programs to reach the public through the media of newspapers, radio and television, and in the schools, with the message that iguana populations of both kinds have been drastically reduced and are threatened with extinction, but that they can be restored and maintained as a valuable renewable natural resource with the cooperation of the public in protection and management. (3) Passage and enforcement of legislation as follows: (a) Prohibition of hunting by any method, from the end of January through April, to permit gravid females to develop normally and lay their clutches assuring a new crop of young. (b) Prohibition of all commercial hunting, and sale in city markets, so that the country people who most need the animals for food will benefit, and the recreational values of hunting these lizards as game animals will be fully realized. (c) Prohibition of hunting with firearms; these reptiles lack the native intelligence and wariness to cope successfully with such deadly long range weapons.

Where populations do not respond to relief from hunting during the breeding season, from commercial hunting, and from hunting with firearms, prohibition of all hunting should be initiated. Refuges should be maintained where hunting is prohibited at all times. Since mobility of the animals is limited, refuges should be small and numerous to be most effective in restocking adjacent depleted areas.

SUMMARY AND CONCLUSIONS

Iguana iguana, Ctenosaura similis, and their near relatives have been hunted and trapped for human food since ancient times. Because of high reproductive potential and adaptability both were long able to withstand such exploitation, but in the last two decades drastic reductions have occurred; iguanas have become scarce or absent in many of the places where they were formerly abundant. In Central America there is a highly organized iguana industry. Skilled professional hunters living in the areas where the animals still occur harvest enormous numbers which are shipped to markets of major cities. The escalating price has not checked demand because the flesh is credited with medicinal properties, and is eaten as a supposed cure for ailments such as impotence. The unlaid eggs are in great demand, also. Because of the demand for eggs and because iguanas are a traditional Lenten meat substitute, hunting is most intense during the breeding season. Habitat destruction and widespread use of firearms have had a part in the decline. Heavy use of agricultrual insecticides is suspected also in some areas, especially for the spiny-tailed iguana.

Most Middle American countries have officials and ecologists who are aware of the need for iguana conservation and most have passed some remedial legislation. In depleted areas it is generally recognized by country people, including hunters, that over-hunting has led to decline. But despite these encouraging trends, the general decline continues and accelerates, and stronger action is essential. More international cooperation is needed in the areas of research-sharing, law enforcement, and perhaps restocking programs. Massive educational programs are needed to convince the general public that a potentially valuable food source is being lost through lack of protection and management. Legislation ought to be passed and rigidly enforced to ban commercial exploitation, to ban hunting during the breeding season, to ban hunting iguanas with firearms, and to set aside certain areas as refuges where the animals may live and reproduce undisturbed. The farming of iguanas may have some promise. Also, large scale hatching and rearing of young for restocking of depleted habitats may have a part in restoration.

Acknowledgments

We extend thanks to the Managua Banco Central, and to Dr. Jaime Incer and Sr. Roberto Incer of that institution for funding our 1976 field study in Nicaragua. Dr. Jaime Villa did much to expedite our field work in Nicaragua and we thank him for the many courtesies extended. Also, we thank the International Fund for Animal Welfare, and especially Ms Donna Hart of that organization for funding our 1979 field work in Mexico and Central America. As a U.S. Peace Corps volunteer, Allen Porter rendered valuable assistance in our 1976 field work. We wish to thank the following residents and informed officials of the host countries visited, each of whom contributed important information to our study.

In Mexico, Dr. Miguel Alvarez del Toro (Director of Zoo and Natural History Museum of Chiapas in Tuxtla Gutierrez).

In Guatemala, Lic. Mario Dary (Director de Jardin Botanico, Universidad de San Carlos, Guatemala City).

In El Salvador, Norma Sandra Azucena (Program Planner, National Parks and Wildlife); Dr. Jaime Jimenez Duran (Department of Biology, University of El Salvador); Dr. Victor Hellebuyck (Director, Natural History Museum, San Salvador); Dr. Hugo Hidalgo (Department of Biology, University of El Salvador): Maria Elisa Sánchez (Herpetoligst, Natural History Museum, San Salvador); Lic. Fransicso Serrano (Head, National Parks and Wildlife, Service of Forestry and Fauna, Renewable Natural Resources, Ministry of Agriculture and Cattle Raising); Victor Zelaya (Wildlife Research Biologist, National Parks and Wildlife).

In Honduras, Lic. Sigfrido Burgos (Director General of Renewable Natural Resources); Donald Hanson (U.S. Peace Corps volunteer working as Wildlife Research Biologist, Renewable Natural Resources); Marcial Erazo Pena (Wildlife Research Biologist, Renewable Natural Resources); Paul C. Purdy (Wildlife Research Biologist and Technical Advisor, on contract from U.S.A. to Renewable Natural Resources).

Finally, Mr. and Mrs. Richard Coffee of Tampico deserve warmest thanks for their hospitality during our visit on the 1979 field trip.

Epilogue

The above account refers to conditions as we found them in 1979. Since then the winds of change have swept over Central America, and many of the statements that applied in 1979 are no longer appropriate as this book goes into press in 1982. Insofar as we are aware, the political upheavals and population changes that have occurred have not in any instance worked to the benefit of the iguanas, and their numbers have continued to dwindle. Nevertheless, some hope for the future can be found in the conservation efforts being made by various governments. In Nicaragua, for instance, the Government of Reconstruction's IRENA (Instituto Nicaraguense de Recursos Naturales y del Ambiente) has generated enlightened legislation banning commercial exploitation of iguanas and protecting them during the breeding season, with serious attempts at enforcement.

23

Behavior, Ecology, and Conservation of South Pacific Banded Iguanas, *Brachylophus*, Including a Newly Discovered Species

John R.H. Gibbons
School of Natural Resources
University of the South Pacific

and

Ivy F. Watkins
Orchid Island Fijian Cultural Centre
Suva, Fiji

INTRODUCTION

Until recently, the biology of the iguanine genus *Brachylophus* of Fiji and Tonga was little known, and studies had been confined to captive animals (Cahill, 1970; Cogger, 1974; Carpenter and Murphy, 1978) or preserved specimens (Avery and Tanner, 1970) of *B. fasciatus*. One purpose of this paper is to provide a summary of our present knowledge of the biology of South Pacific iguanines, including an account of *B. vitiensis*, a newly described species from Fiji. A second purpose is to give an account of the status of South Pacific iguanines as endangered species. For purposes of clarification, we refer to the *Brachylophus* group as South Pacific banded iguanas. At present, two species are recognised, the Banded Iguana, *B. fasciatus* and the Crested Iguana, *B. vitiensis*.

The first author is currently undertaking a field study of banded iguanas, with particular emphasis on their ecology, behavior and evolution within the Fiji Islands. The second author has kept specimens of the Banded Iguana, *Brachylophus fasciatus*, in captivity at Orchid Island Cultural Centre, near Suva, since 1972. Both *B. fasciatus* and *B. vitiensis* (Gibbons, 1981) have now been successfully bred in captivity at Orchid Island.

HISTORICAL ACCOUNT

The species *Brachylophus fasciatus* was first described by Brongniart in

1800 as *Iguana fasciata*. Clarification is needed as to the origin of generic name *Brachylophus*. Burt and Burt (1932) attributed it to Wagler (1830) while Avery and Tanner (1970) briefly mentioned the matter and stated that the name first appeared in Cuvier's "Le Regne Animal" in 1829. In fact, Cuvier (p. 41) referred to the Banded Iguana as belonging to "Les Brachylophes" which, in translation, means the short crested ones. The generic name *Brachylophus* first appears in Wagler (1830), even though he credits it to Cuvier. It also appears in Guérin's (1833) "Iconographie du Regne Animal." Since then the name has stuck, most authors attributing it to Cuvier. Following the finding of the Crested Iguana, this generic name has proved to be an unfortunate choice. It prevented the Crested Iguana being called "*cristatus*," since to have used such a species name would have been a contradiction in terms.

To return to Brongniart, he notes that the species was first reported to him by Riche, a naturalist on board the Esperance, one of two armed transports which set out from France in 1791 under the command of d'Entrecasteaux in search of the missing explorer, La Perouse (Labillardière, 1800; for further details–see Gibbons, 1981). Riche's premature death (Daudin, 1802) prevented him from describing the animal himself and, as a result, much confusion exists regarding the type locality of *Brachylophus fasciatus*. However, examination of Labillardière's account shows that Tongatapu, the largest island in the Tonga Group was the only place visited by the expedition within the Fiji/Tonga area; and also that (relatively large?) lizards were offered as food items for barter on that island. The fact that no European contact took place with Fiji until the early 1800s (Henderson, 1937) is further evidence in support of Tongatapu as the type locality of *Brachylophus fasciatus*. During the period 1774-1800, people from "The Friendly Islands" (Tonga) welcomed passing ships, while the Fijians remained fierce cannibals and were avoided by Cook and other explorers (Poignant, 1976). For these and other reasons (Gibbons, 1981), *Brachylophus brevicephalus* (Avery and Tanner, 1970), whose type locality is Tongatapu, is here considered synonymous with *Brachylophus fasciatus*. Evident failure to check original references led Avery and Tanner (1970) to guess, incorrectly, that the type locality of *Brachylophus fasciatus* was somewhere in Fiji.

In 1929, the Whitney South Sea Expedition collected specimens of *Brachylophus fasciatus* from several islands in the Lau Group (Figure 23.1), including Avea, Moce, Oneata and Vatu Vara; and from Kadavu[1] (Burt and Burt, 1932: 511). The species has from time to time been obtained from other islands, including Viti Levu and Vanua Levu, but in general the Banded Iguana is poorly represented in museum collections.

In 1970, Cahill, working at the Fiji Museum, carried out a brief study of *Brachylophus fasciatus* adding several more locations of the species in Fiji. She mentions that "an old preserved specimen in the Fiji Museum appears to be much more solidly built with a much larger crest than any of the present day specimens collected. This could perhaps point to a marked decrease in size over the past 100 years, or even to the fact that they are no longer reaching their maximum size before extermination." Examination of this specimen (no date or locality) has shown that it conforms with *Brachylophus vitiensis* and not with *B. fasciatus*. In preservative the color patterns of the two species are very subdued, the ground color in both being a dull dark grey, though in live specimens (see below) the color patterns are quite distinct.

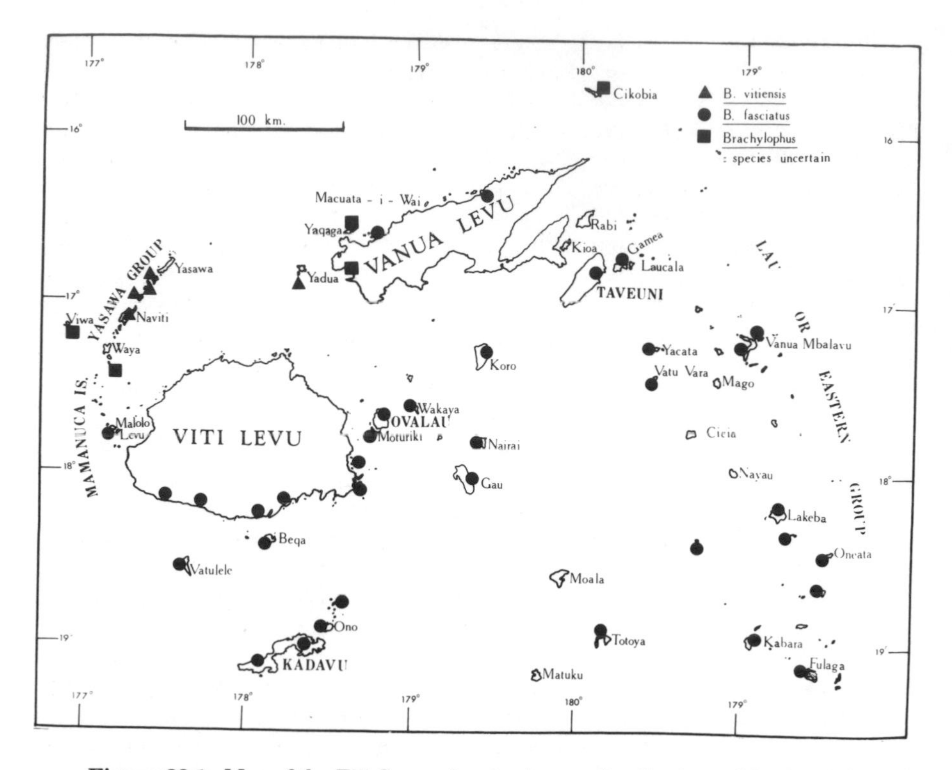

Figure 23.1: Map of the Fiji Group showing known distributions of the Banded Iguana, *Brachylophus fasciatus,* the Crested Iguana, *B. vitiensis* and of iguanines of uncertain species composition. Data are from Cahill (1970)–modified, Burt and Burt (1932), and the present study. The islands of the Tonga Group lie to the Southeast of the Lau Group, approximately equidistant with the Yasawas.

Regarding differences between Tongan and Fijian *Brachylophus fasciatus* as reported by Avery and Tanner (1970), these were very slight, and overlap occurred in all but one morphometric character examined. Criteria for designating species as opposed to subspecies were not discussed, and the erection of a new taxon under such circumstances would seem to serve no useful purpose, even if the errors concerning the type location were rectified.

In 1974, Cogger emphasised the erratic distribution of the Banded Iguana. Its nearest relatives reside in Galápagos Islands and in Central and South America, thousands of kilometers to the east. How did its ancestors reach Fiji and Tonga? According to Cogger, iguanines arrived in the South Pacific by rafting westwards from the Americas on floating vegetation thousands of years ago. All other native land vertebrates in Fiji and Tonga have apparently "island hopped" from South East Asia. Cogger's view has been developed and extended by the first author, and possible trans-Pacific dispersal routes are discussed in detail elsewhere (Gibbons, 1981).

During the last decade, one or both species of *Brachylophus* have been successfully bred in captivity by Cogger (1974) in Australia; at Orchid Island Cultural Centre, Suva, Fiji (see below); and in the United States (Arnette, 1978); but nothing has been reported about their habits in the wild.

Regarding the endangered species status of *Brachylophus,* Bustard (1970), Cahill (1970) and Cogger (1974) all concluded that the introduction of the mongoose had been mainly responsible for the serious decline of the Banded Iguana, which has been decreasing since as early as 1906 (Loveridge, 1945).

Field studies on *Brachylophus* began in July 1978, and have chiefly involved brief visits (4-7 days' duration) to a number of islands within the Fiji Group. The Crested Iguana, *Brachylophus vitiensis,* was found on Yaduatabа, a tiny island close to Yadua (Figure 23.1), during one such trip in January 1979. Subsequently, a more detailed study has been undertaken on the Yaduatabа population, and recently the range of the Crested Iguana has been extended to the Yasawa Group (Figure 23.1).

APPEARANCE OF SOUTH PACIFIC IGUANINES

The Banded Iguana is a slender, green lizard reaching about 70 cm in total length; but more than two-thirds of this value is tail. There is marked sexual dimorphism. The male (Figure 23.2) has two broad pale bands on the trunk which contrast strongly with the emerald green general coloration. Adult females tend to have relatively thinner limbs than males, and have a smaller head, jaws and gular fold. They are considerably smaller in overall size (Table 23.1). They also lack prominent femoral pores. Females from Kadavu Island (Figure 23.1) are commonly uniform green, but specimens from Ovalau, Viti Levu and adjacent islands usually show traces of the banding pattern. There are other differences between Kadavu and Ovalau specimens, but these only become obvious when the animals are aroused or displaying. During display, prominent roaches of skin appear on the backs of Ovalau and Viti Levu males, but not on Kadavu males. Finally, the crest structure, though weakly developed in most *Brachylophus fasciatus* (<0.3 cm), shows its greatest reduction in Kadavu specimens. Previous collections of live *Brachylophus* have been dominated by Kadavu iguanas (e.g., Cogger, 1974) and it is now quite clear that clinal variation exists in *Brachylophus fasciatus.* At present we have a juvenile male from Southern Viti Levu with very well developed crest spines (0.4 cm), but otherwise conforming to *Brachylophus fasciatus.* Few specimens have been available to us from other islands in the Fiji Group, but those that have been examined (from Koro, Kabara, Viwa and Narai) all conform closely to *B. fasciatus.* The exception has been an adult male from Malolo Levu (Figure 23.1). This specimen, now in the Fiji Museum, has a dewlap similar to *B. vitiensis* (see below), a crest of brown spines (0.7 cm), and is intermediate in size between *fasciatus* and *vitiensis.* However, the color pattern is typical of *fasciatus.* Whether this population will be classified as a subspecies of *fasciatus* or given full species status remains to be seen. In Tonga, the Banded Iguana occurs on at least four islands, Tongatapu, Ha 'apai, Va 'vau and Eua (personal communications with Tongan students at the University; various other sources). According to most accounts, Tongan specimens of *B. fasciatus* tend to be even more slender than their Fijian counterparts.

The Crested Iguana looks much more like other iguanine genera, particularly *Iguana,* than does the Banded Iguana. The dorsal crest spines are well developed and prominent, the animal is relatively high at the shoulder and the dewlap is large and almost square in outline. These and other characteristics that separate it from the Banded Iguana are listed in Table 23.1. Apart from femoral counts, all features listed illustrate the gross morphological differences that exist between the two species. During five field trips to Yaduataba, over two hundred and fifty Crested Iguanas have been recorded. The four Crested Iguanas observed at Matacawalevu island in the Yasawas in January 1981 closely conformed with Yaduataba *B. vitiensis,* except that the white bands were somewhat broader (c. 1 cm), and the body seemed more slender.

Figure 23.2: Adult males of Banded Iguana (left) and Crested Iguana (right), showing differences in color pattern and crest structure. Orchid Island, 10 March 1979.

Table 23.1: A Comparison of the External Features of Adult Banded Iguanas with Adults of the New Iguanine

Feature	Banded Iguana (N = 23)* (13 males, 10 females)	Crested Iguana (N = 40)** (25 males, 15 females)
(1) Crest structure	Conical spines <0.4 cm	Conical or recurved soft spines up to 1.2 cm
(2) Color pattern	Male two broad vertical pale blue bands reaching 2 cm width; female with or without bands, or with anterior band poorly developed	Both sexes have two narrow white vertical bands <1.0 cm wide and bordered by dark pigment
(3) Snout vent length, cm mean (range)		
Male	$\overline{X}$ = 16.6 (13.6–19.3)	$\overline{X}$ = 20.5 (19.3–22.3)
Female	$\overline{X}$ = 15.7 (13.4–18.4)	$\overline{X}$ = 20.2 (18.5–21.7)
(4) Adult weight, g mean (range***)		
Male	$\overline{X}$ = 162 (95–207)	$\overline{X}$ = 322 (227–387)
Female	$\overline{X}$ = 123 (105–172)	$\overline{X}$ = 345 (284–404)
(5) Total femoral pores mean (range), male	$\overline{X}$ = 27.1 (23–33)	$\overline{X}$ = 33.1 (31–34)
(6) Sexual dimorphism	Quite obvious for color pattern and size of dewlap	Minimal; restricted to size of femoral pores
(7) Color lability sleeping/display	Males—light green to chocolate brown ground coloration; females exhibit darker color during egg laying	Both sexes—light green to gray or jet black ground coloration
(8) Ventral coloration	Uniform yellowish green	Pale, with green mottling becoming black during display
(9) Appearance of dewlap	Convex in outline; moderately large and pigmented in male, small and pale in female	Large, somewhat square in outline, and pigmented in both sexes
(10) Iris color of eye	Red-orange	Pinkish-gold

*Source of specimens: Kadavu (6 males, 7 females); Viti Levu (3 males); and Ovalau (4 males, 3 females)

**All animals from Yaduataba

***To nearest gram

Within the Yaduataba population there is considerable individual variation. For example, dewlap shape varies considerably in males. It is also our impression that males have slightly longer crest spines than females, and can develop darker (jet black) ground coloration and higher dorsal roaches during display. Some females have a smaller dewlap than that shown in Figure 23.3. Maximum total length recorded for the new iguanine is 87 cm, an adult female. As in *Brachylophus fasciatus,* it is the female which has the longer tail. Reasons for this are discussed below.

Figure 23.3: Adult female Crested Iguana, showing large dewlap. There is very little sexual dimorphism in this species. Orchid Island, 10 March 1979.

HABITAT AND DIET OF SOUTH PACIFIC IGUANINES

Both species of *Brachylophus* occupy coastal and lowland forest, particularly swampy areas. No reliable reports exist of sightings in rainforest. However, whereas the Banded Iguana lives in mesic habitats such as occur throughout most of the Fiji Group, the Crested Iguana is restricted to dry, rainshadow islands like Yaduataba and the Yasawas. Approximate rainfall data for these latter are 180 cm or less per annum. However, most of this rain falls in the hurricane season between October and March, and there is a distinct dry season between May and September. In contrast, most of the remainder of the Fiji Group receives well over 200 cm rainfall per annum, and conditions between May and September can be both wet and cool (<16°C on some nights). Adult Crested Iguanas appear unable to maintain themselves under such conditions, suffering from loss in weight and a variety of other ailments. Significantly, Malolo Levu (Figure 23.1), the island occupied by the 'intermediate' *Brachylophus* is also intermediate in terms of rainfall.

Reports of the Banded Iguana from both Fiji and Tonga indicate that it is most commonly encountered in large broad leaved, leafy trees, particularly the Tahitian chestnut or ivi tree, *Inocarpus fagiferus* (Caesalpiniaceae). It is also found on: mamakara, *Kleinhovia hospita* (Sterculiaceae), dakua, *Agathis vitien-*

sis (Pinaceae); dilo, *Calophyllum inophyllum* (Clusiaceae); vau, *Hibiscus tiliaceus* (Malvaceae); bele, *Abelmoschus manihot* (Malvaceae); and, vesi, *Intsia bijuga* (Caesalpiniaceae). Near inhabited or cultivated areas the Banded Iguana has been frequently observed on introduced trees, particularly mangoes, *Mangifera indica* (Anacardiaceae), guava, *Psidium guajava* (Myrtaceae), and soursop, *Annona muricata* (Annonaceae). These almost certainly represent food trees.

On Ovalau specimens have occasionally been captured in mangroves, though whether mangrove leaves are food items is not yet known. The iguana relies mainly on camouflage for protection and is extremely difficult to find. The great majority of sightings of the animal in the last ten years appear to have been the result of chance encounters rather than active search. Occasionally, Banded Iguanas have been observed on the ground. Probably no photograph has yet been taken of this species in the wild.

Diet of the Banded Iguana is probably a mixed herbivorous/insectivorous one. In captivity, this species often, but not always, shows a strong preference for mealworms and beetle larvae from ivi nuts over most plant foods. Nevertheless, the animal will thrive on a purely vegetarian diet of pawpaw (papaya), banana and *Hibiscus* leaves, and flowers. Fergus Clunie (pers. comm.) reports that some captive Banded Iguanas will eat small caterpillars, and may actively pursue and eat cockroaches. There is, however, great individual variation in food preferences even amongst newly hatched individuals. Some prefer pawpaw to banana or vice versa, and while some will accept mealworms as a choice item, others stick to a vegetarian diet.

Ironically, in view of the short period it has been known to science, the Crested Iguana has been far more thoroughly investigated than the Banded Iguana. Yaduataba, estimated to have a population of about three hundred Crested Iguanas, is a volcanic island 0.7 km^2 in area, and oblong in shape. The animal is found in pockets of primary forest stretching down from the high (c. 50-100 m) central ridge to the coastline. It is absent from both the disturbed semi-open habitat, dominated by *Casuarina equisetifolia,* and the grassland, resulting from repeated burning, on higher parts of the island. The most common tree within this forest is the cevua, *Vavaea amicorum* (Meliaceae), while *Inocarpus* is absent. The dense canopy 5-10 m above ground provides heavy shading, and maintains the rocky soil in a relatively moist and cool condition. Details of perching sites occupied by iguanas on Yaduataba are given in Table 23.2. Of all the animals observed, over one-third were on *Vavaea,* and several iguanas, mainly males, were found wandering on the forest floor. Food identified after dissection of a dead animal (see below) consisted entirely of leaves and flowers, particularly of stamens. No insect remains were found in this animal, and since our captive adult specimens showed little interest in mealworms, there are indications that the Crested Iguana is entirely herbivorous. Unlike the Banded Iguana, this animal has a voracious appetite, and will consume large quantities of banana, pawpaw and flowers. Unlike adults, hatchlings (see below) and juveniles will take mealworms in addition to flowers and fruit.

In January 1981, four Crested Iguanas were recorded on Matacawalevu Island in the Yasawas. All specimens were taken in ivi trees near a village.

Table 23.2: Utilisation of Perching Sites by Iguanas: All Trips

Species of Tree and Family	Fijian Name	Coastal/Rainforest	Number of Iguanas
Vavaea amicorum (Meliaceae)	Cevua	C/F small tree	86
Mallotus tiliifolius (Euphorbiaceae)	Yaqata	C small tree	30
Diospyros spp. (Ebenaceae)	Kau loa	C shrub small tree	21
Hibiscus tiliaceus (Malvaceae)	Vau	C small tree	15
Premna taitenis (Verbenaceae)	Yaro	C small tree	9
Canthium odoratum (Rubiaceae)	Noko ni savu	C small tree	8
Derris trifoliata (Papilionaceae)	Duva	C creeper	8
Eugenia rariflora (Myrtaceae)	Qaqi koro	C cmall tree	4
Ficus obliqua (Moraceae)	Baka	C/F large tree	4
Ervatamia orientalis (Apocynaceae)	Laqaiqai	C/F small tree	3
Fagraea gracilipes (Loganiaceae)	Buabua	C tree	3
Cocos nucifera (Palmae)	Niu	C tree	3
Calophyllum inophyllum (Clusiaceae)	Dilo	C tree	2
Gyrocarpus americanus (Hernandiaceae)	Wiri wiri	C large tree	2
Terminalia littoralis (Combretaceae)	Tavola	C large tree	2
Pongamia pinnata (Papilionaceae)	Karisini	C large tree	2
Morinda citrifolia (Rubiaceae)	Kura	C small tree	2
Entada phasioloides (Mimosaceae)	Wa lai	C/F creeper	2
Thespesia populaea (Malvaceae)	Mulomulo	C tree	1
Messerschmidia argentea (Boraginaceae)	Kau yalewa	C small tree	1
Decaspermum fruticosum (Myrtaceae)	Nuqanuqa	C/dry F shrub	1
Casuarina equisetifolia (Casuarinaceae)	Nokonoko	C tree	1
		TOTAL	210
Others			
Dead trees			3
Forest Floor litter			4
Rocks on forest floor			7
Rocks on beach			4
		GRAND TOTAL	228

PREDATORS

Reports of predation on South Pacific iguanines are scarce. Fergus Clunie (pers. comm.) has informed us that he observed a Banded Iguana being harassed by a cat near Suva, in Viti Levu. To this effect, *Brachylophus fasciatus* does show an alarm reaction to cats and dogs, running away at high speed across open ground, and hiding motionless in thick cover. Cornered, the animal may open the mouth and lower the dewlap. More than half of the dozen Banded Iguanas we have observed from Ovalau had damaged tails and this may reflect attacks (but see below) by feral cats or raptorial birds. Since feral cats infest many islands in the Fiji Group they must be considered important predators.

On Vanuabalavu (Figure 23.1), Birendra Singh (pers. comm.) has reported observation of a Pacific Boa, *Candoia bibronii* swallowing a Banded Iguana, and this snake is found on many islands occupied by *Brachylophus fasciatus*. Attempts by *Candoia* to eat captive adult Banded Iguanas have been observed. When on display to visitors at Orchid Island, the large male "Hercules" was resting on a branch. A Pacific Boa also on display was on another branch nearby, and whilst someone was taking a long time to focus their camera, the snake had edged its way closer to Hercules until it finally grasped Hercules by the head and proceeded to swallow him. It was only when the snake was in the process of taking the final swallow that people noticed what was happening. After grasping either side of the snake's mouth–and the iguana was three-quarters of the way down the throat of the snake by then–one of us (Watkins) held the hind-quarters of the iguana and eased him out. Apart from a few teeth of the snake in his dewlap, Hercules was unharmed and "unmoved" by the experience.

The Swamp Harrier, *Circus approximans,* is also a definite natural predator of South Pacific Iguanines, since their skeletons have been found in Swamp Harrier nests of Yadua Island. The dead iguanine (noted above) on Yaduataba was probably a hawk kill. The animal had chunks of flesh removed from its back and most of the lower jaw was missing. The Grey Goshawk, *Accipiter nifitorques,* may be a predator, particularly of smaller iguanas, while the White Collared Kingfisher, *Halcyon chloris,* a known lizard predator, almost certainly takes hatchlings. The latter is widespread throughout the Fiji Group, including Yaduataba.

To our knowledge there is not a single recorded observation of the Indian Mongoose, widespread on Viti Levu, Vanua Levu and Beqa, killing an iguana. It is therefore surprising that the I.U.C.N. Reptile Red Data book attributes the decline of *Brachylophus fasciatus* mainly to the introduction of the mongoose to Fiji a little over eighty years ago. However, female iguanas do have to descend to the ground to nest, and during this period they would be particularly vulnerable to predation by mongooses.

HABITS OF SOUTH PACIFIC IGUANINES IN THE WILD

Apart from the population at Yaduataba, little is known about the ecology and behavior of South Pacific banded iguanas. This is mainly because the animals occur in remote localities and are extremely well camouflaged. Recent observations have confirmed that, apart from hatchlings which may emerge at night, both species of *Brachylophus* are strictly diurnal. At night, adult and juveniles usually sleep on low horizontal branches, sheltered from the wind. The posture is very characteristic. The head rests on the branch and the front feet encircle it. The longest toe of the hind limb extends horizontally outward and rests upon another branch or a leaf for balance. However, that is not always the case, and a variety of other sleeping postures and positions have been observed. On occasion, animals sleep in high, exposed positions. The posture of the animal is sometimes such that the hind legs grip onto a branch while the body and front legs dangle in space, counterbalanced by the long tail. Animals captured at night by spotlighting struggle only feebly, and resume sleep almost immediately illumination is removed. For the Crested Iguana, at least, this method of capture is far more effective than day-time procedures.

During the first author's initial visit to Yaduataba in January, 1979, a hurricane was taking place within the Fiji Group, and despite heavy rain, all Crested Iguanas observed remained in trees. The majority were inactive and on low branches. Upon close human approach, these animals frequently swung round behind their respective branch to expose as little of the body as possible. During the second visit in fine, warm weather in April, most animals climbed higher in their trees and escaped from view, rather than remaining stationary.

During 1980, the daily activity patterns of Yaduataba iguanas were studied more closely, to the extent that the Fiji Film Unit has now completed a short documentary on the animal's habits in the wild. Shortly after sunrise, the iguana re-orientates its body to gain maximum heat radiation. After several hours of thermoregulation or other activity, the animal moves to a feeding site comprising either young, soft leaves, flowers or flower buds, or less commonly, fruits. Food is eaten quickly in one or two five to ten minutes sessions. In order to consume flowers on long stalks, the front legs are thrust out into space to grab the food, while the hind legs grip onto a tree branch or three trunk. For an arboreal animal like *Brachylophus,* the long tail is of paramount importance in acting as a counterbalance to the head and trunk for activities such as feeding. Females tend to have longer tails than males on account of their often gravid (and hence heavy) condition. Animals with broken tails show reduced arboreal activity, and spend more time on the ground.

There is no doubt that South Pacific banded iguanas are highly selective in their choice of food. For example, young ivi leaves, *Inocarpus fagiferus,* which are thin, pale and soft, are readily taken by captive specimens of both species, but mature leaves, which are relatively thick and hard, are rejected. On Yaduataba, there is a definite movement of animals from higher ground to areas near the main beach in September, as various trees come into flower. At the same time there is some evidence that individual animals have their own home ranges. For example, one adult male with a scarred jaw was observed at the base of a coastal *Messerschmidia argentea* in April 1979 and January 1980, the same tree where it had been recorded in January 1979.

Adult male Crested Iguanas become aggressive and territorial in October, and the method of display and fighting is closely similar to that described for other iguanines (Carpenter, this volume). Severe injuries may result. About forty percent of males sampled had one or more toes missing, while a smaller proportion had damaged tails. On the other hand, few such injuries have been observed in females and juveniles. Territorial behavior continues at least till April. Of the three Crested Iguanas observed wandering on the forest floor in April 1979, two chased the author (Gibbons) for a short distance, dewlap lowered. They then performed several very slow pushups on stiff front legs, presenting laterally, and staring intently at the author and a field assistant companion. Not one made any attempt to retreat when confronted and photographed, and all three assumed the black coloration within a few minutes. Color change always began on the dorsum, and then spread rapidly down the flanks to ventral surface. Subsequently, the limbs and head became darkened. During such display, crest structure was emphasised by the erection of roaches of skin on the dorsum.

From January to March, females are found within male territorial areas, usually on a one to one basis. Mating takes place during this period. By May, all females have left the male areas and move to one part of the island to lay their eggs. The site of this nesting area is believed to be in some low bushes near the main beach (Maika Natera, pers. comm.). The eggs take at least eight months to hatch (see below), hatchling iguanas being observed in January 1980. Like adults and juveniles, these hatchlings are highly arboreal and even more difficult to find. Several observations have been made by Fijians of groups of three or four hatchlings dashing across open ground in close formation. These and other findings suggest that hatchlings have some kind of organised social behavior.

There is also some evidence of parental care in *Brachylophus*. On Yaduataba, an adult and four hatchlings were observed emerging from a hollow tree (Apisai Bogi, pers. comm.) and, likewise, on Tavewa Island in the Yasawas, an adult was seen with four juveniles, sheltering under a rock (pers. comm. with an inhabitant of Tavewa). Apparent parental behavior in the Banded Iguana is noted below.

OBSERVATIONS IN CAPTIVITY

Adult male Banded Iguanas, though normally docile, are extremely aggressive towards other males, and will fight if placed together in the same cage. Severe injury to the neck, head or tail may result. They will display at, and attempt to attack, males in nearby cages. Display involves lowering of the dewlap, head bobbing, and changes in the ground color from light green to darker hues, eventually to a very dark, greenish brown. The dorsal region is the first part of the skin to become dark, as in the Crested Iguana. At the same time, the pale bluish bands develop yellowish tones, and present a vivid contrast with the general body coloration. Head bobbing may involve very rapid, low amplitude movements (>5 per second) with the head held high; this can be termed "shuddering." Alternatively, slower deeper amplitude, movements, with the head held in a lower position, may be exhibited [see Carpenter and Murphy (1978) and Greenberg and Jenssen, this volume, for display-action patterns]. The stimulus that elicits display appears to be a visual one, since male

Brachylophus fasciatus will respond strongly to a mirror. One of us (Gibbons) has observed an adult male, that used to spend most of his time on the living room curtains, perform mirror displays on numerous occasions. The animal would descend, march "purposefully" across the floor in display coloration, enter the bathroom and climb rapidly up a curtain about one metre away from a large mirror. Half way up the curtain, and now visible in the mirror, he would stop and then 'shudder' and 'slow-bob' vigorously and repeatedly. Over a period of weeks, the animal spent more and more time on this bathroom curtain and eventually took up residence there.

Male/male aggression is well-known in the Banded Iguana (see Cogger, 1974) and in captive conditions a dominance hierarchy is established within minutes of introducing the animals to a cage. However, contrary to Cogger, we have found that if a new male is added to the cage of an established resident, it is not always the resident that wins the subsequent encounter. "Samson," who was given to Orchid Island in 1976, attacked and overcame Hercules within thirty minutes of arrival, in spite of the fact that Hercules had occupied the cage since 1973. Though Samson (140 g) was smaller than Hercules (198 g) and though Hercules was removed to another cage and soon established a 'territory' there, Samson always remained a far more aggressive individual than Hercules, and displayed at the slightest provocation. During male/male encounters the combatants try to grip the skin on the neck or at the base of the tail of their opponent. Severe injury can result, and in Hercules' case the wound on his tail became infected, and necessitated amputation.

During male/male encounters female *Brachylophus fasciatus* retreat to the opposite end of the cage and take no part in the interaction. Indeed, they appear non-aggressive apart from the period just prior to and after egg-laying (see below). In response to the head bobbing of males, adult females perform a series of low amplitude nods with the head and trunk held quite close to the ground. They exhibit little other display behavior. Occasionally, they may lower the small, pale dewlap and/or open the mouth on human approach.

The Crested Iguana is quite like the Banded Iguana in social behavior, even though overt aggression and fighting amongst males was not observed during the time they were kept at Orchid Island between January and May 1979. Head bobbing is similar to that seen in the Banded Iguana and involves both 'shuddering' and slower up and down movements of the head. The latter is exhibited by both sexes. Females seem equally assertive in selecting a perching site. Favorite locations are branches partially bathed in sunlight. Early in their confinement at Orchid Island, when more than two individuals were placed in the 1.5 x 0.7 x 0.7 m high cages, it was our impression that in each cage a heterosexual pair amongst the group were consistently occupying these choice positions. The other animals in the cage seemed to become psychologically depressed and lost appetite.

When a heterosexual pair of Banded Iguanas was placed in a cage near the Crested Iguanas the male "Samson" repeatedly displayed at and attempted to attack his neighbours, who, though much larger, failed to retaliate. When placed in his neighbours' cage Samson attacked or threatened both male and female *B. vitiensis*. The latter usually moved slowly away from his advances.

However, after repeated provocation, one of the Crested Iguanas, a male, eventually turned and attacked Samson, forcing him to retreat.

Recent experience with two captive *B. vitiensis* from the Yasawas indicate that they are strongly territoritial and aggressive. The larger specimen, an adult male, was immediately challenged by the resident *B. fasciatus*. During the ensuing fight, the Crested Iguana bit the Banded Iguana on the nape and trampled on its head with the front foot. The Banded Iguana soon fled, and has remained subordinate ever since. Whenever the two meet the larger, dominant *B. vitiensis* attempts to rape the smaller *B. fasciatus*. The latter is dominant over the other Crested Iguana, a young male.

REPRODUCTION

In contrast to earlier reports by Cahill (1970), Cogger (1974) and Arnette (1980), the Banded Iguanas at Orchid Island, which are kept in semi-natural conditions in open air enclosures, show definite seasonal breeding. Mating starts in about November, as the air temperature warms up. The eggs are laid some 6 weeks after the last mating and take 18-30 weeks to hatch when incubated at ambient air temperatures (see Table 23.3). In the single observation of breeding in the Crested Iguana, mating took place on 10 March 1979, eggs were laid on 21 April 1979 and the first hatchling had emerged by 25 December 1979. Time taken for incubation in *Brachylophus* (i.e., 18-30 weeks for *B. fasciatus* and 35 weeks for *B. vitiensis*) presents a striking contrast to the 10-14 weeks given for other iguanines (Wiewandt, this volume).

Table 23.3: Reproductive Data on *Brachylophus*

Species	Clutch Size	Date Laid	Date of First Hatching	Remarks
B. fasciatus	3	Early February 1977	25 July 1977	3 females
	4	31 December 1977 late afternoon	9 May 1978	4 males
	3	5 March 1978 late afternoon	9 October 1978	2 females, 1 male
	4	Early January 1979	5 September 1979	2 males, 2 failed to hatch
	3	3 January 1980	10 July 1980	2 males, 1 female
	6	8 January 1981	–	2 eggs collapsed, 4 still incubating
B. vitiensis	4	21 April 1979 afternoon	24 December 1979	3 eggs hatched sex of hatchlings uncertain
	4	9 June 1980 afternoon	–	1 egg collapsed still incubating

Females of both *Brachylophus fasciatus* and *B. vitiensis* seem to spend most of their adult life in an apparently gravid condition, abdomen distended with eggs. In the Banded Iguana, mating often takes place about two weeks after a clutch of eggs has been laid, and if these observations represent field conditions, one might expect the animals to lay one or two clutches in a season (e.g., in January and perhaps again in March/April, but see below). By May, the air temperature begins to fall rapidly, particularly at night, and one would expect a decrease in sexual activity round this time.

Male Banded Iguanas show courtship by "shuddering," and attempt to grip the skin on the females' back, prior to mounting. Copulation takes 15-20 minutes. Multiple matings occur, at least six having been observed between Hercules and "Letitia" in December 1979.

Only one observation has been made of mating in the Crested Iguana (Figure 23.4). The small male "Lochinvar" had been actively pursuing the large female "Magnifica" off and on for several days, head bobbing and attempting to grip the nape of the neck. But each time she had avoided copulation by walking forward. Approximately half an hour before copulation took place, both animals had head bobbed at a conspecific being photographed outside their cage, and appearing very aroused, had assumed a dark coloration. Lochnivar again commenced courtship and successfully obtained a grip on the female's neck. But on this occasion, instead of moving forward, Magnifica crouched on the ground and raised the base of the tail almost vertically, thereby enabling Lochinvar to insert one hemipenis. The two animals then remained almost motionless until the male released his grip on the neck after ten minutes. After a further fifteen minutes Lochinvar head bobbed three times, and it was then apparent that copulation had ceased. A minute later, head bobbing was repeated, and as the male slowly disengaged from the female, he head bobbed once more, and moved a short distance away. The female remained motionless for another minute, then walked very slowly forward. Upon examination of Lochinvar shortly thereafter, we noticed a white liquid near the cloaca.

Figure 23.4: Copulation of Crested Iguanas. Orchid Island, 10 March 1979.

The Banded Iguana lays eggs in soil and is very selective in choosing a site (see also Cahill, 1970; Cogger, 1974). Clutch size ranges from three to six, but is usually four (see Table 23.3 for details) and the eggs, which are about 3 cm in length, are positioned in a characteristic formation (see Figure 23.5). The female first digs a burrow of about her own body length, and then lays each egg close to the entrance. Considerable exertion is involved and her body color becomes dark green. After laying an egg, she manipulates it down into the burrow using the front foot. The female then tightly packs the soil over the eggs by banging it down with the head. The process is repeated on several occasions as more and more layers of soil are deposited on the eggs. During egg laying, the resident male may watch from nearby, but plays no part in the construction of a nest.

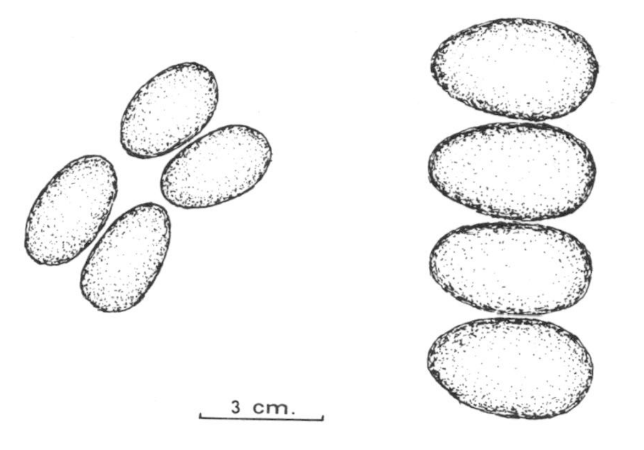

Figure 23.5: Placement of the eggs in burrow in Banded Iguana (left) and Crested Iguana (right). Scale drawing.

Burrowing behavior has been observed in two Crested Iguana females, and is more elaborate than in the Banded Iguana. The female "Cleopatra" spent many days digging a very long burrow. This collapsed and she began digging another. The method of burrowing is very characteristic. The female digs the soil with the front feet, throwing this backwards towards the hind legs. These latter then push the soil away from the burrow. Six weeks after the observed mating, Magnifica, who had in the meantime been observed digging, laid four eggs in a two ended burrow about 10 cm below the soil surface. The placement of these eggs, which measured over 3.5 cm length, was in a line (Figure 23.5), a quite different arrangement from that employed by the Banded Iguana. The eggs were removed for incubation shortly before disastrous flash floods, caused by the building of a new highway, drowned most of the animals at Orchid Island one night in May 1979.

The procedure for incubation was the same in both species. The eggs were buried in moist soil to a depth 2-5 cm in open topped, plastic containers and the soil was moistened by adding a few drops of tap water every 2-3 days. Eggs of both species absorb water and gradually increase in size and weight. Those of the Banded Iguana (c. 10 g) are approximately 3 cm in length initially, and may reach 4 cm shortly before hatching; in the Crested Iguana the eggs are much larger (19 g), initially averaging about 3.5 cm in length and swelling up to 5 cm or more. In both species the egg has a hard parchment-like shell, but this gradually softens and becomes moister about ten days before hatching. At the same time, a brownish oval mark about a centimetre in diameter, begins to appear on the surface of the egg. It is in the vicinity of this mark that the head first appears at the commencement of hatching.

Hatching begins with the appearance of the tiny yellow ringed nostrils through a slit in the egg. About three hours later the head breaks through the shell and, rests on the shell surface with the eyes closed. Within an hour the eyes are opened and the hatchling appears aware of its surroundings, peering around intently. In the Banded Iguana, the front legs usually emerge after a sudden effort some two to five hours later. At this stage, the body can be seen to have a dark green pigmentation, and the breathing is heavy. Within a further hour there is another burst of activity, and this time the animal completely emerges out of the shell and runs, or walks rapidly, to cover, a trace of the yolk sac still visible on the nearly closed umbilicus. Banana and pawpaw subsequently offered to all the hatchling *Brachylophus fasciatus* was refused, even after several days, and therefore they had to be forced fed for about two months.

Evidently, the remaining yolk provides an important source of food for hatchlings, since one individual which (as a result of human disturbance) emerged too quickly and failed to absorb the yolk sac, soon died from starvation in spite of forced feeding. Growth rate in hatchling and juvenile male Banded Iguanas seems to be faster than in females. Both sexes weigh 9-10 g and have a total length of about 21 cm, on emergence. After eighteen months, three juvenile females averaged 41.9 cm in total length and 35 g in weight. On the other hand, four males averaged 42.2 cm by nine months and 42 g in weight.

In our two recorded observations of hatching in the Crested Iguana, just the head is emergent for periods of 13 and 24 hours. However, during this time, occasional movements of the neck caused the opening in the shell to gradually enlarge. As in the Banded Iguana, the ground color of the hatchling is dark green. Both of us observed the first of these two hatchlings (Figure 23.6). Peering around alertly, the animal suddenly jumped out of its shell and raced for cover at high speed. After several hours, the ground color of this hatchling became bright emerald green, and its species-specific pattern of narrow white bands was clearly visible.

The size of hatchlings of this species (three eggs have so far been hatched) is very much larger (total length 27.7-28.8 cm, snout-vent length 8.3-8.5 cm, weight 18.3-20.0 g) than that of hatchling Banded Iguanas (total length 19.8-21.3 cm, snout-vent length 7.7-7.3 cm, weight 8.2-10.5 g), and other features, including the crest and the high ridge (or roach) of skin behind the head, are plainly visible. The skin on the dorsolateral region rapidly develops black pigmentation when the hatchlings are frightened, as in adults. Unlike the Banded Iguana, hatchlings of the new iguanine take food readily without the requirement of forced feeding. So far, they have been fed pawpaw, banana, *Hibiscus* stamens and mealworms. Lactogen, yeast and vitamin B_{12} will later be given as supplements. Sick or distressed individuals assume the black coloration as in adults.

Figure 23.6: Hatching egg of Crested Iguana showing very large pineal eye on the head of the emerging hatchling. 26 December 1979.

Timing of hatching in the Banded Iguana seems fairly well synchronised, a whole clutch often emerging within 24 hours. However, periods of up to five days have been recorded between the first and last emergence. In the Crested Iguana the first two eggs hatched within two days of each other (24 December 1979 and 26 December 1979), but the third individual did not emerge until 2 January 1980, an interval of seven days. There are growing indications that in both species hatching occurs in periods of rain and/or high humidity.

Like adults, *Brachylophus* hatchlings obtain moisture by licking wet leaves. They are attracted by the colors yellow, red, orange and black, but dislike pure white and avoid this color. The iguana stares intently at a color before moving towards or away from it. They immediately react to "sounds," including bird calls and helicopters, turning the head and glancing upwards. Their eyesight is excellent and they will begin head bobbing on seeing individuals in cages 10-20 meters away.

There is now strong evidence that females of both *Brachylophus* species behave aggressively just before and after egg laying, and that this may represent nest defense. Reports of green (female) Banded Iguanas jumping on people from trees have been obtained from Wakaya (David Burness, pers. comm.), Vatulele (Fergus Clunie, pers. comm.) and Viti Levu (Alan Osborne, pers. comm.). Iguanas are greatly feared by Fijians who claim the animal jumps from trees and digs into skin with sharp claws, and that it can only be removed with fire or salt water. The Crested Iguana, Magnifica, leapt against the glass wall of her cage in an attempt to attack spectators shortly after laying eggs. A Fijian villager from Yadua reported that while examining large, soft white eggs in a coconut tree on Yaduataba, he heard a coughing sound. Looking down, he saw an iguana climbing up rapidly below him, and jumped terrified to the ground. There are numerous accounts similar to this one.

At Orchid Island, several juvenile *B. fasciatus* were put in the same cage as their parents at four months of age, and lived together very happily until twenty months later, when their mother laid a further three eggs. Two weeks later she began attacking the youngsters by biting their backs. The juveniles had to be removed from the cage (the three eldest and most severely bitten were half grown females), and within a few days the adults had mated. This took place in March 1979, but no eggs were laid thereafter throughout 1979, even though the female became very swollen with eggs by August 1979. This female, Letitia, was observed mating again with Hercules in November and December 1979, laying a clutch of eggs in January 1980.

These observations suggest that female Banded Iguanas are flexible in the duration of time between fertilisation and egg laying, according to the particular time of year. This conclusion is supported by the fact that no Banded Iguana eggs have ever been laid at Orchid Island between April and November. Furthermore, the female "Delilah" which mated with "Samson" in March 1978 did not lay her eggs at all, but remained gravid until her death in floods in May 1979.

CONSERVATION OF PACIFIC IGUANINES

There is little doubt that South Pacific iguanas have seriously declined in numbers since the onset of the twentieth century (Loveridge, 1945; Bustard, 1970), though the major reasons for this decline are by no means certain. Since both the Banded Iguana and the Crested Iguana, in particular, are restricted to heavily wooded areas, destruction of suitable habitat must have played an important role in their decline, or even extinction, in many localities. Indeed, it is quite possible that iguanines may once have occupied islands to the east of Tonga, becoming extinct before European arrival (but see Gibbons, 1981, indicating arrival of iguanines in Fiji directly from Central America). At the same time, it is becoming more and more apparent that Bustard's (1970) statement that *Brachylophus* is rare throughout its range was wildly inaccurate speculation. To the contrary, several islands, especially uninhabited ones like Vatu Vara (Figure 23.1) may support relatively dense populations, while on others like Koro and Kadavu, iguanas are still quite common. The decline of *Brachylophus* has probably resulted from the following causes:

(1) Habitat destruction has been brought about both directly and indirectly by man's influence. Direct effects include burning and felling of native trees, which has been widespread on many of the islands in the Fiji Group, and even more so in Tonga. Indirect effects include the introduction of domestic animals such as pigs and goats. Both these animals now occur as wild or semi-wild popu-

lations on many islands in Fiji and their effects on vegetation, particularly in drought conditions, can be severe. Goats destroy the understory, rendering iguanas susceptible to predation by raptorial birds and cats. They have had a drastic effect on the ecology of small, uninhabited islands such as Makodroga. Goats were introduced to Yaduataba in 1972 and, before their removal in 1981, their presence severely threatened the continued existence of the iguana population. In addition to their own activities, large areas of bush on that island had been burned to provide the goats with grazing.

(2) A second reason for the decline of the South Pacific iguanines has been the introduction of predators, particularly cats. Feral cats abound on many of the islands in the Fiji Group including Taveuni, where Banded Iguanas are now scarce, and on Yadua. The latter island is also infested with wild pigs, and, according to the villagers that live there, iguanines are believed to be absent. Yadua is separated by only 180 m of shallow water from Yaduataba, where feral cats and pigs are absent. The effect of the mongoose, present for many years on the two largest islands, and recently introduced to Beqa, is not known, though it has definitely led to the extermination of some species of ground nesting bird.

(3) Other reasons cited for the decline of the iguanines have been killing by Fijians and poaching. Neither of these is likely to have had a very widespread effect, but in local areas they have caused a sharp decline. For instance, in 1977, near the airport on Ovalau, some Americans paid the villagers to collect some twenty-four specimens from ivi trees in a nearby swampy area. Eighteen months later, when asked to catch Banded Iguanas, the Fijians searched the same trees and were unable to find a single specimen. Iguanas are usually killed by Fijians when encountered in the bush, mainly as a result of their fearsome reputation. On small, dry islands these killings may have had a very drastic effect on the iguana populations, since ivi trees are usually found in coastal areas close to villages.

Conservation measures to protect the Banded Iguana in Fiji include a $50.00 fine for poaching, and illegal export status. But it is clear that neither of these measures have been very effective. The financial rewards for illegal export of the iguanas are high, particularly in the United States, Austria and West Germany; and, the deterrent effect of penalties for illegally importing the animals has, until recently, been low. Indeed, "Hercules" was donated to Orchid Island in 1973 after a would-be smuggler was apprehended at Nadi airport. Virtually all the estimated 50-100 Banded Iguanas in American Zoos have been obtained without the knowledge or consent of the Government of Fiji, though a wildlife smuggling case involving a number of Banded Iguanas did take place in 1977. Good progress has, however, been made towards conservation of the South Pacific iguanas, particularly in the last few months. This includes both increased public awareness of the problem, and the setting up of Yaduataba as a wildlife reserve. An account of the latter is given below.

Following the finding of the Crested Iguana in 1979, it was clear that Yaduataba required both immediate protection and active conservation measures. A plan was therefore evolved to have the villagers on the nearby island police the area, keeping a record of all visitors. A lump sum of money would be offered for removal of the goats, and for the cessation of burning and tree felling. At the same time, opportunity would still exist for visits to the island by approved researchers, albeit under stringent control. Accordingly, in early 1980 the National Trust for Fiji prepared a submission to the I.U.C.N. for funds to bring about these changes. A joint World Wildlife Fund/IUCN grant totalling US $11,000 was subsequently awarded to the National Trust to carry out the project. In August 1980, a successful agreement was negotiated between the Fijian landowners of Yaduataba, the National Trust, the people of Yadua and two Fiji Government ministers. Under the terms of the agreement, the National Trust would pay for fencing on Yadua prior to the removal of the goats from Yaduataba, which by then numbered over two hundred. This fencing was completed in January 1981, and during the following weeks all but about twenty goats were removed from Yaduataba.

During the past seven years at Orchid Island, it has been stressed to the local people and to overseas visitors that, unless care is taken, the Banded Iguana is in great danger of becoming extinct. It is seldom seen on Fiji's two main, populated islands, although Fijians living in the outer islands report seeing the animal occasionally. They would not go out of their way to kill iguanas unless they were encountered on a path or on nearby overhanging tree branches. Appeals were made to the schoolchildren who visited the Island to do all they could to prevent the killing of the iguana. By allowing the children to handle the iguanas, they could see that the animal was harmless and non-aggressive towards man, and their attitude has changed completely. Through the schools, the message has reached the outer islands that the Banded Iguana must *not* be killed. This already has had some effect; schoolchildren have been known to prevent adults killing specimens seen in the bush, and then to bring the iguanas to Orchid Island for protection. However, at the same time people are requested not to remove iguanas from their native habitat should they find them. Indiscriminate killing still goes on, but there is at least some progress made against this senseless destruction of one of Fiji's most beautiful and interesting animals.

Perhaps the most important long term measure to protect South Pacific banded iguanas is to promote local interest in native fanua. The recently produced documentary film on the Crested Iguana will no doubt help fulfil this need. Also, coins and stamps bearing the likeness of the Banded Iguana have recently been issued as part of the World Wildlife Fund's efforts to protect endangered species. The Crested Iguana is scheduled to appear on a future stamp design. The local news media, particularly "The Fiji Times," have given excellent coverage of both the discovery and hatching of the eggs of the Crested Iguana. It is hoped that Orchid Island Cultural Centre, which suffered severe flooding in both 1979 and 1980, will move to a larger, safer site, and within this complex it is hoped to establish a much more comprehensive collection of Fiji's wildlife. This will help serve local educational needs, and if breeding programs continue to be successful, an outlet for overseas zoos.

DISCUSSION

The study of *Brachylophus* is very much of an ongoing one. During the last two years we have learned more about South Pacific banded iguanas than in all the rest of the time put together, since the original description of the Banded Iguana in 1800. Nevertheless, there are still wide gaps in our knowledge, and many of our earlier conclusions had to be altered after turning out to be incorrect.

Previous literature on *Brachylophus* has been woefully inaccurate in both historical and ecological aspects. Regarding the latter, it should be emphasised that each island is an isolated unit, which may be ecologically quite different from its nearest neighbour. With the recent finding of the Crested Iguana, it is becoming increasingly apparent that the South Pacific Region lags far behind most other parts of the World in terms of biological exploration. Indeed, several islands within the Fiji Group, including Cikobia and parts of the Mamanuca and Yasawa Groups (Figure 23.1) are known to contain iguanas, but no specimens have yet been collected from there. Additional new species are possible, and indeed likely, under such conditions. Also, *Brachylophus fasciatus* has been found in Efate Island, Vanuatu (formerly New Hebrides), but this population is likely to represent a recent introduction (Gibbons, 1981).

What does seem clear is that iguanines have been in the South Pacific for a very long time. The Fiji Group has been in existence since the Eocene, some forty million years ago, and it now appears that Vanua Levu was always separated by deep water from Viti Levu (Howard Plummer, pers. comm.). In view of the great variation that occurs in *Brachylophus* between Yaduataba, the Yasawas and the Mamanuca Group we believe that most of the evolutionary differentiation has occurred within this area. In contrast, most of the *Brachylophus* populations in the remainder of Fiji, and also in Tonga, appear relatively uniform. There are compelling reasons to suggest that *B. vitiensis* resembles the ancestral form of *Brachylophus* far more closely than does *B. fasciatus.* Morphologically, the larger size, well-developed crests, and lack of sexual dichromatism make it far more like other iguanine genera, including *Iguana, Cyclura* and *Conolophus*, than is the Banded Iguana. Ecologically, *B. vitiensis* is restricted to very dry habitats and has an almost entirely herbivorous diet. This is typical of many other iguanines, which are believed to have originated in xeric areas of southern North America (see Etheridge, this volume). Developmentally, *B. vitiensis* exhibits an extremely long egg incubation time, only exceeded by the tuatara, *Sphenodon,* amongst all land vertebrates. Again, this feature is believed to be primitive.

Our account of the ecology and behavior of *Brachylophus* reinforces the view that the South Pacific iguanines are closely related to other members of the subfamily. Recent findings include juvenile gregariousness, female migrations to nesting areas, nest defence and so on. All these are known to occur in other iguanine genera.

In terms of endangered species status, *Brachylophus* is particularly vulnerable. For millions of years it has evolved in an environment free of ground predators–unlike mainland Central and South American iguanines. Clutch sizes are small, egg incubation takes a long time, and females fail to migrate to 'safe areas' as in *Iguana iguana* (see Burghardt and Rand, this volume). The sudden impact of feral cats, mongooses, goats and so on, not to forget man himself, has had a devasting effect on this curious and attractive group of animals.

SUMMARY AND CONCLUSIONS

Differences in morphology, ecology and behavior between the Banded Iguana, *Brachylophus fasciatus,* and a new species, the Crested Iguana, *B. vitiensis* have been noted. While the former is widely distributed throughout Fiji and Tonga, the latter is restricted to dry rainshadow islands, such as Yaduataba and the Yasawas. The Crested Iguana is readily distinguished from the Banded Iguana by its well-developed crest of spines on the nape, by its larger size and by its different color pattern. It also differs from the Banded Iguana in showing little sexual dimorphism, males and females appearing very similar.

Both species are arboreal and found in shaded, forested areas, but whereas the Crested Iguana appears to be almost entirely herbivorous, the Banded Iguana is more omnivorous. Known predators include hawks, Pacific boas (*Candoia* sp.) and feral cats. Males of both species are territorial and possess the ability for extreme color change when aroused. Both sleep in trees at night and seem to have definite home ranges. Breeding is seasonal and occurs mainly between December and March. Females lay a clutch of 3-6 eggs which they bury in moist soil. There are strong indications that each species exhibits nest defence. The extremely long period of egg incubation, 18-35 weeks, is a distinctive feature of the genus *Brachylophus.*

The South Pacific Banded Iguanas are seriously threatened in a number of ways. These include: habitat destruction by man and goats; predation by feral cats; killing by Fijians; and poaching. Recent efforts to protect these animals include (1) the setting up of Yaduataba Island as Fiji's first wildlife reserve, (2) promotion of local interest in nature animals through newspaper articles, radio broadcasts and movie films.

Note

[1]Spelling of place names throughout this text is in Fijian rather than the phonetic English version, e.g., Moce not Mothe, Kadavu not Kandavu.

Acknowledgments

We would particularly like to thank Gwyn Watkins for his help and encouragement in supporting a great many aspects of this project, including photography, arrangements of cages and general administration. Fergus Clunie, Director of the Fiji Museum, and a Member of the National Trust for Fiji, has provided useful advice on numerous occasions, and his sustained efforts to conserve the Crested Iguana, besides being extremely effective, have stimulated increased research effort.

Mick Guinea and David Hassall have accompanied the first author on different field trips, and made valuable observations of the animals at Yaduataba. Munideo Raj helped in drafting Figure 23.1. Funds for this project have been provided, in part, by a University of the South Pacific Research Grant, and more recently, by the New York Zoological Society.

Epilogue: Iguana as Symbol

All important movements and organizations have a common feature; they are always associated with some organism for rallying the emotions. Everyone is familiar with the beloved founder, the honored patriarch, the glorious leader, and others. To illustrate, Communism has Marx, bird lovers have Audubon, and Judaism has Moses.

But this organism need not be human. What could be more American than the bald eagle? Who could deter forest fires as effectively as Smoky the Bear? What could better symbolize the intellectual attainments of Big Ten schools than badgers, wildcats and the like?

. . .[the peace movement] must go beyond an inanimate symbol, free the dove whose relationship with peace is far too subtle and adopt for its cause a meaningful expression of nature's sympathy with us.

Space does not permit enumeration of all the reasons why we should adopt the Iguana of Central and tropical America to satisfy our purpose. It seems relevant, however, to cover briefly a few points of interpretation and natural history.

The Iguana will draw attention to an area especially pertinent to the present world situation.

He is a vegetarian and non-aggressive to the extreme. When personally attacked, he, appropriately enough, lashes out with his long tail.

He clearly represents the "third camp" position, being definitely not found in the U.S., but residing much closer to us than Russia.

Moreover, the correct name of this handsome green creature is *Iguana iguana,* Family Iguanidae. He is obviously from a proud race . . . . Mention could also be made of the Iguana's restraint, his powers of observation, the cruelties he suffers at the hands of his native captors.

But enough–I must now rebuke the arguments that will be brought against *Iguana iguana.* In point of fact, however, I will just list them because of their trivial nature and inform those too naive to agree with me that I have considered them.

Drawing by Neil Greenberg. Adapted from his illustration for *Homage to Fats Navarro* by R. Elman (New Rivers Press, Minneapolis, 1978).

The Iguana engages in combat displays.
He is considered a food delicacy–tastes like chicken.
He can hardly represent [the peace movement] if he can't produce any vocal or written protests.

This then is my proposal. An attempt to go beyond the diverting issues of our time to help create in the Student Peace Union a loyalty and reverance for a living symbolic ideal. I dare not hope I have ended the search; I hope I have at least aroused an awareness of its existence.

Gordon Burghardt
December, 1961

excerpted from "Beyond China"
SPU Discussion Bulletin
1961-1962, Vol. 2 (1): 31-32

Bibliography

Bibliography Chapter Code

Authors	**Chapter Number**
Auffenberg	6
Blanc	2
Boersma	16
Carpenter	12
Case	11
Christian & Tracy	20
Distel & Veasey	14
Drummond & Burghardt	15
Dugan	18
Dugan & Wiewandt	17
Etheridge	1
Fitch, Henderson & Hillis	22
Gibbons & Watkins	23
Greenberg & Jenssen	13
Harris	9
Iverson	4
McBee & McBee	5
Nagy	3
Rand & Greene	8
Ryan	21
Van Devender	10
Werner	19
Wiewandt	7
Introduction	Intro
Section I, II. . .VI	I,II. . .VI

Numbers after references refer to chapters in which the article is cited.

Allen, G.M. 1937. *Geocapromys* remains from Exuma Island. *J. Mamm.* 18:368-370. [VI]

Altmann, J. 1974. Observational study of behavior: sampling methods. *Behaviour* 49:227-265. [18]

Alvarez del Toro, M. 1960. *Reptiles de Chiapas.* Inst. Zool. del Estado Tuxtla Gutiérrez, Chiapas. [22]

Alvarez del Toro, M. 1972. *Los Reptiles de Chiapas.* Gobierno del Estado Chiapas, Tuxtla Gutiérrez, Mexico. [7]

Andrews, R.M. 1971. Structural habitat and time budget of a tropical *Anolis* lizard. *Ecology* 52:262-270. [4]

Andrews, R.M. 1976. Growth rate in island and mainland anoline lizards. *Copeia* 1976: 477-482. [10]

Andriamirado, R. 1976. Quelques remarques sur l'ancienne position de Madagascar par rapport a l'Afrique.–*C.R. Acad. Sci. Paris,* 282, sér. D:2041-2043. [2]

Arnette, J.R. 1980. Breeding the Fiji Island banded iguana, *Brachylophus fasciatus. Int. Zoo Yearbook* 20:78-79 [17, 23]

Auffenberg, W. 1975. The dragon isles: West Indian rock iguanas. *Bahamas Nat.* 1:2-7. [VI]

Auffenberg, W. 1976. Rock iguanas, Part II. *Bahamas Nat.* 2:9-16. [VI]

Auffenberg, W. 1979. A Monitor Lizard in the Philippines. *Oryx* 15:38-46. [Intro]

Auth, D.L. 1979. The thermal biology of the Turks and Caicos Rock Iguana, *Cyclura carinata* Harlan. PhD Diss., Univ. Florida. [4, 6]

Avery, D.F. and Tanner, W.W. 1970. Speciation in the Fijian and Tongan Iguana *Brachylophus* (Sauria, Iguanidae) with the description of a new species. *Great Basin Nat.* 30:166-172. [13, 23]

Avery, D.F. and Tanner, W.W. 1971. Evolution of the iguanine lizards (Sauria, Iguanidae) as determined by osteological and myological characters. *Brigham Young Univ. Sci. Bull. Biol. Ser.* 12(3):1-79. [1, 4]

Avery, R.A. 1973. Morphometric and functional studies on the stomach of the lizard *Lacerta vivipara. J. Zool., Lond.* 169:157-167. [16]

Azevedo, D. da C. 1974. Chuvas no Brasil. Regime, variabilidade e probabilidades de alturas mensais e anuais. Ministerio da Agricultura Departmento Nacional de Meteorologia Brasilia. [8]

Bagnara, J.T. and Hadley, M.E. 1973. *Chromatophores and Color Change.* Prentice-Hall, Englewood Cliffs, N.J. [13]

Bailey, J.W. 1928. A revision of the lizards of the genus *Ctenosaura. Proc. U.S. Natn. Mus.* 73(12):1-55. [1]

Bailey, J. and McBee, R.R. 1964. The magnitude of the rabbit cecal fermentation. *Proc. Montana Acad. Sci.* 24:35-38. [5]

Baldwin, E. 1970. *An Introduction to Comparative Biochemistry.* Cambridge University Press, London. [4]

Barbour, T. 1919. A new rock iguana from Porto Rico. *Proc. Biol. Soc. Wash.* 32:145-148. [VI]

Barbour, T. and Noble, G.K. 1916. A revision of the lizards of the genus *Cyclura. Bull. Mus. Comp. Zool.* 60:139-164. [VI]

Barlow, G.W. 1968. Ethological units of behavior. *In:* D. Ingle (ed.) *The Central Nervous System and Fish Behavior.* Univ. Chicago Press, Chicago, p. 217-232. [13]

Barlow, G.W. 1977. Modal action patterns. *In:* T.A. Sebeok (ed.) *How Animals Communicate.* Univ. Indiana Press, Bloomington, p. 98-134. [13]

Barnes, R.D. 1969. *Invertebrate Zoology.* W.B. Saunders, Philadelphia. [4]

Bartholomew, G.A. 1966. A field study of temperature relations in the Galápagos marine iguana. *Copeia* 1966:241-250. [16]

Bartholomew, G.A. 1977. Energy metabolism. *In:* M.S. Gordon (ed.) *Animal Physiology: Principles and Adaptations,* 3rd Edition. Macmillan Publ. Co., New York. p. 57-110. [3]

Bartholomew, G.A., Bennett, A.F. and Dawson, W.R. 1976. Swimming, diving, and lactate production of the marine iguana, *Amblyrhynchus cristatus. Copeia* 1976:709-720. [7, 16]

Bartholomew, G.A. and Lasiewski, R.C. 1965. Heating and cooling rates, heart rate and simulated diving in the Galápagos marine iguana. *Comp. Biochem. Physiol.* 16:573-582. [16]

Bartholomew, G.A. and Tucker, V.A. 1964. Size, body temperature, thermal conductance, oxygen consumption, and heart rate in Australian varanid lizards. *Physiol. Zool.* 37:341-354. [10]

Bartholomew, G.A. and Vleck, D. 1979. The relation of oxygen consumption to body size and to heating and cooling in the Galápagos marine iguana, *Amblyrhynchus cristatus. J. Comp. Physiol.* 132:285-288. [16]

Bateman, A.J. 1948. Intrasexual selection in *Drosophila. Heredity* 2:349-368. [17]

Batschelet, E. 1965. Statistical methods for the analysis of problems in animal orientation and certain biological rhythms. Amer. Inst. Biol. Sci., Washington, D.C. [15]

Beard, J.S. 1949. The natural vegetation of the Windward and Leeward Islands. *Oxford Forestry Mem.* 21:1-192. [6]

Beebe, W. 1924. *Galápagos, World's End.* G.P. Putnam's Sons, New York. [16]

Beebe, W. 1944. Field notes on the lizards of Kartabo, British Guiana, and Caripito, Venezuels. Part 2. Iguanidae. *Zoologica* 29:195-216, pl. 1-6. [8]

Bennett, A.F. and Dawson, W.R. 1976. Metabolism. *In:* C. Gans and W.R. Dawson (eds.). *Biology of the Reptilia, Vol. 5, Physiology A.* Academic Press, London. p. 127-223. [3, 16, V]

Bennett, A.F. and Nagy, K.A. 1977. Energy expenditure in free-ranging lizards. *Ecology* 58:697-700. [3]

Berry, K.H. 1974. The ecology and social behavior of the chuckwalla, *Sauromalus obesus obesus* Baird. *Univ. Calif. Publ. Zool.* 101:1-60. [4, 7, 10, 11, 12, 13, 17, 19, 21]

Bjorndal, K.A. 1979. Cellulose digestion and volatile fatty acid production in the green turtle. *Chelonia mydas. Comp. Biochem. Physiol.,* 63A:127-133. [4, 5]

Blair, W.F. 1960. *The Rusty Lizard. A Population Study.* Univ. Texas Press, Austin. [15]

Blanc, C.P. 1965. Etudes sur les Iguanidae de Madagascar. I. Le squelette de *Chalarodon madagascariensis* Peters, 1854. *Mém. Mus. Hist. Nat. Paris, sér.* A, 33(3):93-146. [2]

Blanc, C.P. 1969. Etudes sur les Iguanidae de Madagascar. Observation sur l'écologies de *Chalarodon madagascariensis* Peters, 1854. *Oecologia,* 2:292-318. [2]

Blanc, C.P. 1971. Les Reptiles de Madagascar et des îles voisines. *Ann. Univ. Madagascar* (Sci), 8:95-178. [2]

Blanc, C.P. 1977. Faune de Madagascar, *Reptiles Sauriens Iguanidae. CNRS-ORSTOM* Paris. [2]

Boersma, P.D. 1981. An ecological study of the Galápagos marine iguana. *In:* R. Bowman (ed.). *Galápagos Symposium Volume.* AAS, Washington, D.C. (in press). [7, 16, 17]

Borhegyi, S.F. de. 1969. Archaeological synthesis of the Guatemalan Highlands. *In:* G.R. Willey, (ed.). *Handbook of Middle American Indians. Vol. 2. Archaeology of Southern Mesoamerica. Part I.* Univ. of Texas Press, Austin. [22]

Bostic, D.L. Observations on the tumor-like growths on the Chuckawalla, *Sauromalus varius. J. Herp.* 5:76-78. [11]

Boulenger, G.A. 1885. Catalogue of the lizards in the British Museum (Natural History). London, 2:1-497. [1]

Boulenger, G.A. 1890. On the distinctive cranial characters of the iguanoid lizards allied to *Iguana. Ann. Mag. Nat. Hist.,* [6]6(35):412-414. [1]

Bourliere, F. 1963. Observations on the ecology of some large African mammals. *In:* F. Howell and F. Bourliere (eds.) *African Ecology and Human Evolution.* Aldine Publ. Co., Chicago. p. 53-65. [11]

Bowie, L.A. 1974. Comparative study of the gastrointestinal nematodes of two sceloporid lizards in Florida. Master's thesis. University of Florida, Gainesville. [4]

Bradbury, J.W. and Vehrencamp, S.L. 1977. Social organization and foraging in Emballonurid bats. *Behav. Ecol. Sociobiol.* 2:1-17. [17, 18]

Brattstrom, B.H. 1971. Social and thermoregulatory behavior of the bearded dragon, *Amphibolurus barbatus. Copeia* 1971:481-497. [14]

Brattstrom, B.H. 1974. The evolution of reptilian social behavior. *Amer. Zool.* 14:35-49. [11, 17, 21]

Braunwalder, M.E. 1979. Über eine erfolgreiche Zeitigung von Eiern des Grunen Leguans, *Iguana i. iguana,* und die damit verbundene Problematik. *Salamandra* 15:185-210. [VI]

Brett, J.R. 1970. Fish-the energy cost of living. *In:* W.J. McNeil (ed.). *Marine Aquaculture.* Oregon State Univ. Press, Corvallis, pp. 37-53. [3]

Brongniart, A. 1800. Essai d'une classification naturelle des reptiles. *Bull. Soc. Philomath, Paris,* 2:89-91. [1, 23]

Brown, J.L. 1964. The evolution of diversity in avian territorial systems. *Wilson Bull.* 76:160-169. [17, 18, 21]

Brown, J.L. 1975. *The Evolution of Behavior.* W.W. Norton and Co., New York. [17]

Brown, W.S. and Parker, W.S. 1976. Movement ecology of *Coluber constrictor* near communal hibernacula. *Copeia* 1976:225-242. [7]

Buffetaut, E. and Taquet, P. 1975. Les vertébrés du Crétacé et la dérive des continents. *La Recherche,* 6(55):379-381. [2]

Bull, J.J. and Shine, R. 1979. Iteroparous animals that skip opportunities for reproduction. *Amer. Nat.* 114:296-303. [11]

Burghardt, G.M. 1970. Chemical perception in reptiles. *In:* J.W. Johnston, Jr., D.G. Moulton, & A. Turk, eds. *Communication by Chemical Signals,* Appleton-Century-Crofts, New York, p. 241-308. [14]

Burghardt, G.M. 1977. Learning processes in reptiles. *In:* C. Gans and D.W. Tinkle (eds.) *Biology of the Reptilia, vol. 7 Ecology A,* Academic Press, New York, p. 555-681. [18]

Burghardt, G.M. 1977. Of iguanas and dinosaurs: social behavior and communication in neonate reptiles. *Amer. Zool.* 17:177-190. [Intro, 7, 13, 15, 18]

Burghardt, G.M., Greene, H.W. and Rand, A.S. 1977. Social behavior in hatchling green iguanas: Life at a reptile rookery. *Science* 195:689-691. [5, 7, 8, 10, 15, 18, 20]

Burghardt, G.M. 1980. Behavioral and stimulus correlates of vomeronasal functioning in reptiles: Feeding, grouping, sex and tongue use. *In:* D. Muller-Schwarze and R.M. Silverstein (eds.) *Chemical Signals Vertebrate and Aquatic Invertebrates.* Plenum Press, New York, p. 275-301. [II]

Burt, C.E. and Burt, M.D. 1932. Herpetological results of the Whitney South Sea Expedition. VI. Pacific Island amphibians and reptiles in the collection of the American Museum of Natural History *Bull. Amer. Nat. Hist.* 63:461-597. [23]

Bustard, H.R. 1970. Turtles and an iguana in Fiji. *Oryx* 170:317-322. [23]

Cagle, F.F. 1946. A lizard population of Tinian. *Copeia* 1946:4-9. [11]

Cahill, C. 1970. The banded iguana of Fiji. *Fiji Mus. Educ. Ser.,* No. 2:1-14. [7, 13, 17, 23]

Caldwell, D.R. and Bryant, M.P. 1966. Medium without rumen fluid for nonselective enumeration and isolation of rumen bacteria. *Appl. Microbiol.* 14:794-801. [5]

Callard, I.P., Chan, S.W.C. and Callard, G.V. 1973. Hypothalamic-pituitary-adrenal relationship in reptiles. *In:* A. Brodish and E.S. Redgate (eds.). *Brain-Pituitary-Adrenal Interrelationships.* Karger, Basel, pp. 270-292. [13]

Camp, C.L. 1923. Classification of the lizards. *Bull. Amer. Mus. Nat. Hist.* 48(11):289-481. [1]

Carey, W.M. 1966. Observations of the ground iguana *Cyclura macleayi caymenesis* on Caymen Brac, British West Indies. *Herpetologica* 22:265-268. [VI, 4, 12]

Carey, W.M. 1975. The rock iguana, *Cyclura pinguis,* on Anegada, British Virgin Islands, with notes on *Cyclura ricordi* and *Cyclura cornuta* on Hispaniola. *Bull. Florida State Mus. Biol. Sci.* 19:189-24. [6, 7, 17]

Carpenter, C.C. 1961. Patterns of social behavior in the desert iguana, *Dipsosaurus dorsalis. Copeia* 1961:396-405. [12]

Carpenter, C.C. 1962. Patterns of behavior in two Oklahoma lizards. *Amer. Midl. Nat.,* 67:132-151. [12]

Carpenter, C.C. 1963. Patterns of behavior in three forms of the fringe-toed lizards *(Uma-Iguanidae). Copeia* 1963:406-412. [12]

Carpenter, C.C. 1966a. Comparative behavior of the Galápagos lava lizards *(Tropidurus). In:* R.I. Bowman (ed.). *The Galápagos: Proceedings of the Galápagos International Scientific Project.* Univ. California Press, Berkeley, p. 269-273. [12]

Carpenter, C.C. 1966b. The marine iguana of the Galápagos Islands, its behavior and ecology. *Proc. Calif. Acad. Sci.* Series 4, 34:329-376. [7, 12, 14, 16, 17, 20]

Carpenter, C.C. 1967a. Aggression and social structure of iguanid lizards. *In:* W.S. Milstead (ed.). *Lizard Ecology: A Symposium.* Univ. Missouri Press, Columbia 87-105. [12, 14, 19]

Carpenter, C.C. 1967b. Display patterns of the Mexican iguanid lizards of the genus *Uma. Herpetologica* 23:285-293. [12]

Carpenter, C.C. 1969. Behavioral and ecological notes on the Galápagos land iguanas. *Herpetologica* 23:155-164. [12, 14, 20]

Carpenter, C.C. 1977. The display of *Enyaliosaurus clarki* (Iguanidae, Lacertilia). *Copeia* 1977:754-755. [12, 19]

Carpenter, C.C. 1978a. Comparative display behavior in the genus *Sceloporus* (Iguanidae). *Contr. Biol. Geol. Milwaukee Publ. Mus.* No. 18:1-71. [12]

Carpenter, C.C. 1978b. Ritualistic social behavior in lizards. *In:* N. Greenberg and P.D. MacLean (eds.). *Behavior and Neurology of Lizards.* Nat. Inst. Mental Health, Rockville, Maryland. p. 253-267. [12, 20]

Carpenter, C.C., Baldham, J.A. and Kimble, B. 1970. Behavior patterns of three species of *Amphibolurus* (Agamidae). *Copeia* 1970:497-505. [14]

Carpenter, C.C. and G.W. Ferguson. 1977. Variation and evolution of stereotyped behavior in reptiles. *In:* C. Gans and D.W. Tinkle (eds.). *Biology of the Reptilia.* Academic Press, London. Vol. 7 Ecology and Behavior A. 335-554. [12, 13]

Carpenter, C.C. and Murphy, J.B. 1978. Aggressive behavior in the Fiji Island lizard *Brachylophus fasciatus* (Reptilia, Lacertilia, Iguanidae). *J. Herpetol.* 12:215-252. [12, 13]

Carr, A. 1955. *The Windward Road:* Adventures of a naturalist on remote Caribbean shores. A.A. Knopf: New York. [15]

Carr, A. 1963. Orientation problems in the high seas travel and terrestrial movements of marine turtles. *In:* L.E. Slater (ed.). *Bio-telemetry.* p. 179-193. [15]

Carr, A. 1965. The navigation of the green turtle. *Scientific American* 212, 79-86. [15]

Carr, A. 1972. The case for long-range chemoreceptive piloting in Chelonia. *In:* S.R. Galler, K. Schmidt-Koenig, G.J. Jacobs, R.E. Belleville (eds.). *Animal Orientation and Navigation,* National Aeronautics and Space Administration: Washington, D.C. [15]

Carroll, E.J. and Hungate, R.E. 1954. The magnitude of the mirobial fermentation in the bovine rumen. *Appl. Microbiol.* 2:203-214. [5]

Case, T.J. 1975. Species numbers, density compensation, and colonizing ability of lizards on islands in the Gulf of California. *Ecology* 56:3-18. [11]

Case, T.J. 1976a. Body size differences between populations of the chuckawalla, *Sauromalus obesus. Ecology* 57:313-323. [4, 11]

Case, T.J. 1976b. Seasonal aspects of thermoregulatory behavior in the chuckawalla, *Sauromalus obesus* (Reptilia, Lacertilia, Iguanidae). *J. Herp.* 10:85-95 [4]

Case, T.J. 1978. A general explanation for insular body size trends in terrestrial vertebrates. *Ecology* 59:1-18. [11, 21]

Case, T.J. 1979. Optimal body size and an animal's diet. *Acta Biotheor.* 28:54-67. [11]

Case, T.J. and Cody, M.L. (eds.) in prep. *Island Biogeography in the Sea of Cortez.* [11]

Castro, C.S. and Duval, J. 1979. Reproducción en cautividad de iguanas del género *Cyclura. Zoodom* 3:12-18. [7]

Chance, M.R.A. 1962. An interpretation of some agonistic postures; the role of "cut-off" acts and postures. *Symp. Zool. Soc. Lond.* 8:71-89. [14]

Chaplin, S.B. 1974. Daily energetics of the black-capped chickadee, *Parus atricapillus,* in winter. *J. Comp. Physiol.* 89:321-330 [3]

Christian, K.A. 1979. Direct influence of the thermal environment on the fitness of Galápagos Land Iguanas, *Conolophus pallidus.* Abstract. American Society of Ichthyologists and Herpetologists meeting. Orono, Maine. [16]

Christian, K.A. and Tracy, C.R. 1981. The effect of the thermal enviornment on the ability of hatchling Galápagos land iguanas to avoid predation during dispersal. *Oecologia* 49:218-223. [20]

Clarke, R.F. 1965. An ethological study of the iguanid genera, *Callisaurus, Cophosaurus,* and *Hobrookia. Emporia State Res. Stud.* 13:1-66. [12]

Clarke, R.F. and Robison, T.H. 1965. An ethological study of the *Ctenosaura: pectinata, acanthura* and *hemilopha.* Unpubl. manuscript. Emporia State Univ., Emporia, Kansas. [12]

Cliff, F.S. 1958. A new species of *Sauromalus* from Mexico. *Copeia* 1958:259-261. [1]

Cody, M.L. 1966. A general theory of clutch size. *Evolution* 20:174-184. [11]

Cogger, H.C. 1974. Voyage of the banded iguana. *Australian Nat. Hist.* 18:144-149. [13, 17, 23]

Cogger, H.C. 1975. *Reptiles and Amphibians of Australia.* A.H. and A.W. Reed, Sydney. [4]

Cole, C.J. 1966. Femoral glands in lizards: A review. *Herpetologica* 22:199-206. [18]

Congdon, J.D., Ballinger, R.E. and Nagy, K.A. 1979. Energetics, temperature and water relations in winter aggregated *Sceloporus jarrovi* (Sauria: Iguanidae). *Ecology* 60:35-39. [3]

Cope, E.D. 1886. On the species of Iguaninae. *Proc. Amer. Philos. Soc.,* 23:261-271. [1]

Cope, E.D. 1900. The crocodilians, lizards and snakes of North America. *Ann. Rept. U.S. Natn. Mus.,* 1898:153-1270. [1]

Cowles, R.B. 1946. Note on the arboreal feeding habits of the desert iguana. *Copeia* 1946:171-173. [17]

Cox, C.R. and LeBoeuf, B.J. 1977. Female incitation of male competition: a mechanism in sexual selection. *Amer. Natur.* 111:317-335. [18]

Cracraft, J. 1973. Vertebrate evolution and biogeography in the Old World Tropics. *In:* D.H. Traling and S.K. Runcorn (eds.). *Implications of Continental Drift to the Earth Sciences.* Acad. Press, New York. p. 373-393. [2]

Crews, D.P. 1973. Coition-induced inhibition of sexual receptivity in female lizards *(Anolis carolinensis). Physiol. Behav.* 11:463-468. [17]

Crews, D.P. 1974a. Castration and androgen replacement on male facilitation of ovarian activity in the lizard, *Anolis carolinensis. J. Comp. Physiol. Psychol.* 87:963-969. [17]

Crews, D.P. 1974b. The role of group stability and male aggressive and sexual behavior in the control of ovarian recrudescence in the lizard *Anolis carolinensis. J. Zool., Lond.* 172:419-441. [18, 19]

Crews, D.P. 1975. Effects of different components of male courtship behavior on environmentally induced ovarian recrudescence and mating preferences in the lizards, *Anolis carolinensis. Anim. Behav.* 23:349-356. [18]

Cuellar, O. 1966. Oviducal anatomy and sperm storage in lizards. *J. Morphol.* 119:7-20. [17, 19, 21]

Cummins, K.W. 1967. *Caloric Equivalents for Studies in Ecological Energetics.* 2nd Ed., Univ. Pittsburgh Press, Pittsburgh. [6]

Cummins, K.W. and Wuycheck, J.C. 1971. Caloric equivalents for investigations in ecological energetics. Inter. Asso. Theoret. Applied Limnol., Stuttgart. [11]

Curio, E. 1976. *The Ethology of Predation.* Springer-Verlag, Berlin. [6]

Cuvier, G.L.C.F.D. 1829. *Le Regne Animal Distribué d'après son Organisation pour Servir de Base à l'Histoire Naturèlle des Animaux et d'Introduction a l'Anatcmie Comparée.* Nouvelle édition. Tome 2 Dúterville Crouchard Libraire, Paris. [23]

D'Arcy, W.G. 1975. Anegada Island: Vegetation and flora. *Atoll Res. Bull.* 188:1-40. [6]

Darling, F.F. 1937. *A Herd of Red Deer.* Doubleday, N.Y. [18]

Darlington, P.J., Jr. 1957. *Zoogeography: The Geographical Distribution of Animals.* John Wiley, New York. [Intro, 2]

Darwin, C. 1845. *Journal of Researches into the Natural History and Geology of the Countries Visited During the Voyage of H.M.S. Beagle Round the World.* 2nd edition. Ward, Lock and Co., London. [20]

Darwin, C. 1871. *The Descent of Man, and Selection in Relation to Sex.* John Murray, London. [17]

Daudin, F.M. 1802. *Histoire Naturèlle, Générale et Particulière des Reptiles.* Tome 3. Imprimerie de F. Dufart, Paris. [23]

Dawkins, R. and Krebs, J.R. 1978. Animal signals: information or manipulation. *In:* J.A. Krebs and N.B. Davies (eds.) *Behavioral Ecology.* p. 282-309. Sinauer, Sunderland, Mass. [13]

Dawson, W.R., Bartholomew, G.A. and Bennet, A.F. 1977. A reappraisal of the aquatic specializations of the Galápagos marine iguana *(Amblyrhynchus cristatus). Evol.* 31: 891-897. [16]

Deevey, E.S., Jr., 1947. Life tables for natural populations of animals. *Quart. Rev. Biol.* 22:283-414. [10]

DeFazio, A., Simon, C.A., Middendorf, G.A. and Romano, D. 1977. Iguanid substrate licking: a response to novel situations in *Sceloporus jarrovi. Copeai* 1977:706-709. [13]

Derickson, W.K. 1976. Lipid storage and utilization in reptiles. *Amer. Zool.* 16:711-723. [18]

deVries, T. 1973. *The Galápagos Hawk: An Eco-Geographic Study with Special Reference to its Systematic Position.* Vriji Universiteit te Amsterdam. [20]

Diebold, A.R., Jr. 1969. The Huave. *In:* E.Z. Vogt (ed.) *Handbook of Middle American Indians.* Vol. 7, Part 5. Univ. of Texas Press, Austin. [22]

Diefenbach, C.O. Da C. 1975. Gastric function in *Caiman crocodilus.* (Crocodylia: Reptilia) I. Rate of gastric digestion and gastric motility as a function of temperature. *Comp. Biochem. Physiol.* 51A:259-265. [16]

Distel, H. 1976. Behavior and brain stimulation in the green iguana, *Iguana iguana* L. I. Schematic brain atlas and stimulation device. *Brain Behav. Evol.* 13:421-450. [14]

Distel, H. 1978. Behavior and brain stimulation in the green iguana, *Iguana iguana* L. II. Stimulation effects. *Exp. Brain Res.* 31:49-62. [14]

Downhower, J.F. and Armitage, K.B. 1971. The yellow-bellied marmot and the evolution of polygamy. *Amer. Nat.* 105:355-370. [17]

Draper, N.R. and Smith, H. 1966. *Applied Regression Analysis.* John Wiley and Sons, New York. [10]

Drucker, S., Escalante, R. and Weitlaner, R.J. 1969. The Cuitlatec. *In:* E.Z. Vogt (ed.) *Handbook of Middle American Indians. Vol. 7, Part 5, Ethnology, Part I.* Univ. of Texas Press, Austin. [22]

Drummond, H. In press. The adaptive significance of island nest-sites of the green iguana, *Iguana iguana,* and the slider turtle, *Pseudemys scripta. Copeia.* [15]
Drummond, H. and Burghardt, G.M. 1982. Nocturnal and diurnal emergence of neonate green iguanas, *Iguana iguana.* Unpubl. manuscript. [15]
Drury, W.H. and Nisbet, I.C.T. 1964. Radar studies of orientation of songbird migrants in southeastern New England. *Bird Banding* 35:69-119. [15]
Dubuis, A., Faurel, L., Grenot, C. and Vernet, R. 1971. Sur le regime alimentaire du Lezard saharien *Uromastyx acanthinurus* Bell. *C.R. Acad. Sci. Paris,* Ser. D, 273: 500-503. [4]
Duelli, P. 1975. Orientierung ohne richtende Aussenreize bei Reptilien *(Hemidactylus frenatus* Gekkonidae). *Z. Tierpsychol.* 38:324-328. [15]
Duellman, W.E. 1965. Amphibians and reptiles from the Yucatán Peninsula, México. *Univ. Kans. Publ. Mus. Nat. Hist.* 15(12):577-614. [1]
Duellman, W.S. and Duellman, A.S. 1959. Variation, distribution and ecology of the iguanid lizard *Enyaliosaurus clarki* of Michoacan, Mexico. *Occ. Pap. Mus. Zool. Univ. Mich.* 598:1-11. [10, 17]
Dugan, B.A. 1980. A field study of the social structure, mating system and display behavior of the green iguana *(Iguana iguana)* Ph.D. dissertation. University of Tennessee Knoxville. [13, 17, 18]
Dugan, B.A. 1982. A field study of the headbob displays of male green iguanas *(Iguana iguana)*: variation in form and context. *Anim. Behav.* 30:327-338. [17, 18]
Dugan, B.A., Rand, A.S., Burghardt, G.M. and Bock, B.C. 1981. Interactions between nesting crocodiles and iguanas. *J. Herp.* 15:409-415. [15]
Dunham, A.E. 1978. Food availability as a proximate factor influencing individual growth rates in the iguanid lizard *Sceloporus merriami. Ecology* 59:770-778. [10]
Dunn, E.R. 1934. Notes on *Iguana. Copeia* 1934:1-4. [1]
Dunson, W.A. 1976. Salt glands in reptiles. *In:* C. Gans and W.R. Dawson (eds.) *Biology of the Reptilia, Vol. 5. Physiology A.* Academic Press, New York, p. 413-445. [4]
Duvall, D. 1979. Western fence lizard *(Sceloporus occidentalis)* chemical signals. I. Conspecific discriminations and release of a species-typical visual display. *J. Exp. Zool.* 210:321-325. [13, 18]
Edmund, A.G. 1969. Dentition. *In:* C. Gans, A.A. Bellairs and T.S. Parsons (eds.) *Biology of the Reptilia, Vol. 1. Morphology A.* Academic Press, New York. p. 117-200. [4]
Edwards, C.R. 1969. Possibilities of pre-Columbian maritime contacts among New World civilizations. *Mesoamerican Studies,* 4:3-10. [22]
Ehrenfeld, D.W. 1974. Conserving the edible sea turtle: can mariculture help? *Amer. Sci.* 62:23-31. [15]
Eibl-Eibesfeldt, I. 1955. Der Kommentkampf der Meerechse *(Amblyrhynchus cristatus* Bell) nebst einigen Notizen zur Biologies dieser Art. *Z. Tierpsychol.* 12:49-62. [14]
Eibl-Eibesfeldt, I. 1961. *Galápagos: The Noah's Ark of the Pacific.* Doubleday, New York. [17]
Eibl-Eibesfeldt, I. 1962. Neue Unterarten der Meerechse, *Amblyrhynchus cristatus,* nebst weiteren Angaben zur Biologie der Art., *Senckenbergiana Biol.* 43:177-199. [1]
Eibl-Eibesfeldt, I. 1966. Das verteidigen der Eirblageplatze bei der Hood-Meerechse *(Amblyrhynchus cristatus venustissimus), Z. Tierpsychol.* 23:627-631. [7, 20]
Ellis, H.I. and Ross, J.P. 1978. Field observations of cooling rates of Galápagos land iguanas *(Conolophus subcristatus). Comp. Biochem. Physiol.* 59A:205-209. [4]
Emlen, J.M. 1966. The role of time and energy in food preference. *Amer. Nat.* 100:611-617. [6]
Emlen, S.T. and Oring, L.W. 1977. Ecology, sexual selection, and the evolution of mating systems. *Science* 197:215-223. [17, 18, 21]

Enlow, D.H. 1969. The bone of reptiles. *In:* C. Gans, (ed.) *Biology of the Reptilia,* Vol. 1. Academic Press, New York. p. 45-80. [11]

Estes, R. 1970. Origin of the North American lower vertebrate fauna: an inquiry into the fossil record. *Forma Functio* 3:139-163. [Intro]

Estes, R. and Price, L.I. 1973. Iguanid lizard from the upper Cretaceous of Brazil. *Science* 180:748-751. [Intro]

Etheridge, R. 1964. Late Pleistocene lizards from Barbuda, British West Indies. *Bull. Fla. Mus. Biol. Sci.* 9(2):43-75. [VI]

Etheridge, R.E. 1964. The skeletal morphology and systematic relationships of sceloporine lizards. *Copeia* 1964:610-631. [1]

Etheridge, R. 1965. The abdominal skeleton of lizards in the family Iguanidae. *Herpetologica* 21:161-168. [2]

Etheridge, R. 1966. Pleistocene lizards from New Providence. *Quart. J. Fla. Acad. Sci.* 28:349-358. [VI]

Evans, L.T. 1946. The social behavior of the lizard *Sceloporus gramnicus microlepidotus. Anat. Rec.* 95:53-54. [21]

Evans, L.T. 1951. Field study of the social behavior of the black lizard. *Ctenosaura pectinata. Amer. Mus. Nov.* 1943:1-26. [10, 12, 17, 21]

Evans, L.T. 1959. A motion picture study of the maternal behavior of the lizard *Eumeces obsoletus* Baird and Girrad. *Copeia* 1959:103-110. [21]

Fabens, A.J. 1965. Properties and fitting of the von Bertalanffy growth curve. *Growth* 29:265-289. [10]

Feeny, P.P. 1970. Seasonal changes in oak leaf tannins and nutrients as a cause of spring feeding by winter moth caterpillars. *Ecology* 51:565-580. [6]

Ferguson, G.W. 1966. Releasers of courtship and territorial behaviour in the side blotched lizard *Uta stansburiana. Anim. Behav.* 14:89-92. [17]

Ferguson, G.W. 1971. Geographic variation and evolution of stereotyped behavioral patterns of the side-blotched lizards of the genus *Uta. Syst. Zool.* 20:79-101. [13, 14]

Field, C.R. 1976. Palatability factors and nutrient values of the food of the buffalo *(Syncerus caffer)* in Uganda. *E. Afr. Wildl. J.* 14:181-201. [6]

Fischer, K. 1961. Untersuchungen zur sommenkompassorientierung und laufaktivität von Smaragdeidechsen *(Lacerta viridis* Laur). *Z. Tierpsychol.* 18:450-470. [15]

Fisher, R.A. 1958. *The Genetical Theory of Natural Selection.* Dover Publications, New York. [17]

Fitch, H.S. 1970. Reproductive cycles of lizards and snakes. *Univ. Kansas Mus. Nat. Hist. Misc. Publ.* 52:1-247. [7, 8, 10]

Fitch, H.S. 1973a. A field study of Costa Rican lizards. *Univ. Kan. Sci. Bull.* 50:39-126. [7, 10, 22]

Fitch, H.S. 1973b. Population structure and survivorship in some Costa Rican lizards. *Occ. Pap. Mus. Nat. Hist. Univ. Kansas* 18:1-41. [7]

Fitch, H.S., Fitch, A.V. and Fitch, C.W. 1971. Ecological notes on some common lizards of southwestern Mexico and Central America. *Southwest. Nat.* 15:397-399. [22]

Fitch, H.S. and Henderson, R.W. 1977a. Age and sex differences in the ctenosaur *(Ctenosaura similis). Milwaukee Public Mus. Contrib. Biol. Geol.,* 11:1-11. [7, 8, 9, 10, 17]

Fitch, H.S. and Henderson, R.W. 1977b. Age and sex differences, reproduction and conservation of *Iguana iguana. Milwaukee Public Mus. Contrib. Biol. and Geol.,* 13:1-21 [7, 10, 18, 22]

Fitch, H.S. and Henderson, R.W. 1978. Ecology and exploitation of *Ctenosaura similis. Univ. Kansas Sci. Bull.* 51:483-500. [6, 7, 22]

Flanigan, W.J. 1974. Sleep and wakefulness in iguanid lizards *Ctenosaura pectinata* and *Iguana iguana. Brain Behav. Evol.* 8:402-436. [14]

Fleet, R.R. and Fitch, H.S. 1974. Food habits of *Basiliscus basiliscus* in Costa Rica. *J. Herp.* 8:260-262. [4]

Fleming, T.H. 1974. The population ecology of two species of heteromyid rodents. *Ecology* 55:493-510. [10]

Fox, W. 1963. Special tubules of sperm storage in female lizards. *Nature* 195:500-501. [17]

Frakes, L.A. and Kemp. E.M. 1973. Palaeogene continental positions and evolution of climate. *In:* D.H. Tarling and S.K. Runcorn (eds.) *Implications of Continental Drift to the Earth Sciences*. Academic Press, New York. p. 539-558. [2]

Freeland, W.J. and Janzen, D.H. 1974. Strategies in herbivory in mammals: The role of plant secondary compounds. *Amer. Nat.* 108:269-289. [6]

Frick, J. 1976. Orientation and behavior of hatchling green turtles *(Chelonia mydas)* in the sea. *Anim. Behav.* 24:849-857. [15]

Gadgil, M. 1972. Male dimorphism as a consequence of sexual selection. *Amer. Nat.* 106: 574-580. [17, 18]

Gadgil, M. and Bossert, W.H. 1970. Life historical consequences of natural selection. *Amer. Nat.* 104:1-24. [7]

Gasaway, W.C. 1976a. Cellulose digestion and metabolism by captive rock ptarmigan. *Comp. Biochem. Physiol.* 54A:179-182. [4]

Gasaway, W.C. 1976b. Seasonal variation in diet, volatile fatty acid production and size of cecum of rock ptarmigan. *Comp. Biochem. Physiol.* 53A:109-114. [5]

Gasaway, W.C. 1976c. Volatile fatty acids and metabolizable energy derived from cecal fermentation in the willow ptarmigan. *Comp. Biochem. Physiol.* 53A:115-121. [5]

Gastil, G., Minche, J. and Phillips, R.P. 1980. Constraints on the ages of insular isolation in the Gulf of California. *In:* T.J. Case and M.L. Cody (eds.) *Island Biogeography in the Sea of Cortez,* in prep. [11]

Gessaman, J.A. 1973. *Ecological Energetics of Homeotherms*. Utah State Univ. Press, Logan. [3]

Gibbons, J.R.H. 1979. The hind leg pushup display of the *Amphibolurus decresii* species complex (Lacertilia: Agamidae). *Copeia* 1979:29-40. [13]

Gibbons, J.R.H. 1981. The biogeography of the *Brachylophus* (Iguanidae) including a description of a new species, *B. vitiensis* from Fiji. *J. Herp.* 15(3):255-272. [1, 23]

Gibbons, J.W. and Nelson, D.H. 1978. The evolutionary significance of delayed emergence from the nest by hatchling turtles. *Evolution* 32:297-303. [7]

Gibson, A.R. and Falls, J.B. 1975. Evidence for multiple insemination in the common garter snake, *Thamnophis sirtalis. Canad. J. Zool.* 53:1362-1368. [17]

Gillin, J. 1948. Tribes of Guianas. *In:* J.H. Steward (ed.) *Handbook of South American Indians. Vol. 3. The tropical forest tribes.* Smithson. Inst. Bur. Amer. Ethnol. Bull. 143:799-860. [22]

Gilmore, R.M. 1950. Fauna and ethnozoology of South America. *In:* J.H. Steward (ed.) *Handbook of South American Indians. Vol. 6 Physical anthropology, linguistics, and cultural geography of South American Indians.* Smithson. Inst. Bur. Amer. Ethnol. Bull. 143:345-464. [22]

Glander, K.E. 1975. Habitat and resource utilization: an ecological view of social organization in mantled howling monkies. Ph.D. dissertation, The University of Chicago. [10]

Glander, K.E. 1978. Howling monkey feeding behavior and plant secondary compounds: a study of strategies. *In:* G.G. Montgomery (ed.) *The Ecology of Arboreal Folivores.* Smithsonian Inst. Press, Washington, D.C. p. 561-574. [10]

Golley, F.B. 1961. Energy values of ecological materials. *Ecology* 42:581-584. [3]

Golley, F.B., Petrusewicz, K. and Ryszkowski, L. 1975. *Small Mammals: Their Productivity and Population Dynamics.* Cambridge Univ. Press, London. [3]

Gorman, G.C. 1968. The relationship of *Anolis* of the *roquet* species group (Sauria: Iguanidae)-III. Comparative study of display behavior. *Breviora.* 284, 1-31. [14]

Gorman, G.C., Wilson, A.C. and Nakanishi, M. 1971. A biochemical approach towards the study of reptilian phylogeny: evolution of serum albumin and lactic dehydrogenase. *Syst. Zool.* 20:167-185. [2]

Grant, C. 1940. The herpetology of the Cayman Islands. *Bull. Inst. Jamaica, Sci. Ser.* 2:1-65 [VI]

Gray, J.E. 1831. *In:* G. Cuvier (ed.) *The Animal Kingdom, The Class Reptilia.* 9:1-481. [1]

Gray, J.E. 1845. Catalogue of the lizards in the British Museum of Natural History. London. [1]

Greenberg, N. 1976. Thermoregulatory aspects of behavior in the blue spiny lizard, *Sceloporus cyanogenys* (Sauria, Iguanidae). *Behaviour* 59:1-21. [14]

Greenberg, N. 1976. Observations of social feeding in lizards. *Herpetologica* 32:348-352. [6, 18]

Greenberg, N. 1977a. A neuroethological investigation of display behavior in the lizard, *Anolis carolinensis* (Lacertilia, Iguanidae) *Amer. Zool.* 17(1):191-201. [13]

Greenberg, N. 1977b. An ethogram of the blue spiny lizard, *Sceloporus cyanogenys* (Reptilia, Iguanidae). *J. Herpetol.* 11:177-195. [13, 14]

Greenberg, N. and Rodriguez, W.C. 1979. The exploratory behavior of *Anolis carolinensis* Abstract. Proc. Joint Ann. Meeting, Herpetol. League and Soc. Stud. Amphib. Rept.; Knoxville TN. [13]

Greene, H.W., Burghardt, G.M., Dugan, B.A. and Rand, A.S. 1978. Predation and the defensive behavior of green iguanas *(Reptilia, Lacertilia, Iguanidae). J. Herpetol.* 12: 169-176. [15, 20]

Greer, A.E. 1976. On the evolution of the giant Cape Verde scincid lizard *Macroscincus coctei. J. Nat. Hist.* 10:691-712. [4]

Grenot, C. 1976. Ecophysiologie du lezard saharien *Uromastyx acanthinurus* Bell, 1825 (Agamidae herbivore). *Ecole Norm. Super., Publ. Lab. Zool.* 7:1-323. [4]

Griners, J.E. and T.B. Hinton. 1969. The Huichol and Cora. *In:* E.Z. Vogt (ed.) *Handbook of Middle American Indians.* Vol. 8, Ethnology, Part 2. Univ. of Texas Press, Austin. p. 782-813 [22]

Hackforth-Jones, J. and Harker, D.F., Jr. 1979. Oviposition burrows of ctenosaur, *Ctenosaura similis,* in Costa Rica. Unpublished Ms. [7]

Hadley, M.E. and Goldman, J.M. 1969. Physiological color changes in reptiles. *Amer. Zool.* 9:489-504. [13]

Halliday, T.R. 1976. The libidinous newt. An analysis of variations in the sexual behaviour of the smooth newt, *Triturus vulgaris. Anim. Behav.* 24:398-414. [18]

Hamilton, W.J. 1967. Social aspects of bird orientation mechanisms. *In:* R.M. Storm (ed.) *Animal Orientation and Navigation.* Oregon State Univ. Press, Corvallis. [15]

Hamilton, W.J. and Watt, K.E.F. 1970. Refuging. *Ann. Rev. Ecol. Syst.* 1:263-297. [6]

Hansen, R.M. and Sylber, C.K. 1979. Cellulose digestion by young yellow giant chuckwalla lizards. Unpublished Ms. [4, 11]

Harlan, R. 1824. Descriptions of two species of Linnaean *Lacerta* not before described, and construction of the new genus *Cyclura. J. Acad. Nat. Sci. Philadelphia* 4:242-251. [1]

Harlow, H.J., Hillman, S.S. and Hoffman, M. 1976. The effect of temperature on digestive efficiency in the herbivorous lizard, *Dipsosaurus dorsalis. J. Comp. Physiol.* 111:1-6. [4]

Harper, J.L. 1961. The role of predation in vegetational diversity. *In: Brookhaven Symp. Biol. Diversity and Stability in Ecological Systems.* U.S. Atomic Energy Comm. 22:1-12. [6]

Harwood, R.H. 1979. The effect of temperature on the digestive efficiency of three species of lizards, *Cnemidophorus tigris, Gerrhonotus multicarinatus* and *Sceloporus occidentalis. Comp. Biochem. Physiol.* 63A:417-433. [16]

Hastings, J.R. and Humphrey, R.R. 1964a. *Climatological Data for Baja California.* Univ. Ariz. Inst. Atmos. Physics. Tech. Report on the Meteorology and Climatology of Arid Regions. No. 14. [11]

Hastings, J.R. 1964b. *Climatological Data for Sonora and Northern Sinaloa.* Univ. Ariz. Inst. Atmos. Physics. Tech. Report on the Meteorology and Climatology of Arid Regions. No. 15. [11]

Hastings, J.R. 1969a. *Climatological Data and Statistics for Baja California.* Univ. Ariz. Inst. Atmos. Physics. Tech. Report on the Meteorology and Climatology of Arid Regions. No. 18. [11]

Hastings, J.R. 1969b. *Climatological Data and Statistics for Sonora and Northern Sinaloa.* Univ. Ariz. Inst. Atmos. Physics. Tech. Report on the Meteorology and Climatology of Arid Regions. No. 19. [11]

Heath, J.E., Northcutt, R.G. and Barber, R.P. 1969. Rotational optokinesis in reptiles and its bearing on pupillary shape. *Z. vergl. Physiol.* 62:75-85. [14]

Henderson, G.C. 1937. *The Discoveries of the Fiji Islands.* John Murray, London. [13]

Henderson, R.W. 1973. Ethoecological observations of *Ctenosaura similis* (Sauria: Iguanidae) in British Honduras. *J. Herp.* 7:27-33. [10, 12]

Henderson, R.W. 1974. Aspects of the ecology of the juvenile common iguana *(Iguana iguana). Herpetologica* 30:327-332. [8, 9, 10, 14, 15]

Henderson, R.W. and Fitch, H.S. 1978a. Dragons: 25c/lb. *Animal Kingdom* 81(1):12-17. [22]

Henderson, R.W. and Fitch, H.S. 1978b. Plight of the iguana. *Lore* 28(3):2-9. [22]

Henderson, R.W. and Fitch, H.S. 1979. Notes on the behavior and ecology of *Ctenosaura similis* (Reptilia, Iguanidae) at Belize City, Belize. *Brenesia* 16:69-80. [10, 22]

Henke, J. 1975. Vergleichende-morphologische Untersuchungen am Magen-Darm-Trakt der Agamidae und Iguanidae (Reptilia: Lacertilia). *Zool. Jb. Anat.* 94:505-569. [4]

Hernandez de Alba, G. 1948a. Sub-Andean tribes of the Cauca Valley. *In:* J.H. Steward (ed.) *Handbook of South American Islands. Vol. 4. The Circum-Caribbean tribes.* Smithson. Inst. Bur. Amer. Ethnol. Bull. 143:297-327. [22]

Hernandez de Alba, G. 1948b. Tribes of the north Columbia lowlands. *In:* J.H. Steward (ed.) *Handbook of South American Indians. Vol. 4. The Circum-Caribbean tribes.* Smithson. Inst. Bur. Amer. Ethnol. Bull. 143:329-338. [22]

Hernandez de Alba, G. 1948c. The Achagua and their neighbors. *In:* J.H. Steward (ed.) *Handbook of South American Indians.* Smithson. Inst. Bur. Amer. Ethnol. Bull. 143: 399-412. [22]

Hillenius, D. 1959. The differentiation within the genus *Chamaeleo* Laurenti, 1768. *Beaufortia* 8:1-92. [2]

HIMAT. 1979. *Calendario Meteorológico 1979.* Instituto Colombiano de Hidrología, Meteorología, y Adecuación de Tierras; División de Meteorología, Bogotá. [9]

Hirshfield, M.F. and Tinkle, D.W. 1975. Natural selection and the evolution of reproductive effort. *Proc. Nat. Acad. Sci., U.S.A.* 72:2227-2231. [7]

Hirth, H.F. 1963. Some aspects of the natural history of *Iguana iguana* on a tropical strand. *Ecology* 44(3):613-615. [10, 20, 22]

Hladik, C.M. 1978. Adaptive strategies of primates in relation to leaf-eating. *In:* G.G. Montgomery (ed.) *The Ecology of Arboreal Folivores.* Smithsonian Inst. Washington, p. 373-395. [6]

Hoffstetter, R. 1976. Histoire des Mammifères et dérive des continents. *La Recherche* 7:124-138. [2]

Hogan-Warburg, A.J. 1966. Social behavior of the ruff, *Philomachus pugnax* (L.). *Ardea* 54:109-229. [18]

Holdridge, L.R. 1967. *Life Zone Ecology,* revised edition. Tropical Science Center, San José, Costa Rica. [10]

Hoogmoed, M.S. 1973. The lizards and amphisbaenians of Surinam. *Biogeographica.* 4:1-419-[1]

Hoover, W.H. and Heitmann, R.N. 1972. Effects of dietary fiber levels on weight gain, cecal volume and VFA production in rabbits. *J. Nutr.* 102:375-380. [5]

Hoover, W.H. and Clarke, S.D. 1972. Fiber digestion in the beaver. *J. Nutr.* 102:4-16. [5]

Hotton, Nicholas, III. 1955. A survey of adaptive relationships of dentition and diet in the North American Iguanidae. *Amer. Midl. Nat.* 53:88-114. [10]

Houpt, T.R. 1963. Urea utilization in rabbits fed a low-protein ration. *Am. J. Physiol.* 205:1144-1150. [5]

Hover, E.L. and Jenssen, T.A. 1976. Descriptive analysis and social correlates of agonistic displays of *Anolis limifrons* (Sauria, Iguanidae). *Behaviour* 58:173-191. [13]

Howard, R.D. 1978. The evolution of mating strategies in bullfrogs, *Rana catesbeiana* *Evolution* 32:850-871. [17, 18, 21]

Howard, W.E. 1960. Innate environmental dispersal of individual vertebrates. *Amer. Midl. Nat.* 63:152-161. [15]

Huey, R. In press. Temperature, physiology and ecology of reptiles. *In:* C. Gans and F.H. Pough (eds.) *Biology of the Reptilia,* Vol. 12. Academic Press, London and New York. [16]

Hughes, R.N. 1979. Optimal diets under the energy maximization premise: The effects of recognition time and learning. *Amer. Nat.* 113:209-221. [6]

Hungate, R.E. 1950. The anaerobic mesophilic cellulolytic bacteria. *Bacteriol. Rev.* 14:1-49. [5]

Hunsaker, D. 1962. Ethological isolating mechanisms in the *Sceloporus torquatus* group of lizards. *Evolution* 16:62-74. [14]

Hunsaker, D. and Burrage, B.B. 1969. The significance of interspecific dominance in iguanid lizards. *Amer. Midl. Nat.* 81:500-511. [21]

Instituto Meteorologico Nacional. 1974. *Annuario Meteorologico, Ano* 1971. San José, Costa Rica. [10]

Ireland, L.C., Frick, J.A. and Wingate, D.B. 1978. Nighttime orientation of hatchling green turtles *(Chelonia mydas)* in open ocean. *In:* K. Schmidt-Koenig and W.T. Keeton (eds.) *Animal Migration, Navigation and Homing.* Springer Verlag: New York. p. 420-429. [15]

Iverson, J.B. 1977. Behavior and ecology of the rock iguana *Cyclura carinata.* Ph.D. dissertation, Univ. Florida, Gainesville. [7, 10, 12]

Iverson, J.B. 1978. The impact of feral cats and dogs on populations of the West Indian rock iguana, *Cyclura carinata. Biol. Conserv.* 14:63-73. [VI, 6]

Iverson, J.B. 1979. Behavior and ecology of the rock iguana, *Cyclura carinata. Bull. Florida Sate Mus. Biol. Sci.* 24:175-358. [6, 7, 11, 17]

Iverson, J.B. In press. Colic modifications in iguanine lizards. *J. Morph.* 152. [4, 6]

Janzen, D.H. 1973a. Sweep samples of tropical foliage insects: description of study sites, with data on species abundances and size distributions. *Ecology* 54:659-686. [10]

Janzen, D.H. 1973b. Sweep samples of tropical foliage insects: effects of season, vegetation types, elevation, time of day and insularity. *Ecology* 54:687-708. [10, 16]

Janzen, D.H. 1976. Reduction of *Mucuna andreana* (Leguminosae) seedling fitness by artificial seed damage. *Ecology* 57:826-828. [6]

Jenssen, T.A. 1970. Female response to filmed displays of *Anolis nebulosus* (Sauria, Iguanidae). *Anim. Behav.* 18:640-647. [18]

Jenssen, T.A. 1971. Display analysis of *Anolis nebulosus* (Sauria, Iguanidae). *Copeia* 1971:197-209. [13, 14]

Jenssen, T.A. 1977. Evolution of anoline lizard display behavior. *Amer. Zool.* 17:203-215. [18]

Jenssen, T.A. 1978. Display diversity in anoline lizards and problems of interpretation. *In:* N. Greenberg and P.D. Maclean (eds.) *Behavior and Neurology of Lizards.* N.I.M.H., Rockville, MD. DHEW Publ. No. (ADM) 77-491. p. 269-285. [13]

Jenssen, T.A. and Hover, E.L. 1976. Display analysis of the Signature display of *Anolis limifrons* (Sauria, Iguanidae). *Behaviour* 57:227-240. [13]

Jenssen, T.A. and Rothblum, L.M. 1977. Display repertoire analysis of *Anolis townsendi* (Sauria, Iguanidae) from Cocos Island. *Copeia* 1977:103-109. [13]

Johnson, J.L. and McBee, R.H. 1967. The porcupine cecal fermentation. *J. Nutr.* 91:540-546. [5]

Johnson, S.R. 1965. An ecological study of the chuckwalla, *Sauromalus obesus* Biard, in the western Mojave Desert. *Amer. Midl. Nat.* 731:1-29. [7, 10, 11, 21]

Kamil, A. and Sargent, T. 1981. *Foraging behavior: Ecological, ethological, and psychological approaches.* Garland STPM. New York. [II]

Kiester, A.R. 1979. Conspecifics as cues: a mechanism for habitat selection in the Panamanian grass anole *(Anolis auratus). Behav. Ecol. Sociobiol.* 5:323-330. [15]

Kiester, A.R. Gorman, G.C. and Arroyo, D.C. 1975. Habitat selection behavior of three species of *Anolis* lizards. *Ecology* 56:220-225. [15]

King, J.R. 1974. Seasonal allocation of time and energy resources in birds. *In:* R.A. Paynter (ed.) *Avian Energetics.* Nuttall Ornithol. Club Cambridge. p. 4-70. [3]

Kirchhoff, P. 1948a. Food-gathering tribes of the Venezuela llanos. *In:* J.H. Steward (ed.) *Handbook of South American Indians. Vol. 4. The Circum-Caribbean tribes.* Smithson. Inst. Bur. Amer. Ethnol. Bull. 143:445-468. [22]

Kirchhoff, P. 1948b. The tribes north of the Orinoco River. *In:* J.H. Steward (ed.) *Handbook of South American Indians. Vol. 4. The Circum-Caribbean tribes.* Smithson. Inst. Bur. Amer. Ethnol. Bull. 143:481-493. [22]

Kitzler, G. 1941. Die Paarungsbiologie einiger Eidechsen. *Z. Tierpsychol.* 4:353-402. [14]

Knight, D.H. 1968. Ecology of a tropical savanna burned annually by iguana hunters. *Jour. Colorado-Wyoming Acad. Sci.* 61:50. [22]

Kramer, G. 1937. Beobachtungen über Paarungsbiologie und soziales Verhalten von Mauereidechsen. *Z. Morphol. Ökol. Tiere* 32:752-783. [14]

Krebs, J.R. and Davies, N.B. 1981. *An Introduction to Behavioural Ecology.* Sinauer Associates, Sunderland, MA. [17]

Krekorian, C.O. 1977. Homing in the desert iguana, *Dipsosaurus dorsalis. Herpetologica* 33:123-127. [15]

Labillardière, M. 1800. *Voyage in Search of La Perouse.* John Stockdale, Piccadilly, London. [23]

Landberg, H.E. 1965. *World Maps of Climatology.* 2nd Ed. Springer-Verlag, New York, 11:1-102. [6]

Latcham, R.E. 1922. Los animales domesticos de la America precolombiana. *Publ. Mus. Ethnol. Antrop. Chile,* 3:1-199. [22]

Laurent, L.F. 1979. Herpetofaunal-relationships between Africa and South America. *In:* W.E. Duellman (ed.) *The South American Herpetofauna: Its Origin, Evolution, and Dispersal. Monog. Univ. Kansas Mus. Nat. Hist.* 7:55-71. [Intro]

Laurenti, J.N. 1768. Specimen medicum, exhibens synopsin reptilium emendatum cum experimentis circa venena et antidota reptilium Austriacorum. Wien. [1]

Lazell, J.D. 1973. The lizard genus *Iguana* in the Lesser Antilles. *Bull. Mus. Comp. Zool.* 145:1-28. [1, 4, 22]

LeBoeuf, B.J. 1974. Male-male competition and reproductive success in elephant seals. *Amer. Zool.* 14:163-176. [18]

LeGrand, H.E. 1973. Hydrological and ecological problems of karst regions. *Science* 179: 859-864. [7]

Levins, R. 1968. *Evolution in Changing Environments*. Princeton Univ. Press, Princeton, New Jersey. [10]

Lewis, C.B. 1944. Notes on *Cyclura*. *Herpetologica* 2:93-98. [VI, 7]

Licht, P. In press. Seasonal cycles in reptilian reproductive physiology. *In:* E. Lamming (ed.) *Marshall's Physiology of Reproduction*. [8]

Licht, P. and Moberly, W.R. 1965. Thermal requirements of embryonic development in the tropical lizard *Iguana iguana*. *Copeia* 1965:515-517. [7, 8, 9]

Lifson, N., Gordon, G.B., Visscher, M.G. and Nier, A.O. 1949. The fate of utilized molecular oxygen and the source of the oxygen of respiratory carbon dioxide, studied with the aid of heavy oxygen. *J. Biol. Chem.* 180:803-811. [3]

Lifson, N. and McClintock, R. 1966. Theory of use of the turnover rates of body water for measuring energy and material balance. *J. Theoret. Biol.* 12:46-74. [3]

Lipkind, W. 1948. The Caraja. *In:* J.W. Steward (ed.) *Handbook of South American Indians. Vol. 3. The tropical forest tribes*. Smithson. Inst. Bur. Amer. Ethnol. Bull. 143. [22]

Lönnberg, E. 1902. *The Morphological Structure of the Intestine and the Diet of Reptiles. K. Svenska Vet. Acad. Bihang till Handlingar* 28, Afd. IV, No. 8. p. 1-51. [10]

Lothrup, S.K. 1948. The tribes west and south of the Panama Canal. *In:* J.W. Steward (ed.) *Handbook of South American Indians. Vol. 4. The Circum-Caribbean tribes*. Smithson. Inst. Bur. Amer. Ethnol. Bull. 143, p. 253-256. [22]

Loveridge, A. 1945. *Reptiles of the Pacific World*, MacMillan Co., New York. [23]

Low, B.S. 1978. Environmental uncertainty and the parental strategies of marsupials and placentals. *Amer. Nat.* 112:197-213. [20]

Lugo, A.E., Gonzalez-Liboy, J.A., Cintrón, B. and Dugger, K. 1978. Structure, productivity, and transpiration of a subtropical dry forest in Puerto Rico. *Biotropica* 10:278-291. [7]

MacArthur, R. 1972. *Geographical Ecology: Patterns in the Distribution of Species*. Harper and Row, New York. [11]

MacArthur, R. and Pianka, E.R. 1966. On optimal use of patchy environment. *Amer. Nat.* 100:603-609. [6]

MacArthur, R. and Wilson, E.O. 1967. *The Theory of Island Biography*. Princeton Univ. Press, Princeton, New Jersey. [10]

Mackay, R.S. 1964. Galápagos tortoise and marine iguana deep body temperatures measured by radio telemetry. *Nature* 204:355-358. [16]

Mackay, R.S. 1968. Observations on peristaltic activity versus temperature and circadian rhythms in undisturbed *Varanus flavescens* and *Ctenosaura pectinata*. *Copeia* 1968: 252-259. [16]

Mandel, H.G. 1972. Pathways of dung biotransformation: Biochemical conjugations. *In:* B.N. LaDu, H.G. Mandel and E.L. Way (eds.) *Fundamentals of Drug Metabolism and Drug Disposition*. Williams and Wilkins, Baltimore. p. 149-186. [6]

Mares, M.A. and Hulse, A.C. 1978. Patterns of some vertebrate communities in creosote bush deserts. *In:* T.J. Mabry, J.H. Hunziker and D.R. DiFeo (eds.) *Creosote Bush: Biology and Chemistry of Larrea in New World Deserts*. US/IBP synthesis series no. 6. Dowden, Hutchinson and Ross, Stroudsburg, Pennsylvania. p. 209-225. [4]

Mayhew, W.W. 1971. Reproduction in the desert lizard, *Dipsosaurus dorsalis*. *Herpetologica* 27:57-77. [4, 7, 10, 17]

Maynard Smith, J. and Parker, G.A. 1976. The logic of assymetric contests. *Anim. Behav.* 24:159-175. [21]

McBee, R.H. 1971. Significance of intestinal microflora in herbivory. *Ann. Rev. Ecol. Syst.* 2:165-176. [4, 5]

McBee, R.H. 1977. Fermentation in the hindgut. *In:* R.T.J. Clarke and T. Bauchop (eds.) *Microbial Ecology of the Gut*. Academic Press, London p. 185-222. [4, 5]

McBee, R.H. and West, G.C. 1969. Cecal fermentation in the willow ptarmigan. *Condor* 71:54-58. [5]

McElhinny, M.W. 1973. *Palaeomagnetism and Plate Tectonics.* Cambridge Univ. Press, London. [2]

McGinnis, S.M. and Brown, C.W. 1968. Thermal behaviour of the green iguana, *Iguana iguana. Herpetologica* 22:189-198.

McKey, D. 1974. Adaptive patterns in alkaloid physiology. *Amer. Nat.* 108:305-332. [6]

McNab, B.K. 1963. Bioenergetics and determination of home range size. *Amer. Nat.* 97: 133-140. [6, 17]

McNaughton, S.J. 1976. Serengeti migratory wildebeest: Facilitation of energy flow by grazing. *Science* 191:92-94. [6]

Mendelssohn, H. 1980. Observations on a captive colony of Iguana iguana. *In:* J.B. Murphy and J.T. Collins (eds.) *Reproductive Biology and Diseases of Captive Reptiles.* SSAR Contributions to Herpetology No. 1, p. 119-123. [VI]

Merrem, B. 1820. *Versuch eines Systems der Amphibien.* Tentanem systematis Amphibiorum. Marburg. [1]

Mertens, R. 1934. Die Insel Reptilien. *Zoologica* 84:1-205. [11]

Mertens, R. 1946. Die Warn- und Droh-Reaktionen der Reptilien. *Abh. Sencken. Naturf. Ges.* 471:1-108. [14]

Mertens, R. 1960. *The World of Amphibians and Reptiles.* McGraw-Hill, New York. [4]

Merton, L.F.H., Bourn, D.M. and Hnatiuk, R.J. 1976. Giant tortoise and vegetation interactions on Aldabra Atoll. *Biol. Conserv.* 9:293-316. [6]

Metraux, A. 1946. Ethnography of the Chaco. *In:* J.W. Steward (ed.) *Handbook of South American Indians. Vol. 1. The marginal tribes.* Smithson. Inst. Bur. Amer. Ethnol. Bull. 143:197-370. [22]

Meyer, J.R. and Wilson, L.D. 1973. A distributional checklist of the turtles, crocodilians, and lizards of Honduras. *Contrib. Sci. Nat. Hist. Mus. Los Angeles,* 244:1-39. [1]

Miller, G.A. 1918. Mammals and reptiles collected by Theodoor de Booy in the Virgin Islands. *Proc. U.S. Nat. Mus.* 54:507-511. [VI]

Milton, K. 1979. Factors influencing leaf choice by howler monkeys: A test of some hypotheses of food selection by generalist herbivores. *Amer. Nat.* 114:362-378. [6]

Minton, S.A. 1966. A contribution to the herpetology of West Pakistan. *Bull. Amer. Mus. Nat. Hist.* 134:27-185. [4]

Minnich, J.E. 1970. Water and electrolyte balance of the desert iguana, *Dipsosaurus dorsalis,* in its natural habitat. *Comp. Biochem. Physiol.* 35:921-933. [4]

Minnich, J.E. and Shoemaker, J.E. 1970. Diet, behavior, and water turnover in the desert iguana, *Dipsosaurus dorsalis. Amer. Midl. Nat.* 84:496-509. [4]

Mittleman, M.B. 1942. A summary of the iguanid genus *Urosaurus. Bull. Mus. Comp. Zool.* 92:105-181. [1]

Moberly, W.R. 1968. The metabolic responses of the common iguana, *Iguana iguana,* to activity under restraint. *Comp. Biochem. Physiol.* 27:1-20. [4, 14]

Montanucci, R.R. 1968. Comparative dentition in four iguanid lizards. *Herpetologica* 24:305-315. [4, 10]

Montgomery, G.G., Rand, A.S. and Sunquist, M.E. 1973. Post-nesting movements of iguanas from a nesting aggregation. *Copeia* 1973:620-622. [7, 14, 15, 18, 20]

Morgareidge, K.R. and White, F.N. 1969. Cutaneous vascular changes during heating and cooling in the Galápagos marine iguanas. *Nature* 223:587-591. [16]

Morris, D. 1956a. The feather postures of birds and the problem of the origin of social signals. *Behaviour* 9:75-113. [13]

Morris, D. 1956b. The function and causation of courtship ceremonies. *In:* P.P. Grasse (ed.) *L'Instinct dans le Comportement des Animaux et de l'Homme.* Fondation Singer Polignac, Paris. p. 261-286. [18]

Morrison, D.W. 1978. On the optimal searching strategy for refuging predators. *Amer. Nat.* 112:925-934. [6]

Moss, R. 1977. The digestion of heather by red grouse during the spring. *Condor* 79:471-477. [6]

Mrosovsky, N. 1968. Nocturnal emergence of hatchling sea turtles: Control by thermal inhibition of activity. *Nature* 220:1338-1339. [15]

Mrosovsky, N. 1978. Orientation mechanisms of marine turtles. *In:* K. Schmidt-Koenig and W.T. Keeton (eds.) *Animal Migration, Navigation and Homing.* Spring-Verlag: New York. p. 413-419. [15]

Müller, V.H. 1968. Untersuchungen über Wachstum und Altersverteilung einer Population des Grünen Leguans *Iguana iguana iguana* L. (Reptilia: Iguanidae). *Mitt. Inst. Colombo-Alemán Invest. Cient.* 2:57-65. [7, 8, 9, 10, 17, 22]

Müller, V.H. 1972. Ökologische und ethologische Studien an *Iguana iguana* L. (Reptilia: Iguanidae) in Kolumbien. *Zool. Beitr.* 18:109-131. [7, 9, 10, 12, 14, 18, 19, 22]

Murphy, R.W. 1980. Evolution and biogeography of the Baja California herpetofauna. *In:* T.J. Case and M.L. Cody (eds.) *Island Biogeography in the Sea of Cortez.* [11]

Myers, B.C. and Eells, M.M. 1968. Thermal aggregation in *Boa constrictor. Herpetologica* 24:61-66. [16]

Nader, L. 1969. The Zapoetec of Oaxaca. *In:* E.Z. Vogt (ed.) *Handbook of Middle American Indians.* Vol. 7. Univ. Texas Press, Austin. p. 329-359. [22]

Nagy, J.G. and Tengerdy, R.P. 1967. Antibacterial action of essential oils of *Artemesia* as an ecological factor. *Appl. Microbiol.* 16:441-444. [6]

Nagy, K.A. 1973. Behavior, diet, and reproduction in a desert lizard, *Sauromalus obesus. Copeia* 1973:93-102. [4, 7, 10, 17, 21]

Nagy, K.A. 1975. Water and energy budgets of free-living animals: measurement using isotopically labeled water. *In:* N.F. Hadley (ed.) *Environmental Physiology of Desert Organisms.* Dowden, Hutchinson and Ross, Stroudsburg, Penn. p. 227-245. [4]

Nagy, K.A. 1977. Cellulose digestion and nutrient assimilation in *Sauromalus obesus,* a plant-eating lizard. *Copeia* 1977:355-362. [3]

Nagy, K.A. 1980. CO_2 production in animals: analysis of potential errors in the doubly labeled water method. *Amer. J. Physiol.* 238:R466-R473. [3]

Nagy, K.A. and Shoemaker, V.H. 1975. Energy and nitrogen budgets of the free-living desert lizard *Sauromalus obesus. Physiol. Zool.* 48:252-262. [3]

Noble, G.K. and Bradley, H.T. 1933. The mating behavior of lizards; its bearing on the theory of sexual selection. *Ann. New York Acad. of Sci.,* Vol. 35:35-100. [19, 21]

Norris, K.S. 1953. The ecology of the desert iguana *Disposaurus dorsalis. Ecology* 34(2): 265-287. [4, 6, 7, 10]

Norris, K.S. and Dawson, W.R. 1964. Observations on the water economy and electrolyte excretion of chuckwallas (Lacertilia, *Sauromalus). Copeia* 1964:638-646. [13]

Oelrich, T.M. 1956. *The Anatomy of the Head of Ctenosaura pectinata (Iguanidae). Misc. Publ. Mus. Zool. Univ. Michigan* 94:1-122. [14]

Olmsted, D.L. 1969. The Tequistlatec and Tlapanec. *In:* E.Z. Vogt (ed.) *Handbook of Middle American Indians. Vol. 7.* Univ. of Texas Press, Austin. p. 553-564. [22]

Orians, G.H. 1969. On the evolution of mating systems in birds and mammals. *Amer. Nat.* 103:589-603. [17]

Ostrom, J.H. 1963. Further comments on herbivorous lizards. *Evolution* 17:368-369. [6]

Pagden, A.R. 1975. *The Maya: Diego de Landa's Account of the Affairs of Yucatán.* J. Philip O'Hara, Inc., Chicago. [22]

Parker, G.A. 1970. Sperm competition and its evolutionary consequences in the insects. *Biol. Rev.* 45:525-568. [17]

Parker, G.A. 1974. Assessment strategy and the evaluation of fighting behavior. *J. Theor. Biol.* 47:223-243. [13]

Parker, G.A. 1974. Courtship persistence and female-guarding as male time investment strategies. *Behaviour* 48:157-184. [17, 18]

Parker, W.S. 1972. Notes on *Dipsosaurus dorsalis* in Arizona. *Herpetologica* 28:226-229. [7]

Parmenter, R.R. 1978. Effects of temperature and diet on feeding ecology, growth and body size in turtles. *Bull. Ecol. Soc. Amer.* 59:57. [4]

Patton, T.H. 1967. Fossil vertebrates from Navassa Island, W.I. *Quart. J. Fla. Acad. Sci.* 30:59-60. [VI]

Pasteur, G. 1964. Recherches sur l'évolution des Lygodactyles, Lézards afro-malgaches actuels. *Trav. Inst. Scient. Chérif. (Zool.)* 29:1-160. [2]

Peabody, F.E. 1961. Annual growth rings in living and fossil vertebrates. *J. Morphol.* 108:11-62. [11]

Peaker, M. and Linzell, J.L. 1975. *Salt Glands in Birds and Reptiles.* Monographs Physiol. Soc., Cambridge Univ. Press, London. [4, 13]

Pearce, R.C. and Tanner, W.W. 1973. Helmenths of *Sceloporus* lizards in the Great Basin and Upper Colorado Plateau of Utah. *Great Basin Nat.* 33:1-18. [4]

Peracca, M.G. 1891. Osservazioni sulla riproduzione della *Iguana tuberculata* Laur. *Boll. Mus. Zool. Anat. Comp. R. Univ. Torino* 110:1-8. [14, 17, 18, 22]

Perrill, S.A., Gerhardt, H.C. and Daniel, R. 1978. Sexual parasitism in the green tree frog *Hyla cinerea. Science* 200:1179-1180. [18]

Phillips. J.B. and Adler, K. 1978. Directional and discriminatory responses of salamanders to weak magnetic fields. *In:* K. Schmidt-Koenig and W.T. Keeton (eds.) *Animal Migration, Navigation, and Homing.* Springer-Verlag, New York. p. 325-333. [15]

Pianka, E.R. 1970a. Comparative autecology of the lizard *Cnemidophorus tigris* in different parts of its geographic range. *Ecology* 51:703-720. [11]

Pianka, E.R. 1970b. On r and K selection. *Amer. Nat.* 104:592-597. [10]

Pianka, E.R. 1972. r and K selection or b and d selection? *Amer. Nat.* 106:581-588. [10]

Pianka, E.R. 1973. Comparative ecology of two lizards. *Copeia* 1971:129-138. [4]

Pianka, E.R. 1976. Natural selection of optimal reproductive tactics. *Amer. Zool.* 16:775-784. [7, 17]

Pielou, E.C. 1975. *Ecological Diversity.* John Wiley & Sons, New York. [7]

Pitelka, F.A. 1942. Territory and related problems in North American hummingbirds. *Condor* 44:189-204. [6]

Poignant, R. 1976. *Discovery under the Southern Cross.* William Collins Sons and Co., London. [23]

Porter, W.P., Mitchell, J.W., Bechman, W.A. and DeWitt, C.B. 1973. Behavioral implications of mechanistic ecology: Thermal and behavioral modeling of desert ectotherms and their microenvironment. *Oecologia* 13:1-54. [20]

Pough, F.H. 1973. Lizard energetics and diet. *Ecology* 54:837-844. [3, 4, 6, 7, 17]

Pough, F.H. 1980. The advantages of ectothermy for tetrapods. *Amer. Nat.* 115:92-112. [II]

Presch, W. 1969. Evolutionary osteology and relationships of the Horned Lizard genus *Phrynosoma* (Iguanidae) *Copeia* 1969:250-275. [2]

Prestude, A.M. and Crawford, F.T. 1970. Tonic immobility in the lizard, *Iguana iguana. Anim. Behav.* 18:391-395. [14]

Prieto, A.A. and Ryan, M.J. 1978. Some observations on the social behavior of the Arizona chuckwalla, *Sauromalus obesus tumidus* (Reptilia, Lacertilia, Iguanidae). *J. Herpetol.* 12:327-336. [7, 12, 17, 21]

Prieto, A.A. and Sorenson, M.W. 1975. Predator-prey relationships of the Arizona chuckwalla *(Sauromalus obesus tumidus). Bull. New Jersey Acad. Sci.* 20:12-13. [21]

Pulliam, H.R. 1975. Diet optimization with nutrient constraints. *Amer. Nat.* 109:765-768. [6]

Purdue, J.R. and Carpenter, C.C. 1972a. A comparative study of the body movements of displaying males of the lizard genus *Sceloporus* (Iguanidae). *Behaviour* 41:68-81. [12]

Purdue, J.R. and Carpenter, C.C. 1972b. A comparative study of display motion in the iguanid genera *Sceloporus, Uta,* and *Urosaurus. Herpetologica* 28:137-140. [12]

Ramirez, D. 1968. Iguanas y Garrobos Nicas a El Salvador. *La Prensa,* 27 Febrero 1968. [22]

Rand, A.S. 1954. Variation and predator pressure in an island and a mainland population of lizards. *Copeia* 1954:260-262. [11]

Rand, A.S. 1961. A suggested function of the ornamentation of East African forest chameleons. *Copeia* 1961:411-414. [17]

Rand, A.S. 1967a. The adaptive significance of territoriality in iguanid lizards. *In:* W.W. Milstead (ed.) *Lizard Ecology: A symposium.* University of Missouri, Press, Colombia. p. 106-115. [19, 21]

Rand, A.S. 1967b. Ecology and social organization in the iguanid lizard *Anolis lineatopus. Proc. U.S. Nat. Mus.* 122:1-79. [17, 21]

Rand, A.S. 1968a. A nesting aggregation of iguanas. *Copeia* 1968:552-561. [7, 14, 15, 18, 19, 20]

Rand, A.S. 1968b. Desiccation rates in crocodile and iguana eggs. *Herpetologica* 24:178-180. [7]

Rand, A.S. 1972. The temperature of iguana nests and their relation to incubation optima and to nesting sites and season. *Herpetologica* 28:252-253. [8, 10, 18, 19]

Rand, A.S. 1978. Reptilian arboreal folivores. *In:* G.G. Montgomery (ed.) *The Ecology of Arboreal Folivores.* Smithsonian Ins. Press, Washington. p. 115-122. [4, 10, 17]

Rand, A.S., Gorman, G.C. and Rand, W.M. 1975. Natural history, behavior and ecology of *Anolis agassizi. In:* J.B. Graham (ed.) *The Biological Investigation of Mapelo Island, Colombia. Smith. Contrib. Zool.* 176-27-38. [21]

Rand, A.S. and Rand, W.M. 1978. Display and dispute settlement in nesting iguanas. *In:* N. Greenberg and P.D. Maclean (eds.) *Behavior and Neurology of Lizards.* N.I.M.H., Rockville, MD. DHEW Publ. No. (ADM)77-491. p. 245-251. [20]

Rand, A.S. and Robinson, M.H. 1969. Predation on iguana nests. *Herpetologica* 25:172-174. [10, 20]

Rand, A.S. and Williams, E.E. 1970. An estimation of redundancy and information content of anole dewlaps. *Amer. Nat.* 104:99-103. [17, 18]

Rand, W.M. and Rand, A.S. 1976. Agonistic behavior in nesting iguanas: a stochastic analysis of dispute settlement dominated by the minimization of energy cost. *Z. Tierpsychol.* 40:279-299. [7, 14, 18, 20, 21]

Ravicz, R. and Romney, A.K. 1969. The Amuzgo. *In:* E.Z. Vogt (ed.) *Handbook of Middle American Indians. Vol. 7. Ethnology, Part I.* University of Texas Press, Austin. [22]

Regal, P.J. 1977. Behavioral differences between reptiles and mammals: an analysis of activity and mental capabilities. *In:* N. Greenberg and P.D. MacLean (eds.) *Behavior and Neurology of Lizards.* N.I.M.H., Rockville, MD. DHEW Publ. No. (ADM) 77-491. [II]

Renous, S. 1978. Confrontation entre certaines données morphologiques et la répartition géographique des formes actuelles de Sauriens: interprétation phylogénétique et hypothèse paléobiogeographique. *Bull. Soc. Zool. France* 103:219-224. [2]

Rensch. B. and Adrian-Hinsberg, C. 1964. Die visuelle Lernkapazität von Leguanen. *Z. Tierpsychol.* 20:34-42. [15]

Ricklefs, R.E. 1976. *The Economy of Nature.* Chiron Press, Portland. [3]

Ricklefs, R.E. and Cullen, J. 1973. Embryonic growth of the green iguana, *Iguana iguana. Copeia* 1973:296-305. [9]

Robinson, M.D. 1972. Chromosomes, protein polymorphism, and systematics of insular chuckawalla lizards (genus *Sauromalus)* in the Gulf of California. Ph.D. Thesis. University of Arizona. [1, 11]

Rockwood, L.L. 1974. Seasonal changes in the susceptibility of *Crescentia slat* leaves to the flea beetle, *Oedinychus* sp. *Ecology* 55:142-148. [6]

Romer, A.S. and Parsons, T.S. 1978. *The Vertebrate Body.* W.B. Saunders Co., Philadelphia. [4]

Romero Castañeda, R. 1971. *Plantas del Magdalena II (Flora de la Isla de Salamanca) Primera Parte.* Univ. Nac., Facultad de Ciencias, Inst. de Ciencias Naturales, Bogotá, [9]

Rothblum, L.M. and Jenssen, T.A. 1978. Display repertoire analysis of *Sceloporus undulatus hyacinthinus* (Sauria, Iguanidae) from southwestern Virginia. *Anim. Behav.* 26:130-137. [13]

Ruby, D.E. 1976. The behavioral ecology of the viviparous lizard, *Sceloporus jarrovi.* Ph.D. Thesis. Univ. of Michigan, Ann Arbor. [11]

Ruby, D.E. 1977. The function of shudder displays in the lizard, *Sceloporus jarrovi. Copeia* 1977:110-114. [13]

Salt, G.W. 1943. The lungs and inflation mechanism in *Sauromalus obesus. Copeia* 1943:193. [21]

Savage, J.M. 1958. The iguanid lizard genera *Urosaurus* and *Uta* with remarks on related genera. *Zoologica* 43(2):41-54. [1]

Schad, G.A. 1963. Niche diversification in a parasitic species flock. *Nature* 198:404-406. [4]

Schaffer, W.M. 1974. Selection for optimal life histories: The effects of age structure. *Ecology* 55:291-303. [7]

Schein, M.W. and Hale, E.B. 1965. Stimuli eliciting sexual behavior. *In:* F.A. Beach (ed.) *Sex and Behavior.* Wiley, New York. p. 440-482. [17]

Schoener, T.W. 1968a. Sizes of feeding territories among birds. *Ecology* 49:123-141. [17]

Schoener, T.W. 1968b. The *Anolis* lizards of Bimini: Resource partitioning in a complex fauna. *Ecology* 49:704-726. [6]

Schoener, T.W. 1969. Models of optimal size for solitary predators. *Amer. Nat.* 103:277-313. [6, 11]

Schoener, T.W. 1971. Theory of feeding strategies. *Ann. Rev. Ecol. Syst.* 2:369-403. [6]

Schoener, T.W. 1977. Competition and the niche. *In:* C.Gans and D.W. Tinkle (eds.) *Biology of the Reptilia. Vol. 7. Ecology A.* Academic Press, New York. p. 35-136. [17, 19]

Schoener, T.W. and Schoener, A. 1978. Estimating and interpreting body-size growth in some *Anolis* lizards. *Copeia* 1978: 390-405. [10]

Scholes, F.V. and Roys, R.L. 1968. *The Maya Chontal Indians of Acalan-Tixchel.* University of Oklahoma Press, Norman. [22]

Schmidt-Nielsen, K. 1975. *Animal Physiology: Adaptation and Environment.* Cambridge Univ. Press, London. [3]

Schmidt, K.P. and Inger, R.F. 1957. *Living Reptiles of the World.* Doubleday, London. [11]

Schwartz, A. and Carey, W.M. 1977. Systematics and evolution in the West Indian iguanid genus *Cyclura. Stud. Fauna Curacao and other Carib. Is.* 53:1-97. [1, 7, 12]

Sexton, O.J. 1975. Black vultures feeding on iguana eggs in Panama. *Amer. Midl. Nat.* 93:463-468. [20]

Shallenberger, E.W. 1970. Tameness in insular animals: a comparison of approach distance of insular and mainland iguanid lizards. Ph.D. Thesis. University of California, Los Angeles. [11]

Shaw, C.E. 1945. The chuckwallas, genus *Sauromalus. Trans. San Diego Soc. Nat. Hist.* 10:269-306. [1]

Shepard, J.M. 1975. Factors influencing female choice in the lek mating system of the ruff. *Living Bird* 14:87-112. [17, 18]

Shoemaker, V.H., Nagy, K.A. and Costa, W.R. 1976. Energy utilization and temperature regulation by jackrabbits *(Lepus californicus)* in the Mojave Desert. *Physiol. Zool.* 49:364-375. [3]

Siegel, S. 1956. *Nonparametric Statistics for the Behavioral Sciences.* McGraw-Hill Book Company, New York. [18]

Simon, C.A. 1975. The influence of food abundance on territory size in the iguanid lizard *Sceloporus jarrovi. Ecology* 56:993-998. [6, 17, 21]

Smith, D.C. 1977. Interspecific competition and the demography of two lizards. Ph.D. dissertation, University of Michigan, Ann Arbor. [10]

Smith, H.M. 1946. *Handbook of Lizards.* Comstock, Ithaca. [1]

Smith, H.M. 1972. The Sonoran subspecies of the lizard *Ctenosaura hemilopha. Great Basin Nat.* 32:104-111. [1]

Smith H.M. and Taylor, E.H. 1950. An annotated checklist and key to the reptiles of Mexico exclusive of the snakes. *Bull. U.S. Natn. Mus.* 199:v-243. [1]

Smithe, F.E. 1974. *Naturalists's Color Guide Supplement.* Amer. Mus. Nat. Hist., New York. [13]

Smole, W.J. 1976. *The Yanoama Indians.* Univ. Texas Press, Austin. [22]

Sokal, R. and Rohlf, J. 1981. *Biometry.* 2nd ed. W.H. Freeman & Co., San Francisco. [13]

Sokol, D.M. 1967. Herbivory in lizards. *Evolution* 21:192-194. [6]

Sokol, D.M. 1971. Lithophagy and geophagy in reptiles. *J. Herpetol.* 5:69-71. [4]

Soulé, M. and Sloan, A.J. 1966. Biogeography and distribution of the reptiles and amphibians on islands in the Gulf of California, Mexico. *Trans. San Diego Soc. Nat. Hist.* 14:137-146. [1, 11]

Spores, R. 1969. The Zapotec and Mixtec at Spanish contact. *In:* G.R. Willey (ed.) *Handbook of Middle American Indians.* Archaeology of Southern Mesoamerica. Vol. 3, Part 2. Univ. Texas Press, Austin. p. 962-987. [22]

Stamps, J.A. 1973. Displays and social organization in female *Anolis aeneus. Copeia* 1973:264-272. [19, 21]

Stamps, J.A. 1977. Social behavior and spacing patterns in lizards. *In:* C. Gans and D.W. Tinkle (eds.) *Biology of the Reptilia. Vol. 7. Ecology A.* Academic Press, New York. p. 265-334. [19, 21]

Stamps, J.A. and Barlow, G.W. 1973. Variation and sterotypy in the displays of *Anolis aeneus* (Sauria: Iguanidae). *Behaviour* 47:67-93. [14]

Stamps, J.A. and Crews, D.P. 1976. Seasonal changes in reproduction and social behavior in the lizard *Anolis aeneus. Copeia* 1976:467-476. [17]

Stebbins, R.C. 1948. Nasal structure in lizards with reference to olfaction and conditioning of the inspired air. *Amer. J. Anat.* 82:183-222. [1]

Stebbins, R.C. 1966. *A Field Guide to Western Reptiles and Amphibians.* Houghton Mifflin, Boston. [11]

Stout, D.B. 1948. The Cuna. *In:* J.W. Steward (ed.) *Handbook of South American Indians. Vol. 4. The Circum-Caribbean tribes.* Smithson. Inst. Bur. Amer. Ethnol. Bull. 143. [22]

Swain, T. 1976. Angiosperm-reptile co-evolution. *In:* A. d'A. Bellairs and C.B. Cox (eds.) *Morphology and Biology of Reptiles.* Academic Press. London. p. 107-122. [4]

Swanson, P.L. 1950. The iguana *Iguana iguana iguana.. Herpetologica* 6:187-193. [10, 14, 22]

Szarski, H. 1962. Some remarks on herbivorous lizards. *Evolution* 16:528. [6]

Tanner, W.W. and Avery, D.F. 1964. A new *Sauromalus obesus* from the upper Colorado Basin of Utah. *Herpetologica* 20:38-42. [1]

Tarling, D.H. and Runcorn, S.K. 1973. *Implications of Continental Drift to the Earth Sciences.* Academic Press, London. [2]

Taylor, E.H. 1922. The lizards of the Philippine Islands. Bureau of Sci., Manila, Publ. No. 17:1-269. [4]

Thomas, R. 1966. A reassessment of the herpetofauna of Navassa Island. *J. Ohio Herp. Soc.* 5:73-89. [VI]

Thornton, I. 1971. *Darwin's Islands: A Natural History of the Galápagos.* Natural History Press, New York. [20]

Throckmorton, G.S. 1971. Digestive efficiency in the herbivorous lizard *Ctenosaura pectinata. Herp. Rev.* 3(6):108. [6]

Throckmorton, G.S. 1973. Digestive efficiency in the herbivorous lizard *Ctenosaura pectinata. Copeia* 1973:431-435. [4, 6]

Throckmorton, G.S. 1976. Oral food processing in two herbivorous lizards, *Iguana iguana* (Iguanidae) and *Uromastyx aegyptius* (Agamidae). *J. Morph.* 148:363-390. [4]

Tinkle, D.W. 1965. Home range, density, dynamics, and structure of a Texas population of the lizard *Uta stansburiana. In:* W.M. Milstead (ed.) *Lizard Ecology a Symposium.* University of Missouri Press, Columbia. [15]

Tinkle, D.W. 1969. The concept of reproductive effort and its relation to the evolution of life histories of lizards. *Amer. Nat.* 103:501-516. [7, 10]

Tinkle, D.W. and Hadley, N.F. 1973. Reproductive effort and winter activity in the viviparous montane lizard *Sceloporus jarrovi. Copeia* 1973:272-276. [7]

Tinkle, D.W. and Hadley, N.F. 1975. Lizard reproductive effort: Caloric estimates and comments on its evolution. *Ecology* 56:427-434. [7]

Tinkle, D.W., Wilbur, H.M. and Tilley, S. 1970. Evolutionary strategies in lizard reproduction. *Evolution* 24:55-74. [7]

Tosi, J.A., Jr. 1969. *Republica de Costa Rica Mapa Ecologica.* Trop. Sci. Cent., San José, Costa Rica. [10]

Tovar, A.D. 1969. Man's effect on natural fauna. *Florida Nat.* March 9-13, 1969:20. [22]

Trillmich, K. 1979. Feeding behaviour and social behaviour of the marine iguana. *Noticias de Galápagos.* No. 29:19-20. [7]

Trivers, R.L. 1972. Parental investment and sexual selection. *In:* B. Campbell (ed.) *Sexual Selection and the Descent of Man.* Aldine, Chicago. p. 136-179. [17, 19, 21]

Trivers, R.L. 1976. Sexual selection and resource-accruing abilities in *Anolis garmani. Evolution* 30:253-269. [17, 18]

Troyer, K. 1982. Transfer of fermentation microbes between generations in a herbivorous lizard. *Science* 216:540-542. [II, 5, 15]

Turner, F.B. 1970. The ecological efficiency of consumer populations. *Ecology* 51:741-742. [3]

Turner, F.B., Jennrich, R.I. and Weintraub, J.D. 1969. Home ranges and body sizes of lizards. *Ecology* 50:1076-1081. [17, 18]

Turner, F.B., Medica, P.A. and Kowalewsky, B.W. 1976. *Energy Utilization by a Desert Lizard (Uta stansburiana).* US/IBP Desert Biome Monogr. No. 1, Utah State Univ. Press, Logan. [3]

Van Denburgh, J. 1913. The Galápagos lizards of the genus *Tropidurus;* with notes on the iguanas of the genera *Conolophus* and *Amblyrhynchus. Calif. Acad. Sci.* Vol. II:133-202. [19]

Van Denburgh, J. 1914. The gigantic land tortoises of the Galápagos Archipelago. *Proc. Calif. Acad. Sci.* 2:203-374. [6]

Van Denburgh, J. 1922. The reptiles of western North America. *Occ. Papers. Calif. Acad. Sci.* No. 10, 1-1028, 2 vols. [1, 11]

Van Denburgh, J. and Slevin, J.R. 1913. The Galápagoan lizards of the genus *Tropidurus;* with notes on the iguanas of the genera Conolophus and *Amblyrhynchus. Proc. Calif. Acad. Sci.* 11:133-202. [1]

Van Devender, R.W. 1975. The comparative demography of two local populations of the tropical lizard, *Basiliscus basiliscus.* Ph.D. dissertation. University of Michigan, Ann Arbor. [4, 10]

Van Devender, R.W. 1978. Growth ecology of a tropical lizard, *Basiliscus basiliscus. Ecology* 59:1031-1038. [7, 9, 10, 17]

Van Rhijn, J.G. 1973. Behavioral dimorphism in male ruffs, *Philomachus pugnax* (L.) *Behaviour* 47:10-220. [18]

Vaughan, T.A. 1978. *Mammalogy*. W.B. Saunders, Philadelphia. [4]

Villa, J. 1968. Explotación irreflexiva de nuestras iguanas. *El Pensamiento Nacional*, 17 Febrero 1968:2-3. [22]

Vitt, L.J. and Congdon, J.D. 1978. Body shape, reproductive effort and relative clutch mass in lizards: resolution of a paradox. *Amer. Nat.* 112:595-608. [3]

Walcott, C., Gould, J.L. and Kirschvink, J.L. 1979. Pigeons have magnets. *Science* 205: 1027-1029. [15]

Warner, R.R. Roberson, D.R. and Leigh, E.G., Jr. 1975. Sex change and sexual selection. *Science* 190:633-638. [18]

Weber, H. 1957. Vergleichende Untersuchung des Verhaltens von Smaragdeidechsen *(Lacerta viridis)*, Mauereidechsen *(L. muralis)* und Perleidechsen *(L. lepida)*. *Z. Tierpsychol.* 14:448-472. [14]

Webster, T.P. and Burns, J.M. 1973. Dewlap color variation and electrophoretically detected sibling species in a Haitian lizard, *Anolis brevirostris*. *Evolution* 27:368-377. [17]

Weintraub, J.D. 1970. Homing in the lizard *Sceloporus orcutti*. *Anim. Behav.* 18:132-137. [15]

Wells, K.D. 1977a. Territoriality and male mating success in the green frog *(Rana clamitans)*. *Ecology* 58:750-762. [17]

Wells, K.D. 1977b. The social behavior of anuran amphibians. *Anim. Behav.* 25:666-693. [18]

Werner, D. 1972. Beobachtungen an *Ptyodactylus hasselquistii guttatus* (Geckonidae). *Verh. naturf. Ges. Basel* 82:54-87. [14]

Werner, D.I. 1978. On the biology of *Tropidurus delanonis* Baur (Iguanidae). *Z. Tierpsychol.* 47:337-395. [19]

Wernstedt, F.L. 1972. *World Climatic Data*. Climatic Data Press, Lemont, Penn. [8]

Westoby, M. 1974. An analysis of the diet selection by large generalized herbivores. *Amer. Nat.* 108:290-304. [6]

White, F.N. 1973. Temperature and the Galápagos marine iguana–insights into reptilian thermoregulation. *Comp. Biochem. Physiol.* 45A:503-513. [16]

Whittaker, R.H. 1976. *Communities and Ecosystems*. 2nd Ed. MacMillan and Co. New York. [6]

Wiegert, R.G. 1972. Energetics of the nest-building termite *Nasutitermes costas* (Holmgren) in a Puerto Rican forest. *In:* H.T. Odum (ed.) *A Tropical Rain Forest*. U.S. Atomic Energy Comm. p. I57-I64. [6]

Wiegert, R.G. 1976. *Ecological energetics*. Dowden, Hutchinson and Ross, Stroudsburg, Penn. [3]

Wiewandt, T.A. 1977. Ecology, behavior, and management of the Mona Island ground iguana, *Cyclura stejnegeri*. Ph.D. dissertation. Cornell University, Ithaca, New York. [4, 7, 11, 12, 15, 18]

Wiewandt, T.A. 1978. La ecología de comportamiento y conservación de la iguana terrestre de la Isla de Mona, Capitulo 24 (14 pp.). *In: Sesiones de Estudio de Union Iberoamericana de Zoos, IX Congreso*. Compañía Fomento Recreativo, San Juan, P.R. [VI]

Wiewandt, T.A. 1979. La Gran Iguana de Mona. *Nat. Hist.* 88:56-65. [7, 20]

Wiggins, I.L. and Porter, D.M. 1971. *Flora of the Galápagos Islands*. Stanford University Press, Palo Alto. [19]

Wilbur, H.M., Tinkle, D.W. and Collins, J.P. 1974. Environmental certainty, trophic level, and resource availability in life history evolution. *Amer. Nat.* 108:805-817. [7]

Wiley, R.H. 1974. Evolution of social organization and life history patterns among grouse. *Quart. Rev. Biol.* 49:201-227. [18]

Wilhoft, D.C. 1958. Observations on preferred body temperatures and feeding habits of some selected tropical iguanas. *Herpetologica* 14:161-164. [4]

Williams, E.E. and Rand, A.S. 1977. Species recognition, dewlap function and faunal size. *Amer. Zool.* 17:261-270. [18]

Williams, G.C. 1966a. *Adaptation and Natural Selection.* Princeton Univ. Press, N.J. [7, 18, 21]

Williams, G.C. 1966b. Natural selection, the costs of reproduction, and a refinement of Lack's principle. *Amer. Nat.* 100:687-692. [7]

Williams-Smith, H. 1967. Observations on the flora of the alimentary tract of animals and factors affecting its composition. *J. Pathol. Bacteriol.* 89:95-122. [6]

Wilson, E.O. 1975. *Sociobiology: The New Synthesis.* Belknap Press, Cambridge. [18, 19, 20]

Wilson, K.J. and Lee, A.K. 1974. Energy expenditure of a large herbivorous lizard. *Copeia* 1974:338-348. [7]

Wilson, L.D. and Hahn, D.E. 1973. The herpetofauna of the Islas de la Bahía, Honduras. *Bull. Fla. State Mus. (Biol. Sci.)* 17:93-150. [1]

Wolda, H. 1978. Seasonal fluctuations in rainfall, food and abundance of tropical insects. *J. Anim. Ecol.* 47:369-381. [8]

Wood, R.A., Nagy, K.A., MacDonald, N.S., Wakakawa, S.T., Beckman, R.J. and Kaaz, H. 1975. Determination of oxygen-18 in water contained in biological samples by charged particle activation. *Analyl. Chem.* 47:646-650. [3]

Woodbury, R.C., et al. 1977. The flora of Mona and Monita islands. Puerto Rico (West Indies). *Univ. Puerto Rico Agric. Exper. Sta. Bull.* 252:1-60. [6]

Woodley, J.D. 1971. The Jamaican ground iguana in Hellshire. *In:* J.D. Woodley (ed.) *Hellshire Hills Scientific Survey, 1970.* University of West Indies, Inst. Jamaica, Kingston. p. 127-133. [VI]

Woodley, J.D. 1980. Survival of the Jamaican iguana, *Cyclura collei. J. Herp.* 14:45-49. [VI]

Woodward, A.S. 1911. Iguana. *Encyclopaedia Britannica,* 11th Edition. Vol. 14:295-296. [Intro]

Yde, J. 1965. *Material Culture of the Waiwái.* Nationalmuseets Skrifter, Etnografisk Roekke, X., Nat. Mus. Copenhagen. p. 97-134. [22]

Species Indexes

IGUANAS

REPTILES OTHER THAN IGUANAS

Other Noyes Publications

WOLVES OF THE WORLD

Perspectives of Behavior, Ecology, and Conservation

Edited by

Fred H. Harrington
Mount Saint Vincent University
Halifax, Nova Scotia
Canada

Paul C. Paquet
Portland State University
Portland, Oregon

This book brings together the latest worldwide status of the behavior, ecology, and conservation of wolves by authorities from around the world. The North American section presents the latest information available on wolves from Alaska through Canada and Minnesota. The section on the wolves in Eurasia presents a description of these animals poised near the brink of extinction, and covers the wolves in the USSR, Northern Europe, Sweden, Italy, Iran and Israel. Captive wolf behavior and sociology is also discussed separately. The section on conservation examines in detail what must be done to maintain the current range of wolves worldwide. The book provides a telling contrast between wolves as they were in the past throughout the world, and as they may become throughout the world in the future.

It is a most valuable reference work that should be on the book shelf of all those interested in, and studying, animal behavior, sociobiology, wildlife management, ecology, conservation, zoology and animal science.

ISBN 0-8155-0905-7 (1982) **474 pages**

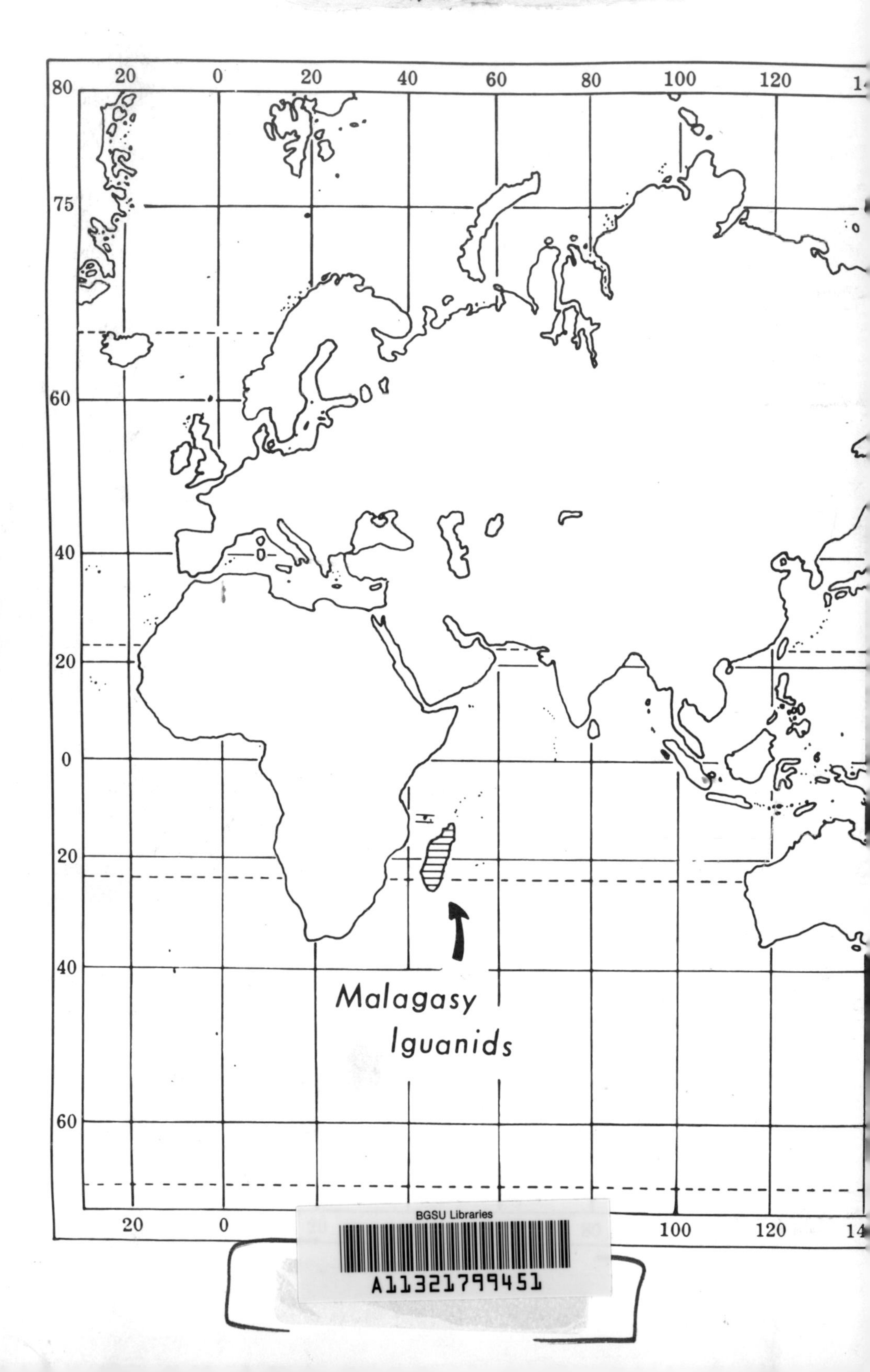
Malagasy
Iguanids